Twentieth Century Physics

Volume II

Twentieth Century Physics
Volume II

Edited by

Laurie M Brown
Northwestern University

Abraham Pais
Rockefeller University
and
Niels Bohr Institute

Sir Brian Pippard
University of Cambridge

Institute of Physics Publishing
Bristol and Philadelphia

and

American Institute of Physics Press
New York

British Library Cataloguing-in-Publication Data
A catalogue record for this book is available from the British Library.

In UK and the Rest of the World, excluding North America:
ISBN   0 7503 0353 0   Vol. I
       0 7503 0354 9   Vol. II
       0 7503 0355 7   Vol. III
       0 7503 0310 7   (3 vol. set)

In North America (United States of America, Canada and Mexico):
ISBN   1-56396-047-8   Vol. I
       1-56396-048-6   Vol. II
       1-56396-049-4   Vol. III
       1-56396-314-0   (3 vol. set)

Library of Congress Cataloging-in-Publication Data are available

Published jointly by Institute of Physics Publishing, wholly owned by The Institute of Physics, London, and American Institute of Physics Press, wholly owned by the American Institute of Physics, New York.

Institute of Physics Publishing, Techno House, Redcliffe Way, Bristol BS1 6NX, UK
Institute of Physics Publishing, Suite 1035, The Public Ledger Building, Independence Square, Philadelphia, PA 19106, USA
American Institute of Physics Press, 500 Sunnyside Boulevard, Woodbury, New York 11797-299, USA

Printed and bound in the UK by Bookcraft Ltd, Bath.

CONTENTS

Chapter 9

ELEMENTARY PARTICLE PHYSICS IN THE SECOND HALF OF THE TWENTIETH CENTURY

Val L Fitch and Jonathan L Rosner

9.1. Introduction

The past 50 years of elementary particle physics have witnessed an explosion of data, followed by simplifications based on classification and solid theory. Attempts to describe the fundamental interactions from a more unified point of view have borne fruit in a combined theory of weak and electromagnetic interactions based on self-interacting quantum fields and a similarly based theory of the strong interactions.

The understanding of the periodic table of the elements bears some similarity to the story of particle physics. An initial systematization of data was followed by firmer theoretical efforts, culminating in the advent of quantum mechanics. The vast variety of atoms and isotopes could be understood in terms of fundamental protons, neutrons and electrons interacting via electromagnetic (well understood) and strong (poorly understood) forces.

In the 1960s, a scheme for classifying the strongly interacting particles based on the group SU(3) began to make sense of the rapidly proliferating spectrum. Eventually, the success of SU(3) and related symmetries was traced to the existence of a few constituents—the *quarks*. Now we are confronted with a proliferation of quarks and *leptons* (the electron, muon, tau and their respective neutrinos) for which a deeper explanation is still lacking. These are summarized in table 9.1.

As more and more fundamental building blocks of matter were being uncovered, the way in which fundamental forces were described also

Table 9.1. *The quarks and leptons as of 1994.*

	Leptons			Quarks	
Symbol	Name	Charge	Symbol	Name	Charge
ν_e	Electron neutrino	0	u	Up	2/3
e^-	Electron	−1	d	Down	−1/3
ν_μ	Muon neutrino	0	c	Charmed	2/3
μ^-	Muon	−1	s	Strange	−1/3
ν_τ	Tau neutrino[a]	0	t	Top	2/3
τ^-	Tau	−1	b	Bottom	−1/3

[a] Not yet directly observed.

evolved. The unification of forces has a long tradition, dating from Newton's synthesis of terrestrial and celestial gravity and Maxwell's synthesis of electricity and magnetism. In this century it included a detailed understanding of the weak interactions and their violation of mirror symmetry, and culminated in the unified theory of weak and electromagnetic interactions of Glashow, Weinberg and Salam, and the discovery of the predicted carriers of the weak force, the W and Z. Still to be understood at the deepest level is the violation of the combined symmetry of charge reversal and mirror reflection, discovered in 1964.

The success of the electroweak theory was particularly heartening because it took place in the context of quantum field theory, previously thought to be useful only for describing electromagnetic processes. A parallel development, also relying on quantum field theory, was the emergence of a theory of strong interactions, now known as *quantum chromodynamics* (QCD). This theory describes why quarks are different from leptons (quarks have a new kind of charge dubbed *colour*, while leptons are colourless) and gives quantitative predictions for their interactions with one another via the exchange of quanta known as *gluons*. It explains, through the dependence of interaction strength on distance, why it makes sense to speak of quarks at all, even though they appear to be permanently bound to one another.

A chart of the carriers of strong and electroweak forces is given in table 9.2. The picture of particle physics as consisting of quarks and leptons interacting via exchanges of photons, gluons, Ws and Zs has come to be known as the 'standard model'.

A symposium has been devoted to the emergence of the standard model [1], and an extensive book [2] treats the whole period with which we are concerned. Specific chapters in particle physics, in the 1930–50s [3] and the period 1947–64 [4], are also the subject of excellent historical reviews. In this chapter, we touch on some high points of the progress made in this fruitful field in the past 50 years. Our hope is to give some

Table 9.2. *Carriers of the strong and electroweak forces.*

Symbol	Name	Force carried	Mass (GeV/c^2)
γ	Photon	Electromagnetic	0
g	Gluon	Strong	0
$W^{\pm}$	W-boson	Weak (charged)	80.3 ± 0.2
Z^0	Z-boson	Weak (neutral)	90.189 ± 0.004

flavour of how far it has come, and where we might expect it to lead in the future.

We do not wish to give the impression that progress in elementary particle physics, any more than in any other field, is an orderly process. In the interest of space, our story omits many blind alleys and wrong experiments. We have chosen to speak of discoveries and ideas that have had some lasting value. At the same time, we cannot claim to be comprehensive in our treatment. There is necessarily some choice of subjects involved, for which we take full responsibility.

We begin in section 9.2 with a few key points of early twentieth century particle physics, to set the stage for our later discussion. This period is dealt with more extensively in reference [5], which may be consulted for citations. We then (section 9.3) describe progress in quantum electrodynamics (QED), which until the advent of electroweak unification and QCD was our only example of a successful, relevant quantum field theory. Except for the treatment of QED, we break our discussion at the mid-1960s, treating properties of matter (section 9.4) and forces (section 9.5) before proceeding further.

The description of strongly interacting particles, or *hadrons*, in terms of quarks (section 9.6) marks a turning point in particle physics in the latter half of this century. With quarks taken seriously, the way was paved for extension of electroweak unification (section 9.7) from its original province of leptons to the whole range of elementary particles. Moreover, the route was now established for the development of QCD (section 9.8). New forms of matter, in the form of the third family of quarks and leptons (section 9.9), could be accommodated without much difficulty in the new framework.

Almost all of the results in particle physics in the past 50 years have been crucially dependent on continued progress in the development of accelerators (section 9.10) and detectors (section 9.11). Elementary particle physics has profited immensely from its overlap with other fields (section 9.12). We mention some puzzles and hopes in section 9.13 and we give our conclusions in section 9.14.

9.2. Preludes (before 1940)

The first 'elementary particle' identified as such was the electron, whose charge-to-mass ratio was first measured by J J Thomson in 1897. The discreteness of the charge itself was demonstrated somewhat later by Millikan.

Experiments by Rutherford in the early part of the twentieth century showed that alpha particles underwent scattering from matter at much greater angles than one might have anticipated. A popular model of the atom at that time envisioned material spread uniformly through it, whereas Rutherford's scattering experiments pointed towards an intense concentration of most of the matter over less than 10^{-4} of the atom's linear size. Niels Bohr, a young visitor at Rutherford's laboratory in Manchester in 1911–12, was inspired by Rutherford's experiments to attempt to construct a model of the atom based on negatively charged electrons orbiting a positively charged nucleus. He was forced to introduce new physics, foreshadowing quantum mechanics, in order to keep the orbits from decaying by emission of radiation. He was not initially motivated by data on the spectra of light emitted by hydrogen, but when he learned of the Balmer spectrum the whole problem became clear to him and his solution was presented within a month.

The Bohr atom made use of an analogy with previously known ideas, such as orbits. A break with the past occurred in the mid-1920s with the fully fledged development of quantum mechanics by Heisenberg, Schrödinger, Born and others. A crucial aspect of quantum mechanics was the scale set by Planck's constant h, with dimensions of (energy) × (time) or (momentum) × (length); another was the identification by de Broglie of the connection between waves and particles: (wavelength) = h/(momentum), confirmed experimentally by Davisson and Germer using electrons. Similarly, though people had been accustomed since Maxwell's time to view electromagnetic radiation in terms of waves, Einstein's explanation of the photoelectric effect in 1905 indicated that light could also be regarded as composed of *quanta*, or discrete units, with (energy) = h (frequency). This idea was confirmed by the discovery of the *Compton effect*, the scattering with change of wavelength of electromagnetic quanta (*photons*) on electrons.

The version of quantum mechanics developed in the mid-1920s applied to particles with velocities small compared with that of light. In seeking an equation of motion for particles not subject to this limitation, Dirac introduced new degrees of freedom. His equation applied to a quantity with a total of four components. A twofold multiplicity allowed one to describe particles such as the electron which have two possible directions of spin. However, an additional twofold multiplicity was a necessary consequence of invariance of Dirac's equation under the transformations of special relativity. Dirac interpreted this additional doubling to imply the existence of *antiparticles*, with opposite charge and the same mass as particles. Thus, there should exist a positively charged version of the electron. This particle, the *positron*, was identified by

Anderson in cosmic radiation in 1932.

The comparison between the charges and masses of atomic nuclei, and the detailed study of their spins, made it clear that one could not build the nucleus merely out of protons, nor of protons and electrons. A new building block was needed, similar in mass to the proton but electrically neutral. This particle, the *neutron*, was discovered by Chadwick in 1932. Its existence made the picture of atomic nuclei fall into place. The charge (Z) counted the number of protons, while the mass number (A) counted the total number of protons and neutrons. The mass of the nucleus was slightly less than the sum of the masses of individual neutrons and protons (*nucleons*), because of the effects of binding energy.

Radiation from outer space had been identified by V Hess and others in the early part of the twentieth century. By the mid-1930s, cosmic rays were a subject of some experimental interest, and were recognized as providing a useful source of highly accelerated particles, as were products of radioactive decay. However, these sources were soon joined by a number of devices invented for artificially accelerating particles, which were then focused with the help of electric and/or magnetic fields. These devices included the Cockcroft–Walton generator, the van de Graaff generator, and the cyclotron. The last was pioneered by Ernest O Lawrence, and was extensively used for particle physics studies until the mid-1950s. At that time, various versions of the *synchrotron* became available.

The prediction of infinite quantities, such as the field energy (self-interaction energy) of the electron, meant that a consistent quantum-mechanical description of the interaction of radiation with matter still had not been found in the 1930s. While it was possible during this period to calculate many processes by an approximation valid to lowest order in the interaction strength, a self-consistent description to all orders was lacking.

The continuous nature of the energy spectrum of electrons or positrons emitted in beta decay, and the balance between initial and final particles, implied that an unseen agent was carrying off momentum and angular momentum in the decay. This particle, dubbed the *neutrino*, always accompanied the electron in beta decay. The fundamental process was then n $\rightarrow$ pe$^-\bar{\nu}_e$, where n is the neutron, p is the proton, e$^-$ is the electron and $\bar{\nu}_e$ is an antineutrino. In a heavy nucleus with a large proton excess, the process p $\rightarrow$ ne$^+\nu_e$, could occur instead, even though forbidden by energy conservation for a free proton and neutron. Both processes were described by an interaction, postulated by Fermi, which was almost, but not quite, correct. It made no provision for the violation, discovered in the 1950s, of mirror symmetry (*parity violation*) in the beta-decay interaction. In its description of the production of the electron–neutrino pair, the Fermi theory was one of the first applications of *quantum field theory*, which makes provision for the production and annihilation of particles.

The extremely short range of the nuclear force led Yukawa to postulate the existence of a new particle, the *meson*, whose exchange would give rise to a short-range interaction. In 1937, a new type of particle was seen in cosmic radiation. Charged, and with a mass very close to that predicted by Yukawa, this new particle (the *muon*) was initially identified as Yukawa's meson. However, if the muon were really the carrier of the strong nuclear force, it should interact strongly with matter. Its persistent failure to do so led to the gradual realization that the muon was *not* Yukawa's meson. The particle predicted by Yukawa was yet to be discovered, as we shall see in section 9.4.

The picture of elementary particle physics as of 1940 was fairly simple and self-consistent. The atom consisted of a nucleus built of neutrons and protons, bound to electrons via electromagnetism. Neutrinos were hypothetical particles emitted in beta decay. The 'four forces of nature' were already in place: strong (holding nuclei together), electromagnetic, weak (associated with beta decay) and gravitational. The elements in the periodic table of the elements up to uranium had almost all been seen and elements heavier than uranium were starting to be discovered. There were few intimations of the rich variety of particles or the progress in understanding of forces that would characterize the next 50 years.

9.3. Quantum electrodynamics

One of the early triumphs of theory in describing elementary particle physics lay in the realm of the purely electromagnetic interactions. This area, *quantum electrodynamics*, or QED, evolved through an interplay of experiment and theory. For many years, its success was regarded as an exception, to be contrasted with much more phenomenological descriptions of the weak and strong interactions. With hindsight, we now know that those theories have followed a route related to that pioneered by QED. Indeed, the weak interactions have now been unified with QED into an *electroweak* theory, and strong forces are described by a theory which could well be unified in the future with the electroweak interactions. For historical purposes, however, it is appropriate to trace the development of QED in its own right. Even today, progress is continuing to be made in calculations, and some interesting puzzles remain for the hardy experimentalist and theorist.

We shall discuss primarily purely electromagnetic processes, not affected by uncertainties of weak- or strong-interaction physics, taking examples mainly from the interaction of photons with electrons or muons. The calculations we shall describe are organized as a series in increasing powers in $\alpha = e^2/\hbar c \approx 1/137$, the *fine-structure constant*.

9.3.1. *Infinite quantities in the theory*

Although the interactions of photons and electrons were described successfully to lowest order in α, as in photon–electron scattering or electron–positron pair production by a photon in an intense external field, it was realized quite early that higher orders in α led to difficulties [6], manifested in a series of calculations which gave infinite answers.

One can see that classical electromagnetic theory is plagued by infinite quantities just by calculating the field energy surrounding a point electron. The energy diverges as $1/r_0$, where r_0 is the minimum distance to the electron taken in the integral over energy density. Could a proper relativistic quantum-mechanical treatment cure this problem?

The description of electrons in a way compatible with special relativity entails also the existence of positrons [7]. With the help of W Furry, Weisskopf [8] showed that the inclusion of contributions from positrons in the electron self-energy calculation reduced the degree of divergence to $\ln(1/r_0)$, where r_0 again represents a minimum cut-off distance. Thus positrons were a partial, but not sufficient, help.

Another infinite quantity occurring in quantum electrodynamics arises as a result of the production of virtual electron–positron pairs by a photon. Like the electron self-energy, this *vacuum polarization* divergence depends on the logarithm of a cut-off parameter. Despite the infinite nature of vacuum polarization, it was possible to calculate its effect on the Coulomb interaction, for instance in a hydrogen atom, by comparing the interaction at large and shorter distances. The result [9] was a prediction that the $2P_{1/2}$ level of hydrogen should lie 27 MHz above the $2S_{1/2}$ level, whereas the Dirac theory predicts them to be degenerate.

9.3.2. *Early experimental developments*

9.3.2.1. *Lamb shift.*

The fine structure in the Balmer Series spectrum of the hydrogen atom was first observed in the H_α line ($n = 3 \rightarrow 2$) by Michelson and Morley in 1887. By the time the Dirac equation was available in the late 1920s more than 15 spectroscopic measurements were available for comparison with the theory. A difficulty was immediately encountered. The intensity ratios of the observed lines were not those expected; even more importantly, the splitting of the lines was different from that predicted. Already in 1933 a letter had been published in *Physical Review* addressing the issue of this discrepancy [10]. It was entitled *On the breakdown of the Coulomb law for the hydrogen atom*. Measurements later in the decade on the D_α line by Houston and Williams [11] sharpened the disagreement with the observation of a third line (deuterium shows less Doppler broadening). This work stimulated Pasternack [12] to observe that the observations could be accounted for if the 2S level were shifted upwards by about 0.03 cm^{-1} (900 MHz in units used subsequently). The vacuum polarization correction alone [9] was much too small and in the wrong direction to account for the discrepancy. This was the situation until after World War II.

In the US during wartime many physicists worked in one of the Radiation Laboratories devoted to radar development, or on the Manhattan Project concerned with nuclear weapons. Willis Lamb, at Columbia University, was originally denied necessary security clearance to work at the Columbia Radiation Laboratory (CRL) because his wife was not a US citizen. Instead, he taught physics, including atomic physics, to Navy students, thus becoming familiar with the problems associated with the H_α and D_α spectra discussed above. Eventually he was allowed to work at the CRL on high-frequency magnetrons in which capacity, though nominally a theoretical physicist, he built one of the first continuous wave magnetrons with his own hands. It operated at 2.7 cm which, not accidentally, was just the frequency of the fine-structure splitting of hydrogen. Immediately after the war, with a graduate student, R C Retherford (who had himself developed an expertise in high-vacuum techniques as well as in the measurement of tiny currents during the war), Lamb proceeded to mount an experiment designed to answer definitively the questions posed by the hydrogen fine structure. He exploited many of the new techniques and instrumentation developed during the war years. It was a brilliant effort and the experiment succeeded beyond all of Lamb's dreams.

The fine-structure splitting between the 2P and 2S levels in hydrogen was in the range of Lamb's 3 cm magnetron. Nominally, the 2S level is metastable with a lifetime sufficiently long to survive a reasonably long path through the apparatus. However, any small stray electric field will mix the 2S and 2P levels and shorten the lifetime of the 2S state, perhaps to the point where no atoms initially in the 2S state could survive the trip through the apparatus. To address these and other problems the apparatus was composed of five distinct elements. First, the source was an oven enclosing a hot tungsten surface which dissociated the molecular hydrogen to atoms. On exiting from the oven, a beam of atomic hydrogen was formed by collimation. Second, an electron beam was positioned to cross the beam of hydrogen atoms to excite at least some of the atoms to the 2S state. This process was very inefficient; only about 1 in 100 million atoms were so excited, but it was enough. Third, the atomic beam was passed through a radio-frequency field to induce transitions from the 2S to various 2P levels. Atoms in the P states decay so rapidly to the ground state that they travel less than 10^{-3} cm before they are lost to the beam of metastable atoms. Fourth, a uniform magnetic field enveloped the whole apparatus to remove the (possible) near degeneracy between the 2P and 2S levels through the Zeeman effect, and thereby minimize the danger of stray electric field which would shorten the lifetime of the beam. Finally, the beam ended with a detector designed to selectively sense the hydrogen atoms in the 2S state and reject all others. Following on the work of Massey and Oliphant, Lamb and Cobas had previously calculated that metastable hydrogen atoms impinging on tungsten would de-excite with the emission of electrons from the tungsten and thereby produce a current. Utilizing this fact, the detector consisted of a plate of tungsten connected to the most sensitive current meter available at the time, an FP 54 electrometer.

The experimental measurement consisted of setting the radio frequency and varying the magnetic field until a dip occurred in the detector current. This corresponded to RF-induced transitions from the 2S to one of the 2P states. By extrapolating to zero magnetic field, Lamb and Retherford found that the transitions occurred at a frequency 1000 MHz less than that deduced from theory, just as expected if the $2S_{1/2}$ level were shifted by this amount. They also reported seeing directly the transitions between the $2S_{1/2}$ and the $2P_{1/2}$ levels at this frequency. In the Dirac theory, these levels have exactly the same energy. The results [13] from this very first experiment, obtained 16 April 1947, are shown in figure 9.1. In this one elegant measurement the speculations of Pasternack had been shown to be correct, but the effect was removed far beyond the realm of speculation and now deserved the most serious attention.

Before World War II there had been considerable theoretical effort directed towards the question of the self-energy of the electron. However, because of the war, interest had remained dormant. Now, with the stimulus of the results of Lamb and Retherford the latent interest developed into a major attack by theoretical physicists, and within a few years the problem was solved to the satisfaction of nearly everyone. (To the end of his life, however, Dirac maintained that any theory involving the subtraction of infinities was ugly, unsatisfactory and surely incomplete.)

Lamb first announced the results, which he had obtained only five weeks earlier, at a conference held on Shelter Island in Peconic Bay, Long Island, New York, 2–4 June 1947. Sponsored by the National Academy of Sciences and organized by Robert Oppenheimer, it was attended by most of the leading theoretical moguls at the time in the US (see figure 9.2, from the second of references [14] p 380). At that conference, not only were the results of Lamb and Retherford made known, but also R Marshak first suggested that there were two kinds of mesons, and H Kramers laid the groundwork for reinterpreting infinite quantities in quantum field theory by means of 'renormalization' [14].

Within a few days after the conference, Bethe had calculated the 'Lamb shift' to be 1040 MHz using old-fashioned non-relativistic methods but with ingenious subtractions of infinite terms [15]. It was a calculation that led to many refinements [16], ending within three years with the fully developed theory of quantum electrodynamics by Feynman, Schwinger and Tomonaga [17].

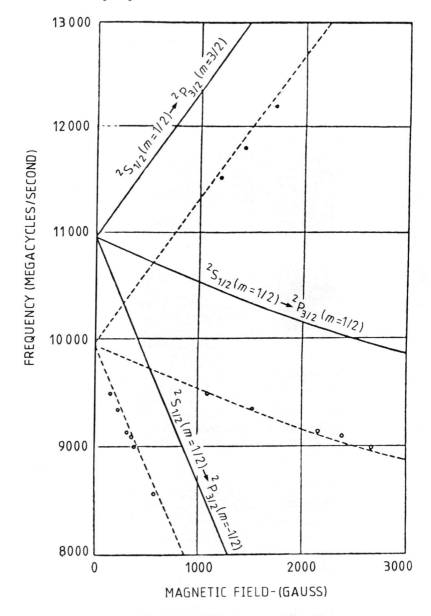

Figure 9.1. *Evidence for the Lamb shift* [13]. *The solid lines are three of the expected values of level splittings as a function of magnetic field in the absence of a level shift; the broken lines are those which would be expected in the presence of a 1000 MHz shift.*

9.3.2.2. Electron magnetic moment. The Dirac theory [7] of the electron predicts the factor g in the relation $\mu = geS/(2m)$ between the spin S and the magnetic moment μ to be exactly 2. Comparison of the fine-structure and hyperfine interactions in sodium and gallium suggested

Figure 9.2. *Participants at the first Shelter Island conference, 2–4 June 1947: (1) I I Rabi; (2) L Pauling; (3) J Van Vleck; (4) W E Lamb Jr; (5) G Breit; (6) D Mac Innes (National Academy of Sciences); (7) K K Darrow; (8) G E Uhlenbeck; (9) J Schwinger; (10) E Teller; (11) B Rossi; (12) A Nordsieck; (13) J von Neumann; (14) J A Wheeler; (15) H A Bethe; (16) R Serber; (17) R E Marshak; (18) A Pais; (19) J R Oppenheimer; (20) D Bohm; (21) R P Feynman; (22) V F Weisskopf; (23) H Feshbach. Photograph provided by D Mac Innes.*

a departure from this value [18]: $g - 2 = 0.00229 \pm 0.00008$. This result was announced at the 1947 Shelter Island conference. Schwinger [19] calculated the effect of the electron self-interaction on this quantity, finding $g - 2 = \alpha/\pi = 0.00232$ in agreement with experiment. The stage was set for a more thorough understanding of such effects. How could one get results in accord with experiment from a theory beset by infinite quantities?

9.3.2.3. Positronium. We have previously discussed ideas lying dormant during the war and blossoming quickly afterwards. Thus as early as December 1945, a few months after the end of the war, Purcell, Torrey and Pound at Harvard, and Bloch and Packard at Stanford had independently discovered nuclear magnetic resonance. In 1946 Wheeler published a paper [20] on an idea he had developed during the war in which he worked out the details of bound states involving one or more electrons (e^-) and positrons (e^+). The simplest system, the e^+e^-

Figure 9.3. *Participants in the Twelfth Solvay Conference, Brussels, 1961. From left to right, front row: S Tomonaga, W Heitler, Y Nambu, N Bohr, F Perrin, J R Oppenheimer, Sir W Lawrence Bragg, C Møller, C J Gorter, H Yukawa, R F Peierls, H A Bethe. Second row: I Prigogine, A Pais, A Salam, W Heisenberg, F J Dyson, R P Feynman, L Rosenfeld, P A M Dirac, L Van Hove, O Klein. Back row: A S Wightman (slightly to front), S Mandelstam, G Chew, M L Goldberger, G C Wick, M Gell-Mann, G Källén, E P Wigner, G Wentzel, J Schwinger, M Cini.*

bound state, we now call positronium. He included the rates of annihilation from the singlet and triplet states and the relative polarization of the two photons from the singlet-state annihilation. This paper was published in an unlikely journal, the *Annals of the New York Academy of Sciences*, because they were offering a prize for the best paper, which Wheeler won. Experimental confirmation required the development of new tools and instrumentation. The predicted bound state was finally discovered in 1951 by Martin Deutsch [21].

9.3.3. Advent of renormalization

The banishment of infinite quantities from quantum electrodynamics has been traced in references [14] and [17]. Many of the participants in this effort are shown in figure 9.3. Early leaders included Heisenberg, Dirac, Oppenheimer and Stückelberg. The procedure of reinterpreting infinite quantities in terms of physical ones was proposed by Kramers at the 1947 Shelter Island Conference [14] and developed systematically by Sin-Itiro Tomonaga, Julian Schwinger and Richard P Feynman (see box). The proof that this method was consistent to every order in perturbation theory was given by Dyson and Salam and Ward.

Three major sorts of infinities occur in quantum electrodynamics. The first, associated with the electron's infinite energy of interaction with its own electromagnetic field, is removed by redefining its mass to be the physical value, order by order, in perturbation theory. The second can be removed by demanding that a free electron produced at a given point in space be detectable with unit probability at some distant point at a later time. The third, related to the polarization of the vacuum pairs by a test charge, can be removed by redefining the electron's charge as its value as seen by a distant observer.

9.3.4. Higher-order corrections and experimental confirmation

The successive improvements in theory and experiment in quantum electrodynamics have been a long and mainly happy story [22].

9.3.4.1. Electron g − 2 factor. The measurement of the electron's anomalous magnetic moment benefited greatly from the ability to investigate magnetically confined free electrons [23]. The most recent experiments make use of traps employing both electric and magnetic fields [24]. A single electron has been kept in solitary confinement for nine months in such a trap!

The results for electrons and positrons for $a \equiv (g - 2)/2$ are [25] $a(e^-) = (1159\,652\,188.4 \pm 4.3) \times 10^{-12}$, $a(e^+) = (1159\,652\,187.9 \pm 4.3) \times 10^{-12}$, where the error is a combination of statistical and systematic uncertainties. The theoretical prediction [26] is

$$a(e) = \alpha/(2\pi) - 0.328\,478\,965\,(\alpha/\pi)^2 + C_3(\alpha/\pi)^3 + C_4(\alpha/\pi)^4 + \delta a(e). \quad (1)$$

The first two terms have been calculated analytically, while only numerical evaluations of $C_3 = 1.176\,11 \pm 0.000\,42$ and $C_4 = -1.434 \pm 0.138$ have been performed. The term $\delta a(e) = 4.46 \times 10^{-12}$ arises from electroweak interactions and from Feynman graphs involving internal muons, τ-leptons and quarks.

Several determinations of α exist, the most precise of which, $\alpha^{-1}(\text{QHE}) = 137.035\,9979 \pm 0.000\,0032$ makes use of the quantum Hall effect [27]. Using this value, one predicts [26]

$$a(e) = (1159\,652\,140 \pm 5.3\,[C_3] \pm 4.1\,[C_4] \pm 27.1\,[\alpha]) \times 10^{-12}. \quad (2)$$

This result agrees with experiment within errors. Until the uncertainty in α is reduced, one is not really testing the $(\alpha/\pi)^4$ term incisively.

9.3.4.2. Muon g − 2 factor. The lowest-order Feynman graphs leading to a difference between $a(e)$ and $a(\mu)$ contribute at second-order in α. Instead of $-0.328\ldots$ for the coefficient of $(\alpha/\pi)^2$, one obtains $+0.754\ldots$ [28].

The earliest measurement of $a(\mu)$ confirmed that it was within 5% of zero [29]. A pioneering series of experiments was begun at CERN

Richard P Feynman

(American, 1918–88)

The contributions of Richard P Feynman to theoretical physics range from the low temperatures of liquid helium to the highest-energy collisions of elementary particles. Feynman developed a relativistic theory of quantum electrodynamics using a path integral approach, aided by diagrams that could be identified with actual physical processes. For this work he shared the Nobel Prize in 1965 with Julian Schwinger and Sin-Itiro Tomonaga. Not only have Feynman diagrams become the standard language of elementary particle theory, but they have been widely applied in other areas, such as nuclear physics and condensed matter theory.

Another contribution of Feynman which gave him great pleasure was a description of the weak interactions (the V–A theory, formulated in collaboration with Murray Gell-Mann and independently by George Sudarshan and Robert E Marshak) which incorporated the newly discovered violation of mirror symmetry. Still another was his recognition, along with J D Bjorken, that experiments at Stanford were probing point-like structures (called 'partons' by Feynman, but soon to be identified as quarks) in the proton.

Feynman was born in Brooklyn, New York, and went to public high school there. He received his undergraduate training at the Massachusetts Institute of Technology and his PhD from Princeton in 1942. During World War II he headed the Scientific Computing division at Los Alamos. After several years on the faculty at Cornell University, he joined the California Institute of Technology in 1950, where he remained intensely active in physics. The *Feynman Lectures in Physics*, based on an introductory course at Caltech, remain one of the foremost guides to novel and simple ways to view our Universe.

in the late 1950s, involving the containment of muons first in a long dipole magnet and then in storage rings [30]. The most recent results of these latter experiments are: $a(\mu^+) = (1165\,910 \pm 11) \times 10^{-9}$, $a(\mu^-) = (1165\,936 \pm 12) \times 10^{-9}$ or, combining the two results, $a(\mu) = (1165\,923 \pm 8.5) \times 10^{-9}$, where the total (statistical and systematic) errors are shown.

The theoretical value contains contributions from QED, intermediate states involving strongly interacting particles ('hadrons') and weak interactions [31]

$$a(\mu)|_{QED} = (1165\,846.955 \pm 0.046 \pm 0.028) \times 10^{-9}$$

$$a(\mu)|_{hadron} = (70.27 \pm 1.75) \times 10^{-9} \qquad (3)$$

$$a(\mu)|_{weak} = (1.95 \pm 0.10) \times 10^{-9}$$

leading to a total of

$$a(\mu)|_{theor} = (1165\,919.18 \pm 1.76) \times 10^{-9}. \qquad (4)$$

The errors in $a(\mu)|_{QED}$ reflect the theoretical uncertainty and that due to the measurement of α. The dominant error in $a(\mu)|_{hadron}$ and hence in (4) comes from uncertainty in hadronic vacuum polarization effects in the $\mathcal{O}(\alpha^2)$ contribution.

An experiment is being mounted at Brookhaven National Laboratory to measure $a(\mu)$ about 20 times more sensitively [30]. Its interpretation will require reduction of the error in the hadronic vacuum polarization contribution by means of more precise e^+e^- annihilation experiments. With that error reduced, one will then be able to test the weak contribution $a(\mu)|_{weak}$ to about 20% of its expected value.

9.3.4.3. Lamb shift. The most recent measurement of the $2S_{1/2} - 2P_{1/2}$ splitting in atomic hydrogen is $(1057\,851.4 \pm 1.9)$ kHz [32], to be compared with the theoretical prediction [33] of $(1057\,853 \pm 14)$ kHz or $(1057\,871 \pm 14)$ kHz, depending on how the proton's structure is described. As in the case of $a(\mu)$, there is satisfactory agreement with experiment, but a hadronic measurement is needed to reduce theoretical uncertainty.

9.3.4.4. Hyperfine interactions. A history of atomic hyperfine structure experiments is given by Ramsey [34]. The most precise value for the splitting between the $^3S_1$ and $^1S_0$ levels in hydrogen is $\Delta \nu_H = (1420\,405\,751.7667 \pm 0.0009)$ Hz. The theoretical prediction is within about 1 kHz of this value, but is affected by unknown proton structure effects. While experimental accuracy far outstrips theory, the agreement is still impressive.

Purely leptonic systems also exhibit hyperfine structure. The experimental value for muonium agrees satisfactorily with theory within the errors associated with uncertainty in the muon mass. In positronium, the observed hyperfine splitting is $\Delta \nu^{\text{exp}} = (203\,389.10 \pm 0.74)$ MHz, to be compared with the prediction $\Delta \nu^{\text{theor}} = (203\,404.5 \pm 9.3)$ MHz, where the dominant source of error comes from uncalculated corrections of order α^2 with respect to the leading result. The calculation of these corrections is a challenge to the most enthusiastic theorist.

9.3.4.5. Positronium decay: a current puzzle. The annihilation of positronium ground states into photons is governed entirely by QED. The $^1S_0$ (singlet) state decays into two photons, while the $^3S_1$ (triplet) state decays into three. The singlet rate [35] is found to be $\lambda_s = (7.994 \pm 0.011)$ ns^{-1}, to be compared with the prediction [20,35,36] $\lambda_s = (7.986\,654 \pm 0.000\,001)$ ns^{-1}, whose error is governed by that on α. The agreement is satisfactory. On the other hand, the value measured in vacuum [37] of the triplet rate, $\lambda_t = (7.0482 \pm 0.0016)$ μs^{-1}, is more than six standard deviations from the calculated value [38] $\lambda_t = (7.03831 \pm 0.00007)$ μs^{-1}. One would need a large $\mathcal{O}(\alpha^2)$ correction to this value to bring theory into accord with experiment. The evaluation of this term represents a frontier of QED.

9.4. New forms of matter up to the mid-1960s
The number of known 'elementary' particles grew by an enormous factor in the two decades after World War II. In this section we describe how that growth took place. The penetrating component of cosmic radiation, the muon, was eventually understood as being distinct from Yukawa's particle, the pion. The antiproton, predicted by Dirac's theory, was eventually found. Many 'strange' particles were discovered, and evidence was gathered for resonant particles, some living less than 10^{-23} s. Classification schemes during this period made widespread use of symmetry principles. The understanding of several hundred particles in terms of a few simple constituents came later.

9.4.1. The 1930s and 1940s: muons, pions and kaons
Modern particle physics starts with the discovery of the deuteron in 1931 by Urey, Brickwedde and Murphy [39]; the neutron in 1932 by Chadwick [40], the positron in 1932 by Anderson [41] (to be confirmed soon after by Blackett and Occhialini in an experiment discussed below) and what is now called the muon (then, the mesotron or meson) by Neddermeyer and Anderson, Street and Stevenson, and the Nishina group in 1937 [42].

Though Dirac's paper [7] predated the discovery of the positron by three years and Yukawa's meson [43] came before the muon by two years, they played no role in stimulating experimental searches. For example, Anderson observed that a person accepting the Dirac theory at face value could have discovered the positron in an afternoon. 'However, history did not proceed in such a direct and efficient manner, probably because the Dirac theory, in spite of its successes, carried with it so many novel and seemingly unphysical ideas, such as negative mass, negative energy, infinite charge density, etc. Its highly esoteric character was apparently not in tune with most of the scientific thinking of the day.' In addition, while today the Dirac theory is held as a monument to deductive reasoning, a splendid example of the 'unreasonable effectiveness of mathematics in physics', it was not accepted for a long time. Leading detractors included the likes of Pauli and Heisenberg. While the discovery of the positron had much to do with the acceptance of the Dirac theory, the theory in no way stimulated the activity leading to the discovery.

In the case of Yukawa's meson theory, Japanese scientific literature was not widely disseminated in the West, and the average physicist simply was unaware of the Yukawa particle. Only after the muon's discovery did Oppenheimer and Serber [44] discuss (with reservations) the possibility of identifying it with the Yukawa particle, making the first reference to Yukawa's paper in Western literature [45].

During the 1930s the only theoretical construct that took hold was the neutrino [46], forced on the community of physicists by energy conservation. This apparent independence of experimental and theoretical work contrasts strikingly with the close communication and interplay that developed between theory and experiment by the 1960s.

Blackett and Occhialini [47] not only beautifully confirmed the observations of Anderson, but gave photographic cloud-chamber evidence of cosmic-ray induced showers of particles. In addition, their experiment initiated one of the most important new techniques of the decade, the use of counter-controlled cloud chambers. The previous practice was to expand the cloud chamber and, after an optimum time, take a photograph. Consequently, most of the photographs were of an empty chamber, or showed old tracks fuzzed out by diffusion. With counter-controlled expansions, more than 75% of the pictures showed ionizing tracks. Never again, relative to the passage of particles, would cloud chambers be expanded randomly in time. The neutral pion excepted, for the next 20 years all of the new particles were discovered through the study of cosmic rays via visual techniques: by the use of counter-controlled cloud chambers and, after World War II, photographic emulsions.

The Yukawa meson theory [43] (1935) and the Fermi theory [48] (1934) of beta decay were patterned on electrodynamics, where the force between charges is viewed as arising from photon exchange. In the case of radiation from atoms, the photons did not previously exist in the atom but were created at the moment of radiation. Likewise, in Fermi's beta-decay theory, the electron and neutrino did not pre-exist in the nucleus but were spontaneously created. This enormous conceptual advance immediately resolved many questions, including the nature of the neutron. From its discovery until the advent of the Fermi theory,

See also
p 119

a hotly debated topic had been whether the neutron was a particle with the same status as the proton or whether it was an electron–proton composite. With the Fermi theory, beta decay was simply the transmutation of the neutron to proton, electron and neutrino. The lifetime of the neutron was not measured until 1948 but in 1934 the Fermi theory estimated it to be $\sim 10^3$ s. (It is now known to be very close to 15 min: 887 ± 2 s [49].)

In the case of the Yukawa theory the particles mediating the interaction were given mass to produce a short-range 'Yukawa potential', $V(r) \sim \exp[-r(mc/\hbar)]/r$. Choosing a range then known to be characteristic of nuclear forces, $\sim 10^{-13}$ cm, one gets a mass around 200 MeV. Oppenheimer and Serber asked whether this was the same particle as the μ-meson (which they prophetically referred to as a heavy electron). However, this association almost immediately ran into conflict with experiment.

If the meson were to account for nuclear forces it should interact strongly with nuclear matter, so that the cross section for interaction should approach the geometric cross section with a mean free path of about 100 g cm^{-2}. It quickly became apparent that the penetrating component of the cosmic radiation interacted with a much smaller cross section. Furthermore, Rossi and Nereson [50] showed in 1940 by studying the flux of this component as a function of altitude that the attenuation through the atmosphere could only be accounted for by including a decay component as well as nuclear interactions. They were able to measure the lifetime to be $(2.15 \pm 0.07) \times 10^{-6}$ s, roughly 100 times longer than the lifetime of the Yukawa particle estimated on theoretical grounds. Shortly afterwards, a meson was seen to decay to an electron plus neutral particles in a cloud chamber [51].

Early on it became apparent that positively and negatively charged mesons could be expected to behave quite differently on coming to rest in matter. The positive particle would very quickly come to thermal energies and decay at its natural rate. The negative particle would come to rest, find itself attracted to a positive nucleus, cascade down through the various atomic levels by emitting radiation or Auger electrons, and then, depending on the nature of the particle, either decay or interact with the nucleus. The whole process, stopping and atomic transitions to the 1S state, was estimated to take less than 10^{-11} s, a time much shorter than the expected decay times.

See also
p 407

The experiment which definitively showed that the meson discovered in 1937 was not Yukawa's particle was performed by Conversi, Pancini and Piccioni [52] (1945). Using magnetized iron as a charge selector, they showed that both positive and negative mesons appeared to decay when coming to rest in carbon, while if the particles came to rest in iron only the positive ones decayed. Since negative mesons with Yukawa-like properties would be expected to interact even in elements as light

Figure 9.4. *The Pic du Midi, site of early cosmic-ray experiments.*

as carbon, the evidence was unmistakable that the mesons could not be the strongly interacting Yukawa type.

In the meantime, significant advances were taking place in the production of photographic emulsions sufficiently sensitive to detect lightly ionizing particles. Following initial applications in nuclear physics, in 1946 Perkins at the Imperial College in London and Powell and his group at Bristol first exposed the new emulsions to cosmic rays [53]. (The history of the photographic emulsion technique and its contributions to particle physics has been beautifully recorded in reference [54].)

Though the first emulsions were not sensitive to minimum ionizing particles, the technique almost immediately showed its value. Perkins found an example of a meson coming to rest and depositing its rest energy into a 'star' of heavily ionizing fragments in an emulsion exposed in an aeroplane flown at 30 000 ft. The Bristol group exposed their emulsions at the Pic du Midi laboratory in the French Alps (figure 9.4), quickly finding examples of π mesons decaying to μ-mesons with characteristic tracks 600 microns long. (This group first labelled the two kinds of mesons as π and μ, the only two Greek letters on Powell's typewriter.) The π-meson decay was clearly two-body. (The emulsions were not yet sufficiently sensitive to see the electron track from the subsequent μ-decay.)

The paper of the Powell Group at Bristol, generally marked as announcing the discovery of the π-meson or pion [55], appeared in

See also p 408

Nature of 24 May 1947. The existence of two types of mesons had been anticipated by Sakata's group in Japan, whose work only became known after several years, and was discussed in the West by R E Marshak and H Bethe [56].

Seven months later in the same journal appeared a paper *Evidence for the existence of new unstable elementary particles* by C D Rochester and C C Butler [57], reporting on the first two examples of 'strange' particle decay in a cloud chamber. We would now call them beautiful examples of K-mesons, the neutral K_S and a charged K^+. Elementary particle physics would be occupied for the next quarter century with the properties of pions, muons and the 'strange' particles.

9.4.2. Pion properties

Nuclear forces were expected to be *charge-independent*, obeying a symmetry which amounted to invariance under rotation in an abstract *isotopic spin* or *isospin* space [5]. As a consequence, Yukawa's theory entailed not only charged but also neutral pions [58], first seen around 1950 in various accelerator and cosmic-ray experiments [59].

9.4.2.1. Spin and parity.
The spin of charged pions was established by comparing the rates for $p + p \to \pi^+ + d$ and $\pi^+ + d \to p + p$. The ratio of the rate for the first process to that for the second is proportional to $2S_\pi + 1$, where S_π is the spin of the pion. The result of measurements of both reactions [60] led to the conclusion that $S_\pi = 0$.

Particles can be characterized by an intrinsic *parity*, describing the behaviour of their fields under space inversion. The negative parity of charged and neutral pions was deduced by comparing [61] rates for the process $\pi^- + p \to n + \gamma$ with $\pi^- + p \to n + \pi^0$, and $\pi^- + d \to 2n$ with $\pi^- + d \to 2n + \gamma$ and $\pi^- + d \to 2n + \pi^0$, and by comparing [62] the yields of neutral and charged pions in proton collisions on light nuclei. An early test of charge-independence by comparing the cross sections for $n + p \to d + \pi^0$ and $p + p \to d + \pi^+$ was performed by R Hildebrand [63].

9.4.2.2. Charged and neutral pion decays.
The lifetime of the charged pion was first measured using pions produced artificially at an accelerator [64]. Within a couple of years, precise electronic timing techniques had yielded a result [65] very close to that known at present (26 ns), and considerably shorter than that of the muon. The dominant channel is $\pi^\pm \to \mu^\pm \nu$.

The nature of the neutral pion was still open to question. It was seen to decay to two photons, but direct measurements of its spin, parity and lifetime were needed. S Sakata, L D Landau and C N Yang [66] showed that a spin-1 particle could not decay to two photons, and showed how to determine the parity by comparing the linear polarizations of the two photons. Parallel polarizations implied even parity, while perpendicular polarizations implied odd parity. Later measurements [67] showed the parity to be odd, in agreement with the result of reference [61].

The lifetime of the neutral pion was determined by direct measurements [68] to be less than 10^{-13} s. Estimates [69] by J Steinberger suggested the possibility of an even shorter lifetime than this upper bound, outside the reach of direct detection methods at the time. It was proposed by H Primakoff [70] to use the rate of neutral pion photoproduction in the Coulomb field of a nucleus to measure the decay rate indirectly. This method was first applied a number of years later, yielding values slightly below 10^{-16} s [71]. Direct methods, now the most accurate, also evolved for measuring such short lifetimes using sandwiches of foils [72].

9.4.3. Antiprotons

The Dirac theory of antiparticles, so stunningly confirmed by the positron's discovery, also predicted a negative version of the proton, the *antiproton* or $\bar{p}$. The antiproton and proton were expected, though not unanimously, to have equal masses. The energy of a new accelerator under consideration at Berkeley was chosen to lie above the threshold for the reaction $p + p \rightarrow p + p + p + \bar{p}$. This accelerator, the Bevatron, not only discovered the antiproton, but a wealth of other particles, ushering in a new era of high-energy physics.

Several groups were searching for antiprotons at the Bevatron. Key players included O Chamberlain, O Piccioni, E Segré, W Wenzel, C Wiegand and T Ypsilantis. The construction of a beam of antiprotons was a crucial feature of the discovery. Protons in the machine struck an internal target, and the resulting negatively charged particles were focused into a well-collimated external beam with a narrow momentum range. Antiprotons were distinguished from the much more abundant negative pions by precise time-of-flight measurements and with the help of focusing Čerenkov counters, a tool that contributed greatly to the experiment's success. Knowing both the momentum and the velocity of particles in the beam, one could measure their masses. A clear signal was seen of negatively charged particles with the mass of a proton [73]. Their identity as examples of antimatter was confirmed by the observation of their annihilations in nuclear emulsions [74].

9.4.4. Strange particles

The story of the discovery and classification of 'strange particles', as they were called, affords a lovely example of order emerging from chaos [75].

9.4.4.1. Discoveries in cosmic rays.

Two techniques for studying cosmic rays were in use during the 1940s: nuclear emulsions, in which the passage of particles exposed a three-dimensional photographic image,

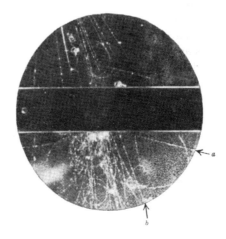

Figure 9.5. *First evidence for a neutral strange particle, from the cloud-chamber experiment of Rochester and Butler [57]. The particle was produced in a lead plate (shown at centre) and decayed to two charged particles (a and b).*

and cloud chambers, in which charged particles left a trail of droplets in a supercooled vapour. Both yielded evidence for new particles besides the positron, muon and pion.

By studying elastic scattering of charged particle tracks on electrons in emulsions, Leprince-Ringuet and Lhéritier [76] deduced the existence of a new particle of mass about 990 times the electron's mass. In retrospect, this appears to have been the first measurement of the mass of the charged K-*meson* or *kaon*.

Using a cloud chamber at Manchester, Rochester and Butler [57] saw two events which we would now characterize as decays of kaons, as mentioned in section 9.4.1. The neutral particle led to the emergence of a 'forked' pair of tracks (figure 9.5), while the charged one was seen decaying to an electron and one or more missing neutral particles. A long 'dry spell' for the Manchester group followed, during which an emulsion experiment [77] saw the decay of a charged kaon to $\pi^+ + \pi^+ + \pi^-$, permitting a decisive measurement of its mass. Meanwhile confirmation of the existence of Rochester and Butler's events came from a cloud chamber operating primarily at high altitude in California [78]. The Manchester group then set up operation on the Pic du Midi in the Pyrenees, where many more events were found.

The neutral kaon was seen to decay to $\pi^+ + \pi^-$. Because of the equal masses of the two pions, the momentum distributions of their tracks were equal, and so were the typical angles they made with the direction of the incident kaon. However, in 1950, an event was seen in which the positive track appeared to be a proton [79]. In the decay of a neutral particle to $p + \pi^-$, the proton, because of its larger mass, tends to carry most of the momentum, leading to an asymmetric configuration. The Pic du Midi group [80], led by R Armenteros, and a cloud chamber group at Indiana [81] led by R Thompson resolved the decay $K^0 \rightarrow \pi^+ + \pi^-$ from what we now understand to be $\Lambda \rightarrow p + \pi^-$. The Λ was the first example of a *hyperon*, a particle heavier than the proton. Cosmic-ray studies also yielded the first evidence for charged hyperons, the Ξ^- and the Σ^+. The Ξ^- was called the 'cascade' particle because it was seen [82] to decay to $\Lambda + \pi^-$, with the subsequent decay of the Λ. The Σ^+ was seen [83] decaying to $n + \pi^+$ and $p + \pi^0$.

In contrast to the neutral kaon, the charged kaon was found to have many different decay modes. Sorting out all these results took some time and effort, in which emulsion studies were of tremendous help.

9.4.4.2. Discoveries in accelerators. The Brookhaven Cosmotron, which began operation in 1952, belonged to a new generation of accelerators constructed to study fundamental interactions at unprecedented energies [84]. Its attention was quickly turned to strange particles. In a series of experiments performed with a hydrogen diffusion cloud chamber, Ralph Shutt and his group [85] were able to demonstrate what cosmic-ray experiments had been unable so far to reveal: the new heavy particles were produced in pairs, as in the reaction $\pi^- + p \rightarrow \Lambda^0 + K^0$, confirming earlier hypotheses by several groups in Japan [86] and by A Pais [87]. In addition to the associated production of Λs just mentioned, Shutt's group was the first to produce and observe neutral and negative Σ hyperons.

9.4.4.3. Associated production and strangeness. The new heavy particles were produced fairly copiously (in about 1% of all cosmic-ray events), but decayed very slowly (with lifetimes about 10^{12} times longer than one might expect if their production and decay were governed by the same interaction). A clear statement of this paradox was given very early by Nambu, Nishijima and Yamaguchi in their second paper of reference [86]

... production and decay are not inverse processes and/or some kind of selection rules (in a very general sense) are at work in the decay reaction).

One option for such a rule was a quantum number ('isotopic parity') taken to be +1 for the nucleon, pion and muon, and −1 for the new ones (corresponding to what we now call the K and Λ). This quantum number was taken to be conserved *multiplicatively* in production processes, but could be violated in the much weaker decay processes. Thus, a kaon always had to be produced in association with a Λ, a rule that came to be known as 'associated production'.

Associated production as formulated required one to assign an odd isotopic parity to the Σ hyperons, but did not prohibit reactions like $p + p \rightarrow \Sigma^+ + \Sigma^+$ or $n + n \rightarrow \Lambda + \Lambda$ that were not seen. Moreover, associated production ran into difficulty with the negative cascade particle, Ξ^-, which decays to $\Lambda\pi^-$ with a typical 'slow' lifetime of the order of 10^{-10} s. By the above scheme we would then assign even isotopic parity to the Ξ^-. But then what would prohibit the decay $\Xi^- \rightarrow n + \pi^-$, a process that has not been seen to this day?

The strong interactions of pions and nucleons conserve isospin symmetry, mentioned at the beginning of section 9.4.2. Pions have isospin $I = 1$ and nucleons have $I = 1/2$. The charge of a pion is the third component of its isospin, while that of the nucleon is displaced by half a unit. For both particles, one can write $Q = I_3 + B/2$, where the *baryon number* B is 0 for pions and 1 for nucleons.

M Gell-Mann [88] and K Nishijima [89] generalized this relation to the new particles by postulating 'displaced' isotopic spin multiplets, with $I = 0$ for the Λ, $I = 1$ for the Σ^+ and Σ^- (thereby predicting the Σ^0), and $I = 1/2$ for charged and neutral kaons. The conservation of charge and I_3 in strong interactions then required every multiplet to be characterized by an additional quantum number, called *strangeness*, also conserved in the strong interactions.

The relation between the charge and the third component of isospin could now be written $Q = I_3 + Y/2$, where the *hypercharge* $Y = B + S$, with a new additive quantum number S assigned to every strongly interacting particle, described the displacement of the isospin multiplet. The K^0 and K^+ were taken to have $S = 1$ and the Λ and Σ were assigned $S = -1$. The reactions $p + p \rightarrow \Sigma^+ + \Sigma^+$ and $n + n \rightarrow \Lambda + \Lambda$ were forbidden by this scheme. Decays such as $K^0 \rightarrow \pi^+ + \pi^-$ and $\Lambda \rightarrow p + \pi^-$, violating S by one unit, were allowed to proceed, but only weakly.

In order that $\Xi^- \rightarrow \Lambda + \pi^-$ be a weak decay violating strangeness by one unit, but to prevent the occurrence of $\Xi^- \rightarrow n + \pi^-$, one had to take $S(\Xi^-) = -2$. Then $K^- + p \rightarrow K^+ + \Xi^-$ should proceed strongly, as was later confirmed. Moreover, $I_3(\Xi^-)$ should equal $-\frac{1}{2}$, entailing the existence of a neutral partner Ξ^0 with $I_3 = +\frac{1}{2}$, a particle eventually seen in a bubble chamber [90].

9.4.4.4. Neutral kaons and their lifetimes. The existence of kaons with both signs of charges had been known for some time before Gell-Mann and Nakano and Nishijima proposed that these mesons had isospin $\frac{1}{2}$. But that proposal implied the existence also of two kinds of *neutral* kaons, the K^0 (with strangeness $S = 1$ like the K^+) and the $\bar{K}^0$ (with strangeness $S = -1$ like the K^-). How could one tell them apart? This question was raised by Fermi in June of 1954 at a seminar by Gell-Mann describing the strangeness scheme.

The resolution of this problem [91] provided a beautiful application of quantum mechanics. The K^0 and $\bar{K}^0$ are two degenerate states; nothing specifies what combination of them corresponds to states of definite mass and lifetime. However, if invariance under charge reflection is assumed in the K decay process (as it was in the mid-1950s), one linear combination of K^0 and $\bar{K}^0$ could decay to $\pi^+ + \pi^-$, while the other could not. These two states were denoted K_1^0 and K_2^0, respectively. The K_1^0 had already been seen, and the K_2^0 was a newly predicted particle that should live much longer than the K_1^0, since its decay to $\pi^+ + \pi^-$ was forbidden.

An experiment at the Brookhaven Cosmotron [92] soon confirmed the existence of the predicted neutral kaon with its characteristic long lifetime and dominant three-body decay mode. Supporting evidence came from four events which, most plausibly, originated from interactions of K_2^0 particles in photographic emulsions [93].

9.4.5. Resonances

9.4.5.1. Pion–nucleon scattering. The isospin of a combined pion–nucleon system can only be $\frac{1}{2}$ or $\frac{3}{2}$, entailing simple relations among pion–nucleon scattering amplitudes [94]. (Other early applications of isotopic spin are noted in references [95].) As artificially produced pions began to be available in the late 1940s because of increased cyclotron energies, attention was turned to the properties of pion–nucleon scattering. The development of electron synchrotrons capable of reaching energies of several hundred MeV led to production of high-energy photons, also useful in producing pions [96]. A crucial result of these experiments was the discovery of the first *isobar*, an excited nucleon state.

9.4.5.2. Nucleon isobars. With increasing pion and photon energies, the cross sections for pion–nucleon elastic scattering and single pion photoproduction grew in a very specific pattern [97]. Theorists [98] proposed that this pattern might signal the existence of a short-lived excited state of the nucleon with $I = J = \frac{3}{2}$. With still higher energies [99], the cross sections traced out a well-defined pion–nucleon resonance, called the (3,3) resonance because of its spin and isospin. Its current appellation is $\Delta(1232)$, where the number (here and later) denotes the mass in MeV/c^2. The phase shift in the corresponding channel was seen to pass through $90°$, as appropriate for a resonance.

Still higher energies led to the discovery of additional peaks in cross sections, corresponding to states with well-defined spins and isospins. By the late 1950s, several of these pion–nucleon resonances had been identified [97].

9.4.5.3. The resonance explosion. Invention of the bubble chamber [100] permitted the study of particle interactions in much greater detail during the mid-1950s. Several groups, including Luis Alvarez and collaborators [101], introduced automated methods to scan photographs of interactions. Pions, nucleons and strange particles all were found to participate in the formation of resonant states with well-defined masses and spins. Widths of these states were typically tens of MeV, as expected for states formed and decaying via the strong interactions [102].

Some of the earliest resonances had been anticipated from spin and isospin properties of nuclear forces. A 'mixture of mesons', not just Yukawa's proposed particle, appeared to be needed [103]. Experiments on the *form factors* of nucleons [104], to be discussed in section 9.6.2.1, could be interpreted as though the photon coupled to nucleons through a spin-1, isoscalar particle [105], decaying to three pions. An $I = J = 1$ resonance decaying to $\pi^+ + \pi^-$ also was suggested [106] on the basis of form-factor experiments. This resonance, the $\rho(776)$, was the first meson resonance to be discovered [107]; the three-pion isoscalar resonance $\omega(783)$ followed shortly thereafter [108].

Resonance searches utilized two major techniques: 'formation', or varying the beam energy and observing a peak in the cross section as in the discovery of the $(3,3)$ resonance, and 'production', or grouping particles in the final state into combinations whose 'effective mass' spectrum could be studied for enhancements. We already noted that formation experiments in pion–nucleon and photon–nucleon collisions led to a fertile yield of resonances. The advent of beams of negatively charged kaons revealed many resonances of negative strangeness formed in K^-p and K^-n reactions. Excited states of the Λ and Σ hyperons thus began to be uncovered. The production experiments had unique ability to see meson resonances, but also were the first to detect several excited baryon states. By the early 1960s, the $\rho(776)$, $\omega(783)$, another three-pion resonance, the $\eta(547)$ (reference [109]), a $J = 1$ 'excited kaon,' $K^*(892)$, and a $J = \frac{3}{2}$ version of the Σ at 1385 MeV had appeared. The 'elementary particles' had thus progressed from a handful to a veritable zoo in the course of 20 years [102].

9.4.6. Unitary symmetry

9.4.6.1. Preludes. By 1960, a wide variety of baryons with spin $\frac{1}{2}$ had been identified. Isospin had proved a successful guide to the classification of all the strongly interacting particles, but the existence of particular multiplets and the masses of the baryonic states remained a mystery. Similarly, it appeared that kaons were strange versions of pions, with the same spin and intrinsic parity.

9.4.6.2. SU(3). Initial efforts to unify non-strange and strange particles [110] included a 'global symmetry' of the strong interactions, composite models of mesons, and searches for various Lie groups containing isospin [SU(2)] as a subgroup. The group SU(3) was proposed [111] as a symmetry of strongly interacting particles in 1959, with the proton, neutron and Λ assigned to a triplet representation. The mesons π, K and an as yet undiscovered eighth meson (now known as the η), all composites of p, n, Λ and their antiparticles, were to form an octet.

Murray Gell-Mann had been searching for several years for higher symmetries containing isospin which could unify the baryons. In 1960 he realized that the octet representation of SU(3) was a perfect home for the spin-$\frac{1}{2}$ states N, Λ, Σ and Ξ. His conclusions on this 'eightfold way' were originally published only in preprint form [112]. Independently, Yuval Ne'eman, an Israeli military attaché in London working for his PhD in physics, was asked by his supervisor, Abdus Salam, to find an appropriate group containing isospin to classify the observed hadrons. He generalized the work of Yang and Mills [113] on self-interacting fields to higher symmetries, and realized the importance of SU(3). He, too, suggested that the spin-$\frac{1}{2}$ baryons belonged to an octet [114]. Gell-Mann's approach was part of a larger scheme in which *currents* (such as the electromagnetic current) played a crucial role [115]. We shall return to that aspect of SU(3) in section 9.5.

9.4.6.3. Consequences for masses; verification. The SU(3) scheme, when combined with an assumption about the nature of symmetry breaking, led Gell-Mann and Okubo [116] to mass relations among the baryons and among the mesons:

$$M(N) + M(\Xi) = [3M(\Lambda) + M(\Sigma)]/2 \qquad M(K) = [3M(\eta) + M(\pi)]/4.$$

The relation for the baryons is well satisfied: with the observed masses, we have 2257 MeV ≈ 2270 MeV. For the mesons, the symmetry breaking is greater, and we have 495 MeV ≈ 446 MeV. The success of the mass formula for baryons and the discovery [109] of the predicted η were notes of encouragement for the SU(3) scheme. Another higher symmetry, based on the group G_2, had room for only seven mesons and seven baryons.

When SU(3) was proposed, an isospin-$\frac{3}{2}$ non-strange multiplet (the Δ(1232)) and an isospin-1 strange multiplet (the Σ(1385)) were known. At the 1962 International Conference on High Energy Physics in Geneva an excited state of the Ξ at 1530 MeV/c^2, decaying to Ξ + π, was announced. The SU(3) scheme and another contender, the G_2 group mentioned above, had very different predictions for the remaining particles if the Δ(1232), Σ(1385) and Ξ(1530) all belonged to the same multiplet.

A ten-dimensional SU(3) representation was the smallest which could accommodate the Δ and its partners. The Gell-Mann–Okubo mass relation predicted equal spacings between the masses of its members. The Ξ(1530) obeyed this prediction. It was expected to have isospin $\frac{1}{2}$. Nine states of the ten would then have been seen: four Δs, three Σs

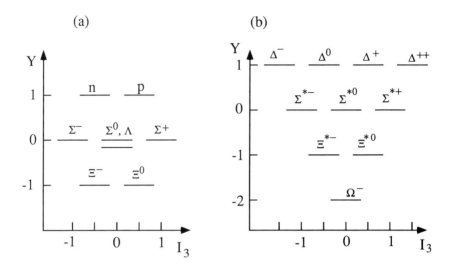

Figure 9.6. *Low-lying baryonic states as classified in SU(3): (a) octet, (b) decimet.*

and two Ξs. The missing tenth state was expected to have strangeness −3 and a mass of about 1675 MeV. With this mass, it would be *stable* with respect to the strong interactions, decaying instead weakly to such channels as $\Lambda + K^-$. The members of the lowest octet and decimet of baryons are shown in figure 9.6.

The next largest representation of SU(3) which could decay to a pair of octets and could hold a Δ is 27-dimensional. It contains resonances of positive strangeness, decaying to such states as $K^+ + p$ and $K^+ + n$. However, no such resonances had been seen [117].

The smallest G_2 multiplet accommodating the Δ was 14-dimensional. Although it had room for a state of strangeness −3 (like the 10-plet of SU(3)), it had an isoscalar positive-strangeness resonance. The absence of such a resonance thus weighed against the G_2 scheme, and favoured the 10-plet (rather than the 27-plet) assignment of the multiplet within SU(3).

The SU(3) prediction of a strangeness −3 state was stressed by Gell-Mann and Ne'eman at the 1962 Geneva Conference [118]. Shortly thereafter, a new 80″ bubble chamber at Brookhaven was exposed to a beam of K⁻-mesons. The story of how the predicted particle, the Ω^-, was discovered [119] in 1964 is one of the exciting chapters in the history of particle physics. Its mass was almost exactly the predicted value, and the first event (figure 9.7) was a particularly clean signature in which all the decay products revealed themselves.

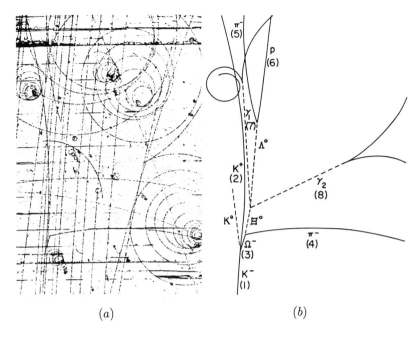

Figure 9.7. *First evidence for the Ω^- baryon* [119]. *(a) Bubble-chamber photograph; (b) interpretation.*

9.5. Interactions up to the mid-1960s

9.5.1. Weak interactions and V–A theory

9.5.1.1. Observation of the neutrino. The neutrino was postulated in the early 1930s to carry off the unseen energy and momentum in the beta-decay process n → p + e⁻ + ν (see section 9.2). However, no direct evidence for it had been found. After World War II, F Reines and some of his colleagues considered using a nuclear explosion as a copious source of neutrinos [120]. Eventually he and Clyde Cowan decided to use neutrinos produced at a nuclear reactor, eliciting an approving letter from Fermi that the new method should be much simpler to carry out and had 'the advantage that the measurement can be repeated any number of times'.

The inverse of the beta-decay reaction, the process $\bar{\nu}+p \rightarrow n+e^+$, was studied. A very large target, well shielded against accidental interactions, was required to observe this process, which is about 10^{18} times less frequent than a proton–proton interaction.

The initial experiment was performed at the Hanford reactor in 1953. Signals with the reactor on and off were compared. The large reactor-off signal, eventually traced to cosmic-ray interactions, prevented a definitive observation [121]. The experiment was then moved to Savannah River, yielding a signal sufficiently convincing [122] for Reines to send a telegram to Pauli in June of 1956 informing him that the particle he predicted more than 25 years earlier had been found.

The observation of the reaction $\bar{\nu} + p \rightarrow e^{+} + n$, without any examples of $\bar{\nu} + n \rightarrow e^{-} + p$, was the first proof of *lepton number conservation* in the weak interactions. The electron and the neutrino, by convention, are assigned lepton number $L = 1$. The neutral lepton accompanying an electron in beta decay (the source of reactor neutrinos) must then be an antineutrino with $L = -1$, giving rise only to positrons (with $L = -1$).

9.5.1.2. Early classification of beta decay. Fermi's 1934 theory of beta decay [5], mentioned in section 9.2, was based on an interaction of the *vector* (V) type between pairs of fermions, one of five possibilities consistent with Lorentz invariance and mirror symmetry. The others are scalar (S), tensor (T), axial vector (A) and pseudoscalar (P), whose names refer to each fermion pair's properties under Lorentz transformation.

The non-relativistic limit of the S and V interactions corresponds to a unit operator evaluated between spinors. Beta-decay transitions of this type were called *Fermi* transitions; many were identified experimentally. The non-relativistic limits of the T and A interactions involve instead the Pauli matrix σ. Transitions of this *Gamow–Teller* type [123] also were identified quite early, through their characteristic change of one unit of nuclear spin, suggesting complements to Fermi's hypothesis.

The energy dependence of the beta-decay spectrum depends on the form of couplings. When S and V or T and A interactions coexist, terms arise of the form m_e/E_e, where m_e and E_e are the electron's mass and energy. Such *Fierz interference* terms [124] were not seen, restricting the allowed interactions.

Electron–neutrino correlations in beta-decay experiments allow one to sort out different interaction types, with distributions proportional to $1 + b\hat{p}_e \cdot \hat{p}_\nu$, with [125] ($b = -1, 1, \frac{1}{3}, -\frac{1}{3}$) for pure (S, V, T, A) interactions, respectively. In the mid-1950s, experiments on $^6\text{He} \rightarrow {}^6\text{Li} + e^- + \bar{\nu}$, $^{19}\text{Ne} \rightarrow {}^{19}\text{F} + e^+ + \nu$ and $^{35}\text{Ar} \rightarrow {}^{35}\text{Cl} + e^+ + \nu$ gave conflicting results, which were soon to be sorted out.

9.5.1.3. Universality hypothesis. The weak interactions, though initially involved only in nuclear beta decay, were soon seen to participate with equal strength in $\mu \rightarrow e\nu\bar{\nu}$ and muon capture by nuclei. This observation led to the hypothesis [126] of *weak universality*, according to which any of the pairs (p, n), (e, ν), and (μ, ν) would interact equally with itself or any other pair. The charged pion's decay proceeds at a rate consistent with weak universality if dominated by a nucleon–antinucleon intermediate state. The decays of strange particles appeared to be due to an interaction similar to but somewhat weaker than the beta-decay interaction.

9.5.1.4. Overthrow of parity conservation. As decays of charged strange mesons were identified in the early 1950s, the masses of particles decaying in various modes all clustered more and more closely about a single value, now known to be $m(K) = 493.6$ MeV/c^2, with lifetimes showing a similar approach to a common value [127]. However, two distinct decays, called $\theta^+ \to \pi^+ + \pi^0$ and $\tau^+ \to \pi^+ + \pi^+ + \pi^-$, could not correspond to the same particle if the weak interactions were invariant under mirror symmetry.

An examination [128] of the kinematic distribution of the three-pion ('τ^+') decays favoured zero relative angular momentum between the two like pions and between them and the unlike pion. In that case, the τ^+ had to have spin J equal to zero and negative intrinsic parity P (since the parity of each pion was identified as negative; see section 9.4.2.1). A mixture of internal relative angular momenta leading to $J^P(\tau^+) = 2^-$ was also possible.

The two-pion ('θ^+') decay can only occur for a particle with $P = (-1)^J$. Thus, since the above analysis favoured $J^P(\tau^+) = 0^-$ or possibly 2^-, the τ^+ and θ^+ could not be the same *if parity was conserved in weak decays.*

The possibility of parity violation in the weak interactions began to be debated informally in seminars early in 1956. At the Sixth International Conference on High Energy Physics in Rochester, New York, in the early spring of that year, Feynman quoted a question by Martin Block asking whether the τ and θ could be different parity states of the same particle with no definite parity, implying that parity is not conserved [129]. Returning from the conference and reviewing all the evidence, T D Lee and C N Yang (figure 9.8) realized that no conclusive test had yet been performed of parity conservation in the weak interactions; such a test should measure the expectation value of an operator odd under spatial reflection, such as the scalar product of a momentum and a spin vector. They suggested several such tests [130]. (For two complementary historical accounts, see references [131].)

Before Lee and Yang's paper was published, R Oehme pointed out to them [132] that one could not see the effects of parity (P) violation in a beta-decay experiment if charge-conjugation invariance (C) and time-reversal symmetry (T) were preserved. Moreover, in the absence of significant final-state interactions, even C invariance alone prevented the observation of a P-violating asymmetry. Thus, Lee and Yang's suggestion of P violation in beta decay implied as well the violation of C. The published work incorporated this suggestion, and was followed by a joint paper [133] which also pointed out that the equality of masses and lifetimes for particles and antiparticles followed just from CPT invariance, a consequence of Lorentz invariance and local field theory [134]. Thus, for example, the observed equality of positive and negative muon lifetimes could not be used as evidence of invariance under charge

665

Figure 9.8. *T D Lee (left) and C N Yang in 1957. Photograph by Alan W Richards.*

conjugation since it followed from much weaker assumptions.

The Lee–Yang suggestion was taken up by C S Wu at Columbia and her collaborators at the National Bureau of Standards in Washington, and by V L Telegdi at Chicago and his student J I Friedman. Wu and her colleagues found that parity was indeed violated [135] in the beta decay of polarized ^{60}Co. Friedman and Telegdi began a search, which ultimately bore fruit [136], for parity violation in the decay $\pi \to \mu \to$ e in emulsions.

Wu returned to Columbia early in 1957 with the preliminary news of her results. Within a weekend Garwin, Lederman and Weinrich [29] were able to devise and perform an experiment searching for parity violation in $\pi \to \mu \to$ e. The story is told in reference [137] by one of the participants.

Lee and Yang suggested that one parity-violating observable could be an up–down asymmetry in the weak decays of polarized Λ hyperons. The observation of this asymmetry [138] was a further confirmation that, indeed, parity was violated in the weak interactions.

The weak interactions as now written down could be described by a theory which violated spatial reflection (P) and charge conjugation (C) symmetries, but was invariant under time reversal (T), the product CP thereby satisfying the requirement of a CPT-invariant theory. The conservation of T by the weak interactions, as newly formulated, explained why the neutron did not appear to have an electric dipole moment. Experiments placed quite stringent limits on such a moment [139]. Its absence had been thought to be evidence against parity violation, but it was now realized that T invariance alone would forbid an electric dipole moment for any elementary particle.

9.5.1.5. V–A theory and conserved vector current (CVC). In the initial experiment demonstrating parity violation in beta decay, the electrons were found to be emitted in a direction correlated with the spin of the initial nucleus. Soon thereafter, a number of experiments found that the spins of beta-decay electrons and positrons were correlated with their own velocities. Electrons were emitted spinning in a predominantly left-handed manner, or with left-handed *helicity*, while positrons emerged primarily with right-handed helicity.

Before the observation of parity violation in the weak interactions, Salam and Landau had proposed that a neutrino could exist in a single helicity state [140]. Indeed, Weyl had considered spinors with two components much earlier, but this possibility was rejected by Pauli in the 1930s for the neutrino because it did not conserve parity [141]. The ordinary spin-$\frac{1}{2}$ particle satisfying the Dirac equation has four degrees of freedom—two for each spin of both particle and antiparticle. When the rest mass is zero, however, the four-component solution breaks apart into two two-component ones, the first describing a left-handed particle and a right-handed antiparticle and the second describing a left-handed antiparticle and a right-handed particle.

A two-component neutrino was tailor-made for the large degree of parity violation observed in beta-decay experiments [142]. If a neutrino had one helicity and its antiparticle the opposite helicity, an electron released with velocity v in beta decay had to have polarization equal to $\pm v/c$, while a positron with the same velocity had to have the opposite polarization. This was indeed verified. Electrons always had polarization $P(e^-) = -v/c$, while positrons had polarization $P(e^+) = v/c$.

In view of the apparent two-component nature of the neutrinos participating in beta decay and the beauty and simplicity of the two-component formalism, Feynman and Gell-Mann and Sudarshan and Marshak proposed [143–145] that *all* particles participate in the weak interactions in the same two-component manner. As a consequence, each particle pair (such as (e, v), (μ, v), and (n, p)) had to participate in the weak interaction with a definite combination of vector (V) and axial (A) strengths, by convention called V–A. The scalar (S), tensor (T) and pseudoscalar (P) interactions vanish under the two-component hypothesis.

The universal V–A interaction agreed with many results at the time it was proposed. In addition to predicting the polarizations of electrons and positrons in beta decay, it reproduced the observed muon lifetime and the electron–neutrino correlations in most beta-decay processes. However, it seemed incompatible with some observations, including the apparent dominance of S and T interactions in ^{6}He beta decay [146] and the apparent absence [147] of the decay $\pi \to e\nu$. In due course, a new ^{6}He experiment [148] confirmed the V–A prediction, and the $\pi \to e\nu$ decay was eventually found at the predicted level [149]. The theory also predicted parity violation in purely *non-leptonic* processes, thus resolving the $\tau - \theta$ puzzle mentioned earlier. It implied that the neutrino emitted in beta decay was left-handed, as was verified in an elegant experiment (see Box 9A) by M Goldhaber, L Grodzins and A Sunyar [150] using the apparatus described in figure 9.9. A more detailed discussion of these and other tests of the V–A theory appears in reference [151].

A general test for Lorentz structure of couplings in muon decay had been proposed in 1950 [152]. The electron energy spectrum in muon decay could be described in terms of a parameter ρ and the ratio x of the electron energy in the muon rest frame to its maximum value. Normalized to unit integral over the range $0 \leqslant x \leqslant 1$, the spectrum has the form $N(x) = 6x^2[2(1 - x) + (4/9)\rho(4x - 3)]$. The V–A theory implied that labelling electrons as particles but muons as antiparticles yields $\rho = 0$, while labelling both muons and electrons as particles yields $\rho = \frac{3}{4}$. Experiments eventually converged on the latter value.

The universal V–A interaction was also generalized to strange particles. Notable among the results [143] was the linkage of strangeness S and charge Q in beta-decay interactions, represented by the empirical rule $\Delta S = \Delta Q$. If $\Delta S = -\Delta Q$ transitions also had been allowed, they would have led to unobserved processes such as $\Xi^- \to \pi^- + n$.

The universal weak (current) × (current) interaction, where each V–A current carries unit charge, implies that all Fermi-type interactions in beta decay stem from the nucleon's V current, while all Gamow–Teller transitions arise from A. A remarkable feature of these currents, particularly of the vector current, is their universal strength. For example, in Fermi transitions between nuclei of zero spin, as in ^{14}O beta decay, the coupling strength appears to be almost identical with that in muon decay, despite the presence of strong interactions which in principle could modify the nucleon's current.

To explain the universality of the weak vector current, Feynman and Gell-Mann [143] and Gershtein and Zel'dovich [153] suggested that the charged weak vector current belonged to an isotopic spin *triplet* of conserved vector currents. This *conserved vector current* hypothesis identified the vector weak current with the generator of the isotopic spin symmetry, whose matrix elements are set by simple isospin considerations, independently of strong interaction details. For example,

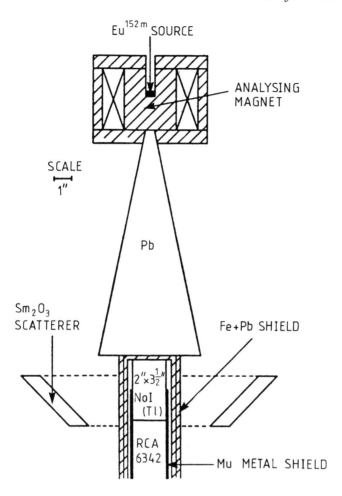

Figure 9.9. *Apparatus in the experiment of Goldhaber, Grodzins and Sunyar [150] to determine the neutrino helicity by measuring circular polarization of resonant scattered γ-rays. The apparatus has cylindrical symmetry about the vertical axis.*

the rate for the decay $\pi^+ \rightarrow \pi^0 + e^+ + \nu_e$ was predicted precisely [143]; the process was observed, with the predicted rate, several years later [154].

9.5.1.6. Two-neutrino hypothesis and its confirmation. The current–current form of the weak interactions, with a Fermi coupling strength $G_F \approx 10^{-5} m_p^{-2}$, leads to cross sections which violate the unitarity of the S-matrix at high energies. One cure for this problem was the proposal, first appearing in Yukawa's theory of the meson and in various forms thereafter [143, 155], that the weak interactions arose from the exchange of a meson, in analogy with the pion for the strong force and the photon for electromagnetism. The V–A nature of the weak interaction required

BOX 9A: THE HELICITY OF THE NEUTRINO

Everyone knows that the neutrino is a most elusive particle requiring tons of material just to detect. Who then would have the daring to even contemplate measuring its spin direction relative to its direction of motion, its helicity? M Goldhaber, L Grodzins and A Sunyar (GGS) (reference [150]) succeeded in doing this in an ingeniously conceived and executed experiment. It was an experiment that required encyclopaedic knowledge of nuclear isotopes as well as cunning experimental technique.

The nucleus ^{152}Eu decays to ^{152}Sm by electron capture from the K-shell with the emission of a neutrino with energy 0.840 MeV. The ^{152}Eu has spin zero but the ^{152}Sm which accompanies the neutrino is in an excited state, Sm* with spin 1. It decays in 3×10^{-14} s to the spin-zero ground state with the emission of a gamma ray. How did GGS put all this together to measure the helicity of the neutrino? By conservation of angular momentum, the spin of the neutrino plus the Sm* must equal the spin of the captured electron, forcing the spin of the Sm* to be opposite to that of the neutrino. Since the Sm* nucleus is recoiling away from the neutrino, the *helicity of the Sm* must be the same as the neutrino*. Measuring the helicity of the neutrino has been reduced to measuring the helicity of the recoiling Sm*.

The angular momentum of the Sm* will be carried away by the emitted gamma ray. It is possible to measure the circular polarization of gamma rays by passing them through magnetized iron since the absorption cross section depends on the spin direction of the scattering electrons. To get the helicity of the Sm* it is still necessary to determine its direction of travel relative to the direction of the gamma. GGS note that the gamma-ray energy is 0.960 MeV, not far from the neutrino energy. If the gamma ray is emitted in the direction the Sm* is travelling, the energy will be boosted just enough so that the gamma will be scattered resonantly (hence, with a large cross section) from Sm in its ground state if the scattering angle is about 90°. This then determines that the gamma ray has been emitted in the direction of the Sm*.

This brilliant, complex scheme was implemented in the apparatus shown schematically in figure 9.9. By measuring the gamma-ray intensity in the NaI crystal as a function of the direction of the magnetization of the iron, the neutrino was found to be *left-handed*!

this particle to have spin 1, i.e. to be a *vector* boson. The modern name for this boson is W.

The vector boson proposal predicted the process $\mu \to e\gamma$ to occur with a branching ratio far above the upper limits set in the early 1960s. The process involved a virtual transition of the muon to some charged intermediate state and a neutrino. The charged intermediate state would then radiate a photon and reabsorb the neutrino to become an electron. However, the predicted decay would not occur if the muon and electron coupled to *separate* neutrinos.

An experiment was performed at Brookhaven National Laboratory by the participants shown in figure 9.10 to study the interactions of neutrinos produced in the decay of pions in the process $\pi^\pm \to \mu^\pm \nu$. Neutrinos distinct from those emitted in beta decay would produce only muons, not electrons, in reactions such as $\nu+n \to \mu^-+p$ and $\bar{\nu}+p \to \mu^++n$. The result was conclusive: the muon and electron neutrinos were different [156].

The history of the two-neutrino experiment is told in reference [157]. The development of neutrino beams had been first proposed by B Pontecorvo and M Schwartz [158], with resulting physics anticipated in a prescient paper by Lee and Yang [159].

One goal of early neutrino experiments was the search for the W-boson mentioned above. The results in experiments on both sides of the Atlantic [160] were negative, placing a lower limit of 2 GeV/c^2 on the mass of the W, which is now known to be much heavier. Neutrino interactions were so rare in those early days (1963) that a bottle of champagne was opened each time an interaction was seen at the European experiment, performed at CERN.

9.5.2. CP violation

9.5.2.1. Implications of P and C violation for neutral kaon systems.

The V–A theory of weak interactions violated both parity (*P*) and charge-conjugation invariance (*C*), while preserving time-reversal invariance and the product *CP*. Now, the original prediction of two types of neutral kaons, the short-lived kaon decaying to $\pi\pi$ and the long-lived one forbidden from doing so [91], was based on the assumption of *C* invariance of the weak interactions. Similar results were shown to follow [161] from combined invariance under the product *CP*. Hence, the K_2^0, mentioned in section 9.4.4, was still expected not to be able to decay to $\pi\pi$ if *CP* invariance was valid in the weak interactions.

9.5.2.2. Initial experiments.

Searches for the decay $K_2^0 \to \pi^+\pi^-$ during the late 1950s and early 1960s failed to find any signal at a level of a part in 300. With the advent of spark chambers permitting selective triggering on events of a given type, a more sensitive search was mounted by

Figure 9.10. *Participants in the two-neutrino experiment at Brookhaven* [156]. *From left to right: J Steinberger, K Goulianos, J-M Gaillard, N Mistry, G Danby, W Hayes, L Lederman, M Schwartz. Top: then; bottom: recently.*

J Christenson, J Cronin, V Fitch and R Turlay at Brookhaven National Laboratory. The early history and motivation for this experiment are described in reference [162]. A signal was indeed seen (figure 9.11), corresponding to one event for every 500 decays [163].

Theoretical analyses [133, 164] had set the stage for a description of CP violation in decays of neutral kaons. After its discovery,

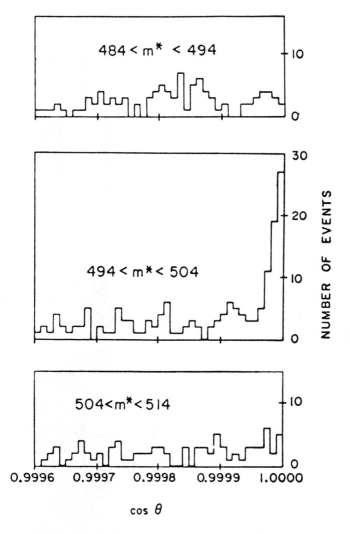

Figure 9.11. *Evidence* [163] *for the decay* $K_L \to \pi^+\pi^-$. *The angular distribution of the two-pion system with respect to the beam is shown for three mass ranges, of which the middle range corresponds to the signal.*

Wu and Yang suggested further experiments [165], establishing many conventions still in use today. Since both K_1^0 and K_2^0 were now seen to decay to pairs of pions, their names were changed to denote states of definite lifetime: K_S for the short-lived neutral kaon, with lifetime 0.09 ns, and K_L for the long-lived neutral kaon, with lifetime 52 ns.

The presence of the decay $K_L \to \pi^+\pi^-$ indicated CP violation in the weak interactions. Assuming the validity of the CPT theorem, this implied also T violation, a result not to be taken lightly. The

673

experimenters spent more than six months analysing their signal before concluding that their data were indeed evidence for CP violation. Detailed analyses (see, for example, reference [166]) without the assumption of CPT invariance later showed that indeed the data were also signalling T violation.

A crucial test confirming that the $\pi^+\pi^-$ final state indeed originated from the long-lived kaon was the observation [167] of interference between the decays $K_L \to \pi^+\pi^-$ and $K_S \to \pi^+\pi^-$. The K_S was produced by *coherent regeneration* [164, 168]. As a K_L beam passes through matter, the phase and magnitude of the beam's K^0 and $\bar{K}^0$ components change in different ways, inducing a mixture of K_S. As a result, one can measure not only the magnitude but also the phase of the ratio of $K_L \to 2\pi$ and $K_S \to 2\pi$ amplitudes.

The evidence for CP violation has so far been confined exclusively to the system of neutral kaons. Reviews of major results since the discovery of CP violation can be found in references [166] and [169].

9.5.2.3. Decays to charged and neutral particles. The origin of CP violation in neutral kaon decay still is not understood in terms of a fundamental theory. A question arose immediately whether the observed effect results entirely from mixing, or stems in part from direct CP violation in the decay process.

The states $K_{S,L}^0$ of definite mass and lifetime can be regarded as mixtures of states $K_{1,2}^0$ with even and odd CP. In a CPT-conserving theory [133], one can write

$$|K_S\rangle = \frac{1}{\sqrt{(1+|\epsilon|^2)}}(|K_1\rangle + \epsilon|K_2\rangle) \qquad |K_L\rangle = \frac{1}{\sqrt{(1+|\epsilon|^2)}}(|K_2\rangle + \epsilon|K_1\rangle)$$

where the complex parameter ϵ denotes the effect of the mixing. The amplitudes for the decays of K_L to $\pi^+\pi^-$ or $\pi^0\pi^0$ may be characterized by their ratios to the corresponding K_S decay amplitudes

$$\frac{A(K_L \to \pi^+\pi^-)}{A(K_S \to \pi^+\pi^-)} \equiv \eta_{+-} \qquad \frac{A(K_L \to \pi^0\pi^0)}{A(K_S \to \pi^0\pi^0)} \equiv \eta_{00}$$

where $\eta_{+-} = \eta_{00} = \epsilon$ if the only CP violation takes place through mixing. In general one can write $\eta_{+-} = \epsilon + \epsilon'$ and $\eta_{00} = \epsilon - 2\epsilon'$. Here ϵ' describes the effect of direct CP violation (i.e. that not due to mixing) in the decay amplitude, leading to a final state with isospin 2.

The leading candidate for a theory of CP violation in the neutral kaon system, the presence of phases in the Cabibbo–Kobayashi–Maskawa matrix (section 9.9), predicts that ϵ' should have a phase very close to that of ϵ, with $\mathrm{Re}(\epsilon'/\epsilon)$ a few parts in 10^4. However, a model with $\epsilon' = 0$, in which CP violation in neutral kaons is due entirely to a $\Delta S = 2$

'superweak' interaction [170] mixing K^0 and $\bar{K}^0$, still is not conclusively excluded by present data.

One experiment [171] finds $Re(\epsilon'/\epsilon) = (7.4 \pm 5.9) \times 10^{-4}$, consistent with zero, while another [172] finds $Re(\epsilon'/\epsilon) = (23 \pm 6.5) \times 10^{-4}$, non-zero at the level of more than three standard deviations. There is not a serious contradiction between the two results if $Re(\epsilon'/\epsilon)$ is around 14×10^{-4}. Improved versions of both experiments are planned. The theory can also be tested by studies of *B*-mesons (section 9.9).

9.5.2.4. Asymmetries in semi-leptonic decays. By comparing rates for neutral kaons to decay to states such as $\pi^- e^+ \nu_e$ and $\pi^+ e^- \bar{\nu}_e$, one can learn the relative admixture of K^0 and $\bar{K}^0$ in each neutral kaon's wavefunction [173]. The $\Delta S = \Delta Q$ rule of strangeness-changing weak interactions mentioned above implies that only negative pions emerge from K^0 and only positive from $\bar{K}^0$. *CP* violation implies an asymmetry proportional to $2Re\epsilon$ in rates of K_L decay to these final states. This asymmetry was searched for and found; the results [174] are in perfect accord with expectations.

9.5.2.5. CP violation and baryon asymmetry in the Universe. Shortly after *CP* violation was discovered, Sakharov [175] pointed out that it was a key ingredient in understanding why the observed Universe contains more baryons than antibaryons. Other crucial features were interactions which violate baryon-number conservation, and a period in the early Universe when reactions governed by these interactions were out of thermal equilibrium. Those conditions have been realized in subsequent models [176], to which we shall return in sections 9.12 and 9.13.

9.5.3. Current algebras

9.5.3.1. The role of the pion. In the newly proposed V–A theory, the weak interactions of strongly interacting particles remained mysterious. While the vector weak current of nucleons seemed to be nearly unaffected by the strong interactions, the axial vector coupling strength, parametrized by $G_A \approx 1.25$, deviated significantly from the value $G_A = 1$ expected if nucleons behaved like leptons. Moreover, decays of strongly interacting particles to purely leptonic final states were characterized by arbitrary constants, such as the pion decay constant f_π for $\pi \to \mu + \nu$.

Armed with experience in both strong and weak interactions, M L Goldberger and S B Treiman [177] embarked on a calculation of f_π. Their result involved a crucial role for the nucleon–antinucleon intermediate state and hence for the pion–nucleon coupling constant $g_{\pi NN}$. They obtained the relation $f_\pi = m_N G_A / g_{\pi NN}$, which is accurate to a few per cent. Here m_N is the mass of the nucleon.

The vector current's time component, when integrated over all space, is the *isotopic charge*, or generator of isotopic spin transformations. The conservation of vector current is thus associated with isotopic spin invariance. The space integral of the *axial* current's time component also performs isotopic spin transformations, rotating left-handed and right-handed fermions oppositely in isospin space. Invariance under separate isospin rotations of left-handed and right-handed fermions is known as *chiral invariance*.

But is the axial current conserved? Apparently not; the divergence of the matrix element of the axial current between one-nucleon states is proportional to m_N, and so cannot vanish unless the nucleon is massless. Under such circumstances, one indeed would have a chiral invariance. Chiral symmetry would be realized in the *Wigner–Weyl* sense, by manifest invariance of both the equations of motion and their solution. One implication of manifest chiral symmetry is the prediction that for every particle with a given parity, there should be a degenerate particle with opposite parity. This certainly did not seem to be so.

In a sequel to their first paper on the pion decay constant, Goldberger and Treiman [178] showed that the requirement of a conserved axial current entails a zero-mass pole in beta-decay amplitudes. Inspired by this result and by an analogy with superconductivity, Y Nambu [179] discovered another mode in which chiral symmetry could be realized. If the *vacuum itself* were not chirally symmetric, so that the part of the symmetry generated by the axial current was *spontaneously broken*, there was no need for parity doubling in the spectrum of states like the nucleon. Instead, a massless pseudoscalar particle, the pion, miraculously appeared; the parity partner of the nucleon was then a state with a nucleon and a pion. This behaviour illustrated a general theorem of J Goldstone [180]: spontaneous breakdown of a global, i.e. space-independent symmetry (such as chiral symmetry), necessarily leads to massless particles in the spectrum. For systems of pions and nucleons, chiral symmetry is said to be realized in the *Nambu–Goldstone* sense: via spontaneous symmetry breaking. In Nambu's approach, the pion satisfied the Goldberger–Treiman relation as a natural consequence of its role in the breakdown of chiral invariance. The zero-mass pole which Goldberger and Treiman had discovered could be, in fact, identified with the pion in the limit of exact chiral invariance.

M Gell-Mann and M Lévy were motivated to understand the deviation of the axial coupling constant G_A from unity by identifying the divergence of the axial current with the pion field itself [181]. They constructed several field-theoretic models in which this was so. The Goldberger–Treiman relation was a welcome consequence. The fact that the divergence of the axial current was non-zero, but identified with the pion field, was called *partial conservation of axial current* (PCAC).

9.5.3.2. Gell-Mann–Lévy discussion of universality. Although the vector current was not expected to be renormalized by the strong interactions, the Fermi constant G as measured in ^{14}O beta decay appeared to be about 3% lower than G_μ as measured from muon decay. Gell-Mann and Lévy proposed [181] that the missing 'strength' was made up by the beta-decay coupling of the proton to the Λ: in modern notation, the hadronic beta-decay current then involved the charge-changing transitions p $\leftrightarrow$ n$\cos\theta$ + $\Lambda\sin\theta$. With $\sin^2\theta \approx 0.06$, one was then able to understand both $G/G_V \approx 0.97$ and the relative suppression of strangeness-changing weak decays (particularly notable for the process $\Lambda \to$ pe$^-\bar{\nu}$ [182]).

9.5.3.3. Gell-Mann's algebra of currents. In search of a dynamical principle underlying the success of unitary symmetry, Gell-Mann [183] (figure 9.12) realized that both isotopic spin (the group SU(2)) and its generalization to SU(3) were generated by *charges*—space integrals of time components of vector currents. These charges F_i ($i = 1, \ldots, 8$) obeyed an algebra of the form $[F_i, F_j] = \mathrm{i}f_{ijk}F_k$, where the f_{ijk} are totally antisymmetric *structure constants*.

The spatial integrals of time components of axial vector currents give rise to charges F_i^A, which necessarily transform as vectors under the symmetry: $[F_i, F_j^A] = \mathrm{i}f_{ijk}F_k^A$. The existence of an octet of light pseudoscalar mesons is the reason the axial charges are not conserved; when acting on the vacuum, they produce a pion, kaon or η.

When one has a set of generators of a symmetry, it is natural to ask for the behaviour of *all* commutators. Gell-Mann postulated that the algebra was completed by the simplest possibility for the commutator of two *axial* charges: $[F_i^A, F_j^A] = \mathrm{i}f_{ijk}F_k$. This is indeed the case for the leptons. It could be true for strongly interacting particles as well if they were made up of more fundamental entities—at least, one could conceive of interactions not affecting the basic relations. These entities could be identified with *quarks* [184]—members of the triplet representation of SU(3). We shall discuss them in section 9.6.

The combinations $(F_i + F_i^A)/2 \equiv F_i^R$ and $(F_i - F_i^A)/2 \equiv F_i^L$ then would obey two independent SU(3) algebras. This SU(3) $\times$ SU(3) structure has led to successful predictions for many quantities previously regarded as the province of intractable strong-interaction physics.

9.5.3.4. Cabibbo theory of strange-particle decays. If vector and axial vector currents transformed as generators of SU(3), their matrix elements between various states of SU(3) multiplets could be related to one another. By 1963, data on beta decays of various hyperons in the baryon octet had been accumulating, and the time was ripe for such an analysis.

N Cabibbo [185] postulated that the charge-changing weak current behaved as a member of an SU(3) octet, consisting of a linear combination of a piece, transforming like a charged pion (with coefficient $\cos\theta$) and a

Figure 9.12. *Murray Gell-Mann (top); Yuval Ne'eman (bottom).*

charged kaon (with coefficient $\sin\theta$). While the angle θ is the one Gell-Mann and Lévy proposed for rescuing universality of the weak hadronic current, we shall henceforth refer to it as θ_C.

The conservation of vector current specifies uniquely the vector current's matrix element between baryon states. However, the nonconservation of axial current allows for two types of matrix elements for the axial current. In SU(3) there are two ways to couple an octet current to initial and final octet baryons, called F (totally antisymmetric) and D (totally symmetric).

One combination of F and D coupling for the axial current is the axial charge of the nucleon, $F + D = G_A \approx 1.25$. The Cabibbo theory described decays such as $K \to \pi e\nu$, $\Lambda \to pe\nu$, $\Sigma^- \to ne^-\nu$, and $\Sigma^- \to \Lambda^0 e^-\nu$ with a single value of θ_C and a self-consistent value of F/D. In modern fits, which also include data on several other hyperon beta decays (and on the ratios of axial to vector couplings in some of them) one finds [186] $\sin\theta_C \approx 0.22$ and $F/D \approx 2/3$. This last value turns out to be close to what one expects in a quark picture of baryons.

9.5.3.5. Adler–Weisberger relation. The PCAC hypothesis permitted calculations of soft pion emission [187] in a manner analogous to calculations of soft photon emission using general principles of electromagnetism [188]. Stephen Adler, a skilled practitioner of this technique, used it to study a series of processes, including low-energy pion–nucleon scattering [189]. Hearing of the proposal by Murray Gell-Mann and his student Roger Dashen to apply the commutator of two axial isospin generators to obtain a relation between the axial–vector coupling constant and pion–nucleon scattering, Adler realized that his experience was ideal for the problem. In very short order he related G_A to the difference in total cross sections of positive and negative pions on nucleons, integrated over energy [190]. At the same time, W I Weisberger produced a similar calculation, and subsequently generalized the result to relate $|\Delta S| = 1$ transitions to kaon–nucleon scattering [191].

The Adler–Weisberger relation in the zero-pion-mass limit may be written

$$G_A^2 = 1 - 2\frac{f_\pi^2}{\pi} \int_0^\infty \frac{d\nu}{\nu} (\sigma^{\pi^- p}(\nu) - \sigma^{\pi^+ p}(\nu))$$

where ν stands for the laboratory energy of the pion. Its prediction for G_A depends to some extent on the treatment of corrections for massive pions, but the resulting value $G_A \approx 1.2$ was sufficiently close to experiment that the power of current algebra was immediately recognized.

9.5.3.6. Other current algebra relations. The Adler–Weisberger relation exploits the nonlinearity of the current commutation relations to normalize axial charges. The commutator between axial and vector charges also contains useful information, providing a relation [192] between the semi-leptonic process $K \to \pi e\nu$ and the purely leptonic decay $K \to \mu\nu$. Even purely hadronic processes such as pion–pion scattering could be attacked by such methods, as shown by Weinberg [193]. Many other successes of current algebra were chronicled in contemporary texts and reviews [194].

9.5.4. Strong-interaction schemes
While the theory of weak interactions enjoyed tremendous progress during the 1950s, the strong interactions remained a mystery. Symmetry arguments sufficed for many results, but the underlying forces were not understood until the advent of quantum chromodynamics in the 1970s. Still, the efforts to understand strong forces bore fruit in a number of areas even in the absence of a fundamental theory.

9.5.4.1. S-matrix theory and dispersion relations. A unitary matrix relating outgoing to incoming scattering states was introduced by Wheeler in the 1930s [195] in the context of non-relativistic nuclear reactions, with a fully relativistic treatment developed independently by Heisenberg [196]. The scattering matrix, or *S*-matrix, as it came to be called, had an interesting history during World War II [197]. News of Heisenberg's work was brought to Japan by German submarine in the form of a letter from Heisenberg to Nishina. A unitary *S*-matrix appeared in Japanese literature of the 1940s in analyses of microwave junctions by Tomonaga and his group [198]. The US wartime microwave work also employed the *S*-matrix in descriptions of junctions [199]. Around this time Stückelberg independently introduced an analogue of the *S*-matrix [200]. In the physics of antenna impedance matching, a transformation very similar to Stückelberg's had been proposed even earlier, forming the basis of the frequently used *Smith chart* [201]. The development of the relativistic *S*-matrix in the post-war years owed much to the work of Møller [202].

As the strong coupling of pions to nucleons became clear in the 1950s, physicists despaired of describing the strong interactions by quantum field theory, so successful for QED. It was hoped [203] that by characterizing the singularities of the *S*-matrix, a theory could avoid expanding amplitudes as a perturbation series, which fails because of the large pion–nucleon coupling constant. Thus began a period of intense study of analytic properties of scattering amplitudes and of 'bootstrap' theories in which the known particles were all viewed as composites of one another. An early success of this programme was the development of dispersion relations for scattering amplitudes [204], relating the real parts of amplitudes to integrals over total cross sections. Precise measurements of pion–nucleon total cross sections and of real parts of amplitudes [205] provided one impressive verification of these relations. Dispersion relations could also be written simultaneously in variables corresponding to relativistic generalizations of energy and momentum transfer [206], providing insight into parameters governing the range of strong forces.

9.5.4.2. Chew–Low theory and isobars. The first pion–nucleon resonance, the $\Delta(1232)$ mentioned above, was described by Chew and Low [207] in a theory that viewed the force between a pion and a baryon (the nucleon

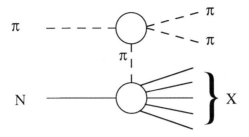

Figure 9.13. *Diagram describing the Chew–Low proposal* [208] *for studying pion–pion scattering.*

or Δ) as due primarily to baryon exchange. Indeed, the singularities in the scattering amplitude most important for its low-energy behaviour are just those associated with baryons in the crossed channel.

9.5.4.3. Pion–pion scattering and the bootstrap programme. The idea that strongly interacting particles could all be viewed as composites of one another (encouraged by Chew and Low's result for the Δ, nucleon and pion) grew in the 1950s into a fully fledged 'bootstrap' programme, dealing only with the self-consistent singularities of the S-matrix actually found in Nature. The key ingredients in this programme were the unitarity of the S-matrix and the analyticity and *crossing symmetry* of the scattering amplitude. This last feature relates such processes as $A + B \rightarrow C + D$ to *crossed reactions* such as $\bar{C} + B \rightarrow \bar{A} + D$.

A simple test of the bootstrap programme was provided by pion–pion scattering. Since pions are not available in the laboratory, how would one employ them as targets? The answer was provided by Chew and Low [208]: use virtual pions (see figure 9.13). In the reaction $\pi + p \rightarrow X + n$, for example, the scattering amplitude has a pole in the invariant momentum transfer, corresponding to exchange of a pion, which dominates the scattering for small enough values of momentum transfer.

Many bootstrap calculations of pion–pion scattering were performed [203], with varying degrees of success in reproducing the dominant low-energy features. These included a prominent P-wave resonance in the $I = 1$ channel, the $\rho(770)$ (the number refers to the mass in MeV/c^2), and a slow increase of the S-wave phase shift in the $I = 0$ channel. Once the low-energy behaviour predicted by current algebra [193] was specified, successful bootstrap calculations were indeed displayed [209]. However, analyticity, unitarity and crossing symmetry proved inadequate to specify scattering amplitudes uniquely.

9.5.4.4. Regge poles. A paper by Tullio Regge [210] dealing with scattering in a non-relativistic potential provided an important step in the description of scattering in particle physics. The S-matrix has poles

681

in complex angular momentum which *change their position as a function of energy*, solving a knotty problem that had plagued field-theoretic descriptions of scattering amplitudes.

Unitarity and the short-range nature of the strong force allowed Froissart to show that total cross sections can grow no more rapidly at high energy than $\sigma_T \sim (\log s)^2$, where s is the square of the centre-of-mass energy [211]. For exchange of any particle of spin J, the total and elastic cross sections should grow as $\sigma_T \sim s^{J-1}$ and $\sigma_{el} \sim s^{2J-2}$, respectively, exceeding the Froissart bound (and violating $\sigma_{el} \leqslant \sigma_T$) for $J > 1$. Particles of high spin had been observed, but cross sections showed no sign of violating the Froissart bound.

Adapting Regge's result to relativistic scattering, Chew and Frautschi proposed [212] that particles lie on *Regge trajectories*, with specific states corresponding to integer or half-integer values of angular momentum $J = \alpha(s)$. Since all the known trajectories had values of $\alpha(t)$ less than or equal to 1 for the scattering regime $t \leqslant 0$, where t represents the invariant momentum transfer variable, there was no conflict with the Froissart bound. Understanding the approach of total cross sections at high energy to an approximately constant value required the introduction of another trajectory with $\alpha(0) = 1$, called the *Pomeranchuk trajectory* in honour of Pomeranchuk's description [213] of the high-energy behaviour of total cross sections.

The derivation of singularities in the complex angular momentum plane from field theory using dispersion relations was performed by Gribov, Bardakci, Barut and Zwanziger, and Oehme and Tiktopoulos in 1962 [214].

The hypothesis of Regge pole exchange also implies a definite phase of the scattering amplitude. Evidence for this phase confirmed the hypothesis. The result relies on the asymptotic energy dependence of the scattering amplitude and its properties with regard to analyticity and crossing. Regge trajectories, an example of which is shown in figure 9.14, display a high degree of linearity in $s = m^2$, a feature to be discussed in section 9.8.3.

9.5.4.5. Duality and forerunners of string theory. Progress in one area of theory frequently leads to results in another. Thus, attempts to understand the strong interactions eventually led to a candidate theory for quantum gravity.

During the 1960s, it was popular to describe scattering processes at high energies by sums of Regge pole exchanges. For processes with no exchange of quantum numbers, the Pomeranchuk trajectory, or *Pomeron*, for short, plays the dominant role. Differences between various such processes are governed by contributions of the non-leading trajectories. If quantum numbers are exchanged, the Pomeron cannot contribute. For example, in pion–nucleon charge exchange, $\pi^- + p \rightarrow \pi^0 + n$,

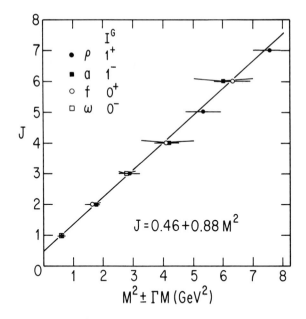

Figure 9.14. *Example of Regge trajectories for non-strange mesons. Points denote values of (squared) mass; error bars denote effects of natural widths. The trajectories for isospins I = 0 and 1 and for positive and negative G-parities (charge-conjugation eigenvalue of the neutral member of the isospin multiplet) are seen to coincide in this case.*

the dominant trajectory is the one on which the ρ-meson lies (see figure 9.14), with energy dependence at high energy well described by ρ trajectory exchange. But what about the behaviour at low energies, when resonances lead to rapid oscillations of the amplitude?

R Dolen, D Horn and C Schmid [215] found that Regge and resonance descriptions of pion–nucleon charge exchange are not to be added to one another, but instead are complementary descriptions of the same physics, the Regge trajectory providing an *average description of the resonant behaviour*. Moreover, the average description of the resonances entails a phase of the scattering amplitude similar to that expected at higher energies from Regge pole exchange. This *duality* between resonances and Regge behaviour applies only to non-Pomeranchuk trajectories [216], and can be visualized in terms of quark graphs [217].

M Ademollo, H Rubinstein, G Veneziano and M Virasoro studied the Dolen–Horn–Schmid duality result for the process $\pi + \pi \rightarrow \pi + \omega$, which is particularly simple under crossing symmetry. Over the course of a year they hammered together a model scattering amplitude with increasing features of duality [218]. Finally Veneziano, *en route* by boat from Israel to Italy and his first postdoctoral job at CERN, found a remarkable formula for a scattering amplitude with both poles and Regge behaviour in two kinematic variables [219]

$$A(s,t) \sim B(1 - \alpha(s), 1 - \alpha(t)) \equiv \frac{\Gamma(1 - \alpha(s))\Gamma(1 - \alpha(t))}{\Gamma(2 - \alpha(s) - \alpha(t))}.$$

The Veneziano amplitude created quite a stir among physicists accustomed to the intractability of the strong interactions. C Lovelace and J Shapiro [220] proposed a model pion–pion scattering amplitude very similar to the one just quoted, with many attractive features, including a low-energy limit in agreement with current algebra results [193]. Attempts to construct a theory yielding the Veneziano amplitude led to the forerunners of string theory [221] and of supersymmetry [222].

String theories of the hadrons were originally viewed as models of the strong interactions, perhaps valid in some limit. One of their seemingly unattractive features was their need to exist in 26 space-time dimensions, rather than the usual four. In 1974 J Scherk and J Schwarz [223] realized that these theories possess a massless spin-2 particle—a good candidate for the graviton—and, together with a few other devotees, began to study them as possible versions of quantum gravity. Then, in 1984, M Green and J Schwarz [224] discovered that supersymmetric versions of certain string theories could exist in 10, rather than 26, dimensions, and the extra six dimensions began to be interpreted in terms of internal symmetries of particles. These *superstring* theories became the subject of intense theoretical activity, which we shall describe further in section 9.13.

All in all, despite the absence of a fundamental theory, much was learned in the 1950s and 1960s from the study of the strong interactions. We shall return to them in section 9.8 in the context of quantum chromodynamics.

9.6. The quark revolution

During the 1960s and early 1970s, particle physics underwent a transformation. Hundreds of strongly interacting resonances and a bewildering hodgepodge of interactions were gradually understood in terms of a few basic building blocks interacting via Yang–Mills fields (to be described in section 9.7). Thus was the physics of quarks and gluons born. In this section we describe the building blocks; we return to the interactions in sections 9.7 and 9.8.

9.6.1. The quark model

9.6.1.1. Triplets as a basis for SU(3). A good deal of the behaviour of a symmetry group such as SU(2) or SU(3) may be learned from its action on its *fundamental* representation, corresponding to a triplet of particles for SU(3). One can build any SU(3) representation out of sufficiently many triplets or antitriplets.

A fundamental triplet, consisting of the proton, neutron and Λ as building blocks, was employed by Sakata [225] for constructing models of elementary particles, and by his colleagues [111] whom we have mentioned earlier in the context of SU(3). However, with the advent of the eightfold way (section 9.4.6.2), the proton, neutron and Λ no longer could be regarded as fundamental; they belonged instead to an *octet* of particles, along with three Σs and two Ξs. Moreover, there appeared to exist a 10-plet of baryons of spin $\frac{3}{2}$, including the $\Delta(1238)$, the $\Sigma(1385)$, the $\Xi(1530)$ and the predicted $\Omega(1675)$. Triplets were employed quite early to construct such states by Goldberg and Ne'eman [226].

Could one regard all the baryons (and the emerging multiplets of spin-0 and spin-1 mesons) as composites of more fundamental entities? During a visit to Columbia University in 1963, Murray Gell-Mann proposed taking SU(3) triplets seriously as fundamental subunits of the hadrons [184, 227]. Calling the subunits *quarks* (as in the passage 'Three quarks for Muster Mark' from *Finnegans Wake* [228]), he found that all baryons could be identified with states of three spin-$\frac{1}{2}$ quarks, while mesons could be represented as quark–antiquark pairs. The members of the SU(3) triplet were an isospin doublet (u, d) and a (strange) singlet s. They had to have fractional charges: $Q(u) = \frac{2}{3}$, $Q(d) = Q(s) = -\frac{1}{3}$, but this was a heavy price, since fractionally charged entities had never been observed in Nature. Working independently at CERN, George Zweig developed a picture of the same fractionally charged constituents of matter [229], which he called 'aces'. In the choice of a name, quarks won out over aces, poetry over poker.

Zweig wished to understand why some decays were allowed and others were forbidden. A spin-1 meson known as the ϕ, conjectured to exist [230] as a mixture of a singlet and octet in SU(3), had been observed [231] decaying to K^+K^- and $\bar{K}^0 K^0$. Its decay to $\rho\pi$, though allowed by the combined symmetry of charge conjugation and isospin known as G-*parity* [232], seemed suppressed. This fact was hard to appreciate group-theoretically but could be comprehended immediately through quark diagrams. (See figure 9.15.) The suppression of decays in which the initial quarks must annihilate one another rather than appearing in the final particles is now known as *Zweig's rule*.

The quark picture immediately explained why baryons appeared only in singlets, octets and 10-plets of SU(3), since these are the states that can be formed of three triplets. Mesons occurred only in singlets and octets, states that can be formed of a triplet and antitriplet. For states without orbital angular momentum, one could make positive-parity baryons of spin $\frac{1}{2}$ and spin $\frac{3}{2}$, and negative-parity mesons of spin 0 and 1. These coincided exactly with the lowest-mass observed families of strongly interacting particles. Additional orbital angular momentum would yield states of higher spin and even or odd parity.

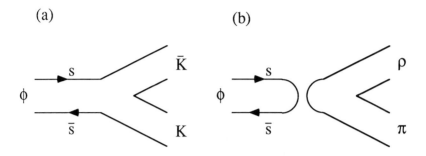

Figure 9.15. *Diagrams illustrating Zweig's rule. (a) Allowed decay; (b) forbidden decay.*

9.6.1.2. The spectrum of resonant particles. The rules for combining quarks into hadrons are simple. For mesons, a quark and antiquark can form states of spin $S = 0$ or 1. For baryons, three quarks can form states of spin $S = \frac{1}{2}$ or $\frac{3}{2}$. These quark spins combine vectorially with an orbital angular momentum L to form states with various total angular momenta J. The intrinsic parities P of these states are also well defined in terms of orbital configurations. Since the relative parities of a quark and antiquark are opposite, the mesons should have parities $P = (-1)^{L+1}$. Indeed, the lowest-lying mesons have $J^P = 0^-$ and 1^-, as predicted for $L = 0$ states. The next-lowest mesons have $J^P = 0^+$, 1^+ and 2^+, as predicted. The lowest-lying baryons (the octet and decimet shown in figure 9.6) indeed have $J^P = (\frac{1}{2})^+$ and $(\frac{3}{2})^+$, as expected for $L = 0$ states, while their first excitations appear to have negative parity and $J = \frac{1}{2}, \frac{3}{2}$ and $\frac{5}{2}$ as expected for $L = 1$ states.

The quark model allowed one to describe masses and magnetic moments of elementary particles with remarkable success [233]. A more algebraic formulation embodying many successes of the quark model was based on the group SU(6) (formed from the product of SU(3) and an SU(2) for quark spin) [234].

There is evidence for a substantial number of orbital excitations and some radial excitations as well. However, the nature of the baryon ground states posed a problem best visualized by referring to the Δ.

9.6.1.3. The quark statistics problem. The Δ(1232) multiplet has isospin $I = \frac{3}{2}$ and spin $J = \frac{3}{2}$. It is a natural candidate for an S-wave state of three quarks. It is symmetric in isospin (containing, for example, the Δ^{++} with three u-quarks). In the absence of orbital angular momentum among the quarks, it is symmetric in spin (containing, for example, the $J_z = \frac{3}{2}$ state with all three quarks aligned along the $+z$ axis). It is symmetric in space if it is the spatial ground state, as is likely, but this behaviour is unacceptable for a state composed of three identical fermions, which must obey the Pauli exclusion principle.

It was shown under certain assumptions [235] that particles of integral spin should obey Bose–Einstein statistics, while particles of half-integral spin should obey Fermi–Dirac statistics. These properties are expressed in terms of commutation relations for boson field operators and anticommutation relations for fermion fields. In 1953, H S Green [236] discovered a generalization of these rules, known as *parastatistics*, based on the assumption of new internal degrees of freedom of the fields. In 1964, Greenberg [237] suggested that quarks obeyed Green statistics, behaving as parafermions of order 3.

The parafermion hypothesis for quarks was equivalent to imagining quarks to occur in three versions, baryons consisting of one of each. A wavefunction antisymmetric in this new degree of freedom was symmetric in all the remaining ones. Greenberg was then able to classify the symmetry of all three-quark wavefunctions, finding a close correspondence with many known baryon states.

A more concrete proposal of an extra degree of freedom for quarks was made by Han and Nambu [238]. The quarks' apparent fractional charges arose from averaging over three species with *integral* charges. Thus, two u-quarks would have charge 1 and one would have charge 0, making an average of $\frac{2}{3}$, while two d-quarks would have charge 0 and one would have charge -1, making an average of $-\frac{1}{3}$. Again, baryons would have one quark of each type.

We shall see the eventual outcome of the quark statistics problem when discussing quantum chromodynamics in section 9.8. The internal 'triplicity' proposed by Greenberg and by Han and Nambu has come to be known as 'colour'. Whereas more 'flavours' of quarks are now known than the u, d and s of the 1960s, the number of colours has remained three. Thus, the quote 'Three quarks for Muster Mark' remains as appropriate as ever.

9.6.2. *Deep inelastic scattering*

9.6.2.1. Elastic scattering and form factors. The advent of new electron accelerators and sensitive detection techniques after World War II enabled the study of electron elastic scattering on protons and neutrons at momentum transfers large enough to probe nucleon structure [104]. The proton and neutron were found to be structures with radii of the order of 1 fm (10^{-13} cm). Their *form factors* (the Fourier transforms of their charge and magnetic moment distributions [239]) were found to fall off at large momentum transfers q roughly as q^{-4}.

9.6.2.2. Design and operation of SLAC. In the 1950s, planning began for a large linear accelerator to be located west of the Stanford University campus. The persistent efforts of W K H Panofsky and others eventually

Figure 9.16. *Aerial view of the Stanford Linear Accelerator Center in 1969.*

led to the construction of the Stanford Linear Accelerator Center (SLAC) (see figure 9.16), which began operation in 1967.

Opinion was divided as to whether the SLAC accelerator was worth the cost and effort [240]. When the machine was conceived, many people thought that the proton form factor would continue to drop with increasing momentum transfer. A similar behaviour was anticipated for each form factor for excitation of specific resonances like the Δ. If true, this behaviour would mean that the high-energy scattering of electrons on protons would prove a barren desert.

9.6.2.3. Discovery and interpretation of scaling. An experiment to study the *inelastic* scattering of electrons on protons was mounted by an MIT–SLAC collaboration. One simply detected the scattered electron, inferring some properties of the final hadronic state from the direction and energy of the electron alone.

688

The results [241] were as surprising as those obtained by Rutherford more than half a century earlier. The number of high-angle scatterings was far in excess of what one expected on the basis of rapidly decreasing form factors, as if the proton itself contained point-like constituents. A 1966 paper by Bjorken [242], predicting such behaviour on the basis of current algebra, had anticipated this viewpoint.

Two relativistically invariant quantities characterize the deep inelastic scattering of a lepton on a target of mass M: the square of the invariant momentum transfer, Q^2, and the product $2M\nu$, where $\nu = E - E'$ is the difference between the initial and final lepton energies E and E' in the laboratory system. Bjorken's result predicted that, aside from well-defined kinematic factors, the scattering could be described in terms of a *scaling* variable $x \equiv Q^2/2M\nu$, i.e. in terms of the *ratio* of the two quantities just mentioned. Observations confirmed this behaviour. The decrease of form factors for elastic scattering and for excitation of specific resonances with increasing Q^2 was compensated by the ability to excite more and more states of high mass.

9.6.2.4. Initial neutrino experiments. In the early plans for Fermilab, it was recognized that deep inelastic scattering of neutrinos on protons and neutrons could play a role complementary to that of the SLAC experiments. Whereas deep inelastic electron scattering is governed by the electromagnetic interactions, and hence is sensitive to charges of the nucleon's constituents, the neutrino reactions known to occur at the time were sensitive to the presence of constituents able to change their charges by ±1 unit.

Two experiments were mounted to study deep inelastic scattering of neutrinos on nucleons in the earliest days of operation of Fermilab: a Harvard–University of Pennsylvania–Wisconsin collaboration [243] (E-1) and a Caltech–Columbia–Fermilab–Rockefeller collaboration [244] (E-21). Experiments at CERN with bubble chambers such as Gargamelle [245], using neutrinos produced at the lower energies of the CERN Proton Synchrotron (PS), also made early studies of deep inelastic neutrino scattering. As in the SLAC experiments, scaling was dramatically confirmed. Moreover, a comparison of the electromagnetic and weak scattering experiments made it possible to determine the charges of constituents [246], confirming the hypothesis advanced several years earlier that the constituents of protons and neutrons had charges of $\pm\frac{1}{3}$ and $\pm\frac{2}{3}$.

9.6.2.5. Parton hypothesis and its successes. Bjorken noted that his current algebra results, which had predicted the scaling behaviour observed at SLAC and in inelastic neutrino interactions, could be interpreted in terms of point-like objects in the proton. Feynman [247] took the idea of point-like constituents seriously, interpreting the scaling

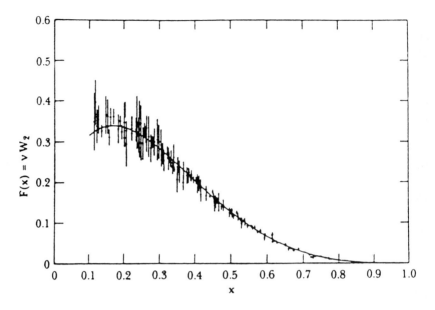

Figure 9.17. *Proton structure function versus the scaling variable x.*

variable $x = Q^2/2M\nu$ as the fraction of a proton's momentum carried by the constituent, or *parton*, that absorbs the momentum and energy from the scattered lepton. A test for the spin of the constituents proposed by C Callan and D Gross [248] soon supported the idea that the point-like objects in the proton indeed were quarks.

A systematic exploration of the parton hypothesis, undertaken by Bjorken and Paschos [249], developed a consistent description of the momentum distributions of partons in protons and neutrons for all deep inelastic scattering data known at the time. Those distributions are parametrized by *structure functions*, an example of which is shown in figure 9.17 (from reference [240] p 152).

Structure functions are also probed when a parton in one hadron annihilates an antiparton in another to produce a lepton pair. That process, first proposed by Drell and Yan in 1971 [250], still plays a major role today in the study of parton distributions and in tests of QCD, and is central to the production of W- and Z-bosons to be discussed in section 9.7. Another effect of point-like constituents in hadrons, predicted by Berman, Bjorken and Kogut [251] in 1971 and observed in a series of experiments at CERN and Fermilab [252], was the production of hadrons at high transverse momenta as a result of parton–parton collisions.

9.6.3. *Electron–positron annihilation*
The cross section for $e^+ + e^- \rightarrow$ hadrons can be normalized by defining a ratio $R \equiv \sigma(e^+ + e^- \rightarrow \text{hadrons})/\sigma(e^+ + e^- \rightarrow \mu^+ + \mu^-)$. In the quark

model this ratio measures just the sums of the squares of the produced charges, and would be expected to increase when the energy crosses the threshold for new kinds of quarks. For example, just above the threshold (about 300 MeV) for the production of pairs of pions which are composed of up and down quarks, this ratio should be $R = 3(q_u^2 + q_d^2)/q_\mu^2$, where q_u, q_d and q_μ (the charge of the up quark, the down quark, and the muon) are $+\frac{2}{3}$, $-\frac{1}{3}$ and -1, respectively. The factor '3' represents the number of quark 'colours', while the muon occurs in only one variety. R should equal $\frac{5}{3}$ if these daring conjectures are correct, i.e. if the assigned quark charges are correct and if quarks have three colours. (The presence of sharp resonances in the cross section prevents this average behaviour from being realized in practice.) As the energy of the colliding particles crosses the threshold for strange quark production as manifested by pairs of kaons, at about 1 GeV, the ratio R should rise to 2. As resonances become broader and start to overlap at higher energies, this average behaviour should be observable.

The study of electron–positron collisions was pioneered by several groups starting in the late 1950s (see section 9.10.9), leading to the construction of machines at Stanford, Orsay (France), Frascati (Italy) and Novosibirsk (USSR). In the late 1960s, a group at Frascati brought into operation the ADONE collider, with a centre-of-mass energy of up to 3 GeV, the first capable of studying the process $e^+ + e^- \rightarrow$ hadrons above the region where prominent resonances led to fluctuations in the cross section. This machine, with an energy sufficient to produce pairs of u, d and s-quarks, should have yielded $R = 2$. Results were consistent with this, though with wide errors [253], as shown in figure 9.18.

In an attempt to reach higher collision energies, the Cambridge Electron Accelerator, or CEA, was adapted for electron–positron collisions in the early 1970s (see section 9.10.9). Above the threshold (slightly below 4 GeV) for 'charmed' quark production ($q_c = +2/3$), the quark model predicted that R should rise to $3\frac{1}{3}$. However, when higher energies of 4 and 5 GeV were reached at CEA, the value of R was found to rise to around 5 [254]! These results were eventually confirmed by the Stanford electron–positron collider SPEAR (sections 9.9.1, 9.10.9). New physics in the form of a τ-lepton (section 9) had unexpectedly appeared, the decay products of the τ being counted as hadrons. (The values of R in figure 9.18 have this contribution subtracted out.) It took some years to straighten out this initially confusing situation [255], but eventually quark counting by means of the ratio R proved a very useful technique.

9.6.4. Searches for free quarks

The absence of free quarks was a continuing source of concern to many. Searches for fractionally charged particles (for a review, see reference [256]) were mounted as soon as proposed; even the early observations of Millikan were examined with quarks in mind. Claims

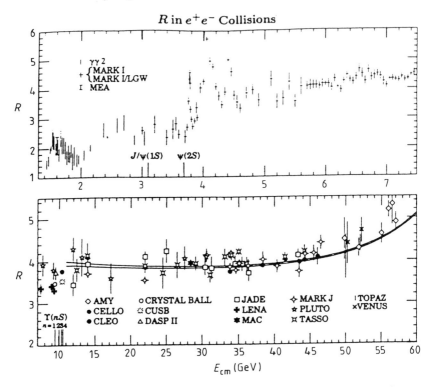

Figure 9.18. *Behaviour of the ratio $R \equiv \sigma(e^+e^- \to hadrons)/\sigma(e^+e^- \to \mu^+\mu^-)$ as a function of centre-of-mass energy E_{CM}.*

for detection of free quarks in cosmic rays [257] were not supported by further evidence [256]. Physicists looked in such diverse places as oysters (where substances with odd chemical properties might be concentrated [258]) and samples from the Moon [259]. Excitement over free quarks peaked in the late 1970s with experiments of Fairbank and collaborators [260] which seemed to show fractional charges on niobium balls heated on a tungsten substrate. However, other experiments [256, 261] failed to see the effect, which was eventually understood to result from inhomogeneities on the surfaces of the balls.

To this day, free quarks have not been observed. Quantum chromodynamics provides a plausible reason why they should be permanently confined in hadrons, as we shall see in section 9.8.

9.7. Electroweak unification

9.7.1. A theory of leptons

9.7.1.1. Yang–Mills fields.
As mentioned in section 9.5.1.6, Yukawa [43] and subsequent authors [143 155] suggested that the weak interactions

stem from exchange of a spin-1 boson, in analogy with the exchange of photons in electromagnetism. A crucial step laid the ground for a self-consistent theory based on this idea. In 1954 C N Yang and R L Mills proposed [113] that isotopic spin symmetry arises in the same way as electromagnetic gauge invariance. Their theory of fields which interact with *one another* as well as with external matter turned out to serve not only to unify electromagnetic and weak interactions, but also to describe strong forces. (In this context see also reference [262].)

The following discussion owes much to an exposition by Yang [263]. In electromagnetism, consider a slowly moving test charge undergoing a virtual displacement by a four-vector dx_μ. The change in the phase of its wavefunction (aside from any $p \cdot x$ contribution, where p is its four-momentum) is $eA^\mu(x)dx_\mu$ (in units with $\hbar = c = 1$), where $A^\mu(x)$ is the vector potential and e is the electric charge of the particle in question.

According to *local gauge invariance*, the phase convention for the particle can be set independently at each space-time point x, corresponding to changing the vector potential by the divergence of a scalar quantity which depends on x. This *local gauge transformation* changes only the difference in phases at the endpoints when a particle is taken along a path from one point to another. When the endpoints coincide, so that the particle is taken on a voyage along a *closed path* in space-time, this phase difference vanishes.

Every experiment corresponds to either taking a particle on a closed path in space-time, or comparing the results of two *different* paths from a point x_1 to another x_2. Thus, one can take a particle through either of two slits or around either side of a solenoid (as in a classic experiment proposed by Aharonov and Bohm [246]). One measures not $A^\mu(x)$, but the spatial integral of the *field strength* $F^{\mu\nu} \equiv \partial^\nu A^\mu - \partial^\mu A^\nu$ over an area enclosed by the particle's closed path in space-time (or by the area between the two paths from x_1 to x_2). The field strength is invariant under a local gauge transformation. Since changes of phase commute with one another, such a gauge theory is said to be *abelian*. The group of phase changes is U(1), the unitary group in one dimension.

Yang and Mills' generalization of local gauge invariance involves not only transformations of phase, but local rotations in isotopic spin space. Thus, the quantity $A^\mu(x)$ induces isotopic spin transformations, and acts as a matrix in isospin space. Its dimension depends on the representation chosen for isotopic spin: two-dimensional for proton and neutron, three-dimensional for π^+, π^0 and π^- and so on. Since rotations about different axes do not commute with one another, the corresponding field strength has an additional term

$$F^{\mu\nu} \equiv \partial^\nu A^\mu - \partial^\mu A^\nu + g[A^\mu, A^\nu].$$

Here g is the gauge coupling strength, analogous to the electric charge e. Such a theory, said to be *non-abelian*, has fields which interact with one another as a result of the additional term in $F^{\mu\nu}$. The isotopic spin group is denoted by SU(2); the 2 stands for the dimension of its smallest non-trivial representation.

9.7.1.2. Glashow's model. The first attempts [143, 155] to unify the charge-changing weak interactions with electromagnetism assumed that charged intermediate bosons (the Ws, mentioned in section 9.5.1.6) together with the (neutral) photon formed an SU(2) triplet. However, the Ws couple only to left-handed fermions, while the photon couples to both left- and right-handed fermions. In order to allow for this difference, Glashow [265] extended the SU(2) to an SU(2) × U(1) group. Gauge bosons corresponding to SU(2) would consist of W^+, W^- and a neutral W^0, while an additional neutral boson (B^0 in today's notation) would correspond to U(1) gauge transformations. The W^0 and B^0 would be mixed by an interaction also giving rise to masses of charged Ws. Two linear combinations of fields then emerged: the massless photon γ, with field $A^\mu(x)$, and a *massive* neutral vector boson Z with field $Z^\mu(x)$

$$A = B \cos\theta + W^0 \sin\theta \qquad Z = -B \sin\theta + W^0 \cos\theta.$$

The mass of the Z was predicted in terms of θ and the mass of the W by $M_Z = M_W / \cos\theta$. Weak processes involving exchange of a Z ('neutral-current interactions') were thus predicted by this theory.

9.7.1.3. Symmetry breaking and the Higgs mechanism. The origin of W and Z masses remained unexplained in Glashow's theory. In an electroweak theory based on a *renormalizable* Yang–Mills theory (one in which divergent quantities could be defined away by redefinitions of fundamental parameters like masses and coupling constants), it appeared that gauge bosons had to remain massless. The *ad hoc* introduction of gauge-boson masses destroys renormalizability, gauge invariance or both. How could one break the SU(2) × U(1) symmetry without spoiling the attractive properties of the theory?

It became apparent in the late 1950s and early 1960s that symmetries could be manifested in two ways (see section 9.5.3.1). In the *Wigner–Weyl* realization, symmetry is manifest through degeneracies in the spectrum. In the *Nambu–Goldstone* mode [179, 180], the vacuum breaks the symmetry but the equations of motion retain it. In this mode, each broken symmetry operation corresponds to a massless, spinless particle, known as a *Nambu–Goldstone boson*.

In a theory with massless vector bosons, such as electromagnetism, only two polarization states are allowed: the two directions transverse to the boson's direction of propagation, or, equivalently, left and right circular polarization. A massive spin-1 particle has, in addition, a longitudinal polarization state. It was noted by J Schwinger,

P W Anderson, Peter Higgs and others [266] that in a gauge theory containing Nambu–Goldstone bosons, such bosons could act as the needed third component of a massive vector boson. One can describe the massless scalar particles as being 'eaten' by the gauge bosons, which then become heavy. This process is now known as the *Higgs mechanism*.

9.7.1.4. The Weinberg–Salam model. The Higgs mechanism was soon employed in the service of electroweak unification by Weinberg and Salam [267, 268] (see figure 9.19). The simplest version of the theory introduced only an SU(2) doublet (ϕ^+, ϕ^0) and its complex conjugate $(\phi^-, \bar{\phi}^0)$. With an appropriate self-interaction of the fields, the combination $\eta \equiv (\phi^0 + \bar{\phi}^0)/\sqrt{2}$ would acquire a non-zero vacuum expectation value v, thereby breaking SU(2) × U(1) down to U(1) of electromagnetism as desired. The fields $\phi^\pm$ and $(\phi^0 - \bar{\phi}^0)/\sqrt{2}$, corresponding to the Nambu–Goldstone bosons of the broken symmetry, would be 'eaten' by the $W^\pm$ and Z, while the difference $H \equiv \eta - v$, often referred to as 'the' Higgs boson, would remain in the spectrum as a massive particle.

9.7.1.5. Renormalizability. The Weinberg–Salam theory remained a curiosity for several years, referred to very little even by its inventors. It predicted unobserved *neutral weak currents*, such as neutrino interactions in which a neutrino rather than a charged lepton emerged, and strangeness-changing decays such as $K^+ \to \pi^+ \nu \bar{\nu}$. Limits on such decays were particularly stringent, lying far below the level of ordinary weak interactions. That is why Weinberg entitled his model 'A theory of leptons'.

It was, furthermore, not clear at the outset that the Higgs mechanism solved the problem of the renormalizability of the electroweak theory, until the proof was supplied in 1971 by Gerard 't Hooft [269], a student of Martinus Veltman in Utrecht (see reference [270] for some historical perspectives on this work) and extended by Benjamin W Lee and Jean Zinn-Justin [271].

The proofs of the self-consistency of the Glashow–Weinberg–Salam electroweak theory were a major factor in its acceptance, and sent the particle physics community into a state of excitement. Weinberg immediately realized the significance of the result [272]. Early calculations of rates for neutral-current processes began to appear [273]. Discussions with experimentalists began in earnest: had neutral currents been overlooked? Variants of the model without neutral currents but with new particles were proposed [274]. Attempts were renewed to extend the theory to hadrons; these will be discussed in section 9.7.3.

9.7.2. Experimental confirmation of neutral currents
With the growth of interest in the Glashow–Weinberg–Salam theory in the early 1970s, the search for weak neutral currents entered a more

695

(*a*)

(*b*)

(*c*)

Figure 9.19. *(a) S Glashow; (b) S Weinberg; (c) A Salam.*

serious phase [275]. Previous upper bounds on such effects were re-examined in many processes. Particularly stringent bounds appeared to exist for reactions initiated by neutrinos.

9.7.2.1. Neutrino scattering. The Gargamelle collaboration at CERN had been investigating deep inelastic neutrino scattering during the late 1960s and early 1970s. Its limits on neutral current events were reported by Perkins at the 1972 International Conference on High Energy Physics [246]. Deep inelastic neutrino scattering was also beginning to be studied by experiments at Fermilab, as mentioned in section 9.6.2.4.

Neutrinos were produced at accelerators mainly from the decays $\pi \rightarrow \mu\nu$ and $K \rightarrow \mu\nu$, and hence were of the muonic type. When interacting in matter via the charged current, they gave rise to hadronic showers and to a clearly identified muon in the final state. The presence of a muon rather than an electron was the basis of the claim that the muon and electron neutrinos were distinct (see section 9.5.1.6). Occasional hadronic showers without an accompanying muon, observed even in the earliest neutrino experiments [276], were usually ascribed to contamination of the beam with neutral particles (particularly neutrons). One could check this possibility by studying in detail the distribution of events with respect to distance along the detector or inward from the lateral boundaries. Neutron-induced events would become rarer as either distance increased, while the rate for neutrino-induced events would be independent of distance.

The likelihood that the Gargamelle Collaboration was seeing neutral-current events was discussed informally at CERN in the early months of 1973. First, however, a detailed check of backgrounds had to be made [277]. Once the experimenters were satisfied that neutrons and other backgrounds could not be responsible for their signal, they announced the discovery of neutral currents in weak interactions of neutrinos. The effects were seen both in deep inelastic scattering [278] and in a 'golden' neutrino–electron elastic scattering event [279]. The E-1A Collaboration at Fermilab also saw deep inelastic neutral-current interactions of neutrinos [280]. Their signal was present at an early stage, but an attempt to understand it better by reconfiguring the detector led to a temporary loss of the effect [281].

How could neutral-current effects have been overlooked for so long? One source of the difficulty in identifying them can be seen in figure 9.20. Define R_ν and $R_{\bar{\nu}}$ as the ratios of neutral-current to charged-current events for deep inelastic neutrino scattering. The predictions of the Glashow–Weinberg–Salam theory as a function of the angle θ (section 9.7.1.2) yield a nose-like curve relating R_ν and $R_{\bar{\nu}}$. The present data, shown as the plotted point [282], sit at the bottom of the 'nose'. The ratio R_ν is expected to be quite small for a large range of values of $\sin^2 \theta$, while the ratio $R_{\bar{\nu}}$, though larger than R_ν, is about as small as it

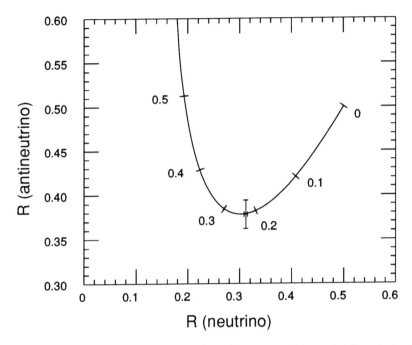

Figure 9.20. *The neutral-current to charged-current ratios R for deep inelastic scattering of neutrinos and antineutrions, plotted parametrically as functions of* $\sin^2 \theta$ *(labels on curve). The plotted point shows an average of recent data.*

can be. Neutral currents were also expected (and eventually found [283]) in *elastic* neutrino–proton scattering.

9.7.2.2. Parity violation in electron–deuteron scattering. The Glashow–Weinberg–Salam theory also predicted neutral-current weak processes involving charged leptons. These processes are normally masked by the much stronger electromagnetic interactions. However, in contrast to electromagnetism, the neutral weak force does not conserve parity. Despite its weakness, its effects can be observed through *interference* of the electromagnetic and weak interactions.

See also p 665

A collaboration led by V W Hughes and C Y Prescott searched at SLAC for the predicted parity violation in deep inelastic scattering of polarized electrons [284]. A difference in interactions of left-circularly polarized and right-circularly polarized electrons would confirm the predicted effect. It was necessary to prepare a source of polarized electrons, and to arrange their arrival at the target with desired (and known) polarization. The choice of target was also important: the predicted effect turns out to be much smaller in hydrogen than in deuterium, so both were investigated.

The results were announced in 1978. The electron had neutral weak interactions, just as the theory had predicted. One could observe the interference of photon and Z^0 exchange. This confirmation was a major source of excitement at the 1978 International Conference on High Energy Physics in Tokyo [285].

9.7.2.3. Parity violation in atoms. A fundamental violation of parity in the electron–nucleus interaction can lead to such effects as optical rotation in atoms [286]. Initial experiments (chronicled in the first of references [284] and the reviews of reference [287]) did not see the expected effects, leading to some bizarre variations on the simplest version of the theory. Eventually laser stability was brought under control, and calculations of atomic physics effects to high orders were performed. The first experiments to claim the observation of parity violation were performed in atomic bismuth, with other groups who had previously not seen a signal then finding one.

More recent experiments [288] have attained ever-improving accuracy in measuring atomic parity violation. The most recent results in caesium by C Wieman's group, when taken in conjunction with atomic physics calculations [289], not only confirm the electroweak theory at better than the 3% level, but provide a useful constraint on physics beyond the standard model [290].

9.7.2.4. Other neutral-current processes. Many other neutral-current effects have been observed over the years since they were first predicted. They include forward–backward asymmetries in electron–positron reactions, parity violation in polarized-electron scattering on nuclei other than hydrogen and deuterium, detailed studies of $\nu_\mu e$ and $\bar{\nu}_\mu e$ elastic scattering and interference between charged and neutral currents in $\nu_e e$ and $\bar{\nu}_e e$ elastic scattering. An excellent source for tracking how these data have steadily improved is the series of reviews of reference [291].

9.7.3. Extension to hadrons and the charm hypothesis
The Cabibbo theory of strange particle decays, discussed in section 9.5.3.4, can be re-expressed in terms of quarks and W-bosons. A charged W can be emitted or absorbed in transitions between a u-quark and the linear combination $d' \equiv d \cos\theta_C + s \sin\theta_C$, with $\sin\theta_C \approx 0.22$. The charge-conjugate couplings also exist.

The charged currents carrying a left-handed u-quark into d' can be taken as part of a triplet. The properties of the neutral member of this triplet can be deduced by recalling the commutation relations for SU(2). The raising and lowering operators $J_\pm$ of angular momentum, for example, are related to the third component J_3 by $[J_+, J_-] = 2J_3$. The normalization of the currents then is specified by the property $[J_3, J_\pm] = \pm J_\pm$. But a neutral current defined in this way contains strangeness-changing pieces. How could these be avoided?

699

Each charged lepton (electron and muon) couples to its own variety of neutrino by the charge-changing current (the analogue of $J_\pm$). The corresponding neutral current is then *diagonal* in neutrino species. Moreover, even if one were to define linear combinations of neutrinos coupled to electrons and muons, neutral currents are avoided as long as these linear combinations are orthogonal to one another.

A quark–lepton analogy, drawn in 1964 by Gell-Mann [184], Bjorken and Glashow [292], Y Hara [293], and Z Maki and Y Ohnuki [294], proposed that as long as each charged lepton had its own neutrino, the d-quark and s-quark should not have to share the u-quark in weak transitions. Instead, one combination $d' = d \cos\theta_C + s \sin\theta_C$ would couple to u-quarks, while the orthogonal combination $s' = -d \sin\theta_C + s \cos\theta_C$ would couple to a new quark, which Bjorken and Glashow called *charmed*. This charmed quark c was to be heavier than the u-quark, but, like it, would have charge $\frac{2}{3}$. The neutral currents as defined by the commutator of two charged currents then are automatically diagonal in flavour.

Several years later, Glashow, J Iliopoulos and L Maiani realized that not only did the Bjorken–Glashow hypothesis guarantee that neutral currents preserve strangeness in lowest order, but strangeness-changing neutral-current effects in higher-order weak processes were also drastically suppressed [295]. The mixing between a neutral kaon and its antiparticle is one such process. Since this mixing appears to be no stronger than the product of two first-order weak transitions (as one would have, for example, in the two-step process $K^0 \leftrightarrow \pi\pi \leftrightarrow \bar{K}^0$), it was possible to deduce an upper limit to the mass of the charmed quark. Glashow, Iliopoulos and Maiani (GIM) predicted that the charmed quark lay below about 2 GeV/c^2 and proposed ways to search for it.

The GIM mechanism was applied by Gaillard and Lee [296] to a systematic study of rare decays of kaons, setting the stage for experimental investigations that continue to this day. Many suggestions for finding charmed particles began to be made [297], particularly after evidence for the predicted neutral currents began to appear. The role of a charmed quark in the cancellation of *triangle anomalies* (appearing in higher-order electroweak calculations) was stressed [298].

9.7.4. Experimental confirmation of charm

9.7.4.1. Early hints. Charmed particles were taken seriously very early in Japan. Some of the initial proposals of quartet models of hadrons had been made by Japanese authors [293, 294]. An event seen in emulsion by K Niu and his colleagues at Nagoya University was interpreted in 1971 as a possible candidate for a charmed particle [299]. It appeared to decay to $\pi^+\pi^0$ with a mass of about 2 GeV/c^2 and a lifetime of about 10^{-14} s. Niu's event (and others like it) suggested that quarks always occurred in pairs: u with d and c with s. Kobayashi and Maskawa [300] recognized that with a third pair one could describe *CP* violation. Both of the quarks they proposed have now been seen (section 9.9).

9.7.4.2. Discovery of the J/ψ and the charmonium spectrum. An early study of the Drell–Yan process (section 9.6.2.4) by Leon Lederman and collaborators observed a shoulder in the effective mass spectrum of muon pairs [301], which could have been interpreted as a new particle viewed with poor mass resolution. Stimulated by experience in the study of lower-mass dilepton pairs and by this result, Samuel C C Ting and collaborators mounted an experiment to measure the dilepton mass spectrum with mass resolution far better than ever achieved before. The experiment consisted in colliding an intense beam of protons from the Brookhaven Alternating-Gradient Synchrotron (AGS) with a beryllium target, and observing electron–positron pairs with a two-arm spectrometer with approximately 30° between the two arms. Čerenkov counters filled with hydrogen served to identify electrons, with bending magnets and eleven planes of proportional chambers providing precise momentum measurements.

By September of 1974, it was clear to Ting's group that they had indeed come upon something either very wrong or very exciting. A peak in the e^+e^- effective mass was showing up at 3.1 GeV/c^2 with almost no background on either side. While a number of cross-checks were performed, Ting spoke of his result discreetly to colleagues, who generally urged him to publish it [240]. Meanwhile, on the West Coast, studies of electron–positron annihilations with the SPEAR detector (section 9.6) had been continuing. Cross sections had been measured in steps of 0.2 GeV in centre-of-mass energy. The values obtained at 3.2 and 4.2 GeV seemed a bit high in comparison with those at other energies. The cross section was remeasured in those regions in June 1974, but no structure was noticed.

The SLAC–LBL Collaboration resumed its scrutiny of the energy-dependence data in October 1974. The cross section at 3.1 GeV, upon closer examination, seemed not to be reproducible. Two out of ten runs at that energy gave much larger values than the others. Measurements were resumed over the weekend of 9 and 10 November in order to check the anomaly.

Ting arrived at SLAC on 11 November for a programme committee meeting with the news of his discovery [302]. The peak he and his colleagues were seeing, dubbed the 'J' (figure 9.21), was almost free of background. He was greeted with news of an electron–positron annihilation cross-section peak seen in the Mark I detector (figure 9.22) by the SLAC–LBL Group, who called their effect [303] the ψ. The dual name J/ψ has persisted for the particle with mass 3.1 GeV/c^2. Within ten days, the SLAC-LBL group had raised the beam energy of SPEAR,

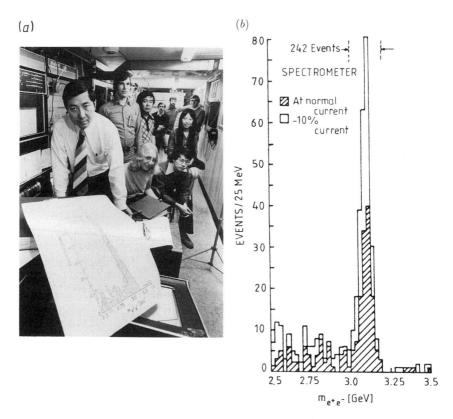

Figure 9.21. *(a) S C C Ting and his group, with a plot of the J resonance. (b) The mass spectrum of e^+e^- pairs, showing the J peak* [302].

finding another bump—the ψ'—at 3.7 GeV [304]. Electron–positron rings at Frascati in Italy and Hamburg in Germany were able to confirm the existence of the ψ in a matter of days [305].

The J/ψ and ψ' could be viewed as the ground state and first radially excited state of a charmed quark and a charmed antiquark with zero relative orbital angular momentum and parallel spins. In analogy with the name *positronium*, for the bound state of an electron and a positron, this system was called *charmonium*. Several sets of authors published interpretations of the new particles based on the charm hypothesis [306]. However, some questions had to be settled before that interpretation could be regarded as established and alternative schemes (see, for example, references [307]) laid to rest.

9.7.4.3. Early confusion in the search for charm. The charm hypothesis predicted P-wave $c\bar{c}$ states lying between the J/ψ and the ψ'. The ψ' should be able to decay to them electromagnetically. A detector at

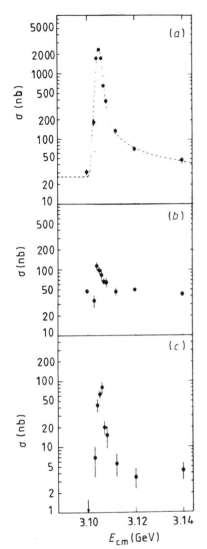

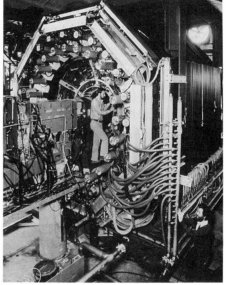

Figure 9.22. *Left: the cross section for e^+e^- annihilations to various final states as measured by the Mark I detector [303]: (a) multihadrons, (b) e^+e^-, (c) $\mu^+\mu^-$. Right: the Mark I Detector with Roy Schwitters.*

SPEAR (located at the other side of the ring from the SLAC–LBL set-up) initially did not see these transitions. They were first found at the DORIS storage ring, and confirmed at SLAC [308].

Mesons were expected containing one charmed quark and one light (u or d) antiquark, decaying to such final states as $\bar{K}\pi$. Early searches were not turning up such decays [309].

The cross section for $e^+e^- \to$ hadrons was indeed increasing above a centre-of-mass energy of 4 GeV in a manner indicating new particle production. However, the increase was actually *too large* for what was expected on the basis of charm. Moreover, an increase in the number of kaons, expected since the charmed quark's dominant decay should be to strange quarks, was not materializing.

The resolution of this confusion is traced in references [255]. Not only were charmed particles being produced in pairs, but a new lepton, the τ, was also being formed in pairs at roughly the same energy. Many of the signals for charm were counterbalanced by signals of the new lepton! It was not until the identification of the new lepton [310] that the 'inclusive' signals for charm could be sorted out.

9.7.4.4. Observation of charmed particles. By the spring of 1976, the absence of a signal for charmed mesons in the SPEAR experiment was becoming a source of concern. Calculations by De Rújula, Georgi and Glashow [311] indicated that the lightest charmed mesons should lie between 1.8 and 1.86 GeV/c^2, and the lightest charmed baryons should have a mass of about 2.2 GeV/c^2. The predictions for baryons were borne out by the discovery of two candidates for charmed baryons, a Σ_c^{++}(= uuc) decaying to a Λ_c^+(= udc) and a pion, in a single neutrino interaction [312]. Observation of additional leptons in neutrino experiments [313] and direct leptons in hadronic reactions [314] also hinted strongly at the existence of charmed particles. Gerson Goldhaber and F M Pierre combed through the Mark I data to see if a charmed-meson signal was lurking there. Indeed it was; the lightest charmed meson, called the D (for doublet [292]) was found [315] in both predicted charge states, $D^0 = c\bar{u}$ and $D^+ = c\bar{d}$, with branching ratios close to those originally estimated in reference [295].

Glashow had exhorted a conference of meson spectroscopists in 1974 to go out and find charm [316]. If they did not do so by the next conference, he would eat his hat. If 'outlanders' (non-meson-spectroscopists) found charm, the spectroscopists were to eat their hats. The organizers of the next (1977) conference graciously distributed candy hats to all participants [317].

9.7.4.5. Spectroscopy of charmonium and charmed particles. The study of charmonium and charmed particles has come a long way since the mid-1970s. The charmonium spectrum (see figure 9.23) is already richer than that of positronium, while numerous non-strange and strange charmed mesons and baryons have been discovered. These systems, the lightest for which ideas of perturbative quantum chromodynamics (section 9.8) begin to hold, thus form one of the testing grounds for the new understanding of the strong interactions that arose in the 1970s.

BOX 9B:
THE DISCOVERY OF CHARMED PARTICLES

Glashow, Iliopoulos and Maiani [295] described how the puzzling absence of strangeness-changing neutral currents could be neatly explained if there existed a new fourth quark, the charmed quark. Immediately experimentalists started to ask: 'How can we find it? What would be the experimental signature of such an object?'. Gradually, the details were fleshed out [297,316]. By the summer of 1974, the masses of charmed mesons and baryons were predicted, as well as their decay modes and their branching ratios. When the J/ψ was discovered in November 1974 it fitted neatly into the scheme as the $c\bar{c}$ analogue of the ϕ_s, the vector meson. The search for charmed mesons and baryons began at laboratories around the world.

At hadron accelerators, searches were started for associated production of charmed mesons and baryons in analogy with strange-particle production more than 20 years earlier, mainly using the predicted decay of the pseudoscalar charmed meson, $D^0 \rightarrow K^-\pi^+$, as an identifying signature. The SLAC group, working with the e^+e^- collider detector, also started a search for charmed particles. However, the first clear-cut example of a charmed baryon came from a bubble-chamber event, produced by a high-energy neutrino [312]. The search for charm at SPEAR was complicated by the appearance of the τ lepton. Not until this confusion was unravelled was the D-meson clearly identified [315] in 1976, a year and a half after the discovery of the J/ψ.

In the meantime, searches for charmed particles at the hadron machines had been singularly unsuccessful, with only upper limits for the production cross sections. The charmed particles, even though produced in great numbers, were swamped by the high attendant backgrounds in hadronic interactions. It was not until the development of silicon strip detectors with their excellent spatial resolution and tolerance for high counting rates that charmed particles were detected at hadron machines [318]. This approach was highly successful. Almost immediately samples of data at the electron–positron colliders, previously counted in tens, were counted in thousands.

At present, experiments at hadron machines are recording millions of events carrying charm. The study of charmonium itself has been extended via proton–antiproton annihilation to states unavailable in e^+e^- collisions.

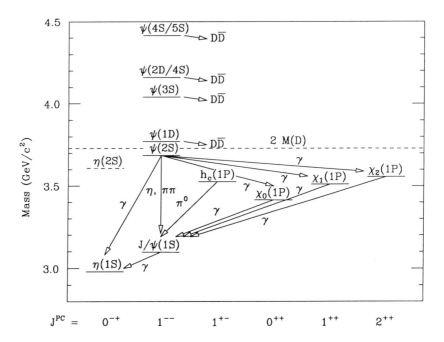

Figure 9.23. *The charmonium spectrum. Observed and predicted levels, as quoted in reference [319], are denoted by solid and broken horizontal lines, respectively. Arrows are labelled by particles emitted in transitions.*

9.7.4.6. Charmed-particle decays. Since charmed quarks couple to strange quarks with a strength proportional to the cosine of the Cabibbo angle (section 9.7.2), the lifetimes of charmed particles are very short, ranging between about 10^{-14} and 10^{-13} s. The *differences* between these lifetimes arising from the strong interactions are a topic of much current interest.

9.7.5. The W and Z

9.7.5.1. Discovery. In a theory with an intermediate vector boson W, the Fermi coupling constant $G_F = (1.166\,39 \pm 0.000\,02) \times 10^{-5}$ GeV^{-2} can be re-expressed as a constant of order one times the square of a coupling constant divided by the square of the W mass. The mass thus follows from the value of the coupling constant.

Early searches for W-bosons, whose coupling constant was unknown, explored mass ranges of a few GeV [160, 320], as mentioned in section 9.5.1.6. Part of the motivation for studies of deep inelastic neutrino scattering at Fermilab was to search for indirect effects of W exchange (see, for example, references [243] and [244]). Sensitivity could thus be achieved to masses up to several tens of GeV.

The Glashow–Weinberg–Salam electroweak theory relates the Fermi coupling constant, the SU(2) coupling strength g, and the W mass M_W by $G_F/\sqrt{2} = g^2/8M_W^2$. Moreover, the relation between g and the electric charge e depends on the angle θ: $e = g\sin\theta$. The combined results predict $M_W = 37.2$ GeV/$\sin\theta$ (with $\alpha = e^2/(4\pi\hbar c) = 1/137$ for the electromagnetic fine-structure constant).

The study of deep inelastic neutrino scattering and parity-violating interactions of polarized electrons in the late 1970s gradually yielded a value of $\sin\theta$ slightly lower than $\frac{1}{2}$, implying a W mass above 75 GeV/c^2. The electroweak theory also predicted the Z mass to be slightly higher than that of the W: $M_Z = M_W/\cos\theta$. With these estimates of W and Z masses, one could anticipate the cross sections for their production in quark–antiquark annihilations. Proton–proton or proton–antiproton collisions with centre-of-mass energies of several hundred GeV would provide quarks with energies sufficient to produce the desired particles. Schemes to collide protons with antiprotons at Fermilab and in the CERN Super Proton Synchrotron (SPS), to be described in greater detail in section 9.10.10.2, began taking shape in the late 1970s. A major step in each project was the collection of sufficient antiprotons in a narrow range of space and momentum. Important steps in this 'cooling' procedure were taken by Simon van der Meer, working at CERN [321]. The effort required to turn the CERN SPS into a proton–antiproton collider was coordinated by Carlo Rubbia [322]. First collisions were seen in 1982. The Fermilab project, requiring the construction of a ring of superconducting magnets, came into operation several years later.

Two detectors were constructed at CERN to study the debris of collisions. The UA1 Collaboration (UA stands for Underground Area) was led by Rubbia, while the spokesman for UA2 was Pierre Darriulat. In January of 1983 the first signals of W production [323] were seen by both groups: an electron of high transverse momentum, with apparent imbalance of transverse momentum in the opposite direction, as expected for a W decaying to an electron and a neutrino. Within six months, both groups also reported the observation of the Z (figure 9.24), decaying to pairs of charged leptons [324].

The W and Z masses are slightly higher than anticipated on the basis of the lowest-order theory (with $\alpha = 1/137$). At the short distances characteristic of the W and Z Compton wavelengths, the vacuum polarization effects mentioned in section 9.3 predict $\alpha \approx 1/128$. With this simple correction, the W and Z masses are in remarkable accord with theory.

9.7.5.2. W and Z properties in hadron colliders. The CERN discoveries of the W and Z were followed by studies of their production and decays with ever increasing accuracy, by the UA1 and UA2 Collaborations and by the Collider Detector Facility (CDF) at Fermilab (figure 9.25).

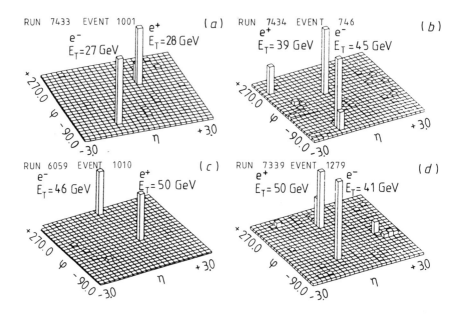

Figure 9.24. *Early evidence for Z production from the first of references [324]. Heights of bar graphs denote energy transverse to the beam deposited in various regions of the UA1 detector.*

A precise determination of the Z mass was made [325], shortly before electron–positron colliders began to refine that quantity further. Production cross sections were measured; comparison of lepton yields in W and Z decays yielded indirect measurements of the ratio of W and Z widths [326], helping to specify the open W decay channels and thereby place an indirect lower bound on the top quark's mass. We shall return to the top quark in section 9.9.

Recently the D0 Detector (figure 9.25) has begun operation at Fermilab. (The name refers to its location around the accelerator ring.) Its fine-grained calorimetry and large angular coverage aim at a precise measurement of the W mass.

9.7.5.3. Results from SLAC and LEP on the Z. The Z mass was predicted by the electroweak theory to be about 90 GeV/c^2. At SLAC and CERN, plans began to develop in the late 1970s and early 1980s to build electron–positron colliders ('Z factories') with at least that energy in the centre of mass.

The Stanford Linear Collider (SLC) involved upgrading the SLAC linear accelerator to a beam energy of 50 GeV, adding a positron source, installing positron and electron cooling, and building two arcs to bring the electrons and positrons into collision (see figure 9.26 (top)). The first results [327], obtained in 1989, included precise measurements of the Z mass and width. More recently, collisions of longitudinally polarized electrons with positrons have provided a precise measurement of $\sin^2 \theta$ through the difference between cross sections for left-handed and right-handed electrons [328].

In the LEP (Large Electron–Positron) Collider, lying on the French–Swiss border at the base of the Jura Mountains (see figure 9.26 (bottom) and reference [330] for an early progress report), positrons and electrons injected from the CERN site are detected at four symmetrically placed locations around the ring. The acronyms for the detectors are ALEPH (Apparatus for LEP Physics), DELPHI (Detector with Lepton, Photon and Hadron Identification), L3 (internal CERN numbering of experiment) and OPAL (Omni-Purpose Apparatus for LEP).

LEP measured the Z width to an accuracy establishing its invisible decay modes to be the three known pairs of neutrinos $\nu_e \bar{\nu}_e$, $\nu_\mu \bar{\nu}_\mu$, and $\nu_\tau \bar{\nu}_\tau$. Precise measurements of the Z mass, of its total width, of branching ratios to various final states and of asymmetries of various sorts, test not only the electroweak theory, but higher-order corrections stemming from the top quark and Higgs boson [331]. The top quark was thereby anticipated to have a mass below 200 GeV/c^2, as indeed transpired [332] (section 9.9).

9.8. Quantum chromodynamics

9.8.1. *Early suggestions of colour-triplicity*
In section 9.6 we mentioned briefly that quarks appear classified by three 'colours', as evidenced by several circumstances.

9.8.1.1. Quark statistics. In baryons, the product of the quarks' spin, space and 'flavour' (u, d, s, etc) wavefunctions appears to be symmetric under interchange of any two quarks. Since quarks have spin $\frac{1}{2}$ and thus should obey Fermi statistics, one would expect their total wavefunction to be antisymmetric under interchange of any two quarks. This goal is achieved by adding a new degree of freedom in which all the quarks in a baryon are antisymmetric.

According to Greenberg [237] and Han and Nambu [238], the three-quark structure of baryons suggested a new SU(3) symmetry, under which quarks would transform as triplets (**3**), while all known hadrons would be singlets. Since

$$3 \times \bar{3} = 1 + 8 \qquad 3 \times 3 \times 3 = 1 + 8 + 8 + 10$$

the known mesons would be the singlets of $3 \times \bar{3}$ (i.e. of quark–antiquark pairs), while the baryons would be the singlets of $3 \times 3 \times 3$ (three-quark states). Nambu [333] showed it plausible that the only states manifest

(a)

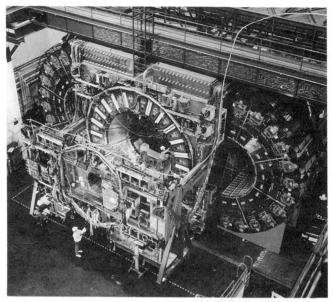

(b)

Figure 9.25. *(a) The CDF detector; (b) the central calorimeter modules of the D0 detector; (c) a recent aerial view of the Fermilab site.*

(c)

Figure 9.25. *Continued*

in Nature are singlets, and discussed the possibility that this new SU(3) corresponded to a gauge theory.

9.8.1.2. Decay of the neutral pion. The decay $\pi^0 \to \gamma\gamma$ was attributed by Steinberger in 1949 [69] to a diagram involving a virtual proton (see figure 9.27). Nearly 20 years later, a similar calculation (involving quark loops instead) was set on a firmer footing by Adler, Bell and Jackiw. The amplitude for this process was found to be proportional to the number of quarks travelling around the loop, indicated to be three by the experimental $\pi^0 \to \gamma\gamma$ rate [334].

9.8.1.3. Cross section for electron–positron annihilations. The study of electron–positron collisions (see section 9.6.3) provided a further confirmation of three quark 'colours'. The cross section $\sigma(e^+ + e^- \to$ hadrons) is proportional to the number N_c of quark colours. Once the charges of quarks produced at any given energy were understood, the data indicated that $N_c = 3$.

9.8.2. Requirements of a gauge theory of strong interactions
Murray Gell-Mann and Harald Fritzsch [335] stressed the colour-triplet nature of quarks in 1972, summarizing the evidence for a threefold degree

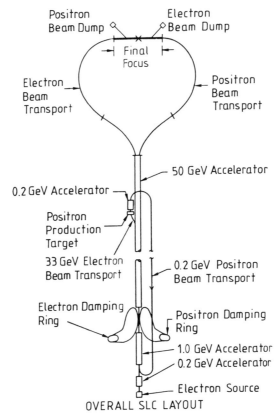

Positron
Beam Dump

Electron
Beam Dump

Final
Focus

Electron
Beam
Transport

Positron
Beam
Transport

50 GeV Accelerator

0.2 GeV Accelerator

Positron
Production
Target

33 GeV Electron
Beam Transport

0.2 GeV Positron
Beam Transport

Electron Damping
Ring

Positron Damping
Ring

1.0 GeV Accelerator

0.2 GeV Accelerator

Electron Source

OVERALL SLC LAYOUT

Figure 9.26. *Top: sketch of the Stanford Linear Collider, from reference* [329]. *Bottom: photograph of the LEP site. The small and large solid circles indicate the position of the SPS and LEP rings, while the dotted line denotes the French–Swiss border.*

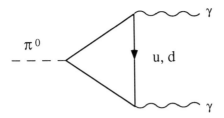

Figure 9.27. *Graph responsible for the decay $\pi^0 \to \gamma\gamma$. The original calculation by Steinberger [69] involved protons instead of u- and d-quarks travelling around the loop.*

of freedom and emphasizing the need for a gauge theory of the strong interactions which would not destroy the successes of current algebra.

A vector-like interaction, similar to that of electromagnetism, was favoured, strong enough at large distances to bind quarks into hadrons yet weak enough at short distances to let the quarks appear quasi-free in deep inelastic scattering. Theories like electromagnetism behave differently. Their vacuum polarization effects (see section 9.3) augment the strength of electromagnetic interactions at short distances. The stage was thus set for a systematic examination of quantum field theories suitable to describe the strong interactions.

9.8.3. Asymptotic freedom and infrared slavery

Vacuum polarization effects in Yang–Mills theories were first calculated correctly, using non-covariant methods, by Khriplovich in 1969 [336]. The self-interaction of the Yang–Mills field led to an important contrast with quantum electrodynamics, the force in a Yang–Mills theory becoming *weaker* at short distances as required for a theory of the strong interactions.

Khriplovich's result seems to have escaped general attention until after its rediscovery in the early 1970s. At that time, several people began examining Yang–Mills theories as candidates for the strong interactions [337]. Two manifestly covariant calculations of the coupling strength's dependence on distance appeared in 1973 [338]. A student of Sidney Coleman at Harvard, H David Politzer, and David Gross and his student Frank Wilczek at Princeton both found a result for SU(N) gauge theories which relates a *running coupling constant* α_N measured at a momentum scale μ_1 to its value at another scale μ_2

$$[\alpha_N(\mu_1)]^{-1} = [\alpha_N(\mu_2)]^{-1} + \frac{1}{4\pi}\left(\frac{11}{3}N - \frac{2}{3}n_f\right)\log\frac{\mu_1^2}{\mu_2^2}.$$

Here n_f is the number of fermion species contributing to the vacuum polarization for a gauge boson. This result made use of the *renormalization group* concept, which specifies the dependence of quantum field theory parameters on changes of scale [339].

The coefficient of the logarithm depending on N stems from gauge-boson loops. Its positive sign means that the interaction in a Yang–Mills theory weakens at large momentum scales (equivalently, at short distances), a property dubbed *asymptotic freedom*. In order for it to hold, the number of fermions n_f in the loop must be small enough for the term proportional to N to dominate.

For an abelian gauge theory such as electromagnetism, the term proportional to N is absent; fermion loops contribute a negative coefficient to the logarithm, and the interaction becomes *stronger* at large momentum scales or short distances.

The *long-distance* behaviour of the coupling constant also differs markedly in abelian and non-abelian theories. In an abelian theory, since the coupling strength ceases to 'run' at scales μ much lower than the mass of the lightest fermion contributing to vacuum polarization, one can measure a well-defined charge at long distances. In non-abelian theories, the logarithmic dependence on μ associated with gauge-boson loops eventually leads the coupling constant to diverge at some small value of μ. The interaction then becomes strong, and perturbation theory ceases to be valid.

It was proposed quite early that the Yang–Mills theory of strong interactions would lead to interaction energies proportional to the interquark separation; such a force between quarks and antiquarks at long distances would keep them from ever being torn apart from one another. This suggestion leads [340] to families with angular momentum J linear in M^2, and is supported by the spectrum of particles lying along the highest Regge trajectories (cf section 9.5.4.4 and figure 9.14). As a counterpart to asymptotic freedom, the confinement of quarks by an ever-increasing potential has sometimes been called *infrared slavery*.

The theory of the strong interactions based on a Yang–Mills quantum field acting on the colour degree of freedom has come to be known as *quantum chromodynamics* or QCD. The quanta of QCD are called *gluons*, since they provide the 'glue' holding hadrons together.

9.8.4. *Scaling violation in deep inelastic scattering*

Although deep inelastic lepton scattering experiments (section 9.6) reveal point-like constitutents in the proton, the details of nucleon structure depend slightly on the momentum transferred to the target by the leptons. Figure 9.28 shows modern data [49] illustrating this behaviour, predicted by QCD. Quarks can emit or absorb gluons, thereby shifting their momenta. A proton probed with very high momentum transfers thus appears to have 'softer' and more numerous constituents. The scaling predicted by Bjorken is violated slightly, to a degree that sheds light on the strong coupling constant $\alpha_s(\mu)$ (the subscript denotes 'strong') at any chosen momentum scale μ. Several quantitative treatments of this feature appeared in the mid-1970s, starting with the

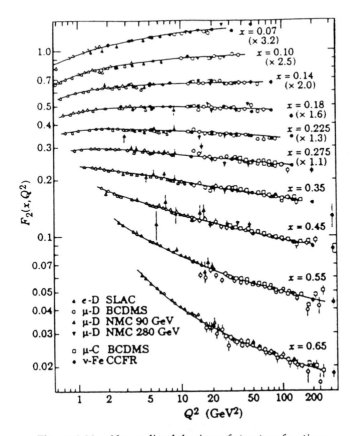

Figure 9.28. *Non-scaling behaviour of structure functions.*

initial discoveries of asymptotic freedom and expressed transparently in the work of Altarelli and Parisi [341]. Since many measurements of α_s are performed at the mass of the Z-boson, a convenient reference point is $\mu = M_Z$, even though the deep inelastic scattering experiments probe lower momentum scales. A recent analysis of deep inelastic scattering [342] finds $\alpha_s(M_Z) \approx 0.12$, consistent with other determinations.

9.8.5. Jets and other high-$p^\perp$ phenomena

An early application of the parton picture, mentioned in section 9.6.2.5, predicted [251] hadron production at high transverse momenta, confirmed by experiments at the CERN ISR (Intersecting Storage Rings) and Fermilab [252]. The behaviour of cross sections at high transverse momenta indicated that the constituents of colliding protons were interacting with one another at a fundamental level.

Direct evidence for gluons emerged from the study of three-jet events in electron–positron annihilations [343], as detailed in reference [240]. In quantum chromodynamics, the process $e^+ + e^- \rightarrow$ hadrons evolves in two stages. First (essentially instantaneously), a quark–antiquark ($q\bar{q}$) pair is produced by the electromagnetic current. Then, over a longer time scale, these quarks materialize into hadrons through production of additional quark–antiquark pairs, with transverse momenta small with respect to the original quarks. The original quark and antiquark thereby define a direction for two *jets* of particles, which become more and more clearly defined with increasing energy. Because QCD interactions grow stronger at momentum scales below 1 GeV/c, it took several GeV in the centre of mass to begin seeing two-jet events. Such jets were identified in e^+e^- annihilations [344] and hadronic collisions [345] in the mid-1970s.

At the 1978 International Conference on High Energy Physics in Tokyo, a much-discussed particle was the Υ (section 9.9), a composite of the fifth (b) quark and its antiquark in the same way that the J/ψ is a charm–anticharm bound state. Had the three-gluon decay of the Υ been seen? It was not clear whether the data were really showing three distinct gluon jets, or just effects of data selection. The higher energies of two new electron–positron colliders, PETRA in Hamburg and PEP at Stanford, allowed a much crisper search for gluon jets. QCD predicted a *third* jet, corresponding to the emission of a gluon by one of the final-state quarks, in a fraction of e^+e^- annihilations into hadrons. This gluon jet should be identifiable if emitted at a sufficient angle with respect to one of the quarks.

A technique for identifying gluon jets [346] was proposed by Wu and Zobernig, members of the TASSO Collaboration at PETRA. A pretty picture of a three-jet event (see figure 9.29) was presented at international conferences during the early summer of 1979 [347]. At the Lepton–Photon Symposium at Fermilab in August 1979, all four groups working at PETRA presented evidence for three-jet events.

As available centre-of-mass energies increased, spectacular particle jets at large angles with respect to the initial beams were observed at the CERN SPS Collider [348] and the Fermilab Tevatron [349]. The decays of Z-bosons to hadrons seen at LEP have yielded vast samples of events with three, four and even more jets. The rate for $n+1$ jets is related to that for n jets by one power of α_s, permitting an estimate $\alpha_s(M_Z) \approx 0.12 \pm 0.01$ from multi-jet production rates compatible with that from deep inelastic scattering.

9.8.6. *Other applications*

9.8.6.1. *Quarkonium decays.* Bound states of charmed quarks such as the J/ψ, mentioned in section 9.7, provided the first laboratory for application of perturbative QCD to decays [350]. These studies were helped greatly by the discovery of the fifth quark b and of the corresponding $b\bar{b}$ bound states such as the Υ (section 9.9). One can

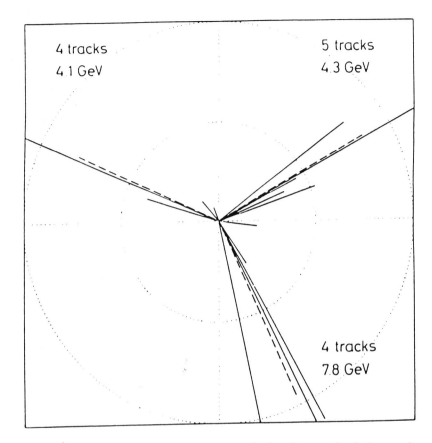

Figure 9.29. *Sketch of particle tracks in an early three-jet event in electron–positron annihilation.*

measure $\alpha_s(m_b)$, for instance, by comparing decays such as $\Upsilon \to 3g$ ($g =$ gluon) and $\Upsilon \to 2g + \gamma$. With higher-order QCD corrections, the result [342] is $\alpha_s(m_b) \approx 0.19$, corresponding to $\alpha_s(M_Z) \approx 0.11$.

9.8.6.2. Hadron production in electron–positron annihilations. The lowest-order QCD correction to the ratio R characterizing hadron production in electron–positron annihilations (section 9.6.3) amounts to a factor $1 + (\alpha_s/\pi)$. Data taken over a wide range of centre-of-mass energies, ranging from several GeV to M_Z, are indeed consistent with this correction.

9.8.6.3. Counting rules. Differential cross sections at high energies and fixed angles [351] behave in ways which may be derived from QCD by tracing the flow of high momentum transfers through diagrams for scattering at the quark level.

717

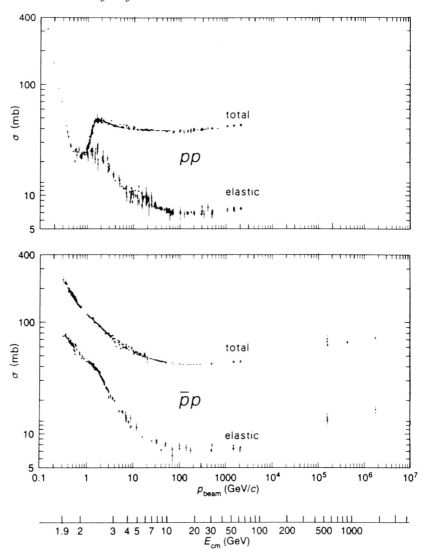

Figure 9.30. *Total and elastic pp and p̄p cross sections as functions of energy.*

9.8.6.4. Non-perturbative effects. Not all phenomena in high-energy collisions are understood quantitatively within the framework of QCD. Scattering at small momentum transfers is still most economically described within the framework of Regge poles (section 9.5.4). The Pomeranchuk trajectory may reflect exchanges of a pair (or more) of gluons [352]. The rise of total cross sections at high energies [353] is a topic of interest, with recent data [354] summarized in figure 9.30.

Multiple particle production [355] in hadronic interactions displays simple regularities understood mainly on a phenomenological basis.

The *rapidity* of a particle may be defined as $y \equiv (\frac{1}{2}) \log[(E + p_z)/(E - p_z)]$, where E is its energy and p_z its momentum along the beam axis. A hadronic collision leads to a number of particles per unit rapidity growing only slightly with increasing energy. A qualitative understanding of this behaviour exists on the basis of quark fragmentation [356].

QCD has thus proved able to describe a wide variety of strong-interaction phenomena. Although QCD only permits perturbative calculations at energies and momenta above one or several GeV, this aspect should not be held against its validity. Methods for dealing with the non-perturbative regime, including lattice gauge theories [357] and QCD sum rules [358], are being actively pursued.

9.9. Three families of quarks and leptons

9.9.1. The τ-lepton

9.9.1.1. Discovery. The repetitive structure of the leptons (electron and its neutrino, muon and its neutrino) stimulated searches for additional 'sequential' doublets, consisting of a heavy charged lepton and its neutrino, with negative results up to the early 1970s [359]. Around that time, several theorists [360] worked out the consequences of heavy charged leptons. The process $e^+ + e^- \rightarrow l^+ + l^-$, under scrutiny at new electron–positron colliders, was one way to produce these leptons.

The Cambridge Electron Accelerator (see section 9.6.3) reported in the summer of 1973 that the ratio $R = \sigma(e^+ + e^- \rightarrow \text{hadrons})/\sigma(e^+ + e^- \rightarrow \mu^+\mu^-)$ at centre-of-mass energies of 4 and 5 GeV exceeded the value of 2 expected for u-, d- and s-quarks [254]. These results were confirmed by the SPEAR Collider later that year [361]. The charmed quark, discovered in 1974 (section 9.7), accounted for part of the rise in R. It contributed $\frac{4}{3}$ to R, leading to a total of $3\frac{1}{3}$. However, the SPEAR results above 4 GeV indicated values of R well in excess of 4.

Other curious features of the SPEAR data at and above 4 GeV appeared not to be attributable to charmed-particle production. Events with an electron, a muon of opposite charge, no other charged particles, and missing energy suggested production of a pair of new leptons: $e^+ + e^- \rightarrow \tau^+ + \tau^-$, with one of them decaying to $e\nu\bar{\nu}$ and the other to $\mu\nu\bar{\nu}$. This signature had been sought earlier at the ADONE electron–positron collider at Frascati [359]. However, in retrospect, the energy was too low for τ pair production.

In 1975, Martin Perl and his collaborators published a paper entitled *Evidence for the existence of anomalous lepton production*, which they concluded with the statement 'A possible explanation for these events is the production and decay of new particles each having a mass in the range of 1.6 to 2 GeV/c^2.'. Two years later they were able to quote a mass of 1.80 ± 0.045 GeV/c^2 for the τ-lepton [310]. Initial failure to detect a crucial decay mode, $\tau \rightarrow \pi\nu$, at the DORIS storage rings at DESY in Hamburg led to some scepticism as late as 1977, but confirmation of this and other modes by the DESY experiments followed soon thereafter [362].

The τ was a truly new discovery in not having been anticipated by any direct theoretical prediction. It was the first member of a third family of quarks and leptons.

9.9.1.2. Properties.

The τ decays to a ν_τ (whose existence remains inferred, having not been detected directly) and a virtual W-boson, materializing into u$\bar{\text{d}}$, u$\bar{\text{s}}$ (with reduced rate), e$\bar{\nu}_e$ or $\mu\bar{\nu}_\mu$. Decay rates for these final states are consistent with the standard weak-interaction theory, leading to an overall lifetime of about 0.3 ps. The ratio of the u$\bar{\text{d}}$ rate to the e$\bar{\nu}_e$ or $\mu\bar{\nu}_\mu$ rate, a factor of three times a small correction, provides further evidence for three colours of quarks. Reference [363] provides a sample of the physics that may be learned from the increasing precision in the measurement of τ properties.

9.9.2. The fifth quark

9.9.2.1. Kobayashi and Maskawa's suggestion and its implications.

Spurred by hints of charmed particles in nuclear emulsions [299] (section 9.7.4.1), M Kobayashi and T Maskawa [300] asked 'As long as there seemed to be two families of quarks and leptons, why not three?'. A third family of quarks would permit the parametrization of CP violation by introduction of a non-trivial phase in the charge-changing weak couplings of quarks.

The search for the charmed quark had been partly motivated by the existence of two families of leptons, a family structure being particularly appropriate for cancelling triangle anomalies [298] (section 9.7.3). Thus, a third doublet of quarks would imply a third doublet of leptons and vice versa. Following the τ's discovery, it was then natural to expect a third quark of charge $-\frac{1}{3}$ (in addition to d and s) named b for 'bottom,' and a third quark of charge $\frac{2}{3}$ (in addition to u and c) named t for 'top' (in analogy to 'down' and 'up'). The quantum number carried by the b-quark has been referred to as 'beauty'.

Some false alarms and harbingers of the true signal marked the search for the third family of quarks. An apparent anomaly in deep inelastic antineutrino scattering [364] suggested that a threshold for production of the b-quark had been crossed. (This effect, not confirmed in other experiments, was later understood in terms of instrumental bias.) A search at Fermilab for particles heavier than the J/ψ decaying to lepton pairs yielded an apparent peak ('Υ') at 6 GeV in the e$^+$e$^-$ channel, not confirmed in the $\mu^+\mu^-$ channel. However, a peak at 9.5 GeV/c^2 in that channel tempted John Yoh, then a postdoctoral fellow working on the experiment, to put a bottle of Mumm's champagne labelled '9.5' (to be opened presently) in the group's refrigerator [365]. Another group, also studying muon pairs at Fermilab, had a single event at 9.5 GeV/c^2, labelled affectionately 'Big Mac'. The $\mu^+\mu^-$ spectrum in the 1976 experiment of reference [366] showed a small anomaly around 10 GeV/c^2.

9.9.2.2. Discovery of the upsilon family. The study of hadronically produced leptons at Fermilab evolved through several stages, culminating in Experiment E-288, dedicated to the study of high-mass lepton pairs, under the leadership of Leon Lederman. After the false alarm of the Υ at 6 GeV in 1976 and a fire in the spring of 1977 whose damage was quickly repaired, the group began running with high intensity and improved resolution in the late spring of 1977. The data soon proved that John Yoh's bottle of champagne had been correctly labelled. Not only was there a peak at 9.5 GeV/c^2, but another smaller one seemed to be riding on its tail, about 0.6 GeV/c^2 higher [367] (see figure 9.31(a)), in a manner reminiscent of charmonium, for which the ψ' lies about 0.6 GeV/c^2 above the J/ψ. In contrast to charmonium, however, there appeared to be *three* narrow peaks [368]. Called $\Upsilon(1S)$, $\Upsilon(2S)$, and $\Upsilon(3S)$, they correspond to the first, second and third S-wave systems of a heavy quark bound to the corresponding antiquark. (Figure 9.32(b) shows a recent spectrum.)

The partial width of the Υ for decay into pairs of leptons, measured at the DORIS storage rings [369], suggested that this particle consisted of quarks with charge $-\frac{1}{3}$, a conclusion strengthened upon observation of the second peak [370], the Υ', whose hadronic parameters were predicted with greater confidence. Once the threshold for production of mesons containing a single heavy quark was passed, the ratio R in e^+e^- rose by $\frac{1}{3}$, confirming the new quark's charge of $-\frac{1}{3}$ and identifying it with the b. The Υ and its excited levels were $b\bar{b}$ states.

9.9.2.3. Beauty mesons. The 12 GeV Cornell Electron Synchrotron, first operated in 1967, was converted in the late 1970s to the Cornell Electron Storage Ring (CESR), with beam energies of up to 8 GeV for electrons and positrons. This machine could not have come at a more opportune time, detecting the $\Upsilon(1S)$, $\Upsilon(2S)$ and $\Upsilon(3S)$ in short order. At higher energy a fourth Υ level was seen, with a natural width [371] indicating its decay into pairs of b-flavoured mesons, now called B-mesons. These new mesons were first identified by the presence of a lepton and additional kaons (corresponding to the weak decay of a b-quark) in their decay products [372]. Reconstruction of specific decay channels [373] yielded a B-meson mass around 5.28 GeV/c^2. Since the threshold for production of a pair of B-mesons is twice this value, the $\Upsilon(4S)$, at a mass of 10.575 GeV/c^2, is thus ideal for producing pairs of B-mesons nearly at rest.

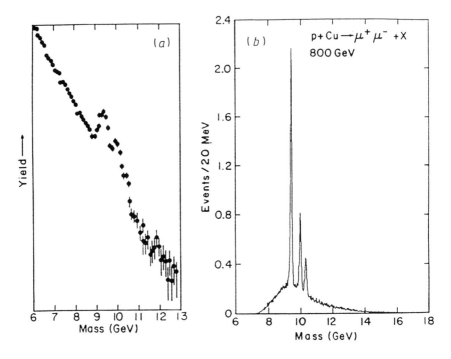

Figure 9.31. *Evidence for the* Υ *resonances in hadronic interactions.* *(a) From reference [367]; (b) a recent spectrum from the second of references [368].*

9.9.2.4. Spectroscopy of the b-quark. The present state of our knowledge about b$\bar{\text{b}}$ levels is summarized in figure 9.32. At least six sets of S-wave levels and two groups of P-wave levels have been identified, all spin-triplets thus far. Their electromagnetic transitions to spin-singlet levels require a b-quark spin to flip, and are expected to occur with rates below present levels of sensitivity.

The b$\bar{\text{b}}$ system exhibits the interplay of short-range and long-range effects in QCD [374, 375]. The interquark potential combines the short-range Coulomb-like behaviour expected from single gluon exchange with the long-range linear behaviour proposed by Nambu [340].

9.9.2.5. Decays of particles with b-quarks. Once the b-quark was identified, its charge-changing weak decays were seen to favour the c-quark and not the lighter u-quark [376], a conclusion reached on the basis of the emitted leptons' momentum spectrum. The decay b $\rightarrow$ ul$\bar{\nu}_l$, initially not detected at all, was eventually identified at the level of about 1–2% of the b $\rightarrow$ cl$\bar{\nu}_l$ process [377]. Considering the different phase space available for the two decays b $\rightarrow$ cl$\bar{\nu}_l$ and b $\rightarrow$ ul$\bar{\nu}_l$, the result implies a ratio of b $\rightarrow$ u and b $\rightarrow$ c couplings of between 0.05 and 0.1.

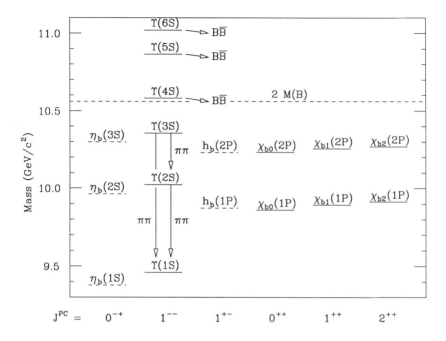

Figure 9.32. *The spectrum of $b\bar{b}$ states. Observed and predicted levels, as quoted in reference [319], are denoted by solid and broken horizontal lines, respectively. In addition to the transitions denoted by arrows, numerous electric dipole transitions have been seen between the Υ and χ_b states, e.g. $3S \to 2P \to 2S \to 1P \to 1S$, $3S \to 1P$ (very weak) and $2P \to 1S$.*

Another peculiar feature of b-quark decays lies in the relative weakness of the b → c weak coupling. Charmed mesons, with masses below 2 GeV/c^2, live for about 0.4 to 1 ps, depending on the light quark they contain. By contrast, mesons with a b-quark live for about 1.5 ps on average [378], in spite of their much larger masses. This curiously long life of the b-quark was first observed by the MAC Collaboration at SLAC under the leadership of David Ritson [379].

9.9.3. *The Cabibbo–Kobayashi–Maskawa (CKM) matrix*

9.9.3.1. *Role in CP violation.* The charge-changing couplings of quarks in the three-family model of Kobayashi and Maskawa [300] can be expressed in terms of a 3×3 matrix with rows pertaining to the charge $-\frac{1}{3}$ quarks d, s, b and columns to the charge $\frac{2}{3}$ quarks u, c, t. The unitarity of this matrix, required by its role in the diagonalization of quark mass matrices, implies the absence of flavour-changing neutral currents, just as for the two-family model.

For *n* quark families, a unitary $n \times n$ matrix has n^2 real parameters. For $n = 2$, a single angle (the Cabibbo angle) suffices to specify the matrix, the remaining three parameters setting relative quark phases. Three families require four real parameters in addition to five arbitrary relative quark phases.

A convenient parametrization of the CKM matrix [380] is

$$
\mathbf{V}_{\text{CKM}} = \begin{bmatrix} V_{\text{ud}} & V_{\text{us}} & V_{\text{ub}} \\ V_{\text{cd}} & V_{\text{cs}} & V_{\text{cb}} \\ V_{\text{td}} & V_{\text{ts}} & V_{\text{tb}} \end{bmatrix}
$$

$$
\approx \begin{bmatrix} 1 - \lambda^2/2 & \lambda & A\lambda^3(\rho - i\eta) \\ -\lambda & 1 - \lambda^2/2 & A\lambda^2 \\ A\lambda^3(1 - \rho - i\eta) & -A\lambda^2 & 1 \end{bmatrix}
$$

with A, ρ and η of order unity. Only the leading orders of the matrix elements are displayed here. The matrix is unitary as written up to order λ^4. The violation of CP invariance is represented by the non-zero value of η, at least three families of quarks being needed for this purpose.

9.9.3.2. Experimental information on matrix elements. The parameter $\lambda \approx 0.22$ is the sine of the Cabibbo angle, well known from the analysis of strange-particle decays. Information on the b-quark lifetime places A around 0.8, while the measurement $|V_{\text{ub}}/V_{\text{cb}}| = 0.05 - 0.1$ implies $(\rho^2 + \eta^2)^{1/2}$ between $\frac{1}{4}$ and $\frac{1}{2}$.

Individual values of η and ρ are less well known. Values of ρ between -0.4 and 0.4 and of η between 0.2 and 0.6 are compatible with present data. Indirect information is provided by CP violation in the kaon system and by mixing of the neutral B-meson $\bar{\text{B}}^0 \equiv \text{b}\bar{\text{d}}$ with its antiparticle $\text{B}^0 \equiv \text{b}\bar{\text{d}}$, first discovered in 1987 [381]. Uncertainties plaguing the estimates of η and ρ from these diagrams include errors in the top quark mass ($m_t = 180 \pm 12$ GeV/c^2; see below), and in various hadronic matrix elements, for which estimates from lattice gauge theories [357] and QCD sum rules [358] are beginning to be useful.

A 'superweak' theory [170] (independent of η in the CKM matrix) can still account for all the CP-violating phenomena observed in the kaon system. Two main checks of the CKM theory of CP violation hinge on rare decays of kaons and on the search for CP-violating B decays [382].

Detection of a difference between the ratios of the amplitudes for decay of K_L and K_S into pairs of charged and neutral pions would provide evidence against the superweak model, as mentioned in section 9.5.2.3. Searches for this difference have achieved a level of better than 0.2%. Two experiments [171, 172] give somewhat different answers on this detection; improved versions of both experiments are planned.

The study of B-meson decays has been pioneered by detectors in e^+e^- colliders (see figure 9.33) such as ARGUS at DESY and CLEO at CESR. More luminous e^+e^- sources of B-mesons, with an energy asymmetry useful in studying time-dependent decay effects, are now being planned at several laboratories. It is also becoming possible to isolate B hadron signals from intense backgrounds in hadron colliders.

9.9.4. On the trail of the top quark; observation

Although the family structure of quarks and leptons is simple and repetitive, the b-quark could have broken away from this pattern, just as the transition metals are associated with variation in the periodic table of the elements. However, several pieces of data indicated that the b-quark had to be a member of a doublet of weak SU(2).

If the b-quark were a singlet, flavour-changing neutral decays such as b $\rightarrow$ s$\mu^+\mu^-$ would occur with a rate much higher than anticipated in standard electroweak theory [383]. No such enhanced rate has been observed. Furthermore, the cross section and forward–backward asymmetry in the reaction e$^+$ + e$^-$ $\rightarrow$ b + $\bar{\text{b}}$, over a range of energies, imply that the b is a doublet member, just like the d and s. The self-consistency of calculations based on box diagrams for mixing of neutral B-mesons with their antiparticles also supports the existence of a heavy top quark.

Ever since the discovery of the b, searches for the top continued at electron–positron colliders such as PEP (Stanford), PETRA (Hamburg), TRISTAN (Japan) and LEP (CERN), and hadron colliders at CERN. With the higher hadronic collision energies available at Fermilab, in a *tour de force* of analysis involving several different decay modes, the CDF Collaboration presented evidence in 1994 for a top quark [332], confirming it in 1995 and measuring its mass to be $m_t = 176 \pm 8 \pm 10$ GeV/c^2. The D0 Collaboration has also observed a statistically significant signal [384], quoting $m_t = 199^{+19}_{-21} \pm 22$ GeV/c^2.

The effect of a large top quark mass on electroweak parameters was first pointed out by Veltman [385]. Virtual top quark contributions to W and Z self-energies alter the lowest-order predictions of the ratio $M_W^2/M_Z^2 = \cos^2\theta$ and the Z mass itself. Since independent (very accurate) determinations of the electroweak angle θ and the Z mass exist, for instance from LEP experiments, the top quark had been anticipated to lie below about 200 GeV/c^2. The measurement of its mass within these bounds was a striking confirmation of the electroweak theory, and indirectly places bounds on other effects such as Higgs bosons.

9.10. Accelerators

Our knowledge of the nucleus and, subsequently, of elementary particles has closely parallelled the development of particle accelerators at higher and higher energies. An excellent and detailed review of accelerator development up to the time of colliding beams (ca 1960) has been given by two of the principal participants in this effort, Livingston and Blewett

See also p 1201

Figure 9.33. *The ARGUS (top) and CLEO (bottom) detectors.*

[386]. For each type of machine, we give here only a few examples of those actually constructed around the world.

9.10.1. *Electrostatic generators*

The first nuclear reactions produced by protons accelerated to high voltages bombarding a target at rest in the laboratory were observed by Cockcroft and Walton [387]. Their work was directly stimulated by the theoretical work of Gurney and Condon and of Gamow on barrier penetration. To produce the required voltage, they developed a scheme to multiply a transformer's output voltage through an ingenious arrangement of rectifying diodes and capacitors. Whereas nearly any multiplying factor can be achieved, the technical advantage lay in the fact that if the transformer's secondary produced a voltage V, no capacitor or rectifier was required to withstand a voltage greater than $2V$.

The more difficult technical problem at the time was posed by the need to provide an ion source at the high voltage and a tube evacuated of air through which the ions could be accelerated. The tube had to be insulated to withstand the high potential between the ion source and the target, with suitable arrangements for a uniform potential gradient along the tube to minimize spurious electrical discharges.

Cockcroft and Walton's incredible technical feat was recognized with a Nobel Prize in 1951. Until the advent of radio-frequency quadrupoles, their accelerator served as the low-voltage stage of every proton accelerator. While theirs was not the first scheme for voltage multiplication, it was superior, their device becoming universally known as the Cockcroft–Walton generator.

The other static voltage source widely used for the acceleration of nuclear particles was the electrostatic generator of van de Graaff [388]. While a Rhodes scholar at Oxford in 1927–28 van de Graaff had become interested in the possibility of producing high voltages by belt-charging. On his return to the United States as a National Research Council Fellow at Princeton University, he constructed his first models (figure 9.34), one for positive and one for negative potentials. Each consisted of a motor at the base turning a pulley which drove an insulating belt. The belt passed vertically about 7 feet into a copper sphere about 24" in diameter which contained the return pulley. The sphere and return pulley arrangement was supported on a glass tube about 2" in diameter. Charge was 'sprayed' on the belt at the bottom and removed in the sphere at the top with brushes.

Van de Graaff's first two generators were used for many years at Princeton in freshman physics demonstration lectures. Eventually one of them was given to the Smithsonian Institution for exhibit on the condition that they supply a duplicate to allow the demonstrations to continue. Subsequently, the duplicate model (made with modern

Figure 9.34. *Robert van de Graaff and one of the earliest versions of his generator.*

materials) has not performed as well as the original, a wonderful testimony to the inventor's skill.

The van de Graaff generators were immediately successful at producing high voltages with relative ease. However, as with the Cockcroft–Walton generator, combining the source of high voltage with an ion source and accelerating column presented a daunting technical problem. With van de Graaff's success, a number of groups became interested in adapting the generator to particle acceleration. Among these were Ray Herb and his group at Wisconsin, Merle Tuve and his group at the Carnegie Institution in Washington, and Compton and Van Atta at MIT, who, joined by van de Graaff [389], produced a machine which developed a potential of 5.1 MV. At more modest voltages the groups at Wisconsin [390] and the Carnegie Institution [391] produced the first practical van de Graaff generators for accelerating particles. With their machine the Carnegie group made the first excitation measurements showing sharp nuclear resonances.

Electrostatic generators have been adapted to accelerate electrons and to produce high-energy x-rays. In this capacity they have extensive industrial and medical usage. In nuclear physics, they have long had the advantage of excellent voltage stability, permitting good control of beam energy in precision measurements of nucleon–nucleon scattering. More recently, the development of negative hydrogen ion sources has permitted the construction of so-called tandem generators. These machines accelerate negative hydrogen ions from ground to a high positive potential where a thin foil strips them of their two electrons; the positive ions are then accelerated back to ground, thereby doubling their energy. Despite these developments, machines relying on high static voltages have been limited to around 50 MeV, restricting them to nuclear physics studies.

9.10.2. The cyclotron

Concurrently with the development of the static high-voltage machines, work began on resonant cyclic accelerators. The first of these, a linear machine designed and constructed by Wideröe [392], consisted of three cylindrical tubes positioned end to end and electrically insulated from each other. A radio-frequency voltage was applied to the central tube, properly phased to accelerate an ion travelling from the first to the second tube. By adjusting the radio frequency to reverse the voltage during the ion's transit down the middle tube, the ion was again accelerated when passing from the second to the third tube. Reading about this device in the library at Berkeley, E O Lawrence realized that the ions might be recycled through the radio-frequency voltages again and again by bending the particles back on their original path with a magnetic field. He quickly saw that the angular frequency of revolution of an ion in a circular path in a magnetic field, $\omega = eB/mc$, remains constant, independent of radius and momentum! In such a way the cyclotron and a whole class of magnetic resonant machines were born [393].

See also p 1201

The first cyclotron applied to physics research had magnetic pole pieces 10″ in diameter, producing protons over 1 MeV in energy. It came into operation in 1932 a few months after the disintegration of lithium by protons was announced by Cockcroft and Walton, confirming their results immediately [394].

A number of features important to successful cyclotron operation were not recognized at the beginning. Happily, the device worked anyway. The cyclotron angular frequency, $\omega = eB/mc$, remains constant only as long as B and m do not change. Countering the requirement of constant B is the need for some vertical focusing to prevent ions with some initial vertical velocity from striking the pole tips of the magnet before undergoing a significant number of rotations. Vertical focusing requires the field to fall off with increasing radius. Fortunately for the cyclotron inventors, any magnet with circular pole tips will, barring heroic attempts at field shaping, naturally show a small decrease of field with radius, thus providing some vertical focusing. At the same time, the change of field with radius should not be so great as to violate the cyclotron equation grossly. This too was the case in the first cyclotron. In addition, relativistic effects cause the mass to increase with energy. Again providence smiled on the inventor, since this effect does not become limiting until a proton's energy reaches about 25 MeV.

After the cyclotron's success at Berkeley, many similar machines were constructed throughout the world. In the 1930s every research university in the country arranged to have a cyclotron for nuclear physics studies. Berkeley, of course, showed the way, with larger and larger cyclotrons. This activity culminated with the construction of a 184" giant, completed only after World War II.

9.10.3. *Phase stability and the synchrocyclotron*

Shortly after the war, McMillan at Berkeley and Veksler in the Soviet Union independently showed that the conflict between requiring the field to decrease with radius for vertical focusing and increase with radius to compensate for the relativistic mass increase could be resolved by changing the frequency during ion acceleration [395]. To quote from McMillan's paper

> The device proposed here makes use of a 'phase stability' possessed by certain orbits in a cyclotron. Consider, for example, a particle whose energy is such that its angular velocity is just right to match the frequency of the electric field. This will be called the equilibrium energy. Suppose further that the particle crosses the accelerating gaps just as the electric field passes through zero, changing in such a sense that an earlier arrival of the particle would result in an acceleration. This orbit is obviously stationary. To show that it is stable, suppose that a displacement in phase is made such that the particle arrives at the gap too early. It is then accelerated; the increase in energy causes a decrease in angular velocity, which makes the time of arrival tend to become later. A similar argument shows that a change of energy from the equilibrium value tends to correct itself. These displaced orbits will continue to oscillate, with both phase and energy varying about their equilibrium values.
>
> In order to accelerate the particles it is now necessary to change the value of the equilibrium energy, which can be done by varying either the magnetic field or the frequency. While the equilibrium energy is changing, the phase of the motion will shift ahead just enough to provide the necessary accelerating force; the similarity of this behaviour to that of a synchronous motor suggested the name of the device ...

The advantages of phase stability far exceeded its single disadvantage of requiring the frequency to change during the time the particles are acquiring energy. The ions are injected at the high-frequency end of the cycle, and the frequency is reduced during the acceleration. The ions are finally brought to an internal target or ejected from the machine at the highest energy (lowest frequency). The frequency cycle is then repeated. Since ions can only be accepted for acceleration during a small part of the whole frequency cycle, the beam intensity will accordingly be much lower, about 1% of that in conventional cyclotrons.

It did not take long for the fundamental soundness of the idea to be lastingly recognized. The ideas expressed by McMillan have been invoked in every subsequent accelerator. Immediately after World War II the 184″ cyclotron, after some modelling studies, was converted to a 'synchrocyclotron'. It first came into operation in November 1946, accelerating deuterons to 190 MeV and, later, protons to 350 MeV. It was thus an enormous success from the beginning, followed between 1948 and 1952 by five additional synchrocyclotrons with energies ranging from 150 to 450 MeV constructed at various universities in the US and a further six elsewhere in the world. The physics pouring from these accelerators was immense, primarily because their energy exceeded in most cases the 290 MeV threshold for creating pions. With these machines the properties of the pion and the muon were elucidated in detail. Pionic and muonic atoms were discovered and quantitative measurements on these new kinds of atoms were made. Parity violation was observed and thoroughly explored in the pion–muon–electron decay chain. It was a fantastically productive period in the history of particle physics.

9.10.4. Electron synchrotrons

The concept of phase stability was also exploited in electron accelerators, leading to the first electron synchrotrons. Synchrotrons accelerate particles in orbits of constant radius. After an electron's energy has reached 4–5 MeV its speed is essentially that of light so that the transit time around an orbit of fixed radius is nearly constant, requiring minimal change in the frequency of the accelerating voltage. Therefore, use of a separate system such as a van de Graaff generator for pre-acceleration to a few MeV simplifies the subsequent synchrotron design enormously.

The acceleration of electrons in circular guide fields is accompanied by significant loss of energy to radiation. This loss implies a smaller orbit and a smaller transit time around the ring. If the particles are accelerated on the falling side of the radio-frequency voltage, those particles arriving earlier will be accelerated by higher voltages thereby compensating for the radiation loss. Under the guidance of McMillan the first electron synchrotron was started in Berkeley in 1946 and completed in 1949, producing electrons of 335 MeV. Experiments using the photons from the electrons impinging on Pb were immediately started and the first photoproduced charged pions were quickly detected [396]. A major result from this machine was the discovery of the neutral pion [68] in 1950 (cf section 9.4.2). As with the synchrocyclotrons, electron synchrotrons were constructed at many universities, most of them in the range of 300 MeV. They were responsible for tracing out the full peak of the famous (3, 3) resonance, mentioned in section 9.4.5.2.

9.10.5. The betatron

Independently of all the activity devoted to electron synchrotrons, in fact back in 1940, the first successful betatron was developed by D W Kerst [397] with an energy of 2.3 MeV. The electric field associated with a changing magnetic field served to accelerate electrons, while the same magnetic field guided the particles, a combination which would seem fairly obvious. However, the attendant technical problems turned out to be severe, thwarting many attempts before Kerst's. His machine followed the detailed magnet design based on elaborate orbit calculations by Kerst and Serber [398]. An interesting technical point about the betatron has become a favourite examination question in courses in electricity and magnetism: it is easily shown that, for a constant orbit radius, the magnetic field at the orbit must be just one-half the average magnetic field linked by the orbit. One problem associated with extrapolating the betatron principle to high energies lies in maintaining this factor of 2 in the face of iron saturation, etc. The development of the betatron culminated with the construction at the University of Illinois of a 300 MeV device, completed in 1950 [399]. We have already noted the desirability of pre-acceleration of electrons in a synchrotron by, for example, van de Graaff generators. Another even more popular method was pre-acceleration of the electrons in the synchrotron itself using the betatron principle.

9.10.6. Proton synchrotrons

By 1948 it had become clear that achieving still higher energy protons than accessible by synchrocyclotrons would hinge on the synchrotron principle. Cyclotrons require the quantity of magnetic iron to increase roughly as the cube of the maximum proton momentum, whereas synchrotron requirements scale only linearly with the momentum. The crossover had already been reached with cyclotrons under construction. The US Atomic Energy Commission planned to construct two machines; one on Long Island at the newly created Brookhaven National Laboratory in the energy range of 2–3 GeV and the other with 5–6 GeV at Berkeley.

The previous considerations of electron synchrotrons make it clear why physicists were more timid in proceeding with proton machines, whose required frequency change was larger. Furthermore, it was necessary to synchronize the frequency closely with the instantaneous magnetic field. At the Brookhaven machine (ultimately named, somewhat grandiosely, the Cosmotron) the protons were injected from a van de Graaff generator with 4 MeV energy. The initial machine design involved a magnetic field of 311 gauss and a frequency of rotation around the magnet ring of 0.4 MHz at injection. At the full energy, 3 GeV, the average field reached 13 kilogauss with a frequency of 4.2 MHz. In contrast to the early electron machines, whose acceleration time was typically 5 ms, the period of acceleration was 1 s, allowing the final energy of 3 GeV to be reached with only about 2 kV of accelerating voltage per turn.

The Cosmotron started operating in 1952 at 2.3 GeV. Almost immediately it began producing noteworthy physics, such as the first observation of associated strange-particle production (cf section 9.4.4). Early in 1954 it started operating at its full design energy of 3 GeV.

The Berkeley synchrotron, called the Bevatron (a name based on the earlier abbreviation for GeV), reached its design energy of 5.4 GeV in October 1954. It could be pushed to 6.4 GeV at a lower cycling rate. Ostensibly the design energy of the Bevatron was chosen to exceed the antiproton production threshold, which is 5.6 GeV for protons striking protons at rest. Appropriately, antiprotons produced in the Bevatron were found by the Chamberlain–Segré group [73] in 1955 (cf section 9.4.3).

9.10.7. Strong focusing

At the 1952 January meeting of the American Physical Society, held on the campus of Columbia University in New York City, a highlight was an invited talk by Enrico Fermi [400], describing research in progress at the University of Chicago's 450 MeV synchrocyclotron which had started operation the preceding year. It was the highest energy accelerator then in operation. Near the end of his talk Fermi described, in jest, the ultimate accelerator of the future, a machine circling the Earth.

Another talk at this same meeting contained the seeds of an idea that made ultra-high-energy accelerators possible. Quoting from the abstract [401] of a paper describing the original pair of quadrupoles devised to focus the external beam of the Princeton cyclotron

> Under these conditions, because of the curvature of the fringing field, their focal lengths in the plane of the field are very nearly equal, but opposite in sign, to their focal lengths in the plane of the pole faces. However, if two of these astigmats are spaced a distance comparable with their focal lengths with their field directions orthogonal, a point image of a point source may be obtained. If the source distance, the separation of the astigmats, and the desired image distance are specified, the focal lengths required for double focusing can be reliably calculated from the thin lens equation.

This is the first published reference to a principle eventually called strong focusing. This idea was quickly exploited by the early summer of 1952 to focus external pion beams at Columbia's Nevis cyclotron [402].

Strong magnetic focusing was in the air. In a highly influential paper, Courant, Livingston and Snyder showed how that principle could be used with great advantage in the design of synchrotrons [403], demonstrating in particular that strong focusing, implemented by alternating the gradient of the magnetic field guiding the particles, could be maintained in the closed orbits required in accelerators, that phase stability would persist, and that the large 'momentum compaction' would result in relatively small beam displacements in spite of rather large momentum spreads. This paper also created a magnetic quadrupole design which would not produce a net bending of the beam, a disadvantage of the original design by the Princeton group.

It developed that Nicholas Christofilos, an electrical engineer working in Greece, had applied in 1950 for a US patent on a conceptual design for an accelerator using alternating-gradient focusing. His seminal contribution went unrecognized because his work was never published, but his patent (awarded 28 February 1956) is contained in reference [404].

Following the prescriptions supplied in the Courant, Livingston and Snyder paper, designs of two alternating-gradient synchrotrons to operate at about 30 GeV, one at Brookhaven and one at CERN, were immediately started. However, the first accelerator to employ the alternating-gradient strong-focusing principle, producing 1.1 GeV electrons in 1954, was constructed under the direction of R R Wilson at Cornell University. The 28 GeV machine at CERN (called the Proton Synchrotron or PS) started operations in 1959, while at Brookhaven experiments were begun in 1960 with the machine (the AGS) operating at 33 GeV and still in service at the time of writing in the mid-1990s.

For many years, proton synchrotrons became the machines of choice in the push towards ever-increasing energies. An accelerator at Serpukov in Russia began operation at 76 GeV in 1971. The Fermilab accelerator, planned for 200 GeV, was constructed by Robert R Wilson (see box) ahead of schedule and under budget to operate at 400 GeV. It revealed the Υ (section 9.9.2) and produced a wide variety of other physics results. It has now been outfitted with a ring of superconducting magnets which permit beam energies of 0.9–1 TeV and can collide protons with antiprotons, yielding centre-of-mass energies of up to 2 TeV (see section 9.10.10.2 below). A machine similar in scope to the original Fermilab accelerator, the SPS, was constructed at CERN, with a beam energy of 400 GeV. It achieved proton–antiproton collisions a few years ahead of Fermilab, making possible the discovery of the W and Z (section 9.7.5).

9.10.8. Linear accelerators

As we have seen, special problems arise with injection and extraction in accelerators such as cyclotrons employing magnetic confinement. These problems are not nearly as challenging for linear machines such as van de Graaff generators. However, large fixed potentials are impossible to deal

Robert R Wilson

(American, b 1914)

When Robert Rathbun Wilson was chosen to lead the construction of a 200 GeV accelerator in 1967 he was given 10 square miles of farmland west of Chicago and a budget of 240 million dollars. By taking high technical risks and eschewing conventional engineering practices, Wilson finished, under budget, a machine yielding twice the energy originally specified. Along with the  machine was a great laboratory which attracted particle physicists from around the world. Wilson, an amateur sculptor, had a passionate interest in architecture as well as accelerator design. Every construction on the site bore his stamp of originality.

Wilson was born in 1914 in Frontier, Wyoming, where his family had a cattle ranch. Educated through high school in local schools, he went to the University of California at Berkeley where he obtained an undergraduate degree followed by a PhD working under E O Lawrence. In 1940 he went to Princeton as an instructor in the physics department (where Richard Feynman was still a graduate student). Almost immediately, because of World War II, he headed a project intended to separate the isotopes of uranium, based on a device of his own invention, the isotron. Moving to Los Alamos in 1943 at the insistence of Robert Oppenheimer who had known him in Berkeley, Wilson became head of the experimental nuclear physics division.

After the war, following a brief period at Harvard, Wilson moved to Cornell University where he was director of the Laboratory of Nuclear Studies. In this capacity he supervised the design and construction of a series of electron accelerators of ever-increasing energy, one of which was the very first machine to use alternating-gradient strong focusing. With his considerable experience in experimental physics and accelerator construction it was natural for him to be selected in 1967 to create, out of the farmlands of Illinois, the Fermi National Accelerator Laboratory.

with beyond about 25 MV. Linear accelerators or linacs, were developed to maintain the ease of beam injection and extraction of fixed potential machines without the limiting high potentials, avoided by using cyclic resonant voltages.

The first proposal for a linear accelerator powered by radio-frequency voltages was by G Ising in Sweden in 1925. However, no attempt was made to implement its general idea until Wideröe succeeded in accelerating potassium and sodium ions in 1928. As noted above, Wideröe's device provided inspiration for the cyclotron of E O Lawrence.

Lawrence was clearly driven to make accelerators to do physics. Concurrently with his development of the cyclotron with his student, M S Livingston, he had another student, D H Sloan, working on a linear device more in keeping with Wideröe's scheme. This early machine consisted of a number of cylindrical tubes arranged in a line with a spacing between tubes sufficient to prevent breakdown of the radio-frequency voltage applied between adjacent sections. Acceleration of the particles occurred only between the tubes. The ions drifted down the axis of each tube for a time sufficient for the alternating voltage to change sign. Progressing down the accelerator, the tubes were made progressively longer to compensate for the ions' acceleration. By 1931 Lawrence and Sloan had, with ten tubes, accelerated mercury ions to 1.25 MeV. Unfortunately, RF generators then available did not permit operation at high frequencies. These early machines were restricted to accelerating slow-moving heavy ions, too slow and highly charged to overcome the Coulomb barrier and initiate nuclear reactions in target materials.

The modern linac stems from the work of William W Hansen at Stanford University starting in the mid-1930s. The initial work centred on developing a single cavity for accelerating electrons, requiring large amounts of RF power at high frequency. Correspondingly, Hansen, with the Varian brothers, set about inventing an appropriate RF generator. This activity yielded the klystron, which played an important role for radar in World War II. In the meantime, Kerst's successful development of the betatron raised doubts whether a linear device could ever compete.

The radar developments during the war, especially with respect to powerful RF sources, raised hopes. After the war, Hansen, with E L Gintzon and J J Woodward, concluded that the magnetrons then available would be suitable for linear accelerators of a few MeV, but that higher energies would require the further development of RF power sources. Hansen pursued the idea of accelerating the electrons with an electromagnetic field travelling in a cylindrical waveguide held in phase with the electrons by 'loading' the waveguide, leading to the first (Mark I) Stanford linac with an energy of 6 MeV.

Concurrently, work initiated by Chodorow and Gintzon was proceeding on a high-power pulsed klystron. After a successful test

Figure 9.35. *Wolfgang K H Panofsky, the builder and first director of the Stanford Linear Accelerator Center.*

in 1949, plans went forward for a GeV electron linac, completed in 1952 (the Stanford Mark III) [405]. Unfortunately, W W Hansen, who had been the inspirational leader of the whole project, died prematurely in 1949 after living to see the successful test of the high-powered pulsed klystron.

The Mark III linac was the progenitor of the two-mile Stanford Linear Accelerator, constructed on the Stanford campus in the early 1960s under the direction of W K H Panofsky (figure 9.35). Initially operated at 20 GeV with an average current of 50 μA, this machine now regularly operates at 50 GeV, corresponding to an average accelerating field of about 150 000 V cm^{-1}.

The post-war development of electron linacs was parallelled by work on proton machines, principally at Berkeley under the direction of L Alvarez and W K H Panofsky [406]. Since protons of equal energy travel much more slowly than electrons, far lower RF frequencies (of the order of 200 MHz) are required, making the accelerators considerably more cumbersome. In fact, the technique of accelerating electrons with a travelling wave in the waveguide cannot be successfully applied to protons, necessitating methods more akin to the old drift tube idea. In particular, what emerged was a resonant cavity 40 ft long and 39" in diameter, operated in the TM$_{010}$ mode at 202.5 MHz and filled with 42 drift tubes. This pioneering work at Berkeley yielded a machine which accepted protons from a 4 MeV van de Graaff generator and accelerated them to 31.5 MeV.

The advantage of a high beam current with low angular divergence has retained for proton linacs a unique place in the accelerator hierarchy. They are used as injectors to the large proton synchrotrons, typically following a Cockcroft–Walton machine or radio-frequency quadrupole. An elaboration of the original standing-wave design of the Berkeley group has been highly successful at the Los Alamos Meson Factory, LAMPF, where protons are accelerated to 800 MeV with an average beam current of 1 mA.

9.10.9. Colliding beams

As noted above, the trail-blazing paper of Courant, Livingston and Snyder showed the feasibility of accelerators using the strong-focusing principle. It also introduced a new era in the sophistication of machine design. In particular, designs could be contemplated providing a new level of precision in beam handling and transport, whether around the closed orbits of an accelerator, or in beam lines external to the machine.

In 1956 it was first recognized that high-intensity beams, positioned and focused with high precision and directed against beams of equal and opposite momenta, could achieve reaction rates sufficient for interesting physics results. Just to contemplate this possibility is somewhat astonishing since the density of target particles in a beam is vastly smaller than in more usual targets such as liquid hydrogen or solid materials. It was Kerst [407] who first recognized that beams stored in fixed orbits could be made to collide with each other repetitively, thereby compensating for their diaphanous nature. All previous accelerators had directed their beams against targets at rest in the laboratory, either within the machine itself, or in an external beam extracted after the completion of the acceleration cycle.

Two beams of equal and opposite energy deposit all of their energy in the centre-of-mass (CM) system, in contrast to the collision of a particle with another at rest, for which the requirement of momentum conservation decreases rather spectacularly the energy available in the CM system at high energies. For example, two protons directed against one another, each with energy E, have an energy $2E$ in the CM system, whereas the CM energy of a proton of energy E striking another at rest is $\sqrt{2m(m + E)}$, where m is the proton mass. Energy is not wasted, of course, in fixed target machines; it yields high-energy beams of secondary particles, such as mesons and neutrinos. However, achieving the maximum energy for the production of new particles hinges on the colliding-beam technique. In 1956, when the notion of colliding beams was being advanced, the 30 GeV proton accelerators at Brookhaven and CERN were under construction. With 30 GeV on a fixed target, the available CM energy is 7.6 GeV; arranging these machines to collide beams with one another would yield almost 8 times as high a CM energy.

The proposal by Kerst *et al* *Attainment of very high energies by means of intersecting beams of particles* [408], recognized that two fixed-field alternating-gradient accelerators [409] could be arranged to circulate their high-energy beams in opposite directions over a common path in a straight section common to the two accelerators. G K O'Neill, at about the same time, published a paper entitled *Storage-ring synchrotron: device for high-energy physics research* [410]. (A non-technical review of the origin of the ideas and the development of the first practical storage rings has been provided in reference [411].) Since beams had already been extracted, highly successfully, from the Cosmotron, he proposed that such an extracted beam be stored in two magnet rings with the stored beams rotating in opposite directions. If the two rings were tangent at one point, the beams could be made to collide. Some details of the O'Neill proposal relating to injection were unworkable, but the general scheme has been followed in every one of the many colliding beam machines constructed subsequently.

Making proposals and suggesting ideas is easy and fun. A rare person is sufficiently convinced of their ultimate value to make a very major commitment of time and effort over a number of years. O'Neill, then an instructor in the Princeton University Physics department, was such a person. He quickly realized that storage rings might be easier to implement with electrons than with protons. With electrons, radiation quickly damps out transverse oscillations, reducing the cross-sectional area of the beams, a major advantage since the beam particles' interaction probability depends inversely on the area. (In those days no mechanism was known for damping the transverse oscillations in proton machines.) Furthermore, radiation automatically provides the energy loss mechanism necessary at injection for the particles to end up in stable orbits.

With a tentative design for an electron–electron collider in hand, O'Neill went to Stanford where the linear electron accelerator already provided an ideal injector. He convinced the director of that facility, W K H Panofsky, as well as Burton Richter and Carl Barber, of the proposal's worth. Somewhat later, O'Neill recruited B Gittelman, also a Princeton faculty member, to work on the project. After a year of detailed design and after funding had been secured from the office of Naval Research, construction was begun in 1959 on a pair of 0.5 GeV electron storage rings at Stanford. The team of Barber, Gittelman, O'Neill and Richter did the pioneering work.

And pioneering work it was. As soon as the first electrons were stored, totally new and unanticipated difficulties arose. To quote O'Neill 'We were crudely reminded that Nature is an ingenious troublemaker.'. For example, the large peak currents induced in the vacuum chamber electromagnetic fields that reacted back on the beam destructively. One by one these problems were solved, with many people contributing to their solution. Finally, in 1965, the first electron–electron scattering results were obtained. The quantum electrodynamics of Feynman, Schwinger and Tomonaga survived an especially clean test. As Feynman has said, 'The test of all knowledge is experiment. Experiment is the sole judge of scientific truth.'.

Electron–electron colliders were limited to studies and tests of electrodynamics. It was clear from the beginning that electron–positron collisions offered a much richer opportunity for interesting physics. Electron–positron annihilations proceed through a virtual photon and thus are ideal for creating vector particles as well as particle–antiparticle pairs. This option had been considered by the Princeton–Stanford group but rejected because of the worry that the positron intensity could never reach the level needed to get physics results.

A most important contributor to the development of storage rings was a group working under the direction of B Touschek at the Frascati laboratory in Italy near Rome. Early in 1960 they decided to make an electron–positron storage ring of 0.25 GeV. An immediate simplification is associated with electrons and positrons which, travelling in opposite directions, can be guided by the same magnetic field and vacuum chamber. However, it was the spirit of the Italian group that guaranteed this electron–positron collider (called ADA, for *aniello d'accumulazione*, or accumulation ring) would be a useful test facility.

Originally constructed and tested at Frascati only for electrons, ADA was loaded on a truck and transported to Orsay near Paris in 1962 with its ultra-high-vacuum chamber intact. The development of storage rings was characterized by the appearance of unexpected instabilities whenever the operation was pushed into a new realm of intensity or energy. Here, true to the pattern, more intense beams could be injected by a 1 GeV linear electron accelerator, and new instabilities (the Touschek effect) associated with the new intensities were found. The source of the new difficulties was eventually identified and the problems brought under control. The first electron–positron annihilations were recorded in ADA in 1963. While their intensity was too low for significant physics to emerge, it was apparent that the problem of low positron intensity could be overcome. Indeed, this was the case and every collider constructed since has been of this type.

ADA had a glorious career as a testing ground for e^+–e^- colliders and was suitably retired in 1965. Early in 1965 the Italian group started the construction of a much larger machine, ADONE, which was to have an energy of 1.5 GeV per beam. The first beam–beam interactions from this machine were observed early in 1967. The maximum energy of ADONE turned out to be a most unfortunate choice. The J/ψ, produced in 1974 at an energy of 3.1 GeV at SLAC, could have been uncovered much earlier at ADONE had that machine been designed to reach, say, 1.6 GeV per beam.

In the early 1960s a French group at Orsay started constructing an electron–positron machine with an energy of 0.385 GeV per beam. Activity in storage-ring design and construction was also initiated in Russia at Novosibirsk under the direction of G Budker. At the Electron–Photon Conference at Stanford University in 1967, experimental results were reported from three storage rings [412]: Orsay (385 MeV), Novosibirsk (510 MeV) and Stanford (550 MeV). The Orsay and Novosibirsk machines were both electron–positron colliders.

Electron–positron colliders were also on the drawing boards in the US. Competitive proposals were received by government agencies from groups at the Cambridge Electron Accelerator (CEA) and the Stanford Linear Accelerator Center (SLAC). The SLAC proposal won out but no funding was forthcoming. Finally, the Stanford Positron–Electron Collider, SPEAR, was constructed as an experiment, albeit a somewhat expensive one, with a maximum energy of 3.5 GeV per beam, starting operation in 1972. In the meantime, the group at CEA, still wanting to build a collider, started by inventing a 'low-β interaction region', a scheme which increases the interaction rates by factors of 10 to 100. This idea, combined with an ingenious and difficult beam handling arrangement, converted the CEA's electron synchrotron into a colliding beam machine, yielding collisions at energies much higher than could be achieved elsewhere and indicating the high value of R already mentioned in sections 9.6.3 and 9.9.1.

The CEA results were largely discounted at the time. A popular impression viewed the collider operation of the CEA as exceedingly difficult and thus unlikely to yield good physics. In the end the CEA group was vindicated; their results were shown to be correct by the more extensive measurements at the SPEAR collider.

We have already shown the behaviour of R as a function of energy in figure 9.18. Below the mass of the top quark the value of R from simple quark and colour counting should approach $3\frac{2}{3}$. This it does, rising at the highest energies as the low-energy tail of the Z^0 is approached.

After SPEAR many electron–positron colliders have been constructed throughout the world, each with a name derived from some appropriate acronym. A listing of those existing near the end of the twentieth century is given in table 9.3.

This activity has culminated in the construction of LEP, the Large Electron Positron collider at CERN, with a circumference of 27 km, operating, in its first phase, at single beam energies up to about 50 GeV. The energy of this machine, completed in 1989, was designed to produce the Z^0, with a mass of 91 GeV. In the second phase of operation, the beam energy will be raised to about 90 GeV, to produce $W^+ - W^-$ pairs. This energy represents the practical limit for electron–positron storage rings, because energy loss to synchrotron radiation is proportional to the fourth power of the beam energy divided by the radius of curvature of

Table 9.3. *Electron–positron colliders with CM energies above 10 GeV.*

Machine	Beam Energy (GeV)	Location
DORIS	4–5.3	DESY (Germany)
VEPP-4	5	Novosibirsk (Russia)
CESR	8	Cornell
PEP	17	Stanford
PETRA	23	DESY (Germany)
TRISTAN	35	KEK (Japan)
LEP	50–100	CERN

the bending magnets. When operating at 50 GeV per beam it amounts to 1.6 MW at the design luminosity, rising to 14 MW at the higher energy. This energy loss must be made up by the considerable radio-frequency power delivered to the beam.

To avoid the limitation of synchrotron radiation a daring proposal was made at SLAC in 1979 [413] to accelerate alternate pulses of electrons and positrons down the length of the two-mile long machine to 46 GeV. Each type of particle is then separately steered to arrange for their head-on collision. Clearly, for these collisions to take place with an appreciable chance of interaction, the beams must be incredibly small (of the order of two wavelengths of visible light) and must be steered with even better accuracy. To appreciate this technical *tour de force*, imagine a needle-shaped bunch of electrons with the diameter of the needle, one-fifth that of a human hair and perhaps a centimetre long, colliding with a similar bunch of positrons moving in the opposite direction, each bunch of particles having travelled through a semi-circle half a mile in radius. This project, the SLC (for Stanford Linear Collider), was successfully completed; while its flux of Z^0s is significantly lower than obtained in LEP, its electrons and positrons can be polarized, adding important information to the results. It is generally agreed that still higher energy electron–positron collisions will require the further development of the idea of the linear collider.

9.10.10. Hadron colliders

9.10.10.1. Proton–proton colliders. As noted above, machines intended to collide protons are intrinsically much more difficult to design. The absence of radiation damping complicates considerably both injection and storage. As has been noted, every proton remembers its history forever in the absence of damping. The first proton storage-ring-collider complex (the ISR or Intersecting Storage Rings) was constructed at CERN using the 28 GeV protons provided by the PS synchrotron, first operating in 1971. It was, technically, a very complex device with elegant solutions to many difficult problems concerning beam handling and proton stacking and storage. Its most significant new technical development was the invention by Simon van der Meer [321] of stochastic cooling, which replaces the radiation cooling in electron machines and has made possible all of the subsequent hadron colliders. This invention recognizes that protons' radial positions oscillate about a central orbit at the so-called betatron frequency. The technique samples the departure of the mean beam position from its nominal radius at a particular point on the ring of magnets, feeding back the amplified signal across a chord to make up for delays in amplification and signal transmission. The amplified signal is applied to correcting electrodes appropriately positioned relative to the pickup electrodes in terms of betatron wavelengths so as to deflect the protons towards their central orbit. Since the pickup electrodes only sense the average position of the beam, some of the protons will be 'heated', but, on the average, the beam is cooled; hence the name 'stochastic cooling'.

Concurrently, a different cooling method was developed by G Budker at Novosibirsk. His method enveloped a beam of protons with an intense beam of electrons moving with the same speed and in the same direction. Any proton with a velocity different from the electrons will scatter from them, losing energy. This technique is most useful for low-energy beams and has been applied successfully to cyclotrons.

9.10.10.2. Proton–antiproton colliders. The cooling method of van der Meer became the crucial feature in the next round of accelerators, the proton–antiproton colliders mentioned in section 9.7.5. Early in 1976 a proposal was prepared by C Rubbia, P McIntyre and D Cline [414] to convert the existing highest-energy accelerators to proton–antiproton colliders, with a physics motivation explicit from their proposal's title 'Producing massive neutral intermediate bosons with existing accelerators'. This proposal was complete with estimates of the production cross sections of $W^{\pm}$ and W^0s (called Z^0s by the time of their discovery). It required beam handling techniques considerably beyond current methods

> The main elements are (1) an extracted proton beam to produce an intense source of antiprotons at 3.5 GeV/c, and (2) a small ring of magnets and quadrupoles that guides and accumulates the $\bar{p}$ beam, (3) a suitable mechanism for damping the transverse and longitudinal phase spaces of the $\bar{p}$ beam (either electron cooling or stochastic cooling), (4) an RF system that bunches the protons in the main ring and in the cooling ring, and (5) transport of the 'cooled' RF bunched $\bar{p}$ beam back to the main ring for injection and acceleration.

> The authors also provided a sketch of the proposal's implementation at the Fermi National Accelerator Laboratory (Fermilab) in Batavia,

Illinois. This proposal implied that the projected production cross sections would permit the construction of a proton–antiproton collider with sufficient luminosity to produce a reasonable rate of detected Ws and Zs. When budgetary constraints at Fermilab did not allow the laboratory to proceed with these plans, despite the importance of the possible results, the authors took their proposal to CERN where a machine of similar energy (the SPS) was also in operation. Here their plans were accepted and construction was started on this grand experiment. As mentioned in section 9.7.5, it resulted in the discovery of the $W^\pm$ early in 1983 and of the Z^0 several months later.

In 1984 Simon van der Meer and Carlo Rubbia were awarded the Nobel Prize 'for their decisive contributions to the large project, which led to the discovery of the field particles W and Z, communicators of weak interaction' [321, 322].

In the meantime, plans at Fermilab were evolving in the late 1970s to make a p$\bar{\text{p}}$ collider of much higher energy using superconducting rings piggy-backed on the original iron magnet ring. This machine is another technical *tour de force* owing to the additional complications associated with the use of superconducting magnets in a pulsed mode around a 6.3 km ring. It was brought into operation in 1985. Each beam has an energy close to 1 TeV, hence its name, the Tevatron. From its initial operation it has been the lead accelerator in the search for the top quark, culminating in its observation (section 9.9.4).

9.11. Detectors: from Rutherford to Charpak

Newtonian mechanics, Maxwell's equations and Einstein's relativity are all recognized as monuments to deductive reasoning. Their place in history both in time and in the development of our understanding of Nature are well known and are part of the textual lore of physics. Conclusions drawn from experimental data which have changed the way we view things—Rutherford scattering, the Bohr atom, the Hubble radius—are likewise part of every textbook. Only rarely do major conceptual developments in the history of science go unnoticed. One example is the recognition that the Sun itself is a star [415]. This was surely a most profound contribution to understanding the Universe, but who first proposed it and how it came to be accepted is totally unknown. Similarly, some developments in instrumentation, in the tools physicists use which have had profound consequences, have had such diffuse origins that it is difficult to identify the individuals who might be credited. Bardeen, Brattain and Shockley are credited for the transistor. Dating from the late 1940s, this is recognized quite properly as a momentous invention. It had a greater impact on the nature of physics research in the last half of the twentieth century than any other invention. But a further development, which truly brought about a revolution not only in physics research but in nearly every aspect of life, was the large-scale integration of transistor circuits on single pieces of silicon. This profound development, occurring in the 1960s, came from incremental advances in many different laboratories, and has a more diffuse history. Near the end of the century commercially available devices are available which have four million transistors on a single piece of silicon!

In a charming essay [416] entitled *The craft of experimental physics*, P M S Blackett begins 'More has been written of what the experimental physicist has discovered than of how he has discovered it. Because he has changed the technique of living by his intense curiosity to find out about obscure things, many of his discoveries have become common knowledge. But his method of experimental discovery, how he works and thinks, is much less known.'.

Also, many of the tools which the experimental physicist has invented and developed for the purpose of doing experiments have not been accorded the attention they deserve. Since these investigations are not an end in themselves, there is nothing which is forgotten faster than an outdated technique no matter how clever or original its conception. From 1920 to 1960 an important device for the physicist was the vacuum tube. Today's students are hard pressed to recognize one, and the inventors are forgotten. In the present section we review the development of some of the tools and techniques which have most contributed to our present knowledge of particle physics.

At the last 'turn of the century', in 1900, the electron had just been discovered. So too with radioactivity. The immediate problem was to unravel the nature of the alpha, beta and gamma rays that had just been discovered. The principal tools consisted of electroscopes and electrometers for measuring ionization.

In 1903 Crookes and also Elster and Geitel discovered the phosphorescence of zinc sulphide on exposure to alpha particles. It had also been discovered rather early that pure zinc sulphide did not fluoresce; some impurity, e.g. copper, was necessary to make it work. A thin layer of finely powdered zinc sulphide spread on a glass plate and exposed to an alpha source such as radium or polonium could be seen to scintillate, with dark-adapted eyes, through a low-powered microscope. The scintillation technique served not only to count particles but also to locate their positions in space. This method became famous later when used to study the scattering of α-particles, leading to Rutherford's discovery of the atomic nucleus. Earlier Rutherford and Geiger had become involved in studying the nature of the α-particle itself. The question of its charge was crucial. By measuring the total charge deposited on an electrometer from a particular source within a set period of time and knowing the flux of particles from the number of scintillations with the same source in the same period of time, the charge on each particle could be obtained. Of course, this method assumes an efficiency of 100% for α-particles producing scintillations, an assumption that had never been tested. In a classic experiment published in 1908 Rutherford and Geiger made an invention to resolve the question [417].

9.11.1. Ionization detectors

The ionization produced by individual α-particles did not provide a sufficient signal in an electrometer. To quote Rutherford and Geiger

> Some preliminary experiments to detect a single α-particle by its direct ionisation were made by us, using specially constructed sensitive electroscopes. As far as our experience has gone, the development of a certain and satisfactory method of counting the α-particles by their small direct electrical effect is beset with numerous difficulties.
>
> We then had recourse to a method of automatically magnifying the electrical effect due to a single α-particle. For this purpose we employed the principle of production of fresh ions by collision. In a series of papers, Townsend [418] has worked out the conditions under which ions can be produced by collisions with the neutral gas molecules in a strong electric field.

As implemented by Rutherford and Geiger, the high electric field was produced by applying voltage on a fine wire running down the centre of a conducting cylindrical tube, 25 cm long and 1.7 cm in diameter. They operated the device with a gain of only a few hundred but it was enough to give distinguishable kicks ('throws') to their electrometer.

The paper ends with six conclusions.

(i) By employing the principle of magnification of ionization by collision, the electrical effect due to a single α-particle may be increased sufficiently to be readily observed by an ordinary electrometer.

(ii) The magnitude of the electrical effect due to an α-particle depends upon the voltage employed, and can be varied within wide limits.

(iii) This electric method can be employed to count the α-particles expelled from all types of active matter which emit α-rays.

(iv) Using radium C as a source of α-rays, the total number of α-particles expelled per second from 1 gram of radium have been accurately counted. For radium in equilibrium, this number is 3.4×10^{10} for radium itself and for each of its three α-ray products.

(v) The number of scintillations observed on a properly prepared screen of zinc sulphide is, within the limit of experimental error, equal to the number of α-particles falling upon it, as counted by the electric method. It follows from this that each α-particle produces a scintillation.

(vi) The distribution of the α-particles in time is governed by the laws of probability.

After these conclusions the authors go on to observe 'Calculation shows that under good conditions it should be possible by this method to detect a single β-particle, and consequently to count directly the number of β-particles expelled from radio-active substances.'.

In connection with this paper two comments are of interest. The first is that Rutherford and Geiger took the charge of the electron to be $e = 3.6 \times 10^{-10}$ esu, vastly different from the currently accepted value of 4.8×10^{-10} esu. This was obtained by averaging the results of J J Thomson (3.4), H A Wilson (3.1) and R A Millikan (4.06) (times 10^{-10} esu), all of which were obtained by observing the effect of an electric field on charged water droplets. In a subsequent paper [419] Rutherford and Geiger report the charge of the α-particle to be 9.3×10^{-10} esu. Independently, they conclude that the charge of the α-particle must be $2e$ and hence $e = 4.65 \times 10^{-10}$ esu, close to the currently accepted value. They then go on to identify the source of the error in the previous determinations of e, i.e. the evaporation of the water droplets during the measurements. Millikan eventually avoided this problem by using oil drops.

The second comment relates to conclusion (vi), which is correct but not supported by their data. Discussing the interval distribution between successive α-particles the authors show, very qualitatively, a curve which starts from zero, rises, and then comes down. They appear to have neglected to allow for the dead-time of the apparatus. We now know that the interval distribution must be, from the Poisson distribution, purely exponential.

And so was born the proportional counter, in general form not dissimilar to that used today. This device continued to be one of the most important detectors of charged particles through the rest of the century. The counter was invented to answer a particular physics question. Necessity is indeed the mother of invention.

It is perhaps surprising that it took 20 more years before the so-called Geiger–Müller (G-M) counter emerged [420]. Of course, World War I removed four years from this period since, unlike in World War II, counting nuclear particles was hardly at the top of the wartime priority list. Nonetheless, it is still surprising that it took so long. In retrospect it is difficult to imagine the circumstances under which experimental physicists laboured and the relatively primitive materials available. It was literally the era of string and sealing wax. It is a sobering experience to read, for example the chapter on Geiger counters in reference [421]. Making G-M tubes was apparently still something of a black art ten years after their invention.

The Geiger–Müller (or 'Zählrohr') counters were an immediate success, with many applications. They were sensitive over the area projected by the outer cylinder, a large area at the time, producing signals sufficiently large to require a minimum amount of vacuum tube amplification to drive mechanical registers for tabulating counts. The G-M counters were indiscriminate, responding to any and all types of ionizing radiation, a serious disadvantage in many applications. However, when arranged in time coincidence the counters could be used to define beams of particles. In this mode, initially in the hands of Bothe [422] and Rossi, these devices became a principal tool in the study of cosmic rays. In 1954, W Bothe was awarded the Nobel Prize in Physics 'for the coincidence method and his discoveries made therewith'. (Rossi's circuit was a very significant improvement: a triple coincidence device, totally symmetric in all the inputs, whereas the original Bothe circuit was twofold and asymmetric. Coincidence circuits, now called logical 'AND' circuits, have continued to be incrementally improved in timing resolution ever since, first in the vacuum tube form and now in transistor versions.)

Somewhat earlier, physicists had started using vacuum tube amplifiers in conjunction with ionization chambers. In 1926 Greinacher [423] showed it possible to amplify the current due to an α-particle sufficiently to register as a click in a headphone. Subsequently, he also detected single protons by the same method. High-voltage power supplies utilizing transformers and vacuum tube diodes became the norm, largely replacing the banks of batteries used previously.

The mechanical counting registers used in conjunction with G-M tubes were limited to a maximum rate of about ten per second. Correspondingly, a counting loss of the order of 10% occurred for rates as low as one per second, stimulating the invention of electronic 'prescaling' of the counts by divide-by-two circuits before registering on mechanical counters. The first of these binary counting circuits, devised by Wynn-Williams in 1932 [424], consisted of two vacuum tubes coupled so that only one could be conducting at a time. The driving pulse would turn off the conducting tube, but at the end of the driving pulse, in the absence of special provisions, either tube could become conducting whereas it was desired that the tubes conduct alternately on successive driving pulses. Memory of the previous conducting state, a necessary intrinsic feature, was provided by an *RC* circuit with a time constant long enough to more than cover the duration of the driving pulse. The divide-by-two circuits could be cascaded to any power of two. Binary division with six stages (by 64) before recording on mechanical registers was common.

A very significant improvement in the reliability of binary dividers was made in the early 1940s by W A Higinbotham using diode coupling between stages. While Higinbotham neither published nor patented his contribution, it is included in reference [425]. The early vacuum tube circuits could respond in the microsecond range, thus avoiding the counting-loss problem of mechanical counters. The essential ingredients of these circuits, bistable elements with intrinsic memory, persist in the highly refined binary transistor circuitry of today.

In the 1960s large-scale integration of transistors was accompanied by a striking decrease in the cost of complex circuits. Their reliability, as measured by the mean time between failures, made it possible to contemplate experiments with massive quantities of electronics. It was in this technological environment that G Charpak, working at CERN, started a revolution in particle detectors by constructing large-area multi-wire proportional chambers. (See also Thompson in the first of references [3], p 274.)

Ever since the advent of scintillation counters, proportional counters had almost become extinct, but now they were resurrected on a grand scale. Multi-wire chambers had been used earlier, but never approaching the magnitude pursued by Charpak. Each wire had its own solid-state amplifier and discriminator to produce the signals necessary to log on magnetic tapes for later computing processing. Charpak further refined the idea of the drift chamber: the time between the passage of a particle through a chamber and the appearance of the signal on the 'sense' wire indicated the distance of the particle's trajectory from the wire. In certain gases the electron drift velocity was remarkably independent of the electric field, making the device very linear. Layers of chambers with the wires at fixed angles to each other provided $x-y$ position information, critically dependent on the position of the sense wires themselves. While the scheme required unprecedented electronic circuitry, such circuitry was inexpensive, and perhaps more importantly, reliable. These devices were fast, acquiring great quantities of data in a short time and providing in ideal form the sort of data required for analysis in a modern digital computer. The drift chamber is now a standard part of the particle physicist's tool kit. For this work Charpak was awarded the 1992 Nobel Prize in Physics.

9.11.2. Scintillation and Čerenkov counters

By 1903 zinc sulphide was known to emit visible flashes of light when bombarded by α-particles, but was relatively insensitive to the gamma and beta radiations most often accompanying the α-particles. The arduous business of counting visible scintillations in dark rooms with dark-adapted eyes was the technique of choice for many years, yielding much information about α-radiating sources, the nature of the α-particles and Coulomb scattering. It was only the invention of photomultipliers capable of detecting the scintillations and converting them to electrical signals that afforded full exploitation of the scintillation phenomenon.

See also p 1200

In the early 1900s certain materials when bombarded with electrons of rather low energy (about 100 volts) showed a propensity to emit more electrons than struck the surface. These surfaces were thus acting as electron multipliers! The first patent on using these materials to amplify small currents, granted in 1919 to Joseph Slepian [426], did not result in practical applications. Only later [427], in 1936, were multiple-stage devices, using both magnetic and electrostatic focusing, invented and applied to the amplification of photoelectric currents. Even though the early photomultipliers were scarcely practical, they were recognized as presenting a unique noise problem, which Shockley and Pierce addressed [428].

The first device resembling those in current use was described by Zworykin and Rajchman in 1939 [429]. With careful attention to electrostatic focusing, these inventors managed to avoid the severe space-charge limitations of the early models. In the nomenclature of the RCA company, the new device became the 931A photomultiplier. This tube had commercial applications in sensing the sound track on movie film, and was also used in World War II, far from its original purpose, as a generator of white noise for masking radio and radar signals. Tubes selected for exceptional sensitivity were labelled as 1P21.

In the meantime a new effect associated with the energy loss of charged particles passing through material had been discovered. In 1934 Čerenkov [430] had observed 'feeble visible radiation' from β-rays passing through a clear liquid. In 1937 Frank and Tamm [431] developed an energy loss formula which included radiation at a particular angle relative to the track of the particle. That this radiation was emitted at a characteristic angle was very shortly afterwards confirmed by Čerenkov [432]. Photomultipliers had still another application: detecting Čerenkov radiation.

While zinc sulphide was almost ideal for detecting slow-moving α-particles, it appeared difficult, if not impossible, to produce the large crystals needed for electron and γ-ray detection. For this reason, the discovery in 1947 by Kallman [433] of scintillation by various organic crystals proved enormously important, immediately stimulating investigations of various materials. P R Bell [434] found that anthracene and later stilbene were good scintillators and could be grown in large sizes. These materials, with very short recovery times, were quickly exploited by experimentalists. Viewed by the RCA 1P21 phototubes, they made sensitive counters with exceedingly fast response that could cover relatively large areas. These devices played a pivotal role in the discovery of the neutral pion, the early lifetime measurements on positronium, and the measurement of the relative polarization of the two γ-rays from positronium annihilation.

Certain fluorescing chemicals were also found which, when added to organic liquids, would result in a solution that would scintillate with good efficiency [435]. Such liquid scintillators proved extremely useful for large volume detectors. Later, in the mid-1950s, it was found that a plastic such as polystyrene could serve as the solvent. Now, towards the end of the century, plastic is the scintillation material of choice in every application except γ-ray spectroscopy, being available commercially in all shapes and sizes.

Parallelling the developments in organic scintillators in the late 1940s, Hofstadter [436] found that sodium iodide activated by thallium impurities proved a very efficient scintillator. Unlike zinc sulphide, it could be grown into very large transparent crystals, and despite its highly deliquescent nature could be packaged into quite stable forms. The high atomic number of the iodine gave the crystal a high sensitivity to γ-rays. Indeed, it became possible, depending on the γ energy and the size of the crystal, to capture the full γ-ray energy. When coupled to a photomultiplier and a device for measuring the resulting spectrum of pulse heights, this crystal led to an enormously valuable spectrometer for γ-rays and mesic x-rays.

Again necessity stimulated invention. The RCA 1P21 phototube had a high sensitivity to photons, but it was difficult to channel the light from a crystal efficiently to photocathodes in the tube's interior. To correct this disadvantage, the RCA company, as well as EMI, developed the first end-window phototubes, the 5819 (RCA) and the 5060 (EMI) (see reference [437]). These new phototubes proved critical for the new spectrometers, essentially making them feasible.

The electronic developments associated with pulse-height analysers (kick-sorters in England) had started at Los Alamos during World War II for use with ionization chambers and proportional counters. In general, they were rather cumbersome and complicated devices, notoriously difficult to maintain. The threshold for each channel would drift relative to its neighbour, which in turn would distort pulse amplitude spectra. It was Wilkinson [438] who first hit on the solution to this problem, by 'laying the pulse on its side', i.e. by converting a pulse height to a pulse time. He charged a capacitor with the input pulse, then discharged it linearly in time, measuring the time between input and the end of the discharge by simply using an oscillator. The resulting time interval was then proportional to the original pulse amplitude. Electronically, it was much easier to keep time intervals constant than to keep voltage amplitude thresholds constant. Today, these circuits, called ADCs ('analogue-to-digital converters'), are ubiquitous throughout the electronics industry. The old technique of setting channel intervals with voltage comparators is still alive because of its speed advantage. When applied to transistor electronics, the device is called a 'flash ADC'.

9.11.3. The visual techniques

We have already mentioned (section 9.4.1) the key role played by photographic emulsions in detecting tracks of elementary particles. Cloud chambers, bubble chambers and spark chambers were also of importance.

9.11.3.1. Cloud chambers. Even before the new century began, C T R Wilson had been studying the process of droplet formation in supersaturated water vapour. It is not a simple process. Even in a supersaturated vapour, tiny droplets will inevitably evaporate in the absence of a formation nucleus, such as a dust particle, which in effect provides a flat surface for condensation. Alternatively, Wilson learned that charged molecular ions also provide centres for droplet formation since the repulsion of the charges causes the droplet to become larger, overcoming the tendency to evaporate away.

Experimentalists were quick to seize on the effects uncovered by Wilson. As we have already noted, in the early 1900s the study of the behaviour of charged water droplets in an electric field provided the first measurements of the electron charge. It was not until 1911, however, that Wilson first observed and photographed [439] the formation of tracks of droplets along the paths of single charged particles in a gas of water vapour supersaturated by a sudden expansion of its volume. Therein was born one of the most important tools in particle physics, the expansion cloud chamber. In 1927 Wilson was awarded the Nobel Prize 'for his method of making the paths of electrically charged particles visible by condensation of vapor'.

The application of this technique to the study of cosmic rays was considerably enhanced when the tracks of ions were found to persist for a time sufficiently long to permit the chamber's expansion and the growth of droplets along the tracks. The expansion could be triggered by signals from counters [47] (cf section 9.4.1). The discoveries of the positron, the muon and strange particles were all made using the expansion cloud chamber.

It occurred to a number of people in the 1930s to arrange for a cloud chamber to be continuously sensitive. A first crude device was constructed by Hoxton [440], but its general principle was elaborated by Langsdorf into a device called a 'diffusion cloud chamber' that became the model for all subsequent chambers [441] (see also reference [442]). A vertical temperature gradient was arranged in a gas such that the vapour was unsaturated at the top but highly saturated at the bottom. In between, droplets would form in a narrow horizontal layer. The sensitive region tended to be rather thin, making the device unsuitable for vertically moving particles such as cosmic rays but presenting clear advantages for the horizontally moving particles from accelerators.

Ralph Shutt's group at Brookhaven exploited the diffusion chamber, initially at the Nevis cyclotron [443] and later at Brookhaven. The chamber could be made of a size ideal for studying the production and decay of the recently discovered strange particles. Traditionally the practice had been to place sheets of material, copper or lead, in the chamber to serve as sources of the new particles when struck by other particles incident from outside the chamber. The ideal target material was hydrogen. To this end, the diffusion chamber of Shutt *et al* contained 21 atmospheres of hydrogen with methanol vapour. When the chamber was exposed to 1.5 GeV negative pions at the Cosmotron at BNL, the associated production of strange particles was discovered.

9.11.3.2. Bubble chambers. Using the diffusion chamber both as a target and as a detector was a clear advantage. However, even when operating at a pressure of 21 atmospheres, the probability of a pion interacting in the chamber was only about one in a thousand. Even with as many as ten incident particles in each photograph, it was still necessary to take 100 pictures per observed interaction. Therefore, when Glaser's invention of the bubble chamber [100] (see section 9.4.5.3) proved that liquid hydrogen (with a density 50 times that of gaseous hydrogen at 21 atmospheres) would work [444], the diffusion chamber was doomed as a tool for doing physics.

The explosive development of the Glaser bubble chamber in its liquid hydrogen version culminated in the 1950s with the Berkeley 72″ chamber. A development of the Alvarez group, it was to be the source of discovery of many unstable particles, as discussed in section 9.4.

Appropriately, the Nobel Prize was awarded in 1960 to Donald Glaser 'for the invention of the bubble chamber', and in 1968 to Luis Alvarez [101] 'for his decisive contributions to elementary particle physics, in particular the discovery of a large number of resonant states, made possible through his development of the technique of using the hydrogen bubble chamber and data analysis'.

9.11.3.3. Spark chambers. An effort was made in the late 1940s to find a counting device that would improve on the rather poor timing resolution characteristic of Geiger counters. Keuffel [445] and Pidd and Madansky [446] constructed 'spark counters' consisting of parallel conducting plates between which high voltages were applied. An electrical discharge would occur whenever a charged particle traversed the gap between the plates. While Keuffel observed that the discharge occurred along the path of the particle, the emphasis at the time was on electronic timing, so the feature of track delineation was ignored. Shortly after, scintillation counters with superb timing characteristics were invented, removing the principal motivation for further spark-counter development. In addition, spark counters with steady voltages were plagued with spurious discharges not associated with the passage of particles. When pulsed voltages were used [447] it was found that spurious discharges could be avoided, but only single tracks could be recorded.

The spark chamber was first proved viable for recording particle tracks by Fukui and Miyamoto [448]. Using pulsed voltages, initiated by scintillation counters, with a neon–argon mixture as the gas between the plates, they showed that tracks of more than one particle could be recorded. The receipt of the preprint of their paper in the US initiated a flurry of activity devoted to further development. Foremost among the leaders in this development were Cronin at Princeton [449] and Cork and Wenzel at Berkeley [450]. By 1960 these groups were doing highly interesting physics using spark chambers. Guided by their work, a Columbia–Brookhaven group built a large, massive spark chamber to detect neutrino interactions at the AGS at Brookhaven (section 9.5.1.6).

Spark chambers had a clear advantage over bubble chambers in that the interaction and tracking of particles could be separated, and the multiple Coulomb scattering, which limited the momentum resolution in bubble chambers, could be reduced significantly. Furthermore, spark chambers could be triggered on interesting events, as determined by accessory information from counters. These advantages were quickly exploited and at least two major discoveries (the existence of two neutrinos [156] and CP violation [163]) were shortly made with this technique.

Initially, information from spark chambers was obtained photographically. The spark positions were digitized from film records and processed on large mainframe computers. Subsequently, in some experiments, the thunder, not the lightning, was recorded using acoustical detectors. Later, the chamber electrodes were made of wire crossing magnetostrictive lines. The discharge current would generate an acoustical signal in the line, to be sensed at its end. The arrival time provided a measure of the spark's position. Such acoustical schemes lent themselves to full automation in a computer environment. Variations developed in the 1960s included the wide gap chamber and the streamer chamber. The spark chamber, operated in a non-visual mode, became the detector of choice in many experiments until replaced by the multi-wire proportional chambers and by Charpak's drift chambers [451], able to acquire data at far higher rates.

9.12. Overlaps with other subjects

As emphasized earlier, particle physics was very different in the periods before about 1960 and after about 1973. Before the 1960s, properties of individual particles were studied without the unifying themes of quarks and leptons. Similarly, the understanding of strong and weak interactions underwent a significant change with the employment of non-abelian gauge theories after the early 1970s.

During these developments, particle physics drew on a number of other areas for insight and experimental results. We mention a few of these.

9.12.1. Nuclear physics

9.12.1.1. Neutron lifetime and decay properties. Nuclear reactors provided copious sources of neutrons, whose mean lifetime and ratio of axial–vector to vector couplings were important ingredients in the evolution of weak-interaction theory. In view of the difficulty of confining a neutron for its mean lifetime (about 15 minutes), an accurate measurement of this

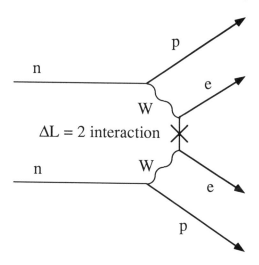

Figure 9.36. *Diagram contributing to neutrinoless double-beta decay. The line labelled by a cross is a Majorana neutrino, containing components with lepton number L = 1 and L = −1.*

quantity remained elusive for many years. 'Magnetic bottle' techniques to contain neutrons led to significant progress. The present value [49] of 887 ± 2 s provides not only useful information on the axial–vector coupling constant [452] (see sections 9.5.1 and 9.5.3), but also constrains the rate at which hydrogen was incorporated into helium in the early Universe (section 9.12.4.4).

9.12.1.2. Neutrinoless double-beta decay. All processes observed so far conserve *lepton number*, an additive quantum number equal to 1 for the negatively charged leptons and their neutrinos and −1 for the corresponding antiparticles. However, since neutrinos can, in principle, mix with their antiparticles [453], an interaction changing lepton number by two units is conceivable. Such a *Majorana mass* can lead to neutrinoless double-beta decay through the graph illustrated in figure 9.36.

Present searches for neutrinoless double-beta decay have observed no signal for this process, leading to an effective upper limit of a few eV on the relevant neutrino Majorana masses [454].

9.12.1.3. Beta-decay limits on neutrino masses. Details of the electron spectrum in nuclear beta decay can reveal distortions stemming from neutrino masses. No such effects have been observed, leading to upper limits of several electronvolts on the electron neutrino mass [455].

9.12.2. Atomic physics

We referred in section 9.7.3.3 to experiments searching for parity violation in atomic physics due to the neutral-current interaction. Many of these results rely on detailed atom-trapping methods developed since the earliest post-World War II days. The trapping methods have also been of great help in measuring anomalous magnetic moments of leptons (section 9.3), in studying neutron decays (see section 9.12.1.1), and in searching for the neutron's electric dipole moment (so far, without success).

9.12.3. Condensed matter

9.12.3.1. Nambu–Jona-Lasinio model. The bound state model of the pion proposed by Nambu and Jona-Lasinio [179] was inspired by analogy with superconductivity [456]. In both cases, a pair of fermions can form a bound state, leading to an energy gap and a zero-mass excitation.

9.12.3.2. Renormalization group. The behaviour of physical quantities under changes in scale was studied not only in elementary particle physics [339], but also in condensed-matter problems, where 'real-space' aspects of scale changes emerged particularly explicitly [457].

9.12.3.3. Lattice gauge theory. Non-relativistic field theories have been formulated on discrete lattices, such as occur in actual solids, yielding many properties of elementary and collective excitations. For example, one can determine the presence or absence of phase transitions, the masses and spatial extent of bound states, and bulk properties such as critical temperatures and specific heats. The methods for describing these phenomena are readily adapted to continuum field theories by approximating space-time with a discrete lattice. This programme [357] has been particularly vigorously applied in constructing a lattice version of quantum chromodynamics (QCD), aimed at understanding all of low-energy hadron physics. The lattice spacing must then be much less than a hadron's size (10^{-13} cm). One current obstacle to this programme is its inability to reproduce the pion well. In the limit of exact (spontaneously broken) chiral symmetry, the pion would be massless, with an infinite Compton wavelength, requiring for its representation a lattice of infinite extent. The actual pion has a Compton wavelength exceeding 10^{-13} cm, demanding many lattice points (and hence larger and faster computers) to reproduce correctly both pions and the rest of hadron physics.

9.12.3.4. String theory and soluble two-dimensional models. We mentioned dual models in section 9.5.4.5, and will come to their present incarnation (string theories) in section 9.13.5. Such theories also have been found relevant to two-dimensional models of critical behaviour in condensed-matter systems [458].

9.12.4. Astronomy, astrophysics, gravitation and cosmology
The interface between elementary particle physics and astrophysics is a vast subject (see, for example, the seminal text of reference [459]), for which we can only touch on a few selected topics. Particle physics is at the heart of many astrophysical processes (such as supernova evolution), while astrophysical constraints have proved useful in anticipating elementary particle properties (such as the number of types of light neutrino). For many problems in astrophysics, solutions based on fundamental particle physics are among several options.

9.12.4.1. Solar neutrinos. The detection of neutrinos from the Sun was proposed long ago by Pontecorvo [460]. The first search for solar neutrinos was mounted by R Davis and collaborators [461], employing the reaction $\nu + {}^{37}Cl \rightarrow e^- + {}^{37}Ar$ with chlorine in the form of cleaning fluid contained in a large tank located in the Homestake Gold Mine in South Dakota to shield against cosmic-ray backgrounds. The detection of a signal relies on the extraction from the tank of less than one atom of argon per day, whose radioactive decay is the signature of its production.

Expectations for Davis' experiment were worked out in detail by Bahcall and collaborators [462]. The rate observed in the chlorine experiment was lower than expected and has remained so over more than 20 years. The present ratio of experiment to theory is about 0.3.

More recently, several other experiments, sensitive to different neutrino energies, have also indicated a solar-neutrino rate lower than expected. An experiment in the Kamioka mine in Japan, sensitive only to the highest-energy neutrinos (above 5 MeV), sees a rate about half that expected. Its detector consists of a large tank of very pure water, viewing neutrino–electron interactions via Čerenkov light [463]. Two experiments sensitive to a lower neutrino energy threshold [464] rely on the reaction $\nu + {}^{76}Ga \rightarrow e^- + {}^{76}Ge$, with extraction of the radioactive germanium and subsequent detection of its decay. These also show a signal somewhat lower than predicted by the standard solar model.

Do these results, if correct, indicate a shortcoming of solar models (with all their attendant details of nuclear physics), or do they point to new elementary particle physics [465]? For example, do electron neutrinos (the type emitted in the Sun) undergo oscillations [466] to other species undetected in the above experiments? Such oscillations could either take place in vacuum or be induced by interaction with the Sun's matter.

Further experiments are in the construction or planning stage [467]. An ideal detector would be sensitive to (i) a wide range of neutrino energies, ranging from less than a few hundred keV to many MeV, (ii) the direction of the neutrinos' source, as in the Kamioka experiment, and (iii) interactions of neutrinos other than ν_e through neutral-current effects.

9.12.4.2. Supernova neutrinos. The Kamioka detector was one of a number set up in the early 1980s for the entirely different goal of searching for proton decay (see section 9.13.3). Another large water Čerenkov detector had been set up in the Morton Salt Mine northeast of Cleveland, Ohio, with the same purpose (see figure 9.37 [468]). After several years of operation, neither detector had seen any signal for proton decay. However, on 23 February 1987, both saw bursts of counts induced by neutrinos emerging from the explosion of a supernova in the Large Magellanic Cloud [469]. The observation by Kamioka was particularly fortunate since the detector was just a minute away from a scheduled shutdown. The supernova was the first seen in 1987, and hence was denoted SN1987A.

This observation confirmed basic predictions of the life cycle of a supernova [469, 470], which in turn relied crucially upon the presence of neutral as well as charged currents in the weak interactions. Furthermore, the neutrinos arrived at the Earth within a few seconds of one another despite travelling about 160 000 light years, allowing estimates of upper limits on their masses [471].

If and when a supernova explodes in our own galaxy (whose diameter is about 60 000 light years), the neutrinos emitted will produce a stronger signal than SN1987A. Several other detectors, to be mentioned in section 9.13.5.5, may be sensitive to this signal.

9.12.4.3. Cosmic microwave background. In 1948, George Gamow, R A Alpher and R C Herman predicted [472] that the black-body radiation remaining from the birth of the Universe should have a temperature of a few degrees K. This expectation was verified by Penzias and Wilson in 1964, whose microwave radiometer revealed a persistent noise signal [473]. A Princeton group that had been looking for the same effect supplied an explanation [474].

The microwave background has proved remarkably homogeneous and isotropic, with a black-body temperature of 2.74 K, but displays a distortion caused by the velocity of our galaxy of some 600 km s^{-1} towards a point in Virgo, as well as fluctuations of order 10^{-5} recently identified by the Cosmic Background Explorer (COBE) Satellite and other experiments [475]. These fluctuations are a relic, looking backwards in time, of the physics generating the much larger fluctuations now seen as galaxies and clusters.

9.12.4.4. Nucleosynthesis and number of neutrinos. The neutron lifetime, as mentioned earlier [452], affects the rate at which hydrogen was incorporated into helium in the early Universe. Another crucial variable in that calculation lies in the number of light neutrino species. The microwave background radiation provides a calibration of the mass density of the Universe contributed by each neutrino species at the time

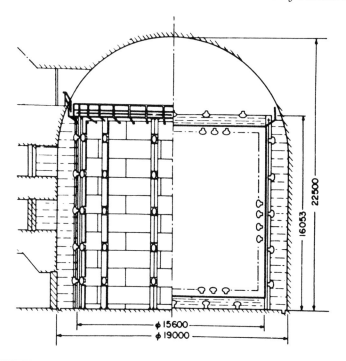

Figure 9.37. *Top: sketch of the Kamiokande II Detector. Bottom: wide-angle photograph of the IMB-3 detector. The hemispheres partially visible in the square plastic sheets are large photomultiplier tubes. The plastic squares, called waveshifters, catch a portion of the light which does not directly strike the photomultipliers. Also visible is a diver. The photograph is taken through about 20 metres of water, showing the remarkable clarity of the purified water.*

759

of helium nucleosynthesis. This density affects the rate of the Universe's expansion and hence the competition between the decay of neutrons and their incorporation into helium. Too many neutrino species would imply too much helium. An upper limit was thus set [476] of no more than four species; even this number started to look unlikely when limits on the neutron lifetime became tighter. Theorists thus breathed a sigh of relief (and at least one theorist won a case of wine) when the study of Z decays at LEP (section 9.7.5.3) indicated that indeed three was the correct number.

9.12.4.5. Baryon asymmetry of the Universe. The observed part of the Universe contains far more baryons than antibaryons. As mentioned in section 9.5.2.5, the groundwork for understanding this asymmetry was laid by A Sakharov [175] in 1967. The necessary ingredients, still important in any current theory [176], included:

(i) CP violation;
(ii) baryon-number violation (to be discussed in section 9.13.3.2);
(iii) a period in which the Universe was out of thermal equilibrium.

The possibilities for realizing these conditions are quite varied, but the form of CP violation incorporated into the phases of the CKM matrix seems insufficient. Rather, the CKM phases may be only one manifestation of a broader role for CP violation.

9.12.4.6. Cosmic-ray physics. Cosmic rays played an important role in elementary particle physics until the advent of accelerators in the mid-1950s, as mentioned in section 9.4. Even today, however, information on fundamental interactions continues to emerge from cosmic-ray studies. As one example, total cross sections for particle interactions appear to rise with increasing energies beyond the limits of terrestrial accelerators [477].

A new field of cosmic-ray physics has been spawned by the possibility of observing point sources of gamma rays at TeV energies and beyond. Čerenkov detection of TeV air showers has pinpointed emissions from the Crab Nebula (the remnants of a supernova seen by Chinese astronomers in 1054), and possibly even from extragalactic sources [478]. The search for point sources of gamma rays at higher energies and for the highest-energy cosmic rays is being undertaken using extensive arrays, some as large as 10^8 m^2 in area [479].

9.12.4.7. Black holes. A star may form with a mass so great that its own gravitational field prevents everything, even light, from escaping from its surface. To quote Laplace [480] (1798)

A luminous star, of the same density as the Earth, and whose diameter should be two hundred and 50 times larger than that of the Sun, would not, in consequence of its attraction, allow any of its rays to arrive at us; it is therefore possible that the largest luminous bodies in the Universe may, through this cause, be invisible.

Astronomical evidence has now been accumulating that such objects, now called [480] 'black holes', do, indeed, exist.

The relation between black holes and elementary particle physics is one of the fascinating unsolved problems of the present century. Stephen Hawking showed a black hole to be capable of producing pairs of particles in its vicinity, one of which is ejected and one of which falls into the black hole [481]. The ejected particle carries off energy which the black hole must supply by losing mass. Thus, a black hole eventually evaporates. Is this behaviour the source of a paradox? A pure quantum-mechanical state is thereby converted to a mixed one, leading to numerous suggestions for modifying the laws of classical gravitation, quantum mechanics, or both. As candidate quantum theories of gravitation, string theories (section 9.13.4) are being used to model such processes.

9.12.4.8. Dark matter. There is ample evidence that not all matter in the Universe takes the form of visible stars, gas or dust. One can measure the mass in our galaxy (or others), for example, by plotting the velocity of rotation of objects around the centre as a function of distance from the centre [482]. Much more mass is inferred than is seen.

Is the 'dark matter' of the Universe in the form of failed stars (Jupiter-size objects) or is it more exotic? Many proposals have been made, including a variety of as yet unseen elementary particles such as very light spinless particles (for example, particles known as 'axions' [483]), particles predicted by supersymmetry (see section 9.13.1.2) and massive neutrinos (see section 9.13.2). Each species seems to be able to account for some but not all of the desired properties of dark matter. For instance, a neutrino with a mass of about 10 eV, stable on the time-scale of the Universe's lifetime, can provide just the right amount of mass for the Universe to neither expand indefinitely nor contract back to a point after a finite time. But a 10 eV neutrino does not form the seeds for small-scale structure seen in galaxies, requiring another mechanism for galaxy formation.

9.12.4.9. Inflation and baryogenesis. The study of the early phases of the Universe led in the early 1980s to a remarkable observation that persisted in essence while undergoing various revisions in detail [484]. It is likely that at one or more stages, the Universe underwent an exponential increase in scale, wiping out all fluctuations. We infer this behaviour from the remarkable isotropy and homogeneity of the Universe today. A consequence of this 'inflationary scenario' is that the Universe should be exactly on the boundary between open (expanding forever) and closed (collapsing back to a point). This observational question remains to be settled by the study of the Universe's dark matter mentioned above.

9.12.4.10. Searches for deviations from Einstein's gravitation theory. The general theory of relativity does not distinguish between the gravitational force felt by any object; all that matters is its mass. Early tests [485] confirming this 'equivalence principle' had been refined to considerable precision by the mid-1960s [486]. A reanalysis [487] of the original experiments seemed to show a departure from the equivalence principle in the mid-1980s, leading to a flurry of improved tests and searches for a 'fifth force'. No evidence for it has survived these improved experiments [488].

9.13. Unsolved problems and hopes for the future

9.13.1. Electroweak theory: symmetry-breaking sector

9.13.1.1. Hunting for the Higgs boson. The electroweak theory described in section 9.7, though immensely successful in reproducing current data, is incomplete. We do not know how the masses of the W, Z, quarks or leptons (all of which break the original symmetry) actually arise.

The Weinberg–Salam mechanism for breaking electroweak SU(2) × U(1) involves the existence of a *Higgs boson H*, corresponding to the fluctuations of a neutral field with respect to its vacuum expectation value v. The non-zero value of v gives rise to the W and Z masses, with $v = 2^{-1/4}G_{\mathrm{F}}^{-1/2} = 246$ GeV fixed by the value of the Fermi coupling constant. Masses of quarks and leptons arise through their Yukawa couplings to the Higgs field.

The *mass* of the Higgs boson, not specified by the electroweak theory, is arbitrary, lying anywhere between the experimentally determined lower limit [489] (about 60 GeV/c^2 at present) and about 1 TeV. The upper limit reflects a requirement that the scattering of two longitudinally polarized gauge bosons preserve *S*-matrix unitarity [490]. A similar requirement for pion–pion scattering is met by the observed spectrum of $\pi\pi$ resonances.

The search for the Higgs boson is one reason often quoted in support of multi-TeV hadron colliders (see section 9.13.5). A straightforward Higgs signature would be a resonance in the W^+W^- and ZZ channels; this and many others are discussed, for example, in references [491] and [492].

9.13.1.2. Alternatives for the Higgs sector: supersymmetry, compositeness. Two main streams of theoretical thought concern the underlying nature of the Higgs boson. One of them views the Higgs boson as an elementary scalar particle whose mass must be 'protected' by some mechanism from acquiring large (and uncontrollable) radiative corrections. *Supersymmetry* [493], the most popular of such schemes, postulates for every particle of spin S the existence of a partner of spin $S \pm \frac{1}{2}$, whose presence in radiative corrections would exactly cancel the contribution of its 'superpartner' if exactly degenerate with it in mass. Supersymmetry, of course, would be verified by observing superpartners, not too different in mass from their ordinary counterparts to facilitate the desired cancellation. Most current versions of supersymmetric theories predict that at least some superpartners should exist below a mass of 1 TeV.

The other class of theories views the Higgs boson as a composite of more fundamental objects [494]. Such theories of *dynamical electroweak symmetry breaking* require the Higgs boson to reveal its structure by an energy of one or two TeV, implying a rich spectrum of resonances in W–W, Z–Z, and W–Z scattering starting at such energies.

9.13.2. Neutrino mass

9.13.2.1. Direct searches. Searches for electron-neutrino masses in beta-decay experiments have substantially improved over the past few years, as mentioned in section 9.12.1.3, with hopes for further advances [495]. Accelerator-based experiments are expected to make modest but not spectacular gains over present bounds on muon and tau neutrino masses.

9.13.2.2. Oscillations. A major hope for future observation of neutrino masses rests upon the phenomenon of neutrino oscillations (section 9.12.4.1), detectable in several ways.

(i) Further experiments could confirm the proposed role of neutrino oscillations in solar neutrino physics by measuring the *spectrum* of neutrinos from the Sun and comparing it with predictions of solar models.

(ii) Studies of the fluxes of different neutrino types generated by cosmic rays in the atmosphere can test for oscillations. At present the observed ratio of muon neutrinos to electron neutrinos seems somewhat smaller than anticipated theoretically [496], but further data are needed.

(iii) Direct searches for oscillations at accelerators [497] are probing new ranges of masses and mixing parameters. Experiments also have been proposed in which an accelerator neutrino beam is directed at an underground target several hundred or more km away [498].

9.13.3. Grand unified theories

9.13.3.1. Particle content. The description of electroweak and strong interactions through two separate Yang–Mills theories in the early 1970s

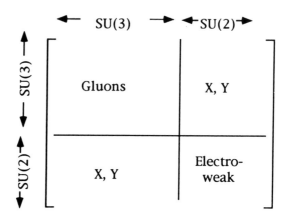

Figure 9.38. *Gauge bosons contained within the SU(5) grand unified group.*

suggested that a *single* such theory might describe particle physics. The 'grand unified' group would have to include $SU(3)_{QCD} \times SU(2)_{weak} \times U(1)_{weak}$ as a subgroup. Whereas the coupling constants of these groups differ at low energies, their different dependences on changes of momentum scale suggest that they approach one another at high energies. At such a 'unification energy', quarks and leptons behave very similarly to one another [499].

Grand unified groups proposed quite early included SU(5) (reference [500]) and SO(10) (reference [501]). The structure of SU(5), in particular, is very easy to visualize, involving the 3×3 matrices of SU(3) and 2×2 matrices of SU(2) arranged in block diagonal form, as shown in figure 9.38. The U(1) fits in naturally as a diagonal SU(5) matrix commuting with both SU(3) and SU(2). A minor blemish on the theory lies in its slightly arbitrary choice of quarks and leptons as members of 5- and 10-dimensional representations of SU(5). This aspect is handled more elegantly in the SO(10) scheme, where a single 16-dimensional representation suffices for each family of quarks and leptons. The presence of small but non-zero masses for neutrinos in SO(10) can be arranged, since the theory includes both left-handed and right-handed neutrinos [502].

Coupling constants in SU(5) do not approach a common value at high energy, a feature which can be bypassed by making the theory supersymmetric [503] or by permitting unification at more than one mass scale [504]. These possibilities are compared with the standard SU(5) scheme in figure 9.39.

9.13.3.2. Proton decay; experiments. Grand unification predicts that the proton can decay, with a pair of quarks turning into an antiquark and a lepton, by the exchange of a gauge boson of the extended group (see figure 9.40). This generic feature [499] was anticipated earlier by

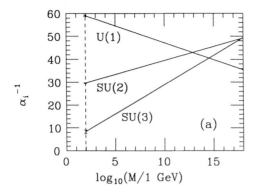

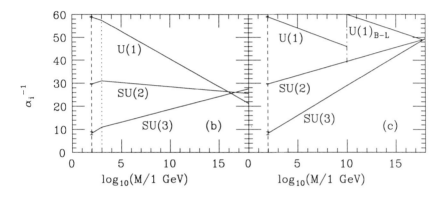

Figure 9.39. *Behaviour of coupling constants predicted by the renormalization group in various grand unified theories [504]. Error bars in plotted points denote uncertainties in coupling constants measured at $M = M_Z$ (broken vertical line). (a) SU(5), showing failure of coupling constants to meet at a point; (b) supersymmetric SU(5) scheme [503] with superpartners at 1 TeV (dotted line); (c) example of a two-scale SO(10) model with an intermediate mass scale (dot–dashed vertical line).*

Sakharov as a way to generate the baryon asymmetry of the Universe [175] in the presence of CP-violating interactions, as mentioned in sections 9.5.2.5 and 9.12.4.5. Many experiments were mounted beginning in the late 1970s to search for proton decay. The simplest SU(5) grand unified theory predicted a proton lifetime lower than about 10^{30} years [176], accessible to experiments with detectors of several tens of tons.

As of now, several multi-kiloton detectors have set limits [505] on the proton lifetime near $\tau_p > 10^{32}$ years. The two largest, the Kamioka and IMB water Čerenkov detectors mentioned in section 9.12.4.2, did record neutrinos from SN1987A. A 50-kiloton version of the Kamioka detector ('Super-Kamiokande') is now under construction [506], aimed at extending the lifetime limit to about 10^{34} years.

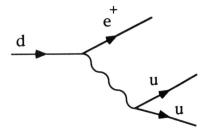

Figure 9.40. *Example of a process contributing to proton decay.*

9.13.3.3. Magnetic monopoles. The transition in the early Universe from a symmetric phase to one with broken symmetry can generate large numbers of magnetic monopoles [507]. Limits on their existence are very stringent [508], thus requiring means to greatly dilute their abundance or to avoid their generation. The apparent dearth of monopoles contributed to the inflationary universe scenario mentioned in section 9.12.4.9.

9.13.4. String theories
Some aspects of dual resonance models and their connection to the physics of strings have already been mentioned in section 9.5.4.5. The connection of string theories to condensed-matter problems was mentioned in section 9.12.3.4. A 'heterotic' version, with both 26- and 10-dimensional properties, gave the promise of realistic grand unification schemes [509]. It was hoped that the observed quarks and leptons could be explained in terms of singularities on surfaces generated when the extra dimensions curled up into unseen structures [510].

The predictive power of string theories for physics at the 100 GeV scale was soon recognized to be limited, as expected for theories whose fundamental scale is the Planck mass $m_P \equiv (\hbar c/G_N)^{1/2} \approx 10^{19}$ GeV/c^2, where G_N is Newton's gravitational constant. Nonetheless, construction of string theories with implications for grand unification schemes continues. Attention has also recently been devoted in string theories to problems of their internal self-consistency and to the construction of models of quantum gravitation. As mentioned in section 9.12.4.7, string theories are being used at present to investigate the behaviour of quantum-mechanical information loss accompanying black-hole evaporation [511].

9.13.5. Future facilities
As questions in elementary particle physics shift to higher energy domains and different particles, new facilities are under construction or under consideration to investigate the new domains. These include the first electron–proton collider (HERA), multi-TeV hadron colliders,

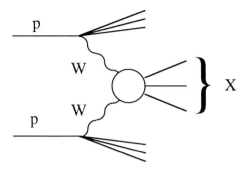

Figure 9.41. *W–W scattering induced by a hadronic collision.*

facilities for producing B-mesons copiously ('B-factories'), and large electron–positron colliders.

9.13.5.1. HERA. The first electron–proton collider began operating in 1991 at the DESY laboratory in Hamburg, Germany [512], colliding 27 GeV electrons with 820 GeV protons at a total centre-of-mass energy of about 300 GeV. So far, HERA has measured the photon–proton total cross section, has produced its first data on proton structure functions, and has observed the expected reaction $e^- + p \rightarrow \nu_e +$ (anything). It will be a welcome instrument for measuring the behaviour of structure functions (section 9.6.2) in new kinematic regions and in searching for new particles.

9.13.5.2. Multi-TeV hadron colliders. In the early 1980s, with the discovery of the W and Z anticipated, plans were laid to explore the mechanism of electroweak symmetry breaking through multi-TeV hadronic collisions. These plans crystallized into a proposal for a Large Hadron Collider (LHC) in the LEP tunnel at CERN and a project to construct the Superconducting Supercollider (SSC) in Texas. A good sampling of the physics accessible to these machines may be found in reference [491]. The LHC is designed for a centre-of-mass energy of 14 TeV and a luminosity of 10^{34} cm^{-2} s^{-1}, affording the study of such processes as W–W scattering (figure 9.41). The SSC was to have 40 TeV in the centre of mass and about ten times less luminosity, but support for it was withdrawn after several years of construction [513].

9.13.5.3. B factories. The standard electroweak theory (described for three families of quarks and leptons in section 9.9) predicts B-meson decays to exhibit CP-violating asymmetries which can be quite large, depending on values of parameters still being determined. However, since B-mesons decay to any given final state with a typical probability of less than 10^{-3}, many B-mesons are needed in order to permit such studies: probably at least 10^8 B$\bar{\text{B}}$ pairs, out of the reach of present electron–positron colliders.

The purest source of B-mesons, lying just above the B$\bar{\text{B}}$ threshold, is the $\Upsilon(4S)$ resonance, produced in e^+e^- collisions with a cross section of about 1 nb. An upgraded version of the CESR e^+e^- collider at Cornell, operating at the $\Upsilon(4S)$ resonance for several years with a luminosity of several times 10^{33} cm^{-2} s^{-1}, could produce 10^8 B$\bar{\text{B}}$ pairs. The most clear-cut CP-violating signatures at the 4S resonance require one to observe the time-dependence of both B-mesons' decays, most easily accomplished using electrons and positrons of different energies. Asymmetric B-factories are being constructed at KEK in Japan and at SLAC in the United States.

9.13.5.4. Large linear electron–positron colliders. The likely existence of a top quark below 200 GeV, the hopes that Higgs bosons and superpartners may exist with masses below several hundred GeV/c^2 and the sheer technical challenge of taking the next step in energy and luminosity have led to planning for a large linear electron–positron collider with a centre-of-mass energy of 500 GeV and a luminosity of 10^{33} cm^{-2} s^{-1}. Workshops on this facility have been held now for several years. In contrast to the Stanford Linear Collider, whose beams are bent in arcs to achieve collisions, this machine would be truly linear, with dimensions set by the attainable accelerating gradients; a total length of about 20 km is typical of plans.

9.13.5.5. Non-accelerator facilities. Several large facilities under construction or under consideration will shed light on particle physics without any need for an accelerator. The 'Super-Kamiokande' detector mentioned earlier will take a further step towards observing a proton's decay or setting improved limits on its lifetime. It will also be sensitive to neutrino interactions, as will a number of other facilities located underground (Soudan [514] in Minnesota, MACRO [515] in Italy, the Sudbury Neutrino Observatory [SNO] [516] in Canada), underwater (DUMAND [517] off the coast of Hawaii), or even under ice (AMANDA [518] at the South Pole). The MACRO detector will also search for monopoles at unprecedented sensitivities.

Searches for 'dark matter' will take on a new dimension with the development of cryogenic techniques, able, in somewhat simplified terms, to detect 'things that go bump in the night', otherwise invisible [519].

Large surface arrays are envisioned to detect cosmic-ray air showers. It may be possible to synchronize electronically the operation of an array as large as 5000 km^2 in area [520]. More modest surface arrays at higher elevations than currently in use are also planned [521].

9.14. Conclusions

Elementary particle physics in the second half of this century has oscillated between complexity and simplicity, between confusion and order.

The number of known 'elementary' particles grew enormously in the years after World War II thanks to new production and detection techniques. While some physicists despaired in the 1950s and 1960s of ever comprehending this burgeoning zoo of particles and their interactions, symmetries such as SU(3) began organizing these particles into families; the advent of the quark model introduced further regularity. However, not until a genuine theory of the strong interactions based on quantum chromodynamics (QCD) appeared in the early 1970s could one understand these successes in a fundamental way.

The weak interactions at the midpoint of the century were in a similar muddle. Failure to identify the most fundamental beta-decay processes made it hard to deduce the structure of the basic interaction. The discovery of parity non-conservation in the weak interactions lifted blinders from physicists' eyes, helped identify some experiments as right and some as wrong, and led almost immediately to a satisfactory description based on the so-called V–A theory. Still, the weak interactions remained incompletely understood. The V–A theory could not be used for calculations of higher precision or at higher energies. It took the synthesis of weak interactions with electromagnetism to construct a theory with these features.

Our present understanding of both strong and electroweak interactions relies on quantum field theory. At the midpoint of the century, field theory appeared to apply to a small area of particle physics, namely, quantum electrodynamics (QED), a limitation that vanished entirely in the course of 20 years.

Now that we understand the strongly interacting particles as consisting of quarks, held together with gluons, and understand the basis for both strong and electroweak forces, new questions have arisen. One of the earliest, arising just seven years after the discovery of parity violation, concerns the violation of combined charge-reflection (C) and parity (P) symmetries. A proposal that this violation requires at least three families of quarks has been supported by the discovery of the third family, but we still do not know whether the observed CP violation arises from this source.

Still mystifying, and possibly related to the origin of CP violation, is the pattern of masses and electroweak couplings of quarks and leptons. Quark masses necessarily break the electroweak symmetry. So do the masses of the weak force carriers, the charged Ws and the neutral Z. The source of this symmetry breaking has been generically termed the 'Higgs mechanism' and vague predictions exist for a 'Higgs boson', whose discovery would undoubtedly shed light on the electroweak symmetry-breaking mechanism. However, lurking in the background of any scenario involving the Higgs boson are many other new particles or effects, to be probed in the next generation of experiments.

The study of elementary particle physics in the past 50 years has been a noble enterprise, complete with its share of successes and setbacks. Let us hope that the progress in the next century is as rich as that in the present one.

9.14.1. Additional sources

In addition to the conference proceedings and historical works mentioned at the beginning of this article [1–5], and in the course of our discussion (see, for example, references [14, 129, 137, 227, 240, 275, 359]), several other useful references are worth mentioning. They include an eminently readable account by Jeremy Bernstein [522], a brief report by the United States Department of Energy [523], a thoughtful book by Steven Weinberg [524], a memoir spanning seven decades by Victor Weisskopf [525], a concise account of results in *Physical Review* by Sam Treiman [526], a new work by Murray Gell-Mann [527], Robert Marshak's last opus [528], and an early history of particle accelerators based on the life and work of Rolf Wideröe [529]. A biennial compilation of relevant data is produced by the Particle Data Group [49], carrying on a long tradition begun in the 1950s by Arthur H Rosenfeld and collaborators.

Acknowledgments

We thank J Bernstein, J D Bjorken, M M Block, L M Brown, U Fano, P Freund, H Frisch, R Hildebrand, Y Nambu, R Oehme, A Pais, M Perl, P Ramond, R Sachs, M M Shapiro, F C Shoemaker, V Telegdi, and A Tollestrup for helpful advice, and Gordon Fraser and Adrienne Kolb for help in obtaining illustrations. Ugo Fano provided indispensable editorial advice in the course of a careful reading of the manuscript. This work was performed in part at the Aspen Center for Physics, and was supported in part by the United States Department of Energy under Grant No. DE FG02 90ER 40560. JLR wishes to thank the Theory Group of Fermilab for their hospitality during the completion of this chapter.

References

[1] Brown L M, Dresden M, Hoddeson L and Riordan M (ed) 1995 *Third Int. Symp. on the History of Particle Physics: The Rise of the Standard Model (Stanford, CA, June 24–27, 1992)* (Cambridge: Cambridge University Press)
[2] Pais A 1986 *Inward Bound* (Oxford: Clarendon)
[3] Brown L M and Hoddeson L (ed) 1983 *Int. Symp. on the History of Particle Physics: The Birth of Particle Physics* (Cambridge: Cambridge University Press)

Colloque International sur l'Histoire de la Physique des Particules (International Colloquium on the History of Particle Physics) (Paris, 21–23 July, 1982) (Les Ulis, France: Les éditions de Physique); 1982 *J. Physique Coll.* **43** C8, supplement 12

[4] Brown L M, Dresden M and Hoddeson L (ed) 1989 *Pions to Quarks: Particle Physics in the 1950s (Proc. Second Int. Symp. on the History of Particle Physics (Fermilab, 1985))* (Cambridge: Cambridge University Press)

[5] Brown L M 1995 Nuclear forces, mesons and isospin symmetry *Twentieth Century Physics* (Bristol: Institute of Physics) Chapter 5

[6] Heisenberg W and Pauli W 1929 *Z. Phys.* **56** 1–61; 1930 *Z. Phys.* **59** 168–90
Oppenheimer J R 1930 *Phys. Rev.* **35** 461–77
Heisenberg W 1934 *Z. Phys.* **90** 209–231
Dirac P A M 1934 *Proc. Camb. Phil. Soc.* **30** 150–63

[7] Dirac P A M 1928 *Proc. R. Soc.* A **117** 610–24; 1928 *Proc. R. Soc.* A **118** 351–61; 1930 *Proc. R. Soc.* A **126** 360–5; 1931 *Proc. R. Soc.* A **133** 60–72

[8] Weisskopf V F 1939 *Phys. Rev.* **56** 72–85

[9] Uehling E A 1935 *Phys. Rev.* **48** 55–63
Serber R 1935 *Phys. Rev.* **48** 49–54

[10] Kemble E C and Present R D 1933 *Phys. Rev.* **44** 1031–2

[11] Houston W V 1937 *Phys. Rev.* **51** 446–9
Williams R C 1938 *Phys. Rev.* **54** 558–67

[12] Pasternack S 1938 *Phys. Rev.* **54** 1113

[13] Lamb W E Jr and Retherford R C 1947 *Phys. Rev.* **72** 241–3
Lamb W E Jr in the first of references [3] ch 20

[14] Schweber S S 1984 *Relativity, Groups and Topology II (1983 Les Houches Lectures)* ed B S de Witt and R Stora (Amsterdam: North-Holland) pp 37–220
See also Marshak R E in the first of references [3] pp 376–401

[15] Bethe H A 1947 *Phys. Rev.* **72** 339–41

[16] Kroll N M and Lamb W E 1949 *Phys. Rev.* **75** 388–98
Feynman R P 1949 *Phys. Rev.* **76** 769–89 (see in particular footnote 13)
Schwinger J 1949 *Phys. Rev.* **75** 898–9
French J B and Weisskopf V F 1949 *Phys. Rev.* **75** 388, 1240–8
Nambu Y 1949 *Prog. Theor. Phys.* **4** 82–94

[17] Schwinger J (ed) 1958 *Selected Papers on Quantum Electrodynamics* (New York: Dover); see also the second of references [3] pp C8-409–23

[18] Foley H M and Kusch P 1947 *Phys. Rev.* **72** 1256–7; 1948 *Phys. Rev.* **73** 412

[19] Schwinger J 1948 *Phys. Rev.* **73** 416–7; Erratum 1949 *Phys. Rev.* **76** 790–817

[20] Wheeler J A 1946 *Ann. NY Acad. Sci* **48** 219–38
See also Pirenne J 1944 *Thesis* University of Paris; 1946 *Arch. Sci. Phys. Nat.* **28** 273; 1947 *Arch. Sci. Phys. Nat.* **29** 121

[21] Deutsch M 1951 *Phys. Rev.* **82** 455–6

[22] Kinoshita T (ed) 1990 *Quantum Electrodynamics* (Singapore: World Scientific)

[23] Schupp A R, Pidd R W and Crane H R 1961 *Phys. Rev.* **121** 1–17

[24] Dehmelt H 1990 *Rev. Mod. Phys.* **62** 525–30
Van Dyck R S Jr 1990 in reference [22] ch 8

[25] Van Dyck R S Jr, Schwinberg P B and Dehmelt H G 1987 *Phys. Rev. Lett.* **59** 26–29

[26] Kinoshita T and Yennie D 1990 in reference [22] ch 1

[27] Cage M E *et al* 1989 *IEEE Trans. Instrum. Meas.* **38** 284–9

[28] Suura H and Wichmann E 1957 *Phys. Rev.* **105** 1930–1
Petermann A 1957 *Phys. Rev.* **105** 1931

[29] Garwin R, Lederman L M and Weinrich M 1957 *Phys. Rev.* **105** 1415–7

[30] Farley F J M and Picasso E in reference [22] ch 11
[31] Kinoshita T and Marciano W J in reference [22] ch 10
[32] Pal'chikov V G, Sokolov Yu L and Yakovlev V P 1983 *Pis. Zh. Eksp. Teor. Fiz.* **38** 347–9 (Engl. Transl. *JETP Lett.* **38** 418–20)
[33] Sapirstein J R and Yennie D R in reference [22] ch 12
[34] Ramsey N F in reference [22] ch 13
[35] Mills A P and Chu S in reference [22] ch 15
[36] Harris I and Brown L M 1957 *Phys. Rev.* **105** 1656–61
[37] Nico J S, Gidley D W, Rich A and Zitzewitz P W 1990 *Phys. Rev. Lett.* **65** 1344–7
[38] Ore A and Powell J L 1949 *Phys. Rev.* **75** 1696–9
      Caswell W G and Lepage G P 1979 *Phys. Rev.* A **20** 36–43
      Adkins G S 1983 *Ann. Phys., NY* **146** 78–128
[39] Urey H, Brickwedde F G and Murphy G M 1932 *Phys. Rev.* **39** 164–5, 864
[40] Chadwick J 1932 *Nature* **129** 312; 1932 *Proc. R. Soc.* A **136** 692–708, 744–8
[41] Anderson C 1932 The positive electron *Science* **76** 238–9; the first of references [3] ch 7
[42] Neddermeyer S H and Anderson C D 1937 *Phys. Rev.* **51** 884–6
      Street J C and Stevenson E C 1937 *Phys. Rev.* **51** 1005
      Nishina Y, Takeuchi M and Ichimaya T 1937 *Phys. Rev.* **52** 1198–9
[43] Yukawa H 1935 *Proc. Phys.-Math. Soc. Japan* **17** 48–57
[44] Oppenheimer J R and Serber R 1937 *Phys. Rev.* **51** 1113
[45] Pais A in reference [2] p 433
[46] Pauli W as described in reference [2] p 315
[47] Blackett P M S and Occhialini G P S 1933 *Proc. R. Soc.* A **139** 699–726
[48] Fermi E 1934 *Nuovo Cimento* **11** 1–19; 1934 *Z. Phys.* **88** 161–71
[49] Montanet L *et al* (Particle Data Group) 1994 *Phys. Rev.* D **50** 1173–1825
[50] Rossi B and Nereson N 1942 *Phys. Rev.* **62** 417–22
      Nereson N and Rossi B 1943 *Phys. Rev.* **64** 199–201
      Rossi B in the first of references [3] ch 11
[51] Williams E J and Evans G R 1940 *Nature* **145** 818–9
[52] Conversi M, Pancini E and Piccioni O 1947 *Phys. Rev.* **71** 209–10
      Piccioni O in the first of references [3] ch 13
      Conversi M in the first of references [3] ch 14
[53] Perkins D H in reference [4] ch 5
      Perkins D H 1947 *Nature* **159** 126–7
      Occhialini G P S and Powell C F 1947 *Nature* **159** 93–4
[54] Powell C F, Fowler P H and Perkins D H 1959 *The Study of Elementary Particles by the Photographic Method* (New York: Pergamon)
[55] Lattes C M G, Muirhead H, Occhialini G P S and Powell C F 1947 *Nature* **159** 694–7
      Lattes C M G, Occhialini G P S and Powell C F 1947 *Nature* **160** 453–6, 486–92
[56] Sakata S and Inoue T 1946 *Prog. Theor. Phys.* **1** 143–50
      Tanikawa Y 1947 *Prog. Theor. Phys.* **2** 220–1
      Marshak R E and Bethe H A 1947 *Phys. Rev.* **72** 506–9
[57] Rochester G D and Butler C C 1947 *Nature* **160** 855–7
[58] Kemmer N 1938 *Proc. R. Soc.* A **166** 127–53
      Sakata S 1941–2 Unpublished correspondence with S Tomonaga (Y Nambu, private communication)
[59] Bjorklund R, Crandall W E, Moyer B J and York H F 1950 *Phys. Rev.* **77** 213–18
      Steinberger J, Panofsky W K H and Steller J 1950 *Phys. Rev.* **78** 802–5
      Steinberger J in reference [4] ch 20

Carlson A G, Hooper J E and King D T 1950 *Phil. Mag.* **41** 701–24

[60] Clark D C, Roberts A and Wilson R 1951 *Phys. Rev.* **83** 649

Durbin R, Loar H and Steinberger J 1951 *Phys. Rev.* **83** 646–8

[61] Panofsky W K H, Aamodt R L and Hadley J 1951 *Phys. Rev.* **81** 565–74

[62] Hales R W, Hildebrand R H, Knable N and Moyer B 1952 *Phys. Rev.* **85** 373–4

[63] Hildebrand R H 1953 *Phys. Rev.* **89** 1090–2

[64] Richardson J R 1948 *Phys. Rev.* **74** 1720–1

[65] Chamberlain O, Mozley R F, Steinberger J and Wiegand C 1950 *Phys. Rev.* **79** 394–5

[66] Sakata S 1942 (unpublished) (Y Nambu, private communication)

Landau L D 1948 *Dokl. Akad. Nauk SSSR* **60** 207–9

Yang C N 1950 *Phys. Rev.* **77** 242–5

[67] Samios N P, Plano R, Prodell A, Schwartz M and Steinberger J 1962 *Phys. Rev.* **126** 1844–9

[68] Kaplon M F, Peters B and Bradt H L 1950 *Phys. Rev.* **76** 1735–6

[69] Steinberger J 1949 *Phys. Rev.* **76** 1180–6

[70] Primakoff H 1951 *Phys. Rev.* **81** 899

[71] Tollestrup A V, Berman S, Gomez R and Ruderman H 1960 *Proc. 1960 Ann. Int. Conf. on High Energy Physics at Rochester (Rochester, August 25– September 1, 1960)* ed E C G Sudarshan *et al* (New York: Interscience [University of Rochester]) pp 27–30

Ruderman H A 1962 *PhD Thesis* Caltech (unpublished)

Bellettini G, Bemporad C, Braccini P L and Foà L 1965 *Nuovo Cimento* A **40** 1139–70

[72] von Dardel G, Dekkers D, Mermod R, Van Putten J D, Vivargent M, Weber G and Winter K 1963 *Phys. Lett.* **4** 51–4

Atherton H W *et al* 1985 *Phys. Lett.* **158B** 81–4

[73] Chamberlain O, Segrè E, Wiegand C and Ypsilantis T 1955 *Phys. Rev.* **100** 947–50

Goldhaber G in reference [4] ch 16

Chamberlain O in reference [4] ch 17

Piccioni O in reference [4] ch 18

[74] Chamberlain O *et al* 1956 *Phys. Rev.* **102** 921–3

Barkas W H *et al* 1957 *Phys. Rev.* **105** 1037–58

[75] Gell-Mann 1982 in the second of references [3] pp C8-395–408; reference [2] ch 20

Perkins D H in reference [4] ch 5

Rochester G D in reference [4] ch 4

[76] Leprince-Ringuet L and Lhéritier M 1944 *C. R. Acad. Sci., Paris* **219** 618–20; 1946 *J. Phys. Radium* **7** 65–9

See, however, the criticism by Bethe H A 1946 *Phys. Rev.* **70** 821–31

[77] Brown R, Camerini U, Fowler P H, Muirhead H, Powell C F and Ritson D M 1949 *Nature* **163** 82–7

See also Fowler P H, Menon M G K, Powell C F and Rochat O 1951 *Phil. Mag.* **42** 1040–9

[78] Seriff A J, Leighton R B, Hsiao C, Cowan E W and Anderson C D 1950 *Phys. Rev.* **78** 290–1

[79] Hopper V D and Biswas S 1950 *Phys. Rev.* **80** 1099–1100

[80] Armenteros R, Barker K H, Butler C C, Cachon A and Chapman A H 1951 *Nature* **167** 501–3

Armenteros R, Barker K H, Butler C C and Cachon A 1951 *Phil. Mag.* **42** 1113–35

[81] Thompson R W, Cohn H O and Flum R S 1951 *Phys. Rev.* **83** 175

Thompson R W, Buskirk A V, Etter L R, Karzmark C J and Rediker R H 1953 *Phys. Rev.* **90** 329–30. An account of subsequent work by this group is given by Thompson R W in the first of references [3] ch 15.

[82] Armenteros R, Barker K H, Butler C C, Cachon A and York C M 1952 *Phil. Mag.* **43** 597–612

Leighton R B, Cowan E W and van Lint V A J 1953 *Proc. Conf. Int. Ray. Cosmique (Bagnères de Bigorre)* (Toulouse: University of Toulouse) pp 97–101

Anderson C D, Cowan E W, Leighton R B and van Lint V A J 1953 *Phys. Rev.* **92** 1089

[83] York C M, Leighton R B and Bjornerud E K 1953 *Phys. Rev.* **90** 167–8

Bonetti A, Levi Setti R, Panetti M and Tomasini G 1953 *Nuovo Cimento* **10** 345–6, 1736–43

[84] Blewett J P in reference [4] ch 10

Livingston M S and Blewett J P 1962 *Particle Accelerators* (New York: McGraw-Hill)

Blewett M H (ed) 1953 *Rev. Sci. Instrum.* **24** 723–870

[85] Fowler W B, Shutt R P, Thorndike A M and Whittemore W L 1953 *Phys. Rev.* **90** 1126–7; 1953 *Phys. Rev.* **91** 1287; 1954 *Phys. Rev.* **93** 861–7; 1955 *Phys. Rev.* **98** 121–30

See also Walker W D 1955 *Phys. Rev.* **98** 1407–10

For an historical account see Fowler W B in reference [4] ch 22.

[86] Nambu Y, Nishijima K and Yamaguchi Y 1951 *Prog. Theor. Phys.* **6** 615–19, 619–22

Miyazawa H 1951 *Prog. Theor. Phys.* **6** 631–3

Oneda S 1951 *Prog. Theor. Phys.* **6** 633–5

[87] Pais A 1952 *Phys. Rev.* **86** 663–72; 1953 *Physica* **19** 869–87

[88] Gell-Mann M 1953 *Phys. Rev.* **92** 833–4; 1953 On the classification of particles (unpublished)

Gell-Mann M and Pais A 1955 *Proc. 1954 Glasgow Conf. on Nuclear and Meson Physics* ed E H Bellamy and R G Moorhouse (London: Pergamon)

Gell-Mann M 1956 *Nuovo Cimento* **4** (Supplement) 848–66

[89] Nakano T and Nishijima K 1953 *Prog. Theor. Phys.* **10** 581–2

Nishijima K 1954 *Prog. Theor. Phys.* **12** 107–8; 1955 *Prog. Theor. Phys.* **13** 285–304

[90] Alvarez L W, Eberhard P, Good M L, Graziano W, Ticho H K and Wojcicki S G 1959 *Phys. Rev. Lett.* **2** 215–9

The history of this and related discoveries is recounted by Alvarez L 1969 *Science* **165** 1071–91.

[91] Gell-Mann M and Pais A 1955 *Phys. Rev.* **97** 1387–9

[92] Lande K, Booth E T, Impeduglia J and Lederman L M 1956 *Phys. Rev.* **103** 1901–4 (an account of this experiment is given by Chinowsky W in reference [4] ch 21)

[93] Fry W F, Schneps J and Swami M S 1956 *Phys. Rev.* **103** 1904–5

[94] Heitler W 1946 *Proc. Ir. Acad.* **51A** 33–9

Nambu Y and Yamaguchi Y 1951 *Prog. Theor. Phys.* **6** 1000–6

[95] Brueckner K A and Watson K M 1951 *Phys. Rev.* **83** 1–9

Watson K M 1952 *Phys. Rev.* **85** 852–7

Adair R K 1952 *Phys. Rev.* **87** 1041–3

[96] Walker R L in reference [4] ch 6

[97] Bishop A S, Steinberger J and Cook L J 1950 *Phys. Rev.* **80** 291

Steinberger J, Panofsky W K H and Steller J 1950 *Phys. Rev.* **78** 802–5

Chedester C D, Isaacs P, Sachs A and Steinberger J 1951 *Phys. Rev.* **82** 958–9

Anderson H L, Fermi E, Long E A, Martin R and Nagle D E 1952 *Phys. Rev.* **85** 934–5

Fermi E, Anderson H L, Lundby A, Nagle D E and Yodh G B 1952 *Phys. Rev.* **85** 935–6

Anderson H L, Fermi E, Long E A and Nagle D E 1952 *Phys. Rev.* **85** 936

[98] Marshak R E 1950 *Phys. Rev.* **78** 346

Fujimoto Y and Miyazawa H 1950 *Prog. Theor. Phys.* **5** 1052–4

Brueckner K A and Case K M 1951 *Phys. Rev.* **83** 1141–7

Brueckner K A 1952 *Phys. Rev.* **86** 106–9

Brueckner K A and Watson K M 1952 *Phys. Rev.* **86** 923–8

Many of these works made reference to earlier schemes by Wentzel G 1940 *Helv. Phys. Acta* **13** 269–308; 1941 *Helv. Phys. Acta* **14** 633–5; 1947 *Rev. Mod. Phys.* **19** 1–18.

Pauli W and Dancoff S M 1942 *Phys. Rev.* **62** 85–108

[99] Walker R L, Oakley D C and Tollestrup A V 1953 *Phys. Rev.* **89** 1301–2

Anderson H L, Fermi E, Martin R and Nagle D E 1953 *Phys. Rev.* **91** 155–68

Yuan L C L and Lindenbaum S J 1953 *Phys. Rev.* **92** 1578–9

Ashkin J, Blaser J P, Feiner F, Gorman J and Stern M O 1954 *Phys. Rev.* **93** 1129–30

[100] Glaser D A 1952 *Phys. Rev.* **87** 665; 1953 *Phys. Rev.* **91** 496, 762–3

[101] Alvarez L A 1969 *Science* **165** 1071–1091; reference [4] ch 19

[102] A brief history of these discoveries is given by Samios N P in reference [1].

[103] Moller C and Rosenfeld L 1940 *Kong. Danske Vid. Selsk., Matt-fys. Medd.* **17** no 8 pp 1–72

Schwinger J 1942 *Phys. Rev.* **61** 387

Rosenfeld L 1948 *Nuclear Forces* (New York: Interscience) p 322

[104] Hofstadter R in reference [4] ch 7 pp 126–43

[105] Nambu Y 1957 *Phys. Rev.* **106** 1366–7

[106] Frazer W R and Fulco J R 1959 *Phys. Rev. Lett.* **2** 365–8

[107] Stonehill D C, Baltay C, Courant H, Fickinger W, Fowler E C, Kraybill H, Sandweiss J, Sanford J and Taft H 1961 *Phys. Rev. Lett.* **6** 624–5

Erwin A R, March R, Walker W D and West E 1961 *Phys. Rev. Lett.* **6** 628–30

[108] Maglic B C, Alvarez L W, Rosenfeld A H and Stevenson M L 1961 *Phys. Rev. Lett.* **7** 178–82

[109] Pevsner A *et al* 1961 *Phys. Rev. Lett.* **7** 421–3

[110] These are documented in the introduction to Gell-Mann M and Ne'eman Y 1964 *The Eightfold Way* (New York: Benjamin).

[111] Ikeda M, Ogawa S and Ohnuki Y 1959 *Prog. Theor. Phys.* **22** 715–24; 1960 *Prog. Theor. Phys.* **23** 1073–99

[112] Gell-Mann M 1961 The eightfold way *Caltech Report* CTSL-20, reprinted in Gell-Mann M and Ne'eman Y 1964 *The Eightfold Way* (New York: Benjamin) p 11

[113] Yang C N and Mills R L 1954 *Phys. Rev.* **96** 191–5

[114] Ne'eman Y 1961 *Nucl. Phys.* **26** 222–9

[115] Gell-Mann M 1962 *Phys. Rev.* **125** 1067–84

[116] Okubo S 1962 *Prog. Theor. Phys.* **27** 949–66

Gell-Mann M 1962 *Proc. 1962 Int. Conf. on High Energy Physics at CERN (Geneva, 4–11 July, 1962)* ed J Prentki (Geneva: CERN) p 805

[117] Goldhaber S, Chinowsky W, Goldhaber G, Lee W, O'Halloran T, Stubbs T F, Pjerrou G M, Stork D H and Ticho H 1962 Presented at the conference by Goldhaber G *Proc. 1962 Int. Conf. on High Energy Physics at CERN (Geneva, 4–11 July, 1962)* ed J Prentki (Geneva: CERN) pp 356–8

[118]   Gell-Mann M 1962 *Proc. 1962 Int. Conf. on High Energy Physics at CERN (Geneva, 4–11 July, 1962)* ed J Prentki (Geneva: CERN)
        Ne'eman Y 1962 Personal conversation with G Goldhaber at 1962 CERN Conference (Geneva, 4–11 July, 1962)
        Goldhaber G in reference [1]
[119]   Barnes V E *et al* 1964 *Phys. Rev. Lett.* **12** 204–6
        See also Barnes V E *et al* 1964 *Phys. Lett.* **12** 134–6
        Abrams G S *et al* 1964 *Phys. Rev. Lett.* **13** 670–2
[120]   Reines F 1979 *Science* **203** 11–16; reference [4] ch 24
[121]   Reines F and Cowan C L Jr 1953 *Phys. Rev.* **92** 830–1
[122]   Cowan C L Jr, Reines F, Harrison F B, Kruse H W and McGuire A D 1956 *Science* **124** 103–4
        Reines F, Cowan C L Jr, Harrison F B, McGuire A D and Kruse H W 1960 *Phys. Rev.* **117** 159–70
[123]   Gamow G and Teller E 1936 *Phys. Rev.* **49** 895–9
[124]   Fierz M 1937 *Z. Phys.* **104** 553–65
[125]   See, for example, Gasiorowicz S 1966 *Elementary Particle Physics* (New York: Wiley) p 502
[126]   Pontecorvo B 1947 *Phys. Rev.* **72** 246–7
        Klein O 1948 *Nature* **161** 897–9
        Puppi G 1948 *Nuovo Cimento* **5** 587–8
        Lee T D, Rosenbluth M and Yang C N 1949 *Phys. Rev.* **75** 905
        Tiomno J and Wheeler J A 1949 *Rev. Mod. Phys.* **21** 144–52, 153–65
[127]   Fitch V and Motley R 1956 *Phys. Rev.* **101** 496–8
        Motley R and Fitch V 1957 *Phys. Rev.* **105** 265–6
        Alvarez L W, Crawford F S, Good M L and Stevenson L 1956 *Phys. Rev.* **101** 503–5. Other references are cited in reference [2] sec 20(c).
[128]   Dalitz R H 1953 *Proc. Conf. Int. Ray. Cosmique (Bagnères de Bigorre)* (Toulouse: University of Toulouse) p 236; 1953 *Phil. Mag.* **44** 1068–80; 1954 *Phys. Rev.* **94** 1046–51
        Fabri E 1954 *Nuovo Cimento* **11** 479–91
        A much fuller account is given by Dalitz R H in reference [4] ch 30.
[129]   Polkinghorne J *Rochester Roundabout: The Story of High Energy Physics* (New York: Freeman) p 57
[130]   Lee T D and Yang C N 1956 *Phys. Rev.* **104** 254–8
[131]   Lee T D 1971 *Elementary Processes at High Energy (International School of Subnuclear Physics, Erice, Italy, 1970)* ed A Zichichi (New York: Academic Press) pp 827–40
        Yang C N *C N Yang: Selected Papers 1945–80, With Commentary* (San Francisco: Freeman) pp 26–31
[132]   Oehme R *C N Yang: Selected Papers 1945–80, With Commentary* (San Francisco: Freeman) pp 32–33
        Lee T D and Yang C N *C N Yang: Selected Papers 1945–80, With Commentary* (San Francisco: Freeman) pp 33–34
[133]   Lee T D, Oehme R and Yang C N 1957 *Phys. Rev.* **106** 340–45
        See also Ioffe B L, Okun' L B and Rudik A P 1957 *Zh. Eksp. Teor. Fiz.* **32** 396–7 (Engl. Transl. 1957 *Sov. Phys.–JETP* **5** 328–30)
[134]   Schwinger J 1953 *Phys. Rev.* **91** 713–28 (see especially p 720ff); 1954 *Phys. Rev.* **94** 1362–84 (see especially equation (54) on p 1366 and p 1376ff)
        Lüders G 1954 *Kong. Danske Vid. Selsk., Matt-fys. Medd.* **28** 5; 1957 *Ann. Phys., NY* **2** 1–15
        Pauli W (ed) 1955 *Niels Bohr and the Development of Physics* (New York: Pergamon) pp 30–51

Bell J S 1955 *Proc. R. Soc. A* **231** 479–95

[135]   Wu C S, Ambler E, Hayward R W, Hoppes D D and Hudson R P 1957 *Phys. Rev.* **105** 1413–15

[136]   Friedman J I and Telegdi V L 1957 *Phys. Rev.* **105** 1681–2; 1957 *Phys. Rev.* **106** 1290–3

[137]   Lederman L M 1993 *The God Particle: If the Universe is the Answer, What is the Question?* (Boston: Houghton Mifflin) pp 256–73

[138]   Crawford F S Jr, Cresti M, Good M L, Gottstein K, Lyman E M, Solmitz F T, Stevenson M L and Ticho H 1957 *Phys. Rev.* **108** 1102–3

Eisler F *et al* 1957 *Phys. Rev.* **108** 1353–5

Leipuner L B and Adair R K 1958 *Phys. Rev.* **109** 1358–63

[139]   Purcell E M and Ramsey N F 1950 *Phys. Rev.* **78** 807

Smith J, Purcell E M and Ramsey N F 1951 (unpublished)

Smith J 1951 *PhD Thesis* Harvard University (unpublished) as quoted by Ramsey N F 1953 *Nuclear Moments* (New York: Wiley) p 8

[140]   Salam A 1957 *Nuovo Cimento* **5** 299–301

Landau L 1957 *Zh. Eksp. Teor. Fiz.* **32** 407–8 (Engl. Transl. 1957 *Sov. Phys.– JETP* **5** 337–8); 1957 *Nucl. Phys.* **3** 127–31

[141]   Pauli W 1933 *Quantentheorie* (*Handbuch der Physik 24/1*) (Berlin: Springer) pp 83–272

[142]   Lee T D and Yang C N 1957 *Phys. Rev.* **105** 1671–5

Jackson J D, Treiman S B and Wyld H W 1957 *Phys. Rev.* **106** 517–21

[143]   Feynman R P and Gell-Mann M 1958 *Phys. Rev.* **109** 193–8

[144]   Sudarshan E C G and Marshak R E 1958 *Proc. Int. Conf. on Mesons and Newly Discovered Particles (Padua–Venice, 22–28 September, 1957)* ed N Zanichelli (Bologna: Società Italiana di Fisica), reprinted in Kabir P K (ed) 1963 *The Development of Weak Interaction Theory* (New York: Gordon and Breach) pp 118–28; 1958 *Phys. Rev.* **109** 1860–2

[145]   Sakurai J J 1958 *Nuovo Cimento* **7** 649–60

[146]   Rustad B M and Ruby S L 1955 *Phys. Rev.* **97** 991–1002

[147]   Anderson H L and Lattes C M G 1957 *Nuovo Cimento* **6** 1356–81

[148]   Allen J S, Burman R L, Herrmannsfeldt W B, Stähelin P and Braid T H 1959 *Phys. Rev.* **116** 134–43

[149]   Fazzini T, Fidecaro G, Merrison A W, Paul H and Tollestrup A V 1958 *Phys. Rev. Lett.* **1** 247–9

Impeduglia G, Plano R, Prodell A, Samios N, Schwartz M and Steinberger J 1958 *Phys. Rev. Lett.* **1** 249–51

[150]   Goldhaber M, Grodzins L and Sunyar A W 1958 *Phys. Rev.* **109** 1015–17

[151]   Telegdi V L in reference [4] ch 32

[152]   Michel L 1950 *Proc. Phys. Soc.* **63** 514–31

[153]   Gershtein S S and Zel'dovich Ya B 1955 *Zh. Eksp. Teor. Fiz.* **29** 698–9 (Engl. Transl. 1956 *Sov. Phys.–JETP* **2** 576–8)

[154]   Depommier P, Heintze J, Mukhin A, Rubbia C, Soergel V and Winter K 1962 *Phys. Lett.* **2** 23–6

Depommier P, Heintze J, Rubbia C and Soergel V 1963 *Phys. Lett.* **5** 61–3

[155]   Klein O 1939 *Les Nouvelles Théories de la Physique* (Paris: Inst. Int. de Coöperation Intellectuelle) pp 81–98

Schwinger J 1957 *Ann. Phys., NY* **2** 407–34

Bludman S A 1958 *Nuovo Cimento* **9** 433–45

Feinberg G 1958 *Phys. Rev.* **110** 1482–3

Glashow S L 1959 *Nucl. Phys.* **10** 107–17

Gell-Mann M 1959 *Rev. Mod. Phys.* **31** 834–8

Lee T D and Yang C N 1960 *Phys. Rev.* **119** 1410–19

[156]   Danby G, Gaillard J M, Goulianos K, Lederman L M, Mistry N,

Schwartz M and Steinberger J 1962 *Phys. Rev. Lett.* **9** 36–44
[157]   Schwartz M 1989 *Rev. Mod. Phys.* **61** 527–32
[158]   Pontecorvo B 1959 *Zh. Eksp. Teor. Fiz.* **37** 1751–7 (Engl. Transl. 1960 *Sov. Phys.–JETP* **10** 1236–40)
Schwartz M 1960 *Phys. Rev. Lett.* **4** 306–7
[159]   Lee T D and Yang C N 1960 *Phys. Rev. Lett.* **4** 307–8
[160]   Block M M *et al* 1964 *Phys. Lett.* **12** 281–5
Bernardini G *et al* 1964 *Phys. Lett.* **13** 86–91
Burns R, Goulianos K, Hyman E, Lederman L, Lee W, Mistry N, Rehberg J, Schwartz M, Sunderland J and Danby G, 1965 *Phys. Rev. Lett.* **15** 42–5
[161]   Landau L 1957 *Zh. Eksp. Teor. Fiz.* **32** 405–6 (Engl. Transl. 1957 *Sov. Phys.–JETP* **5** 336–7); 1957 *Nucl. Phys.* **3** 127–31
[162]   Fitch V L 1981 *Rev. Mod. Phys.* **53** 367–71
[163]   Christenson J H, Cronin J W, Fitch V L and Turlay R 1964 *Phys. Rev. Lett.* **13** 138–40
[164]   Sachs R G 1963 *Ann. Phys., NY* **22** 239–62
[165]   Wu T T and Yang C N 1964 *Phys. Rev. Lett.* **13** 380–5
[166]   Bell J S and Steinberger J 1966 *Proc. Oxford Int. Conf. on Elementary Particles (19–25 September, 1965)* ed T R Walsh (Chilton: Rutherford High Energy Laboratory) pp 193–222
Cronin J W 1981 *Rev. Mod. Phys.* **53** 373–83
[167]   Fitch V L, Roth R F, Russ J S and Vernon W 1965 *Phys. Rev. Lett.* **15** 73–6
[168]   Pais A and Piccioni O 1955 *Phys. Rev.* **100** 1487–9
Case K M 1956 *Phys. Rev.* **103** 1449–53
Good M L 1957 *Phys. Rev.* **106** 591–5
[169]   Lee T D and Wu C S 1966 *Ann. Rev. Nucl. Sci.* **16** 511–90
Kleinknecht K 1976 *Ann. Rev. Nucl. Sci.* **26** 1–50
Kabir P K 1968 *The CP Puzzle: Strange Decays of the Neutral Kaon* (New York: Academic)
Sachs R G 1987 *The Physics of Time Reversal* (Chicago: University of Chicago Press)
Jarlskog C (ed) 1989 *CP Violation* (Singapore: World Scientific)
[170]   Wolfenstein L 1964 *Phys. Rev. Lett.* **13** 562–4
[171]   Gibbons L K *et al* 1993 *Phys. Rev. Lett.* **70** 1203–1206
[172]   Barr G D *et al* 1993 *Phys. Lett.* **317B** 233–42
[173]   Treiman S B and Sachs R G 1956 *Phys. Rev.* **103** 1545–49
Wyld H W Jr and Treiman S B 1957 *Phys. Rev.* **106** 169–70
[174]   Lüth V 1974 *Thesis* University of Heidelberg
Geweniger C *et al* 1974 *Phys. Lett.* **48B** 483–6
[175]   Sakharov A D 1967 *Pis. Zh. Eksp. Teor. Fiz.* **5** 32–5 (Engl. Transl. *JETP Lett.* **5** 24–7)
[176]   Langacker P 1981 *Phys. Rep.* **72** 185–385 and references therein
[177]   Goldberger M L and Treiman S B 1958 *Phys. Rev.* **110** 1178–84; 1958 *Phys. Rev.* **111** 354–61
Treiman S B in reference [4] ch 27
[178]   Goldberger M L and Treiman S B 1958 *Phys. Rev.* **110** 1478–9
[179]   Nambu Y 1960 *Phys. Rev. Lett.* **4** 380–2
Nambu Y and Jona-Lasinio G 1961 *Phys. Rev.* **122** 345–58; 1961 *Phys. Rev.* **124** 246–54
Nambu Y in reference [4] ch 44
[180]   Goldstone J 1961 *Nuovo Cimento* **19** 154–64
Goldstone J, Salam A and Weinberg S 1962 *Phys. Rev.* **127** 965–70
[181]   Gell-Mann M and Lévy M 1960 *Nuovo Cimento* **16** 705–25

[182]   Crawford F S Jr, Cresti M, Good M L, Kalbfleisch G R, Stevenson M L
        and Ticho H K 1958 *Phys. Rev. Lett.* **1** 377–80
        Nordin P, Orear J, Reed L, Rosenfeld A H, Solmitz F T, Taft H T and Tripp
        R D 1958 *Phys. Rev. Lett.* **1** 380–2
[183]   Gell-Mann M 1962 *Phys. Rev.* **125** 1067–84; 1964 *Physics* **1** 63–75
[184]   Gell-Mann M 1964 *Phys. Rev. Lett.* **8** 214–5
[185]   Cabibbo N 1963 *Phys. Rev. Lett.* **10** 531–3
[186]   Bourquin M *et al* 1983 *Z. Phys.* C **21** 27–36
        Leutwyler H and Roos M 1984 *Z. Phys.* C **25** 91–101
        Donoghue J F, Holstein B R and Klimt S W 1987 *Phys. Rev.* D **35** 934–8
[187]   Nambu Y and Lurié D 1962 *Phys. Rev.* **125** 1429–36
[188]   Bloch F and Nordsieck A 1937 *Phys. Rev.* **52** 54–9
        Low F E 1958 *Phys. Rev.* **110** 974–7
[189]   Adler S L 1965 *Phys. Rev.* B **137** 1022–33; 1965 *Phys. Rev.* B **139** 1638–43
[190]   Adler S L 1965 *Phys. Rev. Lett.* **14** 1051–5; 1965 *Phys. Rev.* B **140** 736–47;
        1966 *Phys. Rev.* **149** 1294(E)
[191]   Weisberger W I 1965 *Phys. Rev. Lett.* **14** 1047–51; 1966 *Phys. Rev.* **143** 1302–9
[192]   Callan C G and Treiman S B 1966 *Phys. Rev. Lett.* **16** 153–7
[193]   Weinberg S 1966 *Phys. Rev. Lett.* **16** 879–83; 1966 *Phys. Rev. Lett.* **17** 616–21
[194]   Adler S L and Dashen R F 1968 *Current Algebras* (New York: Benjamin)
        Gasiorowicz S and Geffen D A 1969 *Rev. Mod. Phys.* **41** 531–73
        Lee B W 1972 *Chiral Dynamics* (New York: Gordon and Breach)
[195]   Wheeler J A 1937 *Phys. Rev.* **52** 1107–27
[196]   Heisenberg W 1943 *Z. Phys.* **120** 513–38, 673–702; 1944 *Z. Phys.* **123** 93–112
[197]   Rechenberg H in reference [4] ch 39
[198]   Tomonaga S-I 1947 *J. Phys. Soc. Japan* **2** 151–71; 1948 *J. Phys. Soc. Japan* **3** 93–
        105 (reprinted in Miyazima T (ed) 1971–6 *Scientific Papers of Tomonaga*
        vol 2 (Tokyo: Misuzu Shobo) pp 1–48)
[199]   Dicke R H 1948 *Principles of Microwave Circuits* vol 8, ed C G Montgomery
        (New York: McGraw-Hill) ch 5 pp 130–161
[200]   Stückelberg E C G 1943 *Helv. Phys. Acta* **16** 427–8; 1944 *Helv. Phys. Acta* **17**
        3–26
[201]   Smith P H 1939 *Electronics* **12** 29–31; 1944 *Electronics* **17** 130–133, 318–325
[202]   Møller C 1945 *Kong. Danske Vid. Selsk., Matt-fys. Medd.* **23** no 1 pp 1–48;
        1946 *Kong. Danske Vid. Selsk., Matt-fys. Medd.* **22** 19
[203]   See, for example, Zachariasen F 1964 *Strong Interactions and High Energy
        Physics: Scottish Universities' Summer School 1963* ed R G Moorhouse
        (Edinburgh: Oliver and Boyd) pp 371–409
        Chew G F 1968 *Science* **161** 762–5
[204]   Toll J S 1952 *PhD Thesis* Princeton University (unpublished)
        Gell-Mann M, Goldberger M L and Thirring W E 1954 *Phys. Rev.* **95** 1612–
        27
        Goldberger M L, Miyazawa H and Oehme R 1955 *Phys. Rev.* **99** 986–8
        Goldberger M L 1955 *Phys. Rev.* **97** 508–10; *Phys. Rev.* **99** 979–85
        Goldberger M L, Nambu Y and Oehme 1956 *Ann. Phys., NY* **2** 226–82
        Bremermann H J, Oehme R and Taylor J G 1958 *Phys. Rev.* **109** 2178–90
        Bogoliubov N N, Medvedev B V and Polivanov M V 1958 *Voprosy Teorii
        Dispersionnykh Sootnoshenii* (Moscow: Fizmatgiz)
[205]   Anderson H L, Davidon W C and Kruse U E 1955 *Phys. Rev.* **100** 339–43
        Foley K J, Jones R S, Lindenbaum S J, Love W A, Ozaki S, Platner E D,
        Quarles C A and Willen E H 1967 *Phys. Rev. Lett.* **19** 193–8, 622(E); 1969
        Phys. Rev. **181** 1775–93
[206]   Mandelstam S 1958 *Phys. Rev.* **112** 1344–60; 1959 *Phys. Rev.* **115** 1741–51,
        1752–62

[207]   Chew G F and Low F E 1956 *Phys. Rev.* **101** 1571–9, 1579–87
[208]   Chew G F and Low F E 1959 *Phys. Rev.* **113** 1640–8
[209]   See, for example, Brown L S and Goble R L 1968 *Phys. Rev. Lett.* **20** 346–9;
        1971 *Phys. Rev.* D **4** 723–5
        Basdevant J L and Lee B W 1970 *Phys. Rev.* D **2** 1680–1701
[210]   Regge T 1959 *Nuovo Cimento* **14** 951–76; 1960 *Nuovo Cimento* **18** 947–56
[211]   Froissart M 1961 *Phys. Rev.* **123** 1053–7
[212]   Chew G F and Frautschi S C 1961 *Phys. Rev. Lett.* **7** 394–7
        Blankenbecler R and Goldberger M L 1962 *Phys. Rev.* **126** 766–86
[213]   Pomeranchuk I Ya 1958 *Zh. Eksp. Teor. Fiz.* **34** 725–8 (Engl. Transl. 1958
        Sov. Phys.–JETP **7** 499–501)
[214]   Oehme R 1964 *Strong Interactions and High Energy Physics: Scottish
        Universities' Summer School 1963* ed R G Moorhouse (Edinburgh: Oliver
        and Boyd) pp 129–222 and references [20–23] therein
[215]   Dolen R, Horn D and Schmid C 1967 *Phys. Rev. Lett.* **19** 402–7; 1968
        Phys. Rev. **166** 1768–81
[216]   Freund P G O 1969 *Phys. Rev. Lett.* **20** 235–7
        Harari H 1968 *Phys. Rev. Lett.* **20** 1395–8
[217]   Harari H 1969 *Phys. Rev. Lett.* **22** 562–5
        Rosner J L 1969 *Phys. Rev. Lett.* **23** 689–92
[218]   Ademollo M, Rubinstein H R, Veneziano G and Virasoro M A 1967
        Phys. Rev. Lett. **19** 1402–5; 1968 *Phys. Lett.* **27B** 99–102; 1968 *Phys. Rev.*
        176 1904–25
        See also Mandelstam S 1968 *Phys. Rev.* **166** 1539–52
[219]   Veneziano G 1968 *Nuovo Cimento* **57A** 190–7
[220]   Lovelace C 1968 *Phys. Lett.* **28B** 264–8
        Shapiro J A 1969 *Phys. Rev.* **179** 1345–53
[221]   Nambu Y 1970 *Symmetries and Quark Models, Int. Conf. on Symmetries and
        Quark Models (Wayne State University, 18–20 June, 1969)* ed R Chand
        (New York: Gordon and Breach) pp 269–78
        Susskind L 1969 *Phys. Rev. Lett.* **23** 545–7; 1970 *Phys. Rev.* D **1** 1182–6
        Fubini S, Gordon D and Veneziano G 1969 *Phys. Lett.* **29B** 679–82
[222]   Ramond P 1971 *Phys. Rev.* D **3** 2415–8
[223]   Scherk J and Schwarz J H 1974 *Phys. Lett.* **52B** 347–50
[224]   Green M B and Schwarz J H 1984 *Nucl. Phys.* B **243** 475–536; 1984 *Phys. Lett.*
        149B 117–122
[225]   Sakata S 1956 *Prog. Theor. Phys.* **16** 686–8
[226]   Goldberg H and Ne'eman Y 1963 *Nuovo Cimento* **27** 1–5
[227]   Gell-Mann M 1987 *Symmetries in Physics (1600–1980), Proc. First Int.
        Meeting on the History of Scientific Ideas (Sant Feliu de Guíxols, Catalonia,
        Spain, 20–26 September, 1983)* ed M G Doncel *et al* (Barcelona: Servei de
        Publicacions, Universitat Autònoma de Barcelona) pp 473–97
[228]   Joyce J 1939 *Finnegans Wake* (New York: Viking) p 383
[229]   Zweig G 1964 CERN reports 8182/TH 401 and 8419/TH 412
        (unpublished); second paper reprinted in Lichtenberg D B and
        Rosen S P (ed) 1980 *Devlopments in the Quark Theory of Hadrons* vol 1
        (Nonantum, MA: Hadronic) p 22
[230]   Gell-Mann M in the first of references [183]
        Sakurai J J 1962 *Phys. Rev. Lett.* **9** 472–5
[231]   Bertanza L *et al* 1962 *Phys. Rev. Lett.* **9** 180–3
        Schlein P, Slater W E, Smith L T, Stork D H and Ticho H K 1963
        Phys. Rev. Lett. **10** 368–71
        Connolly P L *et al* 1963 *Phys. Rev. Lett.* **10** 371–6
[232]   Lee T D and Yang C N 1956 *Nuovo Cimento* **3** 749–53

Goldhaber M, Lee T D and Yang C N 1958 *Phys. Rev.* **112** 1796–8
[233]   Dalitz R H 1967 *Proc. XIII Int. Conf. on High Energy Physics* (Berkeley, CA: University of California Press) pp 215–36
Zel'dovich Ya B and Sakharov A D 1966 *Yad. Fiz.* **4** 395–406 (Engl. Transl. 1967 *Sov. J. Nucl. Phys.* **4** 283–90)
Lipkin H J 1973 *Phys. Rep.* **8** 173–268 and references therein
Sakharov A D 1975 *Pis. Zh. Eksp. Teor. Fiz.* **21** 554–7 (Engl. Transl. *JETP Lett.* **21** 258–9)
De Rújula A, Georgi H and Glashow S L 1976 *Phys. Rev.* D **12** 147–62
Lipkin H J in reference [1]
[234]   Gürsey F and Radicati L 1964 *Phys. Rev. Lett.* **13** 173–5
Pais A 1964 *Phys. Rev. Lett.* **13** 175–7
Gürsey F, Pais A and Radicati L 1964 *Phys. Rev. Lett.* **13** 299–301
Bég M A B, Lee B W and Pais A 1964 *Phys. Rev. Lett.* **13** 514–7
Sakita B 1964 *Phys. Rev.* **136** B1756–60; 1964 *Phys. Rev. Lett.* **13** 643–6
[235]   Reference [2] p 528
[236]   Green H S 1953 *Phys. Rev.* **90** 270–3
Greenberg O W and Messiah A M L 1965 *Phys. Rev.* B **138** B1155–67
See also Gentile G 1940 *Nuovo Cimento* **17** 493–7
[237]   Greenberg O W 1964 *Phys. Rev. Lett.* **13** 598–602
[238]   Han M Y and Nambu Y 1965 *Phys. Rev.* **139** B1006–10
[239]   Sachs R G and Wali K C in reference [4] ch 8 pp 144–6
[240]   Riordan M 1987 *The Hunting of the Quark* (New York: Simon and Schuster)
[241]   Bloom E D *et al* 1969 *Phys. Rev. Lett.* **23** 931–4
Breidenbach M, Friedman J I, Kendall H W, Bloom E D, Coward D H, DeStaebler H, Drees J, Mo L W and Taylor R E 1969 *Phys. Rev. Lett.* **23** 935–8
[242]   Bjorken J D, Friedman J I, Kendall H W, Bloom E D, Coward D H, DeStaebler H, Drees J, Mo L W and Taylor R E 1966 *Phys. Rev.* **148** 1467–78; 1967 *Phys. Rev.* **160** 1582(E); see also 1966 *Phys. Rev. Lett.* **16** 408; 1967 *Phys. Rev.* **163** 1767–9; 1969 *Phys. Rev.* **179** 1547–53
[243]   Benvenuti A *et al* 1973 *Phys. Rev. Lett.* **30** 1084–7
[244]   Barish B C *et al* 1973 *Phys. Rev. Lett.* **31** 565–8
[245]   Budagov I *et al* 1969 *Phys. Lett.* **30B** 364–8
[246]   Perkins D H 1972 *Proc. XVI Int. Conf. on High Energy Physics (Chicago and Batavia, IL, September 6–13, 1972)* vol 4, ed J D Jackson *et al* (Batavia, IL: Fermilab) pp 189–247
[247]   Feynman R P 1969 *Phys. Rev. Lett.* **23** 1415–7; 1972 *Photon–Hadron Interactions* (Reading, MA: Benjamin)
[248]   Callan C G and Gross D J 1969 *Phys. Rev. Lett.* **22** 156–9
[249]   Bjorken J D and Paschos E A 1969 *Phys. Rev.* **185** 1975–82
[250]   Drell S D and Yan T-M 1970 *Phys. Rev. Lett.* **25** 316–20, 902(E)
[251]   Berman S M, Bjorken J D and Kogut J 1971 *Phys. Rev.* D **4** 3388–3418
[252]   Büsser F W *et al* 1973 *Phys. Lett.* **46B** 471–6
Cronin J W, Frisch H J, Shochet M J, Boymond J P, Piroué P A and Sumner R L 1973 *Phys. Rev. Lett.* **31** 1426–9; 1975 *Phys. Rev.* D **11** 3105–23
Appel J A *et al* 1974 *Phys. Rev. Lett.* **33** 719–22
Albrow M G *et al* 1978 *Nucl. Phys.* B **135** 461–85; *Nucl. Phys.* B **145** 305–48
[253]   See, for example, Richter B in reference [1]
[254]   Litke A *et al* 1973 *Phys. Rev. Lett.* **30** 1189–92, 1349(E)
Tarnopolsky G, Eshelman J, Law M E, Leong J, Newman H, Little R, Strauch K and Wilson R 1974 *Phys. Rev. Lett.* **32** 432–5
[255]   Gilman F J 1976 *Proc. 1975 Int. Symp. on Lepton and Photon Interactions (Stanford University, August 21–27, 1975)* ed W T Kirk (Stanford, CA:

Stanford Linear Accelerator Center) pp 131–54

Harari H 1976 *Proc. 1975 Int. Symp. on Lepton and Photon Interactions (Stanford University, August 21–27, 1975)* ed W T Kirk (Stanford, CA: Stanford Linear Accelerator Center) pp 317–53

Harari H in reference [1]

[256] Jones L W 1977 *Rev. Mod. Phys.* **49** 717–52

[257] McCusker C B A and Cairns I 1969 *Phys. Rev. Lett.* **23** 658–9

Cairns I, McCusker C B A, Peak L S and Woolcott R L S 1969 *Phys. Rev.* **186** 1394–1400

[258] Rank D M 1968 *Phys. Rev.* **176** 1635–43

[259] Stevens C M, Schiffer J P and Chupka W 1976 *Phys. Rev.* D **14** 716–27

[260] LaRue G S, Fairbank W M and Hebard A F 1977 *Phys. Rev. Lett.* **38** 1011–14

[261] Marinelli M, Gallinaro G and Morpurgo G 1981 *Nucl. Instrum. Methods* **185** 129–40

Morpurgo G 1987 *Fundamental Symmetries: Proc. Int. School of Physics with Low-Energy Antiprotons (Erice, Italy, 27 September–3 October, 1986)* ed P Bloch *et al* (New York: Plenum) pp 131–159

[262] Sachs R G 1948 *Phys. Rev.* **74** 433–41 (see especially eqution (24))

[263] Yang C N 1974 *Phys. Rev. Lett.* **33** 445–7

[264] Aharonov Y and Bohm D 1959 *Phys. Rev.* **115** 485–91

[265] Glashow S L 1961 *Nucl. Phys.* **22** 579–88

[266] Schwinger J 1962 *Phys. Rev.* **125** 397–8

Anderson P W 1963 *Phys. Rev.* **130** 439–42

Higgs P W 1964 *Phys. Lett.* **12** 132–3; 1964 *Phys. Rev. Lett.* **13** 508–9; 1966 *Phys. Rev.* **145** 1156–63

Englert F and Brout R 1964 *Phys. Rev. Lett.* **13** 321–3

Guralnik G S, Hagen C R and Kibble T W B 1965 *Phys. Rev. Lett.* **13** 585–7

Kibble T W B 1967 *Phys. Rev.* **155** 1554–61

[267] Weinberg S 1967 *Phys. Rev. Lett.* **19** 1264–6

[268] Salam A 1968 *Proc. of the Eighth Nobel Symp.* ed N Svartholm (Stockholm: Almqvist and Wiksell/New York: Wiley) pp 367–77

See also Salam A and Ward J C 1964 *Phys. Lett.* **13** 168–71

[269] 't Hooft G 1971 *Nucl. Phys.* B **33** 173–99; 1971 *Nucl. Phys.* B **35** 167–88

't Hooft G and Veltman M 1972 *Nucl. Phys.* B **44** 189–213

[270] Veltman M J G in reference [1]

't Hooft G in reference [1]

Veltman M J G 1994 *Neutral Currents: Twenty Years Later (Paris, 6–9 July, 1993)* ed U Nguyen-Khac and A M Lutz (River Edge, NJ: World Scientific)

[271] Lee B W 1972 *Phys. Rev.* D **5** 823–35

Lee B W and Zinn-Justin J 1972 *Phys. Rev.* D **5** 3121–37, 3137–55, 3155–60

[272] Weinberg S 1971 *Phys. Rev. Lett.* **27** 1688–91

[273] 't Hooft G 1971 *Phys. Lett.* **37B** 195–6

Weinberg S 1971 *Phys. Rev.* D **5** 1412–7

[274] Georgi H and Glashow S L 1972 *Phys. Rev. Lett.* **28** 1494–7

[275] Galison P 1983 *Rev. Mod. Phys.* **55** 477–507; 1987 *How Experiments End* (Chicago: University of Chicago Press); 1994 *Discovery of Weak Neutral Currents: The Weak Interaction Before and After (AIP Conf. Proc. 300)* ed A K Mann and D B Cline (New York: AIP) pp 244–86

[276] Galison P in the second of references [275] p 208

[277] Fry W F and Haidt D 1973 CERN-TCL, Technical memorandum, 22 May, as quoted in reference [275]

Haidt D 1994 *Discovery of Weak Neutral Currents: The Weak Interaction Before and After (AIP Conf. Proc. 300)* ed A K Mann and D B Cline

(New York: AIP) pp 187–206

[278]   Hasert F J *et al* 1973 *Phys. Lett.* **46B** 138–40; 1974 *Nucl. Phys.* B **73** 1–22
[279]   Hasert F J *et al* 1973 *Phys. Lett.* **46B** 121–4
[280]   Benvenuti A *et al* 1974 *Phys. Rev. Lett.* **32** 800–3
        Aubert B *et al* 1974 *Phys. Rev. Lett.* **32** 1454–60
[281]   Cline D B 1994 *Discovery of Weak Neutral Currents: The Weak Interaction Before and After (AIP Conf. Proc. 300)* ed A K Mann and D B Cline (New York: AIP) pp 175–86
        Mann A K 1994 *Discovery of Weak Neutral Currents: The Weak Interaction Before and After (AIP Conf. Proc. 300)* ed A K Mann and D B Cline (New York: AIP) pp 207–243
[282]   Rosner J L 1992 *IV Mexican School of Particles and Fields (Oaxtepec, Mexico, 3–14 December, 1990)* ed M J L Lucio and A Zepeda (Singapore: World Scientific) pp 355–405
        The data are from Bogert D *et al* (FMM Collaboration) 1985 *Phys. Rev. Lett.* **55** 1969–72.
        Allaby J V *et al* (CHARM II Collaboration) 1987 *Z. Phys.* C **36** 611–28
        Blondel A *et al* (CDHSW Collaboration) 1990 *Z. Phys.* C **45** 361–79
        Reutens P *et al* (CCFRR Collaboration) 1990 *Z. Phys.* C **45** 539–50
[283]   Lee W *et al* 1976 *Phys. Rev. Lett.* **37** 186–9
[284]   Prescott C Y in reference [1]
        Prescott C Y *et al* 1978 *Phys. Lett.* **77B** 347–52; 1979 *Phys. Lett.* **84B** 524–8
[285]   Baltay C 1979 *Proc. 19th Int. Conf. on High Energy Physics (Tokyo, 1978)* ed S Homma *et al* (Tokyo: Physical Society of Japan) pp 882–903
        Weinberg S 1979 *Proc. 19th Int. Conf. on High Energy Physics (Tokyo, 1978)* ed S Homma *et al* (Tokyo: Physical Society of Japan) pp 907–18
[286]   Zel'dovich Ya B 1959 *Zh. Eksp. Teor. Fiz.* **36** 964–6 (Engl. Transl. 1959 *Sov. Phys.–JETP* **9** 682–3)
        Bouchiat M A and Bouchiat C 1974 *Phys. Lett.* **48B** 111–4; 1974 *J. Physique* **35** 899–927; 1975 *J. Physique* **36** 493–509
[287]   Commins E D and Bucksbaum P H 1980 *Ann. Rev. Nucl. Part. Sci.* **30** 1–52
        Fortson E N and Lewis L L 1984 *Phys. Rep.* **113** 289–344
        Khriplovich I B 1991 *Parity Nonconservation in Atomic Phenomena* (Philadelphia: Gordon and Breach)
        Stacey D N 1992 *Phys. Scr.* **T40** 15–22
        Sandars P G H 1993 *Phys. Scr.* **T46** 16–21
[288]   Drell P S and Commins E D 1984 *Phys. Rev. Lett.* **53** 968–71; 1985 *Phys. Rev.* A **32** 2196–2210
        Tanner C E and Commins E D 1986 *Phys. Rev. Lett.* **56** 332–5
        Bouchiat M A, Guéna J, Pottier L and Hunter L 1986 *J. Physique* **47** 1709–30
        Noecker M C, Masterson B P and Wieman C E 1988 *Phys. Rev. Lett.* **61** 310–3
        Macpherson M J D, Zetie K P, Warrington R B, Stacey D N and Hoare J P 1991 *Phys. Rev. Lett.* **67** 2784–7
        Wolfenden T M, Baird P E G and Sandars P G H 1991 *Europhys. Lett.* **15** 731–6
        Meekhof D M, Vetter P, Majumder P K, Lamoreaux S K and Fortson E N 1993 *Phys. Rev. Lett.* **71** 3442–5
[289]   Blundell S A, Johnson W R and Sapirstein J 1990 *Phys. Rev. Lett.* **65** 1411–4
        Blundell S A, Sapirstein J and Johnson W R 1992 *Phys. Rev.* D **45** 1602–23
        Dzuba V A, Flambaum V V and Sushkov O P 1989 *Phys. Lett.* **141A** 147–53
[290]   Sandars P G H 1990 *J. Phys. B: At. Mol. Phys.* **23** L655–8
        Marciano W J and Rosner J L 1990 *Phys. Rev. Lett.* **65** 2963–6; 1992 *Phys. Rev. Lett.* **68** 898(E)

Peskin M E and Takeuchi T 1992 *Phys. Rev.* D **46** 381–409

[291]   Kim J E, Langacker P, Levine M and Williams H H 1981 *Rev. Mod. Phys.* **53** 211–52

Amaldi U, Böhm A, Durkin L S, Langacker P, Mann A K, Marciano W J, Sirlin A and Williams H H 1987 *Phys. Rev.* D **36** 1385–1407

Langacker P, Luo M and Mann A K 1992 *Rev. Mod. Phys.* **64** 87–192

[292]   Bjorken B J and Glashow S L 1964 *Phys. Lett.* **11** 255–7

[293]   Hara Y 1964 *Phys. Rev.* **134** B701–4

[294]   Maki Z and Ohnuki Y 1964 *Prog. Theor. Phys.* **32** 144–58

[295]   Glashow S L, Iliopoulos J and Maiani L 1970 *Phys. Rev.* D **2** 1285–92

[296]   Gaillard M K and Lee B W 1974 *Phys. Rev.* D **10** 897–916

[297]   Carlson C E and Freund P G O 1972 *Phys. Lett.* **39B** 349–52

Snow G 1973 *Nucl. Phys.* B **55** 445–54

Gaillard M K, Lee B W and Rosner J L 1975 *Rev. Mod. Phys.* **47** 277–310

[298]   Bouchiat C, Iliopoulos J and Meyer P 1972 *Phys. Lett.* **38B** 519–23

Georgi H and Glashow S L 1972 *Phys. Rev.* D **6** 429–31

Gross D J and Jackiw R 7 *Phys. Rev.* D **4** 7–93

[299]   Niu K, Mikumo E and Maeda Y 1971 *Prog. Theor. Phys.* **46** 1644–6

[300]   Kobayashi M and Maskawa T 1973 *Prog. Theor. Phys.* **49** 652–7

Kobayashi T in reference [1]

[301]   Christenson J H, Hicks G S, Lederman L M, Limon P J, Pope B G and Zavattini E 1970 *Phys. Rev. Lett.* **25** 1523–6; 1973 *Phys. Rev.* D **8** 2016–34

Lederman L M in reference [1]

[302]   Aubert J J *et al* 1974 *Phys. Rev. Lett.* **33** 1404–6

[303]   Augustin J-E *et al* 1974 *Phys. Rev. Lett.* **33** 1406–8

[304]   Abrams G S *et al* 1974 *Phys. Rev. Lett.* **33** 1453–5

[305]   Bacci C *et al* 1974 *Phys. Rev. Lett.* **33** 1408–10

Braunschweig W *et al* 1974 *Phys. Lett.* **53B** 393–6

[306]   Appelquist T and Politzer H D 1975 *Phys. Rev. Lett.* **34** 43–5

De Rújula A and Glashow S L 1975 *Phys. Rev. Lett.* **34** 46–9

Borchardt S, Mathur V S and Okubo S, 1975 *Phys. Rev. Lett.* **34** 38–40

Callan C G, Kingsley R L, Treiman S B, Wilczek F and Zee A 1975 *Phys. Rev. Lett.* **34** 52–6

Appelquist T, De Rújula A, Politzer H D and Glashow S L 1975 *Phys. Rev. Lett.* **34** 365–9

Eichten E, Gottfried K, Kinoshita T, Lane K D and Yan T-M 1975 *Phys. Rev. Lett.* **34** 369–72

Gaillard M K, Lee B W and Rosner J L 1975 *Rev. Mod. Phys.* **47** 277–310

[307]   Goldhaber A S and Goldhaber M 1975 *Phys. Rev. Lett.* **34** 36–7

Schwinger J 1975 *Phys. Rev. Lett.* **34** 37–8

Barnett R M 1975 *Phys. Rev. Lett.* **34** 41–3

Nieh H T, Wu T T and Yang C N 1975 *Phys. Rev. Lett.* **34** 49–52

Sakurai J J 1975 *Phys. Rev. Lett.* **34** 56–8

[308]   Braunschweig W *et al* 1975 *Phys. Lett.* **57B** 407–12

Feldman G J *et al* 1975 *Phys. Rev. Lett.* **35** 821–4

Tanenbaum W *et al* 1975 *Phys. Rev. Lett.* **35** 1323–6; 1978 *Phys. Rev.* D **17** 1731–49 and references therein

[309]   Boyarski A M *et al* 1975 *Phys. Rev. Lett.* **35** 196–9

[310]   Perl M L *et al* 1975 *Phys. Rev. Lett.* **35** 1489–92; 1976 *Phys. Lett.* **63B** 466–70; 1977 *Phys. Lett.* **70B** 487–90

[311]   De Rújula A, Georgi H and Glashow S L 1976 *Phys. Rev.* D **12** 147–62

[312]   Cazzoli E G, Cnops A M, Connolly P L, Louttit R I, Murtagh M J, Palmer R B, Samios N P, Tso T T and Williams H H 1975 *Phys. Rev. Lett.* **34** 1125–8

[313] Benvenuti A *et al* 1975 *Phys. Rev. Lett.* **34** 419–22
Blietschau J *et al* 1976 *Phys. Lett.* **60B** 207–10
von Krogh J *et al* 1976 *Phys. Rev. Lett.* **36** 710–3
Barish B C *et al* 1976 *Phys. Rev. Lett.* **36** 939–41

[314] See, for example, Boymond J P, Mermod R, Piroué P A, Sumner R L, Cronin J W, Frisch H J and Shochet M J 1974 *Phys. Rev. Lett.* **33** 112–5
Appel J A *et al* 1974 *Phys. Rev. Lett.* **33** 722–5

[315] Goldhaber G *et al* 1976 *Phys. Rev. Lett.* **37** 255–9
Peruzzi I *et al* 1976 *Phys. Rev. Lett.* **37** 569–71

[316] Glashow S L 1974 *Experimental Meson Spectroscopy—1974* ed D A Garelick (New York: AIP) pp 387–392

[317] Riordan M in reference [240] p 321

[318] Anjos J C *et al* 1989 *Phys. Rev. Lett.* **62** 513–6

[319] Eichten E and Quigg C 1994 *Phys. Rev. D* **49** 5845–56

[320] Lederman L M and Pope B G 1971 *Phys. Rev. Lett.* **27** 765–8

[321] van der Meer S 1985 *Rev. Mod. Phys.* **57** 689–97

[322] Rubbia C 1985 *Rev. Mod. Phys.* **57** 699–722

[323] Arnison G *et al* 1983 *Phys. Lett.* **122B** 103–116; 1983 *Phys. Lett.* **129B** 273–82
Banner M *et al* 1983 *Phys. Lett.* **122B** 476–85

[324] Arnison G *et al* 1983 *Phys. Lett.* **126B** 398–410
Bagnaia P *et al* 1983 *Phys. Lett.* **129B** 130–40

[325] Abe F *et al* (CDF Collaboration) 1989 *Phys. Rev. Lett.* **63** 720–3

[326] Abe F *et al* (CDF Collaboration) 1993; D0 Collaboration 1993, as reported by Swartz M 1994 *Lepton and Photon Interactions: XVI Int. Symp. (Ithaca, NY, August, 1993) (AIP Conf. Proc. 302)* ed P Drell and D Rubin (New York: AIP) pp 381–424

[327] Abrams G S *et al* (Mark II Collaboration) 1989 *Phys. Rev. Lett.* **63** 724–7

[328] Abe K *et al* 1993 *Phys. Rev. Lett.* **70** 2515–20

[329] Seeman J T 1991 *Ann. Rev. Nucl. Part. Sci.* **41** 393

[330] Schopper H 1985 *Proc. Int. Symp. on Lepton and Photon Interactions at High Energy (Kyoto, August 19–24, 1985)* ed M Konuma and K Takahashi (Kyoto: Kyoto University) p 769

[331] Swartz M 1994 *Lepton and Photon Interactions: XVI Int. Symp. (Ithaca, NY, August, 1993) (AIP Conf. Proc. 302)* ed P Drell and D Rubin (New York: AIP) pp 381–424

[332] Abe F *et al* (CDF Collaboration) 1994 *Phys. Rev. Lett.* **73** 225–31; 1994 *Phys. Rev. D* **50** 2966–3026; 1995 *Phys. Rev. Lett.* **74** 2626–31
Abachi S *et al* (D0 collaboration) 1995 *Phys. Rev. Lett.* **74** 2632–7

[333] Nambu Y 1966 *Preludes in Theoretical Physics in Honor of V F Weisskopf* ed A De-Shalit A *et al* (Amsterdam: North-Holland/New York: Wiley) pp 133–42

[334] Adler S L 1969 *Phys. Rev.* **177** 2426–38
Bell J S and Jackiw R 1969 *Nuovo Cimento* **60A** 47–61
Okubo 1970 *Symmetries and Quark Models, Int. Conf. on Symmetries and Quark Models (Wayne State University, 18–20 June, 1969)* ed R Chand (New York: Gordon and Breach) pp 59–79

[335] Fritzsch H and Gell-Mann M 1972 *Proc. XVI Int. Conf. on High Energy Physics (Chicago and Batavia, IL, September 6–13, 1972)* vol 2, ed J D Jackson *et al* (Batavia, IL: Fermilab) pp 135–65
See also Gell-Mann M 1972 *Acta Phys. Austriaca Suppl.* **IX** 733–61
Bardeen W A, Fritzsch H and Gell-Mann M 1973 *Scale and Conformal Symmetry in Hadron Physics* ed R Gatto (New York: Wiley) p 139

[336] Khriplovich I B 1969 *Yad. Fiz.* **10** 409–24 (Engl. Transl. 1970 *Sov. J. Nucl. Phys.* **10** 235–42)

[337]   't Hooft G in reference [1]
        Gross D in reference [1]
[338]   Gross D J and Wilczek F 1973 *Phys. Rev. Lett.* **30** 1343–6; 1973 *Phys. Rev.*
        D **8** 3633–52; 1974 *Phys. Rev.* D **9** 980–93
        Politzer H D 1973 *Phys. Rev. Lett.* **30** 1346–9; 1974 *Phys. Rep.* **14C** 129–80
[339]   Stückelberg E C G and Petermann A 1953 *Helv. Phys. Acta* **26** 499–520
        Gell-Mann M and Low F E 1954 *Phys. Rev.* **95** 1300–12
        Bogoliubov N N and Shirkov D V 1955 *Dokl. Akad. Nauk SSSR* **103** 203–6;
        1956 *Nuovo Cimento* **3** 845–63
        Wilson K 1969 *Phys. Rev.* **179** 1499–1512; 1970 *Phys. Rev.* D **2** 1438–72; 1971
        Phys. Rev. D **3** 1818–46
        Callan C G Jr 1970 *Phys. Rev.* D **2** 1541–7
        Symanzik K 1970 *Commun. Math. Phys.* **18** 227–46
        Wilson K G and Kogut J 1974 *Phys. Rep.* **12C** 75–199
[340]   Nambu Y 1974 *Phys. Rev.* D **10** 4262–8
[341]   Altarelli G and Parisi G 1977 *Nucl. Phys.* B **126** 298–318
[342]   Bethke S 1993 *Proc. XXVI Int. Conf. on High Energy Physics (Dallas, TX,*
        August 6–12, 1992) ed J R Sanford (New York: AIP) pp 81–113
        Voss R 1994 *Lepton and Photon Interactions: XVI Int. Symp. (Ithaca, NY,*
        August, 1993) (AIP Conf. Proc. 302) ed P Drell and D Rubin (New York:
        AIP) pp 144–171
[343]   Newman H *et al* (Mark J Collaboration) 1979 *Proc. Int. Symp. on Lepton*
        and Photon Interactions at High Energies (Fermilab, August 23–29, 1979)
        ed T B W Kirk and H D I Abarbanel (Batavia, IL: Fermilab) pp 3–18
        Berger Ch *et al* (PLUTO Collaboration) 1979 *Proc. Int. Symp. on Lepton and*
        Photon Interactions at High Energies (Fermilab, August 23–29, 1979) ed T
        B W Kirk and H D I Abarbanel (Batavia, IL: Fermilab) pp 19–33
        Wolf G *et al* (TASSO Collaboration) 1979 *Proc. Int. Symp. on Lepton and*
        Photon Interactions at High Energies (Fermilab, August 23–29, 1979) ed T
        B W Kirk and H D I Abarbanel (Batavia, IL: Fermilab) pp 34–51
        Orito S *et al* (JADE Collaboration) 1979 *Proc. Int. Symp. on Lepton and*
        Photon Interactions at High Energies (Fermilab, August 23–29, 1979) ed T
        B W Kirk and H D I Abarbanel (Batavia, IL: Fermilab) pp 52–69
[344]   Hanson G J *et al* 1975 *Phys. Rev. Lett.* **35** 1609–12
[345]   Selove W 1979 *Proc. 19th Int. Conf. on High Energy Physics (Tokyo, 1978)* ed
        S Homma *et al* (Tokyo: Physical Society of Japan) pp 165–70
        McCarthy R L 1979 *Proc. 19th Int. Conf. on High Energy Physics (Tokyo,*
        1978) ed S Homma *et al* (Tokyo: Physical Society of Japan) pp 170–1
        Cool R L 1979 *Proc. 19th Int. Conf. on High Energy Physics (Tokyo, 1978)* ed
        S Homma *et al* (Tokyo: Physical Society of Japan) pp 172–3
        Clark A G 1979 *Proc. 19th Int. Conf. on High Energy Physics (Tokyo, 1978)*
        ed S Homma *et al* (Tokyo: Physical Society of Japan) pp 174–6
        Hansen K H 1979 *Proc. 19th Int. Conf. on High Energy Physics (Tokyo, 1978)*
        ed S Homma *et al* (Tokyo: Physical Society of Japan) pp 177–81
        Nakamura K 1979 *Proc. 19th Int. Conf. on High Energy Physics (Tokyo,*
        1978) ed S Homma *et al* (Tokyo: Physical Society of Japan) pp 181–
        3; summarized by Sosnowski R 1979 *Proc. 19th Int. Conf. on High*
        Energy Physics (Tokyo, 1978) ed S Homma *et al* (Tokyo: Physical Society
        of Japan) pp 693–705.
[346]   Wu S L and Zobernig G 1979 *Z. Phys.* C **2** 107–10
[347]   Wiik B 1979 *Proc. Neutrino 79, Int. Conf. on Neutrinos, Weak Interactions,*
        and Cosmology (Bergen, June 18–22, 1979) ed A Haatuft and C Jarlskog
        (Bergen: University of Bergen) pp 113–54
        Söding P 1980 *Proc. European Physical Society Int. Conf. on High Energy*

Physics (Geneva, 27 June–4 July, 1979) vol 1, ed W S Newman (Geneva: CERN) pp 271–81

[348] Banner M *et al* (UA2 Collaboration) 1982 *Phys. Lett.* **118B** 203–10

Bagnaia P *et al* (UA2 Collaboration) 1983 *Z. Phys.* C **20** 117–34

[349] Abe F *et al* (CDF Collaboration) 1993 *Phys. Rev. Lett.* **70** 1376–80

[350] Novikov V A, Okun L B, Shifman M A, Vainshtein A I, Voloshin M B and Zakharov V I 1978 *Phys. Rep.* **41C** 1–133

[351] Gunion J F, Brodsky S J and Blankenbecler R 1972 *Phys. Rev.* D **6** 2652–8; 1973 *Phys. Rev.* D **8** 287–312

Brodsky S J and Farrar G R 1973 *Phys. Rev. Lett.* **31** 1153–6; 1975 *Phys. Rev.* D **11** 1309–30

Blankenbecler R, Brodsky S J and Gunion J F 1975 *Phys. Rev.* D **12** 3469–87

[352] Nussinov S 1975 *Phys. Rev. Lett.* **34** 1286–9

Low F E 1976 *Phys. Rev.* D **12** 163–73

[353] Block M M and Cahn R N 1985 *Rev. Mod. Phys.* **57** 563–98; 1990 *Czech J. Phys.* **40** 164–75

[354] Hikasa K *et al* (Particle Data Group) 1992 *Phys. Rev.* D **45** S1–S584

[355] Dremin I M and Quigg C 1978 *Science* **199** 937–41

[356] Field R D and Feynman R P 1978 *Nucl. Phys.* B **136** 1–76

[357] Wilson K 1974 *Phys. Rev.* D **10** 2445–59

Mackenzie P B and Kronfeld A S 1993 *Ann. Rev. Nucl. Part. Sci.* **43** 793–828

[358] Shifman M A, Vainshtein A I and Zakharov V I 1979 *Nucl. Phys.* B **147** 385–447, 448–518, 519–534

Shifman M A 1983 *Ann. Rev. Nucl. Part. Sci.* **33** 199–233

Reinders L J, Rubinstein H and Yakazi S 1985 *Phys. Rep.* **127** 1–97

[359] Alles-Borelli V, Bernardini M, Bollini D, Brunini P L, Massam T, Monari L, Palmonari F and Zichichi A 1970 *Lett. Nuovo Cimento* **4** 1156–9

Bernardini M, Bollini D, Brunini P L, Fiorentino E, Massam T, Monari L, Palmonari F, Rimondi F and Zichichi A 1973 *Nuovo Cimento* **17A** 383–9

Perl M in reference [1]; 1994 *Stanford Linear Accelerator Center Report* SLAC-PUB-6584 (to be published in *Proc. Int. Conf. on the History of Original Ideas and Basic Discoveries in Particle Physics (Erice, Sicily, 29 July–4 August, 1994)*)

[360] Tsai Y-S 1971 *Phys. Rev.* D **4** 2821–37

Thacker H B and Sakurai J J 1971 *Phys. Lett.* **36B** 103–5

[361] Augustin J-E *et al* 1975 *Phys. Rev. Lett.* **34** 764–7

[362] Alexander G *et al* (PLUTO Collaboration) 1978 *Phys. Lett.* **78B** 162–6

[363] Drell P and Patterson J R 1993 *Proc. XXVI Int. Conf. on High Energy Physics (Dallas, TX, August 6–12, 1992)* ed J R Sanford (New York: AIP) pp 3–32

Schwarz A S 1994 *Lepton and Photon Interactions: XVI Int. Symp. (Ithaca, NY, August, 1993) (AIP Conf. Proc. 302)* ed P Drell and D Rubin (New York: AIP) pp 671–94

[364] Benvenuti A *et al* 1976 *Phys. Rev. Lett.* **36** 1478–82

[365] Lederman L in reference [1]

[366] Kluberg L, Piroué P A, Sumner R L, Antreasyan D, Cronin J W, Frisch H J and Shochet M J 1976 *Phys. Rev. Lett.* **37** 1451–4

[367] Herb S W *et al* 1977 *Phys. Rev. Lett.* **39** 252–5

Innes W R *et al* 1977 *Phys. Rev. Lett.* **39** 1240–2, 1640(E)

[368] Ueno K *et al* 1979 *Phys. Rev. Lett.* **42** 486–9

Lederman L 1989 *Rev. Mod. Phys.* **61** 547–60

[369] Berger Ch *et al* (PLUTO Collaboration) 1978 *Phys. Lett.* **76B** 243–5

Darden C W *et al* (DASP Collaboration) 1978 *Phys. Lett.* **76B** 246–8

[370] Bienlein J K *et al* 1978 *Phys. Lett.* **78B** 360–3

C W Darden *et al* (DASP Collaboration) 1978 *Phys. Lett.* **78B** 364–5

[371]   Andrews D A *et al* (CLEO Collaboration) 1980 *Phys. Rev. Lett.* **45** 219–20
        Finocchiaro G *et al* (CUSB Collaboration) 1980 *Phys. Rev. Lett.* **45** 222–5
[372]   Bebek C *et al* (CLEO Collaboration) 1981 *Phys. Rev. Lett.* **46** 84–7
        Chadwick K *et al* (CLEO Collaboration) 1981 *Phys. Rev. Lett.* **46** 88–91
        Spencer L J *et al* (CUSB Collaboration) 1981 *Phys. Rev. Lett.* **47** 771–4
        Brody A *et al* (CLEO Collaboration) 1982 *Phys. Rev. Lett.* **48** 1070–4
        Giannini G *et al* (CUSB Collaboration) 1982 *Nucl. Phys.* B **206** 1–11
[373]   Behrends S *et al* (CLEO Collaboration) 1983 *Phys. Rev. Lett.* **50** 881–4
        Giles R *et al* (CLEO Collaboration) 1984 *Phys. Rev.* D **30** 2279–94
[374]   Buchmüller W and Cooper S 1988 *High Energy Electron–Positron Physics*
        ed A Ali and P Söding (Singapore: World Scientific) pp 410–87
[375]   Rosner J L 1991 *Testing the Standard Model (Proc. 1990 Theoretical Advanced
        Study Institute in Elementary Particle Physics (Boulder, CO, 3–27 June,
        1990))* ed M Cvetič and P Langacker (Singapore: World Scientific) pp
        91–224
[376]   Spencer L J *et al* (CUSB Collaboration) 1981 *Phys. Rev. Lett.* **47** 771–4
[377]   Fulton R *et al* (CLEO Collaboration) 1990 *Phys. Rev. Lett.* **64** 16–20
        Albrecht H *et al* (ARGUS Collaboration) 1990 *Phys. Lett.* **234B** 409–16; 1991
        Phys. Lett. **255B** 297–304
[378]   Venus W 1994 *Lepton and Photon Interactions: XVI Int. Symp. (Ithaca, NY,
        August, 1993) (AIP Conf. Proc. 302)* ed P Drell and D Rubin (New York:
        AIP) pp 274–91
[379]   Fernandez E *et al* (MAC Collaboration) 1983 *Phys. Rev. Lett.* **51** 1022–5
[380]   Wolfenstein L 1983 *Phys. Rev. Lett.* **51** 1945–7
[381]   Albrecht H *et al* (ARGUS Collaboration) 1987 *Phys. Lett.* **192B** 245–52
[382]   Ellis J, Gaillard M K, Nanopoulos D V and Rudaz S 1977 *Nucl. Phys.* B
        131 285–307
        Carter A B and Sanda A I 1980 *Phys. Rev. Lett.* **45** 952–4; 1981 *Phys. Rev.*
        D **23** 1567–79
        Bigi I I and Sanda A I 1981 *Nucl. Phys.* B **193** 85–108
        Winstein B and Wolfenstein L 1993 *Rev. Mod. Phys.* **65** 1113–47
[383]   Kane G L and Peskin M E 1982 *Nucl. Phys.* B **195** 29–38
[384]   Abachi S *et al* (D0 Collaboration) 1995 *Phys. Rev. Lett.* **74** 2632–7
[385]   Veltman M 1977 *Nucl. Phys.* B **123** 89–99
[386]   Livingston M S and Blewett J P 1962 *Particle Accelerators* (New York:
        McGraw-Hill)
[387]   Cockcroft J D and Walton E T S 1932 *Proc. R. Soc.* A **136** 619–30; 1932
        Proc. R. Soc. A **137** 229–42
[388]   van de Graaff R J 1931 *Phys. Rev.* **38** 1919–20
        For earlier work using pulsed electrostatic generators see Breit G and Tuve
        M A 1928 *Nature* **121** 535–6 and other authors noted in reference [2] pp
        405–6.
[389]   van de Graaff R J, Compton K T and Van Atta L C 1933 *Phys. Rev.* **43**
        149–57
        Van Atta L C, Northrop D L, Van Atta C M and van de Graaff R J 1936
        Phys. Rev. **49** 761–76
[390]   Herb R G, Parkinson D B and Kerst D W 1935 *Phys. Rev.* **48** 118–24
[391]   Tuve M S, Hafstad L R and Dahl O 1935 *Phys. Rev.* **48** 315–37
[392]   Wideröe R 1928 *Arch. Elektrotech.* **21** 387, 486 (Engl. Transl. Livingston M S
        (ed) 1966 *The Development of High-Energy Accelerators* (New York: Dover)
        pp 92–114)
[393]   Lawrence E O and Livingston M S 1931 *Phys. Rev.* **37** 1707; 1931 *Phys. Rev.*
        38 834; 1931 *Phys. Rev.* **40** 19–35
[394]   Lawrence E O, Livingston M S and White M G 1932 *Phys. Rev.* **42** 150–1

[395]   McMillan E M 1945 *Phys. Rev.* **68** 143–4
        Veksler V 1945 *J. Phys. (USSR)* **9** 153–8, reprinted in Livingston M S (ed) 1966 *The Development of High Energy Accelerators* (New York: Dover) pp 202–10
[396]   McMillan E M, Peterson J M and White R S 1949 *Science* **110** 579–83
[397]   Kerst D W 1940 *Phys. Rev.* **58** 841; 1941 *Phys. Rev.* **60** 47–53
[398]   Kerst D W and Serber R 1941 *Phys. Rev.* **60** 53–8
[399]   Kerst D W, Adams G D, Koch H W and Robinson C S 1950 *Phys. Rev.* **78** 297
[400]   Fermi E 1952 *Phys. Rev.* **86** 611
[401]   Shoemaker F C, Britton R J and Carlson B C 1952 *Phys. Rev.* **86** 582
[402]   Fitch V L and Rainwater J 1953 *Phys. Rev.* **92** 789–800
[403]   Courant E D, Livingston M S and Snyder H S 1952 *Phys. Rev.* **88** 1190–6
[404]   Christofilos N 1956 *US Patent* No 2,736,799, reprinted in Livingston M S (ed) 1966 *The Development of High-Energy Accelerators* (New York: Dover) pp 270–80
[405]   Chodorow M, Ginzton E L, Hansen W W, Kyhl R L, Neal R B and Panofsky W K H, 1955 *Rev. Sci. Instrum.* **26** 134–204
[406]   Alvarez L W, Bradner H, Franck J V, Gordon H, Gow J D, Marchall L C, Oppenheimer F, Panofsky W K H, Richman C and Woodyard J R 1955 *Rev. Sci. Instrum.* **26** 111–33
[407]   Kerst D W 1955 (unpublished). This was first discussed at a conference sponsored by the Midwest Universities Research Association, MURA, in late 1955 (Shoemaker F C, private communication).
[408]   Kerst D W, Cole F T, Crane H R, Jones L W, Laslett L J, Ohkawa T, Sessler A M, Symon K R, Terwilliger K M and Nilsen N V 1956 *Phys. Rev.* **102** 590–1
[409]   Symon K R, Kerst D W, Jones L W, Laslett L J and Terwilliger K M 1956 *Phys. Rev.* **103** 1837–59
[410]   O'Neill G K 1956 *Phys. Rev.* **102** 1418–9
[411]   O'Neill G K 1966 *Sci. Am.* **215** 107–116
[412]   Marin P 1967 *Proc. Third Int. Symp. on Electron and Photon Interactions (SLAC, 1967)* ed S M Berman (Stanford, CA: SLAC) pp 376–86
[413]   Schwarzschild B M 1980 *Phys. Today* **33** January pp 19–21
[414]   Rubbia C, McIntyre P and Cline D 1977 *Proc. Int. Neutrino Conf. (Aachen, 1976)* ed H Faissner (Braunschweig: Vieweg) pp 683–7
[415]   See, for example, Manchester W R 1992 *A World Lit Only by Fire* (Boston: Little Brown) p 294
[416]   Blackett P M S 1933 *Cambridge University Studies* ed H Wilson (London: Nicholson and Watson) pp 67–96
[417]   Rutherford E and Geiger H 1908 *Proc. R. Soc.* A **81** 141–61
[418]   Townsend J S 1901 *Phil. Mag.* **1** (Ser. 6) 198–227; 1902 *Phil. Mag.* **3** 557–76; 1903 *Phil. Mag.* **5** 389–98; 1903 *Phil. Mag.* **6** 358–61, 598–618
[419]   Rutherford E and Geiger H 1908 *Proc. R. Soc.* A **81** 162–73
[420]   Geiger H and Müller W 1928 *Phys. Z.* **29** 839–41
[421]   Neher H V 1938 (seventeenth printing in 1952) *Procedures in Experimental Physics* ed J Strong *et al* (New York: Prentice-Hall) pp 259–304
[422]   Bothe W 1929 *Z. Phys.* **59** 1–5
        Rossi B 1930 *Nature* **125** 636
[423]   Greinacher H 1926 *Z. Phys.* **36** 364–73
[424]   Wynn-Williams C E 1932 *Proc. R. Soc.* A **136** 312–24
[425]   Elmore W C and Sands M 1949 *Electronics* (New York: McGraw-Hill) p 210
[426]   Slepian J 1919 *US Patent* No 1,450,265 (April 3, 1919)
[427]   Zworykin V K, Morton G A and Mather L 1936 *Proc. IRE* **24** 351–75

[428] Shockley W and Pierce J R 1938 *Proc. IRE* **26** 321–32
[429] Zworykin V K and Rajchman J A 1939 *Proc. IRE* **27** 558–66
[430] Čerenkov P A 1934 *C. R. Acad. Sci. URSS* **2** 451–4
[431] Frank I and Tamm I 1937 *C. R. Acad. Sci. URSS* **14** 109–14
[432] Čerenkov P A 1937 *Phys. Rev.* **52** 378–9
[433] Kallman H 1947 *Nat. Tech.* July
[434] Bell P R 1948 *Phys. Rev.* **73** 1405–6
[435] Reynolds G T, Harrison F B and Salvini G 1950 *Phys. Rev.* **78** 488
[436] Hofstadter R 1948 *Phys. Rev.* **74** 100–1; reference [4] ch 7 pp 126–143
[437] Morton G A 1949 *RCA Rev.* **10** 525–53
[438] Wilkinson D H 1950 *Proc. Camb. Phil. Soc.* **46** 508–18
[439] Wilson C T R 1911 *Proc. R. Soc.* A **85** 285–8; 1912 *Proc. R. Soc.* A **87** 277–92
[440] Hoxton L G 1933–4 A continuously operating cloud chamber *Proc. Virginia Acad. Sci.* **9** 23
[441] Langsdorf A Jr 1936 *Phys. Rev.* **49** 422; 1939 *Rev. Sci. Instrum.* **10** 91–103
[442] Vollrath R E 1936 *Rev. Sci. Instrum.* **7** 409–10
[443] Shutt R P, Fowler E C, Miller D H, Thorndike A M and Fowler W B 1951 *Phys. Rev.* **84** 1247–8
[444] Hildebrand R H and Nagle D E 1953 *Phys. Rev.* **92** 517–8
[445] Keuffel J W 1948 *Phys. Rev.* **73** 531
[446] Pidd R W and Madansky L 1949 *Phys. Rev.* **75** 1175–80
[447] Cranshaw T E and DeBeer J F 1957 *Nuovo Cimento* **5** 1107–17
[448] Fukui S and Miyamoto S 1959 *Nuovo Cimento* **11** 113–5
[449] Cronin J W 1967 *Bubble and Spark Chambers* vol 1, ed R P Shutt (New York: Academic) pp 315–405
[450] Wenzel W A 1964 *Ann. Rev. Nucl. Sci.* **14** 205–38
[451] Charpak G 1993 *Rev. Mod. Phys.* **65** 591–8
[452] Freedman S J 1990 *Comments Nucl. Part. Phys.* **19** 209–20
[453] Majorana E 1937 *Nuovo Cimento* **14** 171–84
[454] Kayser B, Gibrat-Debu F and Perrier F 1989 *The Physics of Massive Neutrinos* (Teaneck, NJ: World Scientific)
       Boehm F and Vogel P 1987 *Physics of Massive Neutrinos* (Cambridge: Cambridge University Press)
       Bilenky S M and Petcov S T 1987 *Rev. Mod. Phys.* **59** 671–754; 1989 *Rev. Mod. Phys.* **61** 169(E)
[455] Kawakami H *et al* 1991 *Phys. Lett.* **256B** 105–11
       Robertson R G H, Bowles T J, Stephenson G J Jr, Wark D J, Wilkerson J F and Knapp D A 1991 *Phys. Rev. Lett.* **67** 957–60
       Decman D and Stoeffl W 1992 *Bull. Am. Phys. Soc.* **37** 1286
       Holzschuh E, Fritschi M and Kündig W 1992 *Phys. Lett.* **287B** 381–8
       Weinheimer Ch *et al* 1993 *Phys. Lett.* **300B** 210–6
[456] Bardeen J, Cooper L N, and Schrieffer J R 1957 *Phys. Rev.* **108** 1175–1204
       Anderson P W 1958 *Phys. Rev.* **112** 1900–16
[457] See, for example, Kadanoff L P 1966 *Physics* **2** 263–72; 1975 *Phys. Rev. Lett.* **34** 1005–8
       Kadanoff L P, Götze W, Hamblen D, Hecht R, Lewis E A S, Palciauskas V V, Rayl M, Swift J, Aspnes D and Kane J 1967 *Rev. Mod. Phys.* **39** 395–431
       Kadanoff L P and Houghton A 1975 *Phys. Rev.* B **11** 377–86
       Wilson K G 1971 *Phys. Rev.* B **4** 3174–83, 3184–3205; 1975 *Rev. Mod. Phys.* **47** 773–840
       Wilson K G and Kogut J 1974 *Phys. Rep.* **12C** 75–199
       Fisher M E 1974 *Rev. Mod. Phys.* **46** 597–616

[458] Friedan D, Qiu Z and Shenker S 1984 *Phys. Rev. Lett.* **52** 1575–8; 1985 *Phys. Lett.* **151B** 37–43

[459] Weinberg S 1972 *Gravitation and Cosmology: Principles and Applications of the General Theory of Relativity* (New York: Wiley)

[460] Pontecorvo B 1946 *Chalk River Laboratory Report* PD-205 (unpublished)

[461] Davis R Jr, Mann A K and Wolfenstein L 1990 *Ann. Rev. Nucl. Part. Sci.* **39** 467–506

[462] Bahcall J N 1989 *Neutrino Astrophysics* (Cambridge: Cambridge University Press)

[463] Hirata K S *et al* 1991 *Phys. Rev.* D **44** 2241–60; 1992 *Phys. Rev.* D **45** 2170(E) Totsuka Y 1992 *Rep. Prog. Phys.* **55** 377–430

[464] Abazov A I *et al* 1991 *Phys. Rev. Lett.* **67** 3332–5
Gavrin V *et al* 1993 *Proc. XXVI Int. Conf. on High Energy Physics (Dallas, TX, August 6–12, 1992)* ed J R Sanford (New York: AIP) pp 1101–10
Anselmann P *et al* 1992 *Phys. Lett.* **285B** 376–89, 390–7; 1993 *Phys. Lett.* **314B** 445–8

[465] Bludman S, Hata N, Kennedy D C and Langacker P 1993 *Phys. Rev.* D **47** 2220–33

[466] Pontecorvo B 1967 *Zh. Eksp. Teor. Fiz.* **53** 1717–25 (Engl. Transl. 1968 *Sov. Phys.–JETP* **26** 984–8)
Wolfenstein L 1978 *Phys. Rev.* D **17** 2369–74
Mihkeev S P and Smirnov A Yu 1985 *Yad. Fiz.* **42** 1441–8 (Engl. Transl. *Sov. J. Nucl. Phys.* **42** 913–7); 1986 *Nuovo Cimento* **9C** 17–26; 1987 *Usp. Fiz. Nauk* **153** 3–58 (Engl. Transl. 1987 *Sov. Phys.–Usp.* **30** 759–90)

[467] Norman E B *et al* 1992 *The Fermilab Meeting DPF 92 (Proc. 1992 Division of Particles and Fields Meeting (Fermilab, 10–14 November, 1992))* ed C H Albright *et al* (Singapore: World Scientific) pp 1450–2
Raghavan R S and Pakvasa S 1988 *Phys. Rev.* D **37** 849–57

[468] Hirata K S *et al* 1988 *Phys. Rev.* D **38** 449

[469] Hirata K *et al* 1987 *Phys. Rev. Lett.* **58** 1490–3
Bionta R M *et al* 1987 *Phys. Rev. Lett.* **58** 1494–6

[470] Colgate S A and White R H 1966 *Astrophys. J.* **143** 626–81
Arnett W D 1982 *Astrophys. J. Lett.* **263** L55–7

[471] Bahcall J and Glashow S L 1987 *Nature* **326** 476–7
Arnett W D and Rosner J L 1987 *Phys. Rev. Lett.* **58** 1906–9
Abbott L F, De Rújula A and Walker T P 1988 *Nucl. Phys.* B **299** 734–56

[472] Gamow G 1948 *Phys. Rev.* **74** 505–6; 1948 *Nature* **162** 680–2
Alpher R A and Herman R C 1948 *Nature* **162** 774–5; 1949 *Phys. Rev.* **75** 1089–95 and references therein

[473] Penzias A A and Wilson R W 1965 *Astrophys. J.* **142** 419–21

[474] Dicke R H, Peebles P J E, Roll P G and Wilkinson D T 1965 *Astrophys. J.* **142** 414–9
Dicke R H and Peebles P J E 1966 *Nature* **211** 574–5

[475] Smoot G F *et al* 1992 *Astrophys. J. Lett.* **396** L1–5

[476] Steigman G, Schramm D N and Gunn J E 1977 *Phys. Lett.* **66B** 202–4
Walker T P, Steigman G, Schramm D N, Olive K A and Kang H S 1991 *Astrophys. J.* **376** 51–69

[477] Yodh G B *First Aspen Winter Physics Conf.* ed M M Block (*Ann. NY Acad. Sci.* **461** 239–59)

[478] Weekes T C 1988 *Phys. Rep.* **160** 1–121
Weekes T C *et al* 1989 *Astrophys. J.* **342** 379–95
Punch M *et al* 1992 *Nature* **358** 477–8

[479] Alexandreas D E *et al* 1991 *Phys. Rev.* D **43** 1735–8; 1992 *Nucl. Instrum. Methods* A **311** 350–67

Cronin J W *et al* 1992 *Phys. Rev.* D **45** 4385–91
Nagano M *et al* 1992 *J. Phys. G: Nucl. Phys.* **18** 423–42
Chiba N *et al* 1992 *Nucl. Instrum. Methods* **A311** 338–49
Bird D J *et al* 1993 *Phys. Rev. Lett.* **71** 3401–4

[480] Misner C W, Thorne K S and Wheeler J A 1973 *Gravitation* (San Francisco: Freeman) p 872

[481] Hawking S 1975 *Commun. Math. Phys.* **43** 199–220; *Quantum Gravity: An Oxford Symposium, 1975* ed C J Isham *et al* (Oxford: Clarendon) pp 219–67; 1981 *Encyclopedia of Physics* ed R G Lerner and G L Trigg (Reading, MA: Addison-Wesley) pp 81–83

[482] Trimble V 1987 *Ann. Rev. Astron. Astrophys.* **25** 425–72
Primack J R, Seckel D and Sadoulet B 1988 *Ann. Rev. Nucl. Part. Sci.* **38** 751–807

[483] Peccei R D and Quinn H R 1977 *Phys. Rev. Lett.* **38** 1440–3; 1977 *Phys. Rev.* D **16** 1791–7
Weinberg S 1978 *Phys. Rev. Lett.* **40** 223–6
Wilczek F 1978 *Phys. Rev. Lett.* **40** 279–82

[484] Guth A 1980 *Phys. Rev.* D **23** 347–56
Linde A 1983 *Phys. Lett.* **129B** 177–81 and references therein

[485] Eötvös R v, Pekár D and Fekete E 1922 *Ann. Phys., Lpz* **68** 11–66

[486] Dicke R H, Roll P G and Krotkov R 1964 *Ann. Phys., NY* **26** 442–517

[487] Fischbach E, Sudarsky D, Szafer A, Talmadge C and Aronson S H 1986 *Phys. Rev. Lett.* **56** 3–6, 1427(E)

[488] Adelberger E G, Heckel B R, Stubbs C W and Rogers W F 1991 *Ann. Rev. Nucl. Part. Sci.* **41** 269–320

[489] For a recent reference see, for example, Buskulic D *et al* (ALEPH Collaboration) 1993 *Phys. Lett.* **313B** 299–311.

[490] Veltman M 1977 *Acta Phys. Pol.* **B8** 475–92; 1977 *Phys. Lett.* **70B** 253–4
Lee B W, Quigg C and Thacker H B 1977 *Phys. Rev. Lett.* **38** 883–5; 1977 *Phys. Rev.* D **16** 1519–31

[491] Eichten E, Hinchliffe I, Lane K and Quigg C 1984 *Rev. Mod. Phys.* **56** 579–707; 1986 *Rev. Mod. Phys.* **58** 1065(E)

[492] Dawson S, Gunion J F, Haber H E and Kane G L 1990 *The Higgs Hunter's Guide* (Redwood City, CA: Addison-Wesley)

[493] Wess J and Bagger J 1983 *Supersymmetry and Supergravity* (Princeton, NJ: Princeton University Press)
Freund P 1986 *Introduction to Supersymmetry* (Cambridge: Cambridge University Press)

[494] Weinberg S 1976 *Phys. Rev.* D **13** 974–96; 1979 *Phys. Rev.* D **19** 1277–80
Susskind L 1979 *Phys. Rev.* D **20** 2619–25

[495] Decman D and Stoeffl W in reference [455]

[496] Hirata K S *et al* 1992 *Phys. Lett.* **280B** 146–52
Beier E W *et al* 1992 *Phys. Lett.* **283B** 446–53
Fukuda Y *et al* 1994 *Phys. Lett.* **335B** 237–45

[497] Arik E *et al* (CHORUS Collaboration) 1991 CERN Experiment WA-95, approved September 1991, K Winter, spokesperson
Kadi-Hanifi M *et al* (NOMAD Collaboration) 1991 CERN Experiment WA-96, approved September 1991, F Vannucci, spokesperson

[498] Bernstein R H and Parke S J 1991 *Phys. Rev.* D **44** 2069–78 and references therein

[499] Pati J C and Salam A 1974 *Phys. Rev.* D **10** 275–89

[500] Georgi H and Glashow S L 1974 *Phys. Rev. Lett.* **32** 438–41

[501] Georgi H 1975 *Proc. 1974 Williamsburg DPF Meeting* ed C E Carlson (New York: AIP) pp 575–82

Fritzsch H and Minkowski P 1975 *Ann. Phys., NY* **93** 193–266

[502] Gell-Mann M, Ramond P and Slansky R 1979 *Supergravity* ed P van Nieuwenhuizen and D Z Freedman (Amsterdam: North-Holland) pp 315–21

Yanagida T 1979 *Proc. Workshop on Unified Theory and Baryon Number in the Universe* ed O Sawada and A Sugamoto (Tsukuba, Japan: National Laboratory for High Energy Physics)

[503] Amaldi U, Bohm A, Durkin L S, Langacker P, Mann A K, Marciano W J, Sirlin A and Williams H H 1987 *Phys. Rev.* D **36** 1385

Amaldi U, de Boer W and Fürstenau H 1991 *Phys. Lett.* **260B** 447–55

Langacker P and Polonsky N 1993 *Phys. Rev.* D **47** 4028–45

[504] Rosner J L 1994 *DPF 94 Proc. DPF 94 Meeting (Albuquerque, NM, August, 1994)* (Singapore: World Scientific)

[505] Seidel S *et al* 1988 *Phys. Rev. Lett.* **61** 2522–25

Hirata K S *et al* 1989 *Phys. Lett.* **220B** 308–16

[506] Totsuka Y 1992 *Rep. Prog. Phys.* **55** 377–430

[507] Preskill J P 1979 *Phys. Rev. Lett.* **43** 1365–8

[508] See, for example, Adams F C, Fatuzzo M, Freese K, Tarlé G, Watkins R and Turner M S 1993 *Phys. Rev. Lett.* **70** 2511–14

[509] Gross D J, Harvey J A, Martinec E and Rohm R 1985 *Phys. Rev. Lett.* **54** 502–5; 1985 *Nucl. Phys.* B **256** 253–84; 1986 *Nucl. Phys.* B **267** 75–124

[510] Candelas P, Horowitz G T, Strominger A and Witten E 1985 *Nucl. Phys.* B **258** 46–74

Witten E 1985 *Nucl. Phys.* B **258** 75–100

[511] Callan C G Jr, Giddings S, Harvey J A and Strominger A 1992 *Phys. Rev.* D **45** 1005–9

[512] See, for example, Derrick M *et al* (ZEUS Collaboration) 1992 *Phys. Lett.* **293B** 465

Ahmed T *et al* (H1 Collaboration) 1993 *Phys. Lett.* **299B** 374–84, 385–93

[513] Ritson D 1993 *Nature* **366** 607–10

[514] Ayres D S *et al* 1991 *Proc. 25th Int. Conf. on High Energy Physics (Singapore, August 2–8, 1990)* ed K K Phua and Y Yamaguchi (Singapore: World Scientific [South East Asia Theoretical Physics Association and the Physical Society of Japan]) pp 480–1

Thron J L 1993 *Proc. XXVI Int. Conf. on High Energy Physics (Dallas, TX, August 6–12, 1992)* ed J R Sanford (New York: AIP) pp 1232–7

[515] Calicchio M *et al* 1988 *Nucl. Instrum. Methods* A **264** 18–23

Ahlen S P *et al* 1993 *Nucl. Instrum. Methods* A **324** 337–62

[516] Norman E B *et al* 1992 *The Fermilab Meeting DPF 92 (Proc. 1992 Division of Particles and Fields Meeting (Fermilab, 10–14 November, 1992))* ed C H Albright *et al* (Singapore: World Scientific) pp 1450–2

[517] Roberts A 1992 *Rev. Mod. Phys.* **64** 259–312

[518] Barwick S *et al* 1993 *Proc. XXVI Int. Conf. on High Energy Physics (Dallas, TX, August 6–12, 1992)* ed J R Sanford (New York: AIP) pp 1250–3

[519] Sadoulet B 1990 *Proc. First Int. Symp. on Particles, Strings, and Cosmology (Boston, MA, 27–31 March, 1990)* ed P Nath and S Reucroft (Teaneck, NJ: World Scientific) pp 147–84

[520] Cronin J W 1992 *Proc. Symp. on the Interface of Astrophysics with Nuclear and Particle Physics (Zuoz, Switzerland, 11–18 April, 1992)* ed M P Locher (Villigen: Paul Scherrer Institute) pp 341–3

[521] Klein S *et al* 1992 *The Fermilab Meeting DPF 92 (Proc. 1992 Division of Particles and Fields Meeting (Fermilab, 10–14 November, 1992))* ed C H Albright *et al* (Singapore: World Scientific) pp 1364–6

[522] Bernstein J 1989 *The Tenth Dimension: An Informal History of High Energy*

Physics (New York: McGraw-Hill)

[523]   United States Department of Energy 1990 *The Ultimate Structure of Matter: The High Energy Physics Program from the 1950s through the 1980s* (Washington, DC: USDOE Office of Energy Research)

[524]   Weinberg S 1992 *Dreams of a Final Theory* (New York: Pantheon)

[525]   Weisskopf V 1991 *The Joy of Insight* (New York: Basic Books)

[526]   Treiman S B 1993 *A Century of Particle Theory*

[527]   Gell-Mann M 1994 *The Quark and the Jaguar: Adventures in the Simple and the Complex* (New York: Freeman)

[528]   Marshak R E 1993 *Conceptual Foundations of Modern Particle Physics* (River Edge, NJ: World Scientific)

[529]   Waloschek P (ed) 1994 *The Infancy of Particle Accelerators: Life and Work of Rolf Wideröe* (Braunschweig: Vieweg)

Chapter 10

FLUID MECHANICS

Sir James Lighthill

10.1. Yet another great success for twentieth century physics

10.1.1. A parallel revolution in fluid mechanics

Much of this book has been concerned with those profound changes in the pure and applied physical sciences that were initiated at the outset of the twentieth century through revolutionary discoveries which included radioactivity, quantum theory, the nuclear atom and relativity, together with thermionics and x-rays and their many applications. These were discoveries that transmuted the nineteenth century world-view of physics while addressing with scintillating success many of its recognized failures.

This chapter recalls yet another success story of twentieth century physics which, like the others, had profound implications for the human condition. It was based on brilliant experiments but depended above all on revolutionary approaches to interpretation, which both accounted for some spectacular failures of earlier approaches and, for the first time, placed upon sound foundations a certain major branch of physics: fluid mechanics.

Our environment on Earth confers a special human importance on two fluids: air and water; and for this principal reason we limit the present history to the mechanics of these two fluids. Indeed, the failures of nineteenth century fluid mechanics were particularly notorious for these familiar fluids of low viscosity—whereas, for example, the lubricating action of liquids of much higher viscosity in bearings under load had by 1886 been reliably accounted for with a sound fluid-mechanical theory [1].

The low viscosity of air, on the other hand, had tempted nineteenth century physicists to try to relate its dynamics to that of an 'ideal' fluid without any viscosity at all. For such a fluid there existed a most

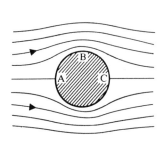

Figure 10.1. *Contrast between the potential field of flow around a circular cylinder (left) and an experimentally observed field [3] (right). Note: in both fields the oncoming flow is steady (not varying with time) but in the observed field the 'wake' (behind the cylinder) is highly unsteady.*

elegant and extensive theory [2], constructed—sometimes, in response to suppositions about the ether!—by many of the century's finest physicists. Yet the true motions of air, when disturbed by the movement of a typical body through it, bore in reality no relation whatsoever to the predictions of this ideal-fluid theory.

Admittedly, in acknowledging the theory's failure, the scientists concerned paid excessive attention to just one erroneous prediction—the famous d'Alembert's Paradox, that a body moving at uniform velocity through fluid would experience no resistive force—instead of admitting that the predicted flow field as a whole was quite different from that actually occurring. This predicted flow field was just such a regular 'potential' field, derived from a scalar potential, as is familiar from electrostatics or magnetostatics, and was in total contrast (see figure 10.1) to the real field, with its swirling wake behind the body [3]. And yet another false prediction of the theory was that fixed-wing aircraft would be unable to fly, because the air could not exert any force at all—resistive or lifting force—on a steadily moving body.

Just one scientist, Ludwig Prandtl, must be given the main credit for the brilliant discovery [4] which resolved these anomalies. Indeed his revolutionary discovery of the boundary layer in 1904 had the same transforming effect on fluid mechanics as Einstein's 1905 discoveries had on other parts of physics.

Such a very special level of recognition is due to Prandtl for two principal reasons. In theoretical terms his solution, besides representing an extremely early example of a singular perturbation (see section 10.1.2 below), was successfully applied to field equations of a fully non-linear type. In practical terms, Prandtl's new insights brought about the

Ludwig Prandtl

(German, 1875–1953)

Ludwig Prandtl was born on 4 February 1875 in Freising, Bavaria and studied Mechanical Engineering in Munich under August Foeppl. His doctor's thesis appeared in 1900 with some immediately acclaimed discoveries on the torsion of beams. However, by the following year, when appointed Professor of Mechanics in Hannover, he had already begun his path-breaking career in fluid mechanics. It was to the Third International Congress of Mathematicians (Heidelberg, 1904) that his epoch-making discovery of the boundary layer was first presented. That year, the famous Göttingen mathematician Felix Klein secured for Prandtl a chair in applied physics (later renamed applied mechanics) at Göttingen; which was to become the base from which he progressively revolutionized the understanding of aerodynamic drag and lift and of fluid flows at high Mach number. Many of his later contributions were made as leader of a fine team whose other members (including A Betz, W Tollmien, M Munk, J A Ackeret, H Schlichting, A Busemann) would themselves all win international recognition. In 1909 he married Gertrude Foeppl, by whom he had two daughters. His superb guide to fluid mechanics ultimately appeared in an extended 1952 English-language edition as *Essentials of Fluid Dynamics*. He died in Göttingen on 15 August 1953. His complete works [4] were published by Springer in 1963.

introduction of 'streamlined' shapes which, by permitting just modest deviations from potential flow fields, would experience very low resistive forces; while Prandtl himself went on to generate the first quantitative understanding of lift (and drag) forces on fixed-wing aircraft based on sound physical principles [5].

The twentieth century history of many crucially important developments that were to evolve from boundary-layer theory is sketched in section 10.2 below. These included developments in which Prandtl and scientists of his Göttingen school (A Betz, H Schlichting and above all Theodore von Kármán, who would soon set up new schools first at Aachen and later at Caltech) combined the new ideas with ideas from elsewhere, including the work of Zhukovski in Russia, to develop knowledge on boundary layers and wakes, and on the aerodynamic forces associated with them. Also, they included important studies on the instability of such shear layers (where early advances were made by physicists distinguished in other contexts, including Rayleigh and Sommerfeld) and on their tendency to develop the chaotic form of fluid motion known as turbulence—on which key papers by Osborne Reynolds, concentrating on turbulence in pipe flow, had first appeared in the 1880s and where important later advances would be made not only by Prandtl but also by H L Dryden in the United States, G I Taylor in England and A N Kolmogorov in Russia.

Here all of that extended historical account is preceded by a succinct introduction which begins with a quite simple explanation (section 10.1.2) of the idea of a singular perturbation. This leads on to a brief indication (section 10.1.3) of what the very earliest discoveries on boundary layers and wakes meant for those new aeronautical developments that were, in due course, to transform the human condition in ways which are outlined in section 10.4.

Then a first sketch is given (section 10.1.4) of that parallel revolution in knowledge of non-linear effects in the generation and propagation of waves in fluids—including sound waves and water waves—which began from elucidation of the physics of shock waves. These, indeed, offered another outstanding example of a singular perturbation successfully applied to field equations of a non-linear type, where Rayleigh [6] and Taylor [7] made independently the key discovery in two papers published in 1910. The whole scientific study constituted an early example of what would later be given the generic name of 'catastrophe theory'—just as the fluid mechanics of turbulence became an important forerunner to later generic concepts of chaos. Moreover, the same consistent involvement of twentieth century exponents of fluid mechanics in non-linear field theories led them to pioneer many new concepts—from solitons to uses of non-linear Schrödinger equations—that, after first appearing in the advanced study of water waves, would contribute later to other areas of the physical sciences.

The two great revolutions in twentieth century fluid mechanics were combined in many later developments that are outlined in section 10.4. For example, aircraft flying at speeds near, or greater than, the speed of sound may exhibit complex interactions of boundary layers with shock waves. Again, the mechanics of bodies near the ocean surface—whether ships or offshore structures—may often involve a similar mixture of considerations from the modern non-linear studies of both water waves and shear layers.

Geoffrey Ingram Taylor

(British, 1886–1975)

Geoffrey Ingram Taylor—a grandson of the mathematical originator George Boole —was born in London on 7 March 1886. He studied Mathematics and Physics at Cambridge; where, after successful early experimental researches within quantum optics, he produced in 1910 his revolutionary paper on the physics of shock waves. After his 1911 appointment to a newly established Readership in Dynamical Meteorology at Cambridge, he began to publish pioneering experimental investigations of turbulence in the atmosphere.

Moreover, he demonstrated the importance of turbulent dissipation of energy in tidal streams (1919) and pursued fundamental investigations on the motion of rotating fluids (1922) and on hydrodynamic stability (1923); while his 1921 paper *Diffusion by continuous movements* crucially contrasted turbulence with apparently analogous kinetic-theory phenomena and led in 1935 to massive developments in statistical theory of turbulence. Appointment in 1923 to a Royal Society Research Professorship allowed him thereafter to concentrate his energies on personal researches. Alongside crucial discoveries in solid mechanics (including the beginnings of dislocation theory), he continued to expand knowledge of shock waves and of atmospheric processes (e.g. buoyant plumes and thermals) while making also key initial studies (1951–52) on the fluid mechanics of the biosphere. He married Stephanie Ravenhill in 1925. His complete works in 4 volumes were published by Cambridge University Press (1958–71). He died in Cambridge on 27 June 1975.

Beyond our history of responses of fluid mechanics to such engineering challenges, we offer in section 10.5 an enormously larger-scale view of the mechanics of the Earth's whole fluid envelope including oceans, rivers and the atmosphere. The twentieth century has seen great advances in the successful application of fluid mechanics to the understanding, and to the prediction, both of disasters such as storms and floods, and moreover of weather and climate in general. Yet, many of these advances, too, can trace their lineage back to the revolution in fluid mechanics achieved in 1904 by Ludwig Prandtl.

10.1.2. An extremely simple example of a singular perturbation
Prandtl's introduction of what we now call a singular perturbation [8] was applied to field equations of a fully non-linear type. However, the essential idea can be illustrated with an exceedingly simple application to an ordinary differential equation which is linear with constant coefficients, and where the correctness of the method is easily verified by comparison with an exact solution.

Suppose that, in seeking to solve the differential equation

$$\varepsilon y'' + y' + ky = 0 \quad \text{for } 0 < x < 1 \tag{1}$$

subject to the boundary conditions

$$y = 0 \quad \text{at } x = 0, \qquad y = 1 \quad \text{at } x = 1 \tag{2}$$

we know that the second differential coefficient y'' is multiplied by an extremely small factor ε (which might be the viscosity in a fluid-mechanical application). The ordinary perturbation theory which is used throughout physics would look first for the solution with $\varepsilon = 0$ and then improve it by a process of successive approximation; for example, one involving an expansion in powers of ε.

This approach seems at first to make good progress because when $\varepsilon = 0$ equation (1) takes the still simpler form

$$y' + ky = 0 \tag{3}$$

which has the well known general solution

$$y = ae^{-kx} \tag{4}$$

with a an arbitrary constant. However, it is impossible to find a value of a which satisfies both the boundary conditions given in (2); for example, the choice $a = e^k$ satisfies the second condition, giving

$$y = e^{k(1-x)} \quad \text{with } y = 1 \text{ at } x = 1 \tag{5}$$

but this solution fails to satisfy the first condition since

$$\text{the value of } y \text{ at } x = 0 \text{ is } e^k \tag{6}$$

instead of being equal to zero. Moreover it is easily verified that no procedure of moving to higher approximations through the usual expansion in powers of ε yields any improvement at all.

The difficulty arises of course because, although two boundary conditions as in (2) are appropriate to a second-order differential equation (1), nevertheless the solution of a first-order equation (3) is uniquely determined by just a single boundary condition as in (5). The 'extra boundary condition' $y = 0$ at $x = 0$ (which might be the no-slip condition at a solid boundary in the fluid-mechanical application) makes completely impossible the usual perturbation approach.

If, however, we probe just what has gone wrong, we must recognize that the solution (5) which meets the second boundary condition should in general satisfy the true differential equation (1) to quite a close approximation because the term $\varepsilon y''$ is very small compared with the other terms in the equation. By contrast, things become very different as the solution (5) approaches $x = 0$ because it must encounter a region of very rapid change between the value (6) and the true value of $y = 0$ at $x = 0$.

In such a region the rate of change y' can become very large and its own rate of change y'' can become enormously larger so that even the term $\varepsilon y''$ in equation (1) takes a large value. If $\varepsilon y''$ and y' are both far bigger than the term ky, then equation (1) can be approximated as

$$\varepsilon y'' + y' = 0 \tag{7}$$

so that $\varepsilon y' + y$ has zero rate of change and must take a constant value c. But the solution of

$$\varepsilon y' + (y - c) = 0 \tag{8}$$

is

$$y - c = Ae^{-x/\varepsilon} \tag{9}$$

where A is another arbitrary constant; here A replaces a while $1/\varepsilon$ replaces k in the former general solution (4) of equation (3).

Now, given that $y = 0$ at $x = 0$ (the first boundary condition), the value of A is determined as $-c$, so that

$$y = c(1 - e^{-x/\varepsilon}) \tag{10}$$

in the region of rapid change. The broken line in figure 10.2 shows, however, that such rapid change soon disappears as y tends to an asymptotic value c.

In its simplest form, the idea of singular perturbation theory is that equation (7) with its solution (10) is just appropriate to the 'boundary layer' region of very rapid change with x extremely small, while equation

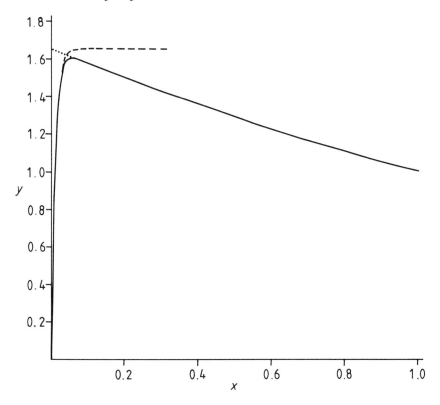

Figure 10.2. *Solid line, exact solution of equation (1) subject to conditions (2). Dotted line, 'external' solution (5). Broken line, 'boundary-layer' solution (10).*

(3) with its solution (4) is appropriate to all other regions. Moreover the two solutions are readily matched together provided that

$$c \text{ is given the value } e^k \qquad (11)$$

so that the value c of y just outside the region of rapid change coincides with the limit (5) of y as it approaches the region around $x = 0$.

Figure 10.2 assesses the accuracy of this approach. The solid line shows the exact solution of equation (1) subject to the conditions in (2) while the broken line shows the 'boundary-layer' solution (1), with c as in (11), and the dotted line shows the solution (5) for the region outside the boundary layer. The idea that the true solution makes, quite simply, a smooth transition between its two approximate forms is well substantiated in this case†.

† *More advanced forms of singular perturbation theory [8] are able to achieve any desired degree of accuracy by matching a series expansion for y in powers of ε with coefficients functions of x/ε in the region of very rapid change to another expansion in powers of ε with coefficients functions of x outside that region.*

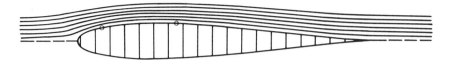

Figure 10.3. *For specially designed shapes, potential flow can be achieved (as illustrated here for flow, symmetrical about the broken line, around a blade whose cross-section is a Zhukovski [10] aerofoil) except in a very thin boundary layer. For magnified illustrations of the fluid motion in the two circular regions near the flow boundary, see figure 10.4 below.*

The above introductory account of a singular perturbation has concentrated on an extremely simple example, in order that readers may easily appreciate the idea and why it can work so well. It is a case where an exact solution (figure 10.2) was readily obtained; on the other hand, we shall see in the next two sections that the revolution in fluid mechanics came from applications of the same idea to field equations (that is, to partial differential equations) of a fully non-linear type where progress would have been impossible without the use of such a radically new approach.

10.1.3. How d'Alembert's Paradox became d'Alembert's Theorem

Of the two remaining introductory sections this first is devoted to outlining Prandtl's early discoveries about the boundary layer, and to sketching the resulting transformation of ideas about how a body moving with uniform velocity disturbs the air around it. In section 10.1.1 we recalled d'Alembert's Paradox, that the potential flow field suggested by theories of an 'ideal' fluid without any viscosity would generate no resistive force on such a body [9]. Prandtl's work explained why, for typical body shapes, the true flow field is so vastly different (see figure 10.1) in spite of the extremely low viscosity of air; but, above all, it gave the first indications that for specially designed body shapes the flow field could be much closer to a potential field—in such a way that the resistive force while not zero would become very small (see figure 10.3).

This was a development in which, effectively, d'Alembert's Paradox became d'Alembert's Theorem: an encouraging affirmation that, if ever the flow around a steadily moving body could be made quite close to a potential flow, then the resistive force should likewise become quite close to that zero force which an exactly potential flow would exert. These considerations prompt, of course, the question 'just what may be needed for the flow of an only slightly viscous fluid to become close to a potential flow?'.

Although (as has been indicated more than once) the field equations for the flow of such a fluid are 'of a fully non-linear type'—with non-linear terms as large and important as the linear terms in those equations—nevertheless it turns out that a potential flow field does exactly satisfy these full non-linear equations for a viscous fluid. On the other hand a potential flow field can satisfy only one boundary condition at a solid surface instead of the two which a real fluid satisfies. The genius of Prandtl showed itself in his bold introduction of a revolutionary approach—to which the reader has already been introduced, under its modern name singular perturbation theory—in recognition of this deficiency in the number of boundary conditions that a potential flow field can satisfy.

Actually, the flow field around a body moving with uniform velocity through still air is easiest to describe in a frame of reference with the body at rest, in which the problem becomes the exactly equivalent problem of how a stationary body disturbs a uniform wind. The classical potential-flow solution to this problem satisfies just one boundary condition; namely that, because there can obviously be no flow across the solid surface, the flow velocity at that surface must be directed along it (that is, tangentially).

A real fluid, however, must satisfy a second boundary condition, stating that the magnitude of that flow velocity tends to zero as the surface is approached. At the solid surface, indeed, departures from local thermodynamic equilibrium must be just as small as in the rest of the fluid (owing—in both locations—to the huge frequency of molecular collisions); so that fluid in contact with the surface satisfies the equilibrium conditions of zero velocity relative to the surface and temperature equal to that of the surface.

It is impossible for this second boundary condition to be satisfied by any potential flow field; for which, rather, the velocity magnitude is always substantial on the solid surface (and, actually, attains its maximum at a certain point of the surface). As in the simple example of section 10.1.2, however, it remains possible that the external potential flow is separated from the body's surface by a thin boundary layer [4, 11] in which the velocity falls steeply from its potential-flow value to a zero value at the solid surface (see figure 10.4).

Simplified forms of the field equations can be expected to apply in such a boundary layer (if it exists) just as equation (1) within its boundary layer takes the simplified form (7). Here, we merely describe in physical terms the nature of these simplified field equations for the boundary layer; which are interesting primarily because

(i) their solutions for certain shapes of body represent boundary layers which remain thin and attached to the surface—these are the 'streamlined' shapes which experience low resistive force because the flow around them differs only moderately from a potential flow— whereas

(ii) for a much wider range of shapes, solutions to these boundary-layer equations simply cease to exist downstream of a certain point on

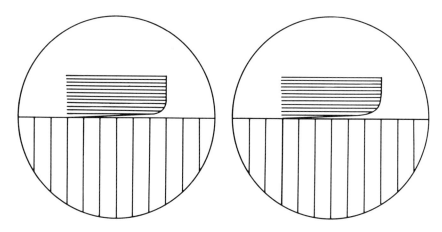

Figure 10.4. *Magnified illustration of flow (from left to right) in the two circular regions of figure 10.3. Each horizontal line near the flow boundary has length equal to the local flow speed. The boundary layer, within which the speed tends to zero, is very thin in the left-hand region. Moreover, in the right-hand region it remains attached to the boundary even though diffusion of momentum has caused it to grow a little thicker.*

the body surface, near which the flow along that surface can be described as undergoing 'separation' from the solid surface.

It is of course this flow separation that in case (ii) may produce a prominent swirling wake behind a bluff body; whereas the attached boundary layer in case (i) may emerge as a thin, far less prominent wake behind a streamlined body.

These distinctions make plain the crucial importance of those boundary-layer equations [4, 11] which Prandtl derived already in 1904. The physical effects represented in those equations are as follows.

First of all, the very steep gradient of velocity across a boundary layer (from a potential-flow value, say V, at its outer edge down to zero at the solid surface) allows viscous effects to become important even for fluids of very low viscosity; indeed, the thickness δ of the layer adjusts itself to the viscosity μ of the fluid. Velocity gradients of the order of V/δ produce viscous stresses of the order $\mu V/\delta$; here, a stress is of course a force per unit area, so that forces per unit volume in a layer of thickness δ are of the order $\mu V/\delta^2$. Simply from a balance of these viscous forces per unit volume against any other effects (see below), it is already clear that the small thickness δ of a boundary layer for fluids of different viscosity μ must vary as the square root $\mu^{1/2}$ of the viscosity.

For extremely small values of μ, then, boundary layers may be very thin indeed. At the same time, the actual viscous stresses in these layers (of order $\mu V/\delta$) may themselves be small (of order $\mu^{1/2}$); so that, in cases where the boundary layer remains attached, these stresses may contribute only a modest augmentation to the zero resistive force on the body suggested by d'Alembert's Theorem.

Those other effects, independent of viscosity, against which the viscous forces per unit volume have to be balanced are two in number. Newton's second law of motion, of course, equates the total force on any particle to its mass times its acceleration, so that force per unit volume on a fluid is equal to its density ρ times its acceleration. This is the non-linear term in the field equations of fluid mechanics; thus, in steady flow around a body, the acceleration of any particle of fluid along its (curved) path has to be written as $v\,dv/ds$ in terms of the distance s along that path—because its velocity v is the rate ds/dt of increase of s—and this gives a mass-acceleration $\rho v\,dv/ds$ per unit volume.

At the same time, the total force per unit volume which this has to balance includes not only the viscous force but also any gradient $(-dp/ds)$ of the fluid pressure† along the path (with negative sign because pressures increasing with s oppose the motion). Outside the boundary layer, where v assumes its potential-flow value V, this gradient $(-dp/ds)$ is the sole force balancing the mass-acceleration term $\rho V\,dV/ds$, giving the famous Bernoulli equation

$$p + \tfrac{1}{2}\rho V^2 = \text{constant} \tag{12}$$

explained in 1738 (for such steady motion without viscous effects) by Daniel Bernoulli [12].

Prandtl brilliantly recognized that the pressure would be given throughout the boundary layer by this same equation (12) depending on the velocity V outside it (briefly, any pressure change across the very thin boundary layer must be negligible since pressure gradients at right angles to the flow are limited to their centrifugal values). It followed that the pressure gradient in the flow direction, $(-dp/ds)$, takes a value $\rho V\,dV/ds$ which, in combination with viscous forces, has to balance the mass-acceleration $\rho v\,dv/ds$ within the boundary layer. It is this balance

$$\rho v\frac{dv}{ds} - \rho V\frac{dV}{ds} = \text{viscous force per unit volume} \tag{13}$$

which—in physical terms—constitutes Prandtl's boundary-layer equations, governing how the velocity v falls from its value V just outside the thin boundary layer to a zero value at the solid surface. (Here, because the viscous stress takes a value $\mu\,dv/dn$, proportional to a gradient of v normal to the flow, each particle of fluid is subjected to a viscous force per unit volume $\mu\,d^2v/dn^2$, associated with the difference of stresses

---

† *Here, more strictly, p should be taken as the 'excess pressure' (any excess of fluid pressure over its purely hydrostatic value), so that the gradient $(-dp/ds)$ accounts also for any component of gravity along the path. In the case of air, however, that distinction is generally unimportant.*

acting on both sides of it—and, as foreshadowed earlier, with order of magnitude $\mu V/\delta^2$.)

These equations were of revolutionary significance because they focused attention on that thin layer whose dynamics was really determining the character of the whole flow. In mathematical terms they are quite complicated non-linear field equations of 'parabolic' type which would be subjected to over half a century's intensive study because of their complexity and because of their importance—see (i) and (ii) above—for a wide range of body shapes [11]. Although section 10.2 includes some account of these analyses that confirmed how the equations can be solved by a systematic progression in the downstream direction, which may either proceed regularly as in (i) or else abruptly cease as in (ii), we reproduce here just the essential physical insight into those alternatives that was included in Prandtl's 1904 paper.

Viscosity, of course, is the physical process by which fluid momentum is diffused down its gradient. This has different effects in three different types of boundary layers: (a) when the external flow speed V is increasing in the direction of flow (dV/ds positive); (b) when it is slightly decreasing (dV/ds negative but small); and (c) when it is falling more steeply (dV/ds negative but not small). In case (a) the pressure gradient $\rho V dV/ds$ is accelerating the fluid in the boundary layer (which therefore remains thin) while any excess momentum is transferred by diffusion to the solid surface. In case (b) the pressure gradient is gradually reducing the momentum of fluid near the surface, which would be brought to rest but for the fact that diffusion from the external flow replenishes that momentum. In case (c) such diffusion is insufficient to keep this fluid moving and a stagnant region forms near the wall from which the main flow separates.

Briefly, then, the shapes that can benefit from d'Alembert's Theorem by retaining a thin, attached boundary layer are those shapes for which the external potential flow field combines regions of types (a) and (b) only and avoids regions of type (c). The flow speed V increases over the front part of the body to a maximum where flow lines are most packed together, and is then reduced only gradually over the smoothly tapering rear portion (see figure 10.3). For example, when a herring after a burst of swimming glides forward rigidly, its rate of loss of speed is low because its shape satisfies this condition.

Widespread engineering application of these ideas is described in section 10.2, where their further development takes, in essence, three forms. First, imprecise phrases like 'very low viscosity' are abandoned in favour of that quantitative comparison of a product of flow quantities with μ which 'Reynolds number' gives; so that boundary layers become thinner (relative to body size) with increasing Reynolds number.

Next, the tendency as Reynolds number becomes still larger for the interior of a boundary layer to become turbulent is explained along with several implications. Above all, increased diffusion in the boundary layer assists in countering the tendency to separation—as Prandtl himself was to find—so that a rather wider range of boundary layers with retarded external flow becomes comprised within case (b) above.

Finally, boundary-layer theory was also developed by Prandtl in another, even more dramatic, form [5]. This explained how the nature of thin wakes from well designed wing shapes could allow wings to experience large forces ('lift') at right angles to the flow, even while resistive forces would continue to be low in the general spirit of d'Alembert's Theorem.

10.1.4. The physics of shock waves

This last introductory section is devoted to one further topic, completely different from that of boundary layers, where a long-standing enigma in fluid mechanics was resolved early in the twentieth century by an application of, yet again, the essential idea (section 10.1.2) that is now called singular perturbation theory. It was the resolution of that enigma which first elucidated the physics of shock waves.

The present brief account of this major achievement may serve to introduce that wide subject—non-linear effects in the generation and propagation of waves in fluids—to which section 10.3 is devoted. There are many different kinds of waves in fluids; amongst which sound waves in air may be most familiar to our ears, and gravity waves on a water surface to our eyes. Non-linear effects include interaction of such waves with fluid flows, and indeed this interaction is the principal feature of waves in fluids which makes their physics so different from that of other types of waves.

Linear theories of waves in general became well developed during the nineteenth century, and Rayleigh's superb treatise *The Theory of Sound* [13] expounded comprehensively an enormous body of knowledge obtained from linear wave theory as applied to sound generation and propagation. Yet in section 253 of this great work Rayleigh brilliantly showed how consideration of just the simplest possible problem in the non-linear theory of sound posed an enigma which it was quite impossible to resolve with the knowledge then available. Briefly, in any large-amplitude sound wave, higher values of the pressure may propagate faster than lower values and 'catch up' with them so that, apparently, continuous motion ceases to be possible.

Here, the enigma is explained for air, treated as a perfect gas with constant specific heats in a ratio γ which is close to 1.4. Ever since a famous 1816 paper of Laplace [14] it had been appreciated that, in sound waves, those changes of pressure p and density ρ whose ratio $dp/d\rho$ is the square of the sound speed c are 'adiabatically' related; this means that a particle of fluid experiences no heat input (or, indeed, output) so that when it expands it cools—by doing work—and the pressure falls more than it would in an isothermal process. For air with undisturbed pressure p_0 and density ρ_0 this adiabatic relationship takes the form

$$\frac{p}{p_0} = \left(\frac{\rho}{\rho_0}\right)^{\gamma} \text{ giving } c^2 = \frac{\mathrm{d}p}{\mathrm{d}\rho} = c_0^2 \left(\frac{\rho}{\rho_0}\right)^{\gamma-1} \text{ with } c_0^2 = \frac{\gamma p_0}{\rho_0}. \tag{14}$$

On linear theory sound is propagated at the undisturbed speed of sound c_0, whereas on non-linear theory higher pressures travel at an increased sound speed, amounting on a first approximation to

$$c = c_0 + \frac{\gamma - 1}{2} \frac{p - p_0}{\rho_0 c_0}. \tag{15}$$

But, this is the propagation speed relative to motions of the air, whose velocity u in the direction of propagation is given on linear theory by a well known relation

$$p - p_0 = \rho_0 c_0 u. \tag{16}$$

(Indeed, in a wave travelling at the linear-theory speed c_0, the acceleration of any particle of fluid is $-c_0 \mathrm{d}u/\mathrm{d}x$ so that equation (16) allows the mass-acceleration per unit volume $(-\rho_0 c_0 \mathrm{d}u/\mathrm{d}x)$ to balance the pressure force $-\mathrm{d}p/\mathrm{d}x$.)

These arguments suggest that the speed $c + u$ at which pressure changes may be propagated on non-linear theory can, by equations (15) and (16), be written

$$c_0 + \frac{\gamma + 1}{2} u \quad (= c_0 + 1.2u \quad \text{for } \gamma = 1.4). \tag{17}$$

Note that the excess propagation speed for air is about $1.2u$, out of which just one-sixth is due to the increase (15) in c while five-sixths is due to convection of sound at the air velocity u; here, already, there is an important interaction of a fluid flow with waves.

Although readers have just been introduced by crude approximate arguments to expression (17) for the propagation speed, a brilliant 1859 analysis by the great mathematician Bernhard Riemann [15] had proved it to be absolutely accurate for plane sound waves of any amplitude propagated one-dimensionally into still air under adiabatic conditions. Briefly, the relationship $c = c_0 + \frac{1}{2}(\gamma - 1)u$ is exact; the expressions for pressure and density may be derived from c by equations (14); and, most important of all, each value of u is propagated at precisely the speed (17).

These conclusions were well known to Lord Rayleigh, who recognized also their sensational implications. Figure 10.5 shows these in the case of a single pulse of positive excess pressure, represented as an initial plot (solid line) of fluid velocity u against distance. On a linear theory of one-dimensional sound waves, each value of u would

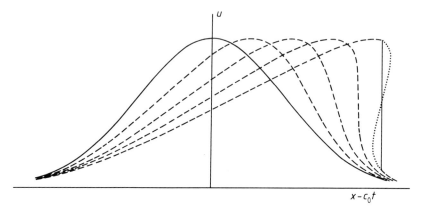

Figure 10.5. *Solid line, initial waveform (graph of fluid velocity u against $x - c_0 t$). Broken lines, changed waveforms at later times, expected since values of u are propagated not at speed c_0 but at speed $c_0 + 1.2u$. Dotted line, a still later waveform similarly derived, yet manifestly impossible. Vertical line, tentative resolution of the enigma proposed by Riemann [15].*

be propagated at speed c_0, so that the shape of the pulse would remain unchanged when plotted (as here) against $x - c_0 t$.

On the exact non-linear theory, however, each value of u is propagated at speed $c_0 + 1.2u$, so that after a time t it has travelled a distance $(c_0 + 1.2u)t$. In the representation of figure 10.5, then, where the values of u are plotted against $x - c_0 t$, each value of u has been shifted a distance $1.2ut$ to the right. Small values of u have hardly moved at all, while large values have moved much more—permitting them, as suggested earlier, to 'catch up' with smaller values.

These distorted pulse shapes are shown in figure 10.5 (broken lines) for a sequence of increasing values of t until a time has been reached where the pulse shape has a vertical tangent. Pulse shapes at still later times continue to be predicted by the theory; however, parts of these shapes are shown in figure 10.5 as dotted lines in recognition of the clear impossibility of the fluid velocity u taking three different values at one and the same point!

A tempting idea for resolving this enigma is to suppose that the solution develops a discontinuity. Riemann himself noticed that a discontinuity—indicated in figure 10.5 by the vertical solid line— could be inserted in place of the dotted line in such a way that total mass and momentum for the system are conserved, while all continuous parts of the curve (broken lines) still satisfy exact equations for sound propagation under adiabatic conditions. Rayleigh, however, objected that this idea not only (i) left unexplained how a discontinuity would arise but also, still more seriously, (ii) failed to satisfy overall conservation of energy.

810

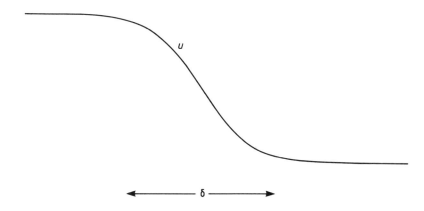

Figure 10.6. *The abrupt change of velocity u within the extremely small 'Taylor thickness' δ of a (not too strong) shock wave takes up this distribution, such that diffusion balances non-linear effects.*

Much later, in 1910, the enigma was resolved both by Rayleigh himself [6] and by the young G I Taylor [7] in two independent studies published in a single number of the Proceedings of the Royal Society. Essentially, the idea that would later be called 'singular perturbation theory' was again used so as to overcome both the objections (i) and (ii) above.

In our first simple illustration (section 10.1.2) and in the aerodynamic application (section 10.1.3) of this idea, it was a discontinuity between external trends and a boundary condition that necessitated the introduction of 'a region of very rapid change' at the boundary, with gradients large enough for 'diffusion' effects to become important. In the present application no such 'boundary layer' can appear; instead, a discontinuity is emerging in the midst of the fluid; yet it once more gives rise to a region with gradients large enough for diffusion to become important (and the analyses of Rayleigh and Taylor showed that diffusion of momentum as in section 10.1.3 is of quite comparable importance with thermal diffusion).

It is because the effects of diffusion (being proportional to gradients) can, if the region of very rapid change is thin enough, attain any required level that they are able to cancel out those effects of excess propagation speed which tend to produce the unrealistic 'overturning' of the waveform represented by the dotted curve in figure 10.5. Instead, there is formed something close to the discontinuous solution shown as a solid vertical line; where, however, the discontinuity possesses a definite thickness δ ('the Taylor thickness') and an associated internal structure (figure 10.6) which allow those 'overturning' and 'diffusion' effects to be in precise balance [16].

While countering Rayleigh's objection (i) by explaining how this nearly discontinuous wave—called a shock wave—arises, the above indication of important diffusion effects within it also demolishes any idea that the process is adiabatic. On the contrary, the fluid traversed by a shock wave experiences a departure from adiabatic conditions, which of course—in accordance with the second law of thermodynamics—involves an increase in entropy (so that p/p_0 rises above the value given by (14) for given ρ/ρ_0). This counters objection (ii) because the shock wave achieves energy conservation by abandoning entropy conservation—instead of the other way round.

Shock waves are much thinner than boundary layers; essentially, because the main non-linear term $\rho v \mathrm{d}v/\mathrm{d}s$ in the field equations of fluid mechanics includes a rate of change $\mathrm{d}/\mathrm{d}s$ in the flow direction—that is, across the shock wave's small thickness δ. Therefore, while the balance (13) in a boundary layer equates a diffusion term of order $\mu V/\delta^2$ to other terms independent of δ (the flow being directed along the boundary), a similar balance in a shock wave equates a diffusion term to a non-linear term of order $\rho V^2/\delta$. Accordingly, the shock wave thickness δ comes out to be directly proportional to the relevant diffusivity rather than to its square root, and typical values of δ in atmospheric air vary from a millimetre for quite weak shock waves to a few micrometres for much stronger ones.

For practical engineering purposes, then, it often suffices to treat shock waves as sharp discontinuities satisfying conservation equations for mass, momentum and energy. These equations had actually been written down in 1889—for fluids in general—by A Hugoniot [17] who, on the other hand, had not been in a position to appreciate why, in most fluids including air, the physics of shock-wave formation (see above) permits only the appearance of compressive discontinuities.

Important applications of such approaches to supersonic aerodynamics are outlined in section 10.4. They are above all founded on a deep study of supersonic flow in shaped nozzles [18] which Ludwig Prandtl published in the same year as his boundary-layer theory: 1904, the *annus mirabilis* that transformed fluid mechanics.

10.2. Boundary layers and wakes; instability and turbulence; heat and mass transfer

10.2.1. The liveliest of non-dimensional parameters

This section 10.2 traces the further development of many ideas—and especially those related to boundary layers and wakes—that have been briefly introduced in section 10.1.3. The historical developments here described were centred around problems of how stationary solid objects disturb a steady wind. (As stated on p 804, these problems are identical to questions of how still air is disturbed by the steady movement of a solid body through it; when, indeed, air motions are most conveniently considered as relative to the movement of the body.)

Non-dimensional parameters proved useful in the twentieth century development of several branches of physics. Their particular importance in fluid mechanics may be illustrated by two contrasting examples, related to assessments of the significance for airflows of different physical properties: compressibility and viscosity.

The air density ρ, according to equation (14), should change by just a very small fraction of its undisturbed value ρ_0—so that compressibility might be neglected—provided that the pressure p changed by a very small fraction of p_0. But, if the wind speed is U, Bernoulli's equation (12) suggests that changes in pressure may be of the order of changes in $\frac{1}{2}\rho V^2$ where flow velocities V are of the same order of magnitude as U; so that changes in pressure are very small compared with the undisturbed pressure p_0 if the ratio

$$\frac{1}{2}\frac{\rho_0 U^2}{p_0} = \frac{1}{2}\gamma \left(\frac{U}{c_0}\right)^2 \tag{18}$$

is very small.

Extensive experimental studies confirmed the importance of the ratio of wind speed to sound speed

$$\frac{U}{c_0} = M \tag{19}$$

as a non-dimensional parameter small values of which (say, values less than 0.2) could be used to diagnose the absence of any influence of compressibility. The flow field is then 'solenoidal'—like a magnetic field—with flow speeds increasing wherever flow lines come together because the volume flux through a tube of flow lines remains constant.

The name 'Mach number' given to M honours Ernst Mach who in 1887 interpreted photographs of bullets in flight, showing (see section 10.4.2 below) how the pattern of airflow undergoes an exciting change [19], with the prominent appearance of shock waves, when M exceeds 1. Here, however, we are entirely concerned with the 'unexciting' tendency of compressibility to lose all importance for those flows, with small values of M, to which section 10.2 is devoted.

See also p 873

Actually, various other non-dimensional parameters of fluid mechanics [11] share this relatively unexciting property that their smallness implies the negligibility of a certain physical property. By contrast there are, as explained in section 10.1.3, absolutely no airflows around solid bodies for which the effects of viscosity can be neglected, since these are always important in that boundary layer whose development critically influences the character of the entire flow.

Nevertheless a non-dimensional parameter involving viscosity does prove important for a wide variety of reasons other than any misguided

813

aim of diagnosing when viscosity might be negligible. This parameter depends not only on the wind speed U but also on the body's linear dimension ℓ.

Thus, for all bodies of a well-defined shape but different sizes (commonly described as 'geometrically similar' bodies) the dimension ℓ in the direction of the wind is usually taken as a measure of size. Then the parameter

$$\frac{\rho U \ell}{\mu} = R \qquad (20)$$

is non-dimensional since the viscosity μ is defined as the ratio of a stress (which, like a pressure, has the dimension ρU^2) to a velocity gradient (with dimension U/ℓ).

Whenever the Mach number (19) is small, so that compressibility is negligible, the pattern of disturbances to the wind caused by the presence of the stationary body can depend only on the (effectively uniform) air density ρ, on the viscosity μ, on the wind speed U and on the body size ℓ. But the general principles of dimensional analysis tell us that this apparent dependence on four variables is really only a dependence on the single non-dimensional parameter (20). (Briefly, any changes in ℓ, U, ρ and μ which preserve the value of R are readily seen to be equivalent to simple changes in the fundamental units of length, time and mass, which leave unaffected both the laws of mechanics and the relation of viscous stress to velocity gradient.)

Accordingly, the airflow around a large body is just a scaled-up version of the flow around a smaller, but geometrically similar, model of the body at the same value of R. This principle has many applications to wind-tunnel experiments on airflow; for example, a scale-model with a reduced value of ℓ may be tested in a wind-tunnel that uses air at high pressure or low temperature (respectively, increasing ρ or decreasing μ) so that the ratio R is unchanged.

An even more important use of R, the non-dimensional parameter (20), is to characterize which out of several possible types of flow will occur in a particular case. Thus the flows loosely described in section 10.1 as motions of 'a fluid of very low viscosity' are more correctly described as flows with high values of R; indeed, it is for $R \geqslant 10^3$ (about) that a thin boundary layer can occur, and any dependence (section 10.1.3) of its thickness δ on the square root $\mu^{1/2}$ of viscosity is better expressed as the non-dimensional statement [11] that a thickness-to-length ratio

$$(\delta/\ell) \text{ tends to vary as } R^{-1/2}. \qquad (21)$$

However, such thinning of a boundary layer as R grows larger and larger is by no means indefinitely continued.

For his fine 1880s studies of flow in pipes, the Manchester engineer Osborne Reynolds [20] had used a ratio just like (20) but with the length

ℓ replaced by the pipe's dimension normal to the flow (the internal diameter), and had discovered that the flow tended to become turbulent (that is chaotic) when that ratio exceeded about 2000. In his honour, the ratio (20) is named Reynolds number by aerodynamicists, who moreover use R_δ to mean the corresponding ratio

$$R_\delta = \frac{\rho U \delta}{\mu} = \left(\frac{\delta}{\ell}\right) R \tag{22}$$

based on the boundary layer's dimension δ normal to the flow. This flow within a boundary layer tends (see section 10.2.3) to become turbulent when such a ratio (22) reaches a critical value whose rough order of magnitude is again 10^3; however, since R_δ tends—by equation (21)—to vary as $R^{1/2}$, it may be for values of R in the broad neighbourhood of 10^6 rather than of 10^3 that a boundary layer becomes turbulent.

Prandtl recognized very early [21] that such a transition to turbulence in boundary layers played three important roles, all associated with the increased diffusion of momentum which results from chaotic motions. Thus,

(i) the boundary layer becomes considerably thicker, while
(ii) resistance receives a substantial increase in consequence of greater diffusion of momentum towards the solid surface, and yet
(iii) this same diffusion makes the layer relatively less prone to separation;

that is, able (see p 807) to remain attached to the surface at relatively higher values of the rate of decrease $(-dV/ds)$ of external flow velocity with distance s along the surface.

Prandtl saw, too, that (ii) and (iii) are in an interesting conflict. For example, an 'aerofoil' shape may need to be as thin as that in figure 10.3 if separation is to be avoided at $R = 10^4$, when the boundary layer takes a steady, 'laminar' form (figure 10.4). Yet, at Reynolds numbers around 10^6 or more, aerofoil sections almost twice as thick (with $(-dV/ds)$ almost twice as large) will avoid separation as a consequence of transition to turbulence in the boundary layer; and then the associated increase (ii) in resistance, though substantial, is far less than would have been the excess resistance associated with separated flow.

Moreover [21], the resistance to flow over an aerodynamically 'bad' shape like a sphere is reduced by a large factor (around 3) when the boundary layer makes such a transition—from laminar to turbulent—before it separates (figures 10.7 and 10.8). This transition may be generated either (a) by an increase in R to rather large values (in excess of 10^5); or, at lower values of R, (b) by turbulence artificially generated, as shown in some dramatic photographs [22] of flow around a sphere with and without a 'trip wire' which promotes transition. It is, of course,

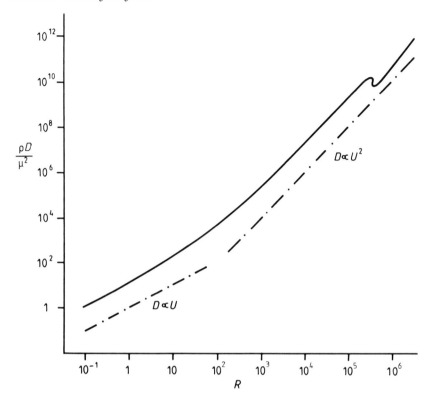

Figure 10.7. *A first illustration of the 'liveliness' of dependence on Reynolds number $R = \rho U \ell / \mu$. The drag force D on a smooth sphere of diameter ℓ in a steady wind of speed U, while broadly varying as U^2 (due to pressures of order ρU^2) for $R > 10^3$ and yet as U (due to viscous stresses of order $\mu U / \ell$) for $R < 10$, experiences [29] a threefold drop, due to transition to turbulence in the boundary layer, at around $R = 3 \times 10^5$.*

well known that many ball games are made interestingly complicated by such sensitive dependence of forces on a moving sphere to both speed (or, more correctly, Reynolds number) and various surface disturbances.

Reynolds number has been described [11] as 'the liveliest of non-dimensional parameters' because régimes of flow vary so strikingly over an extended range of orders of magnitude of R. Such variations are found both for increases in R (as briefly illustrated above) from 10^3 to a sequence of higher orders of magnitude and also for decreases to low values, even less than 1, as revealed under the microscope in the marvellous world of the swimming of small organisms [23].

10.2.2. New roles for vorticity

It is a paradox that the most efficacious twentieth century tool for understanding the forces (including the aeronautically crucial lift force) that may act on a stationary body in a wind was a concept, VORTICITY,

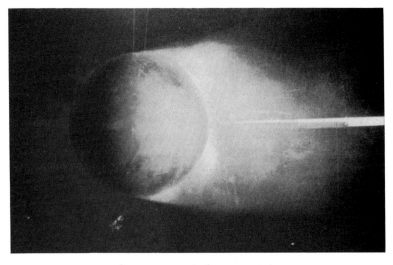

(a)

(b)

Figure 10.8. *Flow (from left to right) around a smooth sphere (a) without, and (b) with, a trip wire* [22].

which—far from having been previously neglected—was a specially favoured tool of nineteenth century fluid-mechanics theorists! Obsessed with vorticity, they nevertheless had used it to pursue many fruitless quests (for example, the representation of atoms as vortices in the ether) which need not be recalled here, whilst one of their most penetrating general results—Kelvin's Circulation Theorem, destined to play a key role in aerodynamic lift theory—was mistakenly viewed by its author as incompatible with any possibility of fixed-wing aircraft sustaining a lift force.

Some quite new roles for vorticity emerged, then, in the twentieth century. These were above all derived from reinterpretation of discoveries on boundary layers and wakes in terms of vorticity distributions that possessed both well defined momentum and associated consequences for aerodynamic forces.

The importance of vorticity derives from the idea that, wherever viscous stresses can be neglected, a small sphere of fluid is subjected only to pressure forces [11]; which, acting through its centre, are not altering its angular momentum (figure 10.9). The sphere's motion can be divided into three parts: (i) uniform translation with the velocity v of the centre, (ii) rigid rotation with angular velocity $\frac{1}{2}\omega$ where ω is the vorticity—parts (i) and (ii) carry, respectively, all the sphere's momentum and angular momentum—and (iii) a symmetrical squeezing or 'straining' motion in which the sphere is instantaneously changing its shape into an ellipsoid of the same volume. Part (iii) involves elongations in some directions and foreshortenings in others, which are, respectively, decreasing or increasing its moment of inertia about these directions. Conservation of angular momentum implies, then, that components of vorticity along axes which are being elongated or foreshortened are, respectively, increased or decreased; and, indeed, the vorticity vector itself is subject to precisely the same changes in length and direction as a line of particles of fluid through the centre of the sphere undergoes in the course of the sphere's straining motions.

This leads to the idea of a 'vortex line' as a line, or 'necklace', of moving particles of fluid that—wherever viscous stresses can be neglected—continues to point always in the direction of the vorticity vector; a vector whose magnitude, moreover, varies in proportion as the necklace is locally stretched. The familiar smoke ring is a visible bundle of vortex lines.

A typical shearing motion, such as a boundary layer (figure 10.4), is a region of strong vorticity, with magnitude ω equal to the gradient dv/dn of velocity. Indeed, pure rotation with angular velocity ω is obtained by vectorially adding such a shearing motion to another at right angles, and this indicates $\frac{1}{2}\omega$ as the rotary component (ii) for each (figure 10.10).

The above considerations lead to the classical vector relationship $\omega = \text{curl } v$ expressing vorticity components as a combination of gradients of velocity components. This has two major implications, of which the first is that the vorticity field itself is solenoidal (div $\omega = 0$) so that vortex lines can never end in the fluid. Furthermore, just as the corresponding magnetostatic relationship curl $H = J$ allows us to regard a current distribution J as precisely determining its associated magnetic field H in accordance with the Biot–Savart law, so too the vorticity distribution ω completely determines the fluid flow field v.

Where viscous stresses can by no means be neglected, as in a boundary layer, their effect is to produce diffusion, not only of velocity

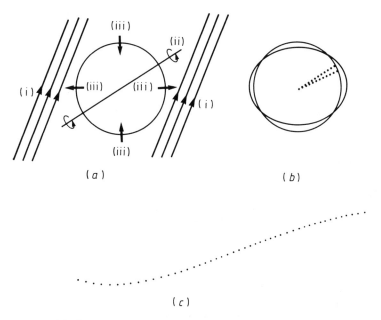

(a)   (b)

(c)

Figure 10.9. *(a) The instantaneous motion of a small sphere of fluid may be divided into three parts, with part (ii)—rotation at angular velocity $\frac{1}{2}\omega$—carrying all the angular momentum. Now, pressure forces on the sphere act through its centre, and so are not changing this angular momentum. (b) Yet ω may be changing because the 'straining' motion (iii) is altering moments of inertia about different axes, in such a way that the vorticity vector is subject to the same straining motions as a line of fluid particles. (c) Thus a 'necklace' of fluid particles which coincides with a vortex line at one instant will continue to do so.*

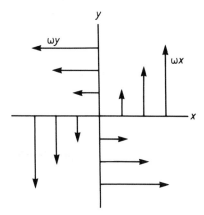

Figure 10.10. *Since these two shearing motions combine vectorially to give pure rotation with angular velocity ω, the rotary component (ii) in each has angular velocity $\frac{1}{2}\omega$.*

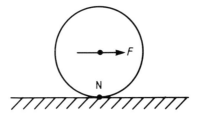

Figure 10.11. *Rotation is generated in a small sphere of fluid, touching a solid surface at N—where it satisfies a no-slip condition—by a force per unit mass F, where (see section 10.1.3) $F = -(1/\rho)(dp/ds) = V(dV/dS)$.*

components (section 10.1.3) but also of velocity gradients, so that the vorticity too undergoes diffusion with just the same value (μ/ρ) of the diffusivity. Vortex lines, then, are subject not only to convection by the fluid motions but also to such diffusion.

When a stationary body disturbs a uniform wind, the solid surface itself is the only available source of vorticity. From this viewpoint, a boundary layer arises when diffusion of the vorticity generated at the surface, together with convection downstream, allows that vorticity to remain confined in a thin boundary layer and wake—outside of which a potential flow (that is, a flow with zero vorticity) is to be found.

The rate of vorticity generation at a solid surface (per unit area) is VdV/ds in the notation of section 10.1.3. A schematic diagram (figure 10.11) readily suggests why this pressure-gradient force per unit mass tends to generate rotation in a small sphere of fluid which—like a rubber ball!—satisfies a no-slip condition at the surface.

From this point of view, boundary-layer regions defined on p 807 as (a) with dV/ds positive, (b) with dV/ds negative but small and (c) with dV/ds negative but not small are characterized by (a) positive rates of vorticity injection, (b) small negative rates (vorticity abstracted gradually enough to be counterbalanced by diffusion), and (c) large negative rates, leading to backflow ($dv/dn < 0$) near the surface. Out of these, only (c) makes the boundary layer separate.

Prandtl, about a decade after his initial explanation of flows that avoid separation (because no region (c) is present), made another extraordinary discovery about such flows [5]. This was that the vorticity shed behind the body exercises a determining influence on both airflow and aerodynamic forces. Here, a nineteenth century theorem proved invaluable: a vorticity distribution ω not only determines the flow field v but also specifies the momentum of that flow field very directly (in mathematical terms, as $\frac{1}{2}\rho$ times the moment of the vorticity distribution).

Admittedly, these effects may be relatively insignificant for a purely symmetrical flow like that of figure 10.3, simply because boundary layers

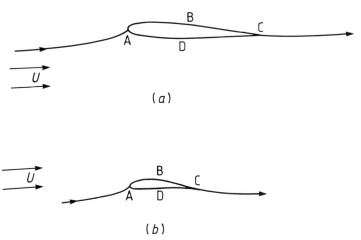

Figure 10.12. *Lift-generating airflows around wing cross-sections. (a) Flow about a symmetrical shape (the same as in figure 10.3) when the 'angle of attack' between the wind vector U and the 'chord' (line joining leading and trailing edges) is 6° . The circulation (amount by which the integral ∫ Vds taken along ABC exceeds its value taken along ADC) is positive because values of the velocity V just outside the boundary layer are greater on the upper surface than on the lower surface. (b) Flow at the same angle of attack about a cambered cross-section. The circulation now includes a contribution due to camber in addition to that due to angle of attack.*

on the body's upper and lower surfaces shed equal and opposite vorticity (clockwise and anticlockwise, respectively). In a unit length of wake, then, there is no resultant vorticity, although it does have a (very small) moment corresponding to that rate of creation of forward momentum in the wake fluid which Newton's third law tells us must accompany the (very small) resistive force acting on the body.

But things are quite different wherever some departure from flow symmetry—arising either when a symmetrical shape like that of figure 10.3 is placed in the wind at a positive 'angle of attack', or when the centreline of its cross-section is not straight but 'cambered' (arched)— removes any such tendency for vorticity on the upper and lower surfaces to be equal (figure 10.12). Then, the wake may carry a large and growing downward momentum, whose rate of increase represents the aerodynamic lift on the body (such an upward force of the air on the body being equal and opposite to the body's rate of communication of downward momentum to the air).

Twentieth century developments in the aerodynamics of aircraft wings—at those relatively low Mach numbers to which section 10.2 is devoted—have concentrated on winning ever greater quantitative precision for each element within Prandtl's fundamental picture [5] of the associated vorticity pattern, which critically involves 'bound' vorticity (any resultant vorticity attached to the wing) as part of a continuous field of vortex lines with the 'trailing' vorticity in the wake. Here this pattern can be sketched in just the simplest terms.

Evidently, the expression $\omega = dv/dn$ for vorticity in a boundary layer, where the velocity v rises from zero at the surface to V at the edge of the boundary layer, implies a resultant (total) vorticity $+V$ per unit area of a wing's upper surface, alongside a similar resultant $-V$ on the lower surface. Bound vorticity, representing a positive resultant (per unit span) for the wing section as a whole, takes therefore the form of any difference between values of the integral $\int V ds$ along the upper and the lower surfaces. Such a difference, called the 'circulation' Γ around the wing section, appears whenever the extent and magnitude of clockwise values of V on its upper surface exceed those for its anticlockwise values on its lower surface (figure 10.12).

Circulation is important because it generates lift; Bernoulli's equation (12) already suggests this—since it makes the pressure greatest where V is least—but a rather general mathematical analysis known as Zhukovski's theorem [10] established the lift as $\rho U \Gamma$ per unit span for potential flow of 'two-dimensional' character [9] over a wing of arbitrary (though uniform) section and very large span. Prandtl's more physical analysis [5], on the other hand, employed fundamental properties of vortex lines

(i) to infer the lift per unit span on each cross-section as $\rho U \Gamma$ for rather general 'three-dimensional' wing shapes; and, moreover,

(ii) to quantify the associated 'penalty', in the form of an extra resistance now commonly described as 'drag due to lift'.

The insight essential to these achievements was a recognition that, since vortex lines cannot end in the fluid, any bound vorticity must be linked in a continuous system with trailing vorticity in the wake.

Figure 10.13(*a*) offers a crude schematic view of how the bound vorticity Γ, varying with position z along the wing span (and of course falling to zero in the region beyond the wing tips $z = \pm b$ where no source of vorticity exists), is incorporated within such a pattern. (Note: dotted lines in this diagram are deliberately oversimplified indications that the vortex lines must be closed; also, they constitute a recognition—which can be linked to the above-mentioned Kelvin's Circulation Theorem—that, around any wing with positive camber and/or angle of attack, it is possible to establish a smooth flow with attached boundary layers incorporating bound vorticity Γ only if equal and opposite vorticity—sometimes referred to as a 'starting vortex'—has been shed earlier when the motion was initiated. Actually, patterns of trailing vorticity far behind an aircraft tend to 'roll up' into a pair of concentrated vortices such as are often observed on humid days as 'condensation trails'.)

The vortex wake close behind the wing (solid lines) has a flow pattern as indicated in figure 10.13(*b*)—due to Prandtl [5]—with a downward momentum, given by $\frac{1}{2}\rho$ times the moment of the vorticity distribution,

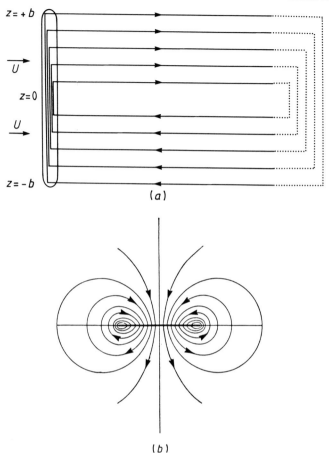

$z = +b$

U

$z = 0$

U

$z = -b$

(a)

(b)

Figure 10.13. *(a) Here, solid lines show schematically the overall pattern of vortex lines (incorporating both bound vorticity $\Gamma(z)$—attached to the wing, and greatest where the vortex lines are most thickly packed—and also trailing vorticity) for a lifting wing-pair in a steady airflow. (For dotted lines, see text.) (b) Prandtl's diagram [5] of the pattern of wake flow in a vertical plane just downstream of the wing associated with this vorticity distribution. Forces on the wing arise from the continued shedding of this pattern of airflow; the lift L being its additional downward momentum per unit time while its kinetic energy per unit length of wake is the drag due to lift, D_L.*

which increases at a rate

$$L = \rho U \int_{-b}^{b} \Gamma \, \mathrm{d}z \tag{23}$$

representing the lift on the wing—each section of which contributes as in (i) above. It is important, too, that each unit length of the growing wake has a calculable kinetic energy, D_L; which, being necessarily supplied from work done in maintaining the wing's steady relative

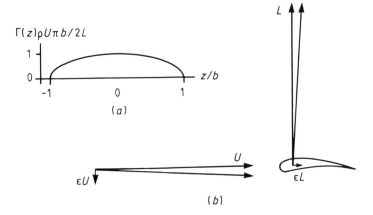

Figure 10.14. (a) This 'elliptic' distribution of circulation was shown by Prandtl to reduce drag due to lift to its minimum value εL, while the associated trailing vorticity would generate at the wing a uniform downward motion εU. (b) Indeed, where a horizontal wind of speed U produces an upward force L, its downward deflexion by the vertical motion εU must give an extra horizontal force εL.

motion through the air per unit distance travelled, is—exactly as in (ii) above—the drag due to lift.

For a given span $2b$, Prandtl successfully demonstrated that the least possible drag due to lift is

$$D_L = \varepsilon L \quad \text{where } \varepsilon = \frac{L}{2\pi \rho U^2 b^2} \qquad (24)$$

this minimum being achieved—amongst all distributions of Γ across the span—by the 'elliptic' shape of distribution actually shown in figure 10.14. Prandtl also uncovered yet another instructive way of looking at this drag due to lift: the wake vorticity ω produces a velocity field v which, at the wing itself, takes the form of a downward motion εU. Locally, this is equivalent to the wind being inclined downwards through an angle ε, and then the expected force L normal to this effective wind is inclined backwards at an angle ε and so includes a 'drag' component εL (figure 10.14).

D'Alembert's Theorem does not, of course, apply to these flows which—far from being nearly potential flows—are strongly influenced by a complicated distribution of vorticity. Yet ε is a small number, of the order of magnitude 10^{-2}, for aircraft wing pairs whose span $2b$ greatly exceeds their other dimensions. Then, as remarked at the end of section 10.1.3, the general spirit of d'Alembert's Theorem is retained in these airflows where huge lift forces, capable (see section 10.4) of sustaining hundreds of tonnes of metal in the air, are achievable with only a modest drag penalty.

10.2.3. Types of transition, types of turbulence: (i) struggles before 1940
However justifiably a historian of twentieth century fluid mechanics may adopt a heroic narrative style for outlining key discoveries on boundary layers and wakes made in the century's first quarter, he must acknowledge that such a style would be out of place for describing the course of twentieth century developments in turbulence research. This has been a field where progress in the twentieth century was much more gradual—notwithstanding the benefits of starting from the strong position of Reynolds's 1880s researches [20]. Only during the 1940s, indeed, did understanding begin substantially to progress beyond the point which Reynolds had reached in those researches and in a powerful theoretical analysis [24] dated 1895.

Although (section 10.2.1) his 1880s studies had concentrated on turbulence in pipe flow, they had recognized too the existence of some different, sharply contrasted types of turbulent flow and of transition to turbulence. Yet investigators during the next half century showed an excessive tendency to regard 'turbulence' as a single, integrated phenomenon, research on any aspect of which would illuminate the field as a whole. This led to prolonged struggles up blind alleys; which may have yielded some refinement of techniques, but need not be chronicled in a brief survey.

Famous papers are rarely recalled in their entirety! For pipe flow itself, Reynolds had not just determined an easily remembered minimum Reynolds number for transition to turbulence. A meticulous observer, he had also uncovered two extraordinary features—intermittency, and abruptness—of that transition; which involved the intermittent, sudden appearance of abrupt 'flashes' of intensely chaotic motion in the midst of an otherwise laminar flow. Only in 1951 were these two features rediscovered, by Howard Emmons, in brilliant experiments at Harvard on transition in boundary layers on flat plates [25]; comprising the intermittent, abrupt initiation of growing 'spots' of highly chaotic flow, each separated by a sharp boundary from a laminar-flow environment, but all merging downstream into a fully turbulent boundary layer (figure 10.15).

For a long time, again, little attention was paid to Reynolds's lists of other types of flow which

(i) show a much greater proneness to transition (by comparison with pipe flow), and yet

(ii) make that transition in more restrained ways which he described as 'sinuous'.

These lists included, among parallel flows, those where the vorticity has a maximum in the flow itself (instead of at a solid boundary); and, among curved flows, those with velocity greatest 'on the inside'. Broadly, they are flows where fluid mechanics theory has proved rather

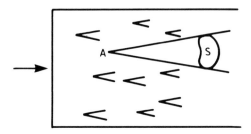

Figure 10.15. *Emmons's observations [25] of turbulent 'spots' in the boundary layer on a flat plate. Different 'spots' originate randomly, each at some point A which is the apex of a wedge within which the spot S subsequently grows as shown. Other wedges (all with about the same included angle) within which spots grow as they are swept downstream are also shown. Turbulent motion is fully developed within each spot. Spots thus randomly initiated merge together, forming ultimately a fully turbulent boundary layer.*

readily able to forecast flow instability (as Rayleigh did first in each case [26, 27]) and where, initially, disturbances predicted to develop unstably are rather similar to those observed. Yet it was only in the second half of the twentieth century that distinctions between 'abrupt' and 'restrained' transition would be properly recognized, and interpreted by ideas from non-linear stability theory—ideas of which the great Russian physicist Landau [28] had sketched the first in 1944.

In the meantime, differences in kind between the diffusion of momentum by molecular viscosity and by turbulence had been well brought out in Reynolds's 1895 paper [24], and may here be illustrated by the case of a parallel airflow with velocity U in the x-direction which varies as a function $U(y)$ of the coordinate y in a perpendicular direction (figure 10.16). If the airflow is turbulent, this $U(y)$ represents the average of a randomly fluctuating fluid velocity, with components

$$U(y) + u, \quad v, \quad w \quad \text{in the } x, \ y, \ z \text{ directions.} \tag{25}$$

Now, although the velocity fluctuations u, v, w have zero means, a product like uv can have a non-zero mean $\langle uv \rangle$, in which case excess x-momentum (ρu per unit volume) is transported in the y-direction at an average rate

$$\rho \langle uv \rangle \tag{26}$$

now known as a Reynolds stress. Energy is then being extracted from the mean flow, and fed into the turbulence, at a rate

$$-\rho \langle uv \rangle \frac{dU}{dy} \tag{27}$$

per unit volume.

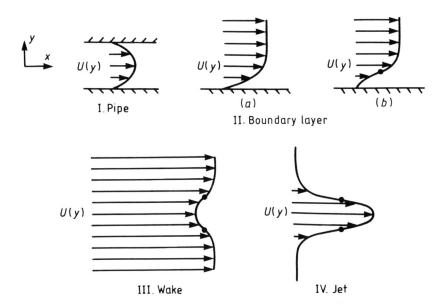

Figure 10.16. *Some (closely) parallel flows: (I) pipe; (II) boundary layer on a flat plate in cases (a) dV/ds positive and (b) dV/ds small and negative; (III) wake; (IV) jet. Vorticity maxima are found at the solid boundary in cases I and II(a) but appear in the flow itself (black circles) in cases II(b), (III) and IV.*

In a laminar flow, by contrast, the only source of momentum transport is viscous diffusion, interpreted of course in an apparently similar way on the kinetic theory of gases—where equations like (25), (26) and (27) may arise (with (25) representing velocities of molecules while angle brackets in (26) and (27) signify weighted means based on molecular weights). These superficial similarities are, however, gravely misleading; molecular velocities u, v, w are enormous (of order the sound speed) but a mean like (26) takes a very small value (the viscous stress, $-\mu dU/dy$) because molecular collisions act constantly to restore thermodynamic equilibrium and it is only over the mean free path between collisions (10^{-4} mm in atmospheric conditions) that a molecule's momentum shows, in a statistical sense, any effective 'persistence'.

Such contrasts not only demonstrate an absolute scale separation between the statistics of molecular velocities in kinetic theory and the statistics of fluid velocities in turbulence, but suggest also that attempts at finding anything analogous to a mean free path in turbulence may be fruitless. Ultimately, those attempts—known as 'mixing length theories' and pursued throughout the century's second quarter [29]— were abandoned in recognition that velocity fluctuations on length scales comparable with the thickness of the entire parallel flow contribute to the Reynolds stress (26).

Interspersed, however, with many less productive attacks on

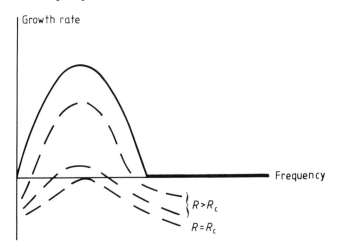

Figure 10.17. *Growth rate of small sinusoidal disturbances to a parallel shear flow: typical results when vorticity maximum is in the midst of the fluid* [11]. *Solid line, vorticity neglected. Broken lines, growth rates for finite Reynolds number R (there are growing disturbances when R exceeds a 'critical' value R_c).*

problems of turbulence and transition, a moderate number of relatively valuable additions to knowledge were made during the period before the 1940s. Tollmien in 1935 completed Rayleigh's work on parallel flows in the limit of zero viscosity by establishing [30] that the existence of a vorticity maximum in the midst of the fluid is not only a necessary, but also a sufficient, condition for their instability to small disturbances. Figure 10.17 shows in a typical case how the rate of growth of disturbances varies with frequency (solid line); while results for finite Reynolds number (broken lines) indicate how, as might be expected, viscous effects diminish that rate of growth [11]. This is a flow where transition to turbulence starts from an initial development of regular, 'sinuous' disturbances at or around the frequency for maximum growth rate.

In (slightly) curved flows, where the condition 'greater velocity on the inside of the bend' for a similar instability to small disturbances in the zero-viscosity limit had in 1916 been demonstrated [27] by Rayleigh, G I Taylor showed seven years later [31] that the stabilizing effect of viscosity could be calculated in an important special case—where experiment and theory were in agreement about the stability boundary. Near the boundary, moreover, the disturbances predicted (rings of vorticity, aligned with the flow direction, of axially alternating sign) were 'robustly' detected in the experiments (figure 10.18).

Instability, of course, was interpreted by practically all scientists before 1940 as just such an instability to small disturbances. It was a paradox for them, therefore, that common flows without any vorticity

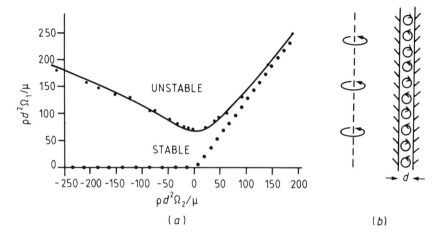

Figure 10.18. *Instability of fluid motion in the annular gap between an inner solid cylinder and an outer hollow cylinder rotating about the broken-line axis at angular velocities Ω_1 and Ω_2, respectively. (a) Solid line, stability boundary calculated by G I Taylor [31], whose experimental points are also shown. Dotted line, stability boundary with viscosity neglected. (b) Form (both predicted and observed) of disturbances.*

maximum in the midst of the fluid—including flow in a pipe or in a boundary layer on a flat plate—are (see above) stable to small disturbances in the zero-viscosity limit; especially, since such stability was presumed to be reinforced by any viscous effects.

The resolution of this paradox emerged in two halves. Prandtl used yet another singular perturbation theory to initiate the first half in 1921; showing [32] that, when the undisturbed vorticity attains its maximum at a solid surface, the effect of viscosity on small disturbances is felt above all within an 'inner surface layer'—considerably thinner than the main boundary layer (or pipe as the case may be)—where viscous action gives the fluctuations in surface frictional force a 45° phase lead over local velocity fluctuations which can tend to destabilize the disturbances. (This phase lead arises from the fact that, for sinusoidal disturbances of frequency ω, an extra term $\rho i \omega v$ on the left-hand side of equation (13) is capable of becoming dominant in such an inner surface layer.) Prandtl's 1921 paper is important in a wider context because it foreshadowed the general 'triple deck' theory (see section 10.4.2) of inner surface layers within main boundary layers.

But once again the powerful analytical skills of W Tollmien were needed to achieve (in 1929) a complete stability calculation [33], taking viscous effects fully into account, for the boundary layer on a flat plate. The unstable disturbances take the form of progressive waves that are exponentially amplified, and the analysis had to allow for viscous effects not only in the inner surface layer of Prandtl but also in a thin 'critical layer' around where the wave speed coincided with the local flow speed. Tollmien's brilliant analysis was given some important detailed refinement four years after by H Schlichting [34], and the waves— whose full significance would emerge only afterwards, with the second half of the resolution of the paradox—are now referred to as Tollmien– Schlichting waves.

Beyond all this work on how transition to turbulence may be initiated, a stride forward in analysing fully developed turbulence was taken by G I Taylor [35] in his 1922 study 'Diffusion by continuous movements'. Here a potent tool, for characterizing how diffusion by chaotic yet continuous fluid flows differs from that achieved by free motions of gas molecules between collisions, was created by extending that idea of a mean product of two velocity fluctuations which appears in the Reynolds stress (26).

The full tensor of Reynolds stresses includes the mean values, not only of uv as in (26) but also of u^2, v^2, w^2, wu and vw; with each mean product (on multiplication by ρ) representing a component of momentum transport. Taylor's innovation was to consider the mean product of a component (u, v or w) of the velocity fluctuation at a point P with its value (or the value of one of the others) at a nearby point Q. For example, if Q lies at a distance r downstream of P, the mean product

$$\langle u_P u_Q \rangle \tag{28}$$

(or 'covariance') coincides with the mean square $\langle u_P^2 \rangle$ (or 'variance') when r is very small; yet the chaotic nature of turbulence implies that, when r becomes large, u_P and u_Q become essentially uncorrelated so that the covariance (28) tends to zero.

The curve of (28) as a function of r gives, then, a first indication of the spatial scales of coordinated motions (or 'eddies') within the turbulence (figure 10.19). Another such indication is given by its Fourier transform $\Phi(k)$ which, by a general statistical theorem, indicates the spectrum of turbulence as a function of the downstream wavenumber k—in the sense that $\Phi(k)\,dk$ represents that part of the variance $\langle u^2 \rangle$ which is contributed to it by sinusoidal components $a\sin(kx + \alpha)$ of u with wavenumbers k in the interval dk.

One type of turbulence to which Taylor's innovations were applied was that studied (in 1929–36) by Hugh Dryden and his colleagues [36] in the course of their development of closed-circuit wind-tunnels. These would become the main test facilities for tackling the central problem (see section 10.2.1) of how stationary objects disturb a steady wind; however, great care was needed to ensure that the wind would indeed be practically steady in the working section of the wind-tunnel (figure 10.20). After the flow downstream of that section is accelerated by a fan, a gradual expansion of cross-section retards it slowly enough (case (b), dV/ds negative but small, of p 820) for separation from the tunnel walls to be avoided. Then its slow passage through a so-called honeycomb

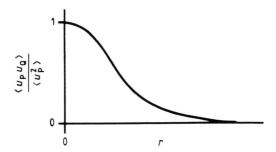

Figure 10.19. *Typical dependence [39] of the covariance (28) on the point Q's distance r downstream of P.*

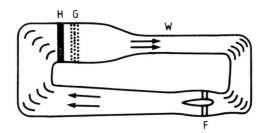

Figure 10.20. *A typical closed-circuit wind-tunnel (schematic). W, working section; F, fan; H, honeycomb; G, gauze screens. Note: the airflow is assisted in turning corners by guide vanes.*

(at Reynolds numbers low enough to retain laminar flow) brings it to a constriction carefully shaped to achieve condition (a), dV/ds positive, so that the wall boundary layers in the working section remain thin while the flow at restored speed between them becomes very nearly uniform. But the ability to demonstrate this demanded the development of refined 'hot-wire' techniques for measurement of the extremely low levels of residual turbulence left in the flow.

Wollaston wire, with a platinum core of radius 5 μm embedded in a silver wire of ten times the radius, can have a very small length of platinum core uncovered (by etching) so that, in an electric circuit, it provides the main resistance—with a value sensitively dependent on the temperature of the platinum. In a wind, this temperature is determined by a balance between ohmic production of heat and its removal by the airflow, whose instantaneous velocity may thus be measured with a suitable circuit. Dryden showed how, provided that this circuit includes compensation for the wire's thermal inertia, such a measurement can have excellent accuracy.

In the meantime the frequency spectrum obtained from the measured velocity fluctuations allows a direct derivation of Taylor's spectrum $\Phi(k)$

with respect to downstream wavenumber (since a sinusoidal component $a\sin(kx + \alpha)$ on being swept past the hot wire at velocity U becomes $a\sin(kx - kUt + \alpha)$, with frequency kU). Also, the use of two hot wires at points P and Q permits the covariance (28) to be measured as a function of the distance r between them.

Turbulence of the type which may remain in the working section of a wind-tunnel was found to be 'isotropic'; that is, with statistical properties which were invariant on any rotation (or reflection) of axes. This permitted valuable determinations—to which both Taylor [37] and von Kármán [38] contributed—of the entire tensor of covariances between different components of velocity at nearby points P and Q in terms of the vector separation between them [39], provided that just one component (28) was known as a function of r when Q is a distance r downstream of P. Similarly, the complete energy spectrum $E(k)$ of velocity fluctuations as a function of the magnitude k of the three-dimensional wavenumber vector could be expressed in terms of Taylor's one-dimensional spectrum $\Phi(k)$ by a differential relationship

$$E(k) = k^3 \left[\frac{\Phi'(k)}{k} \right]'. \tag{29}$$

Yet isotropic turbulence seemed merely to be just one more 'type of turbulence', by no means generic; particularly as the component (26) of Reynolds stress is zero in the isotropic case so that no acquisition (27) of energy from a mean flow is possible and the turbulent energy must always decay as a function of time.

10.2.4. *Types of transition, types of turbulence: (ii) the new taxonomy*

Many of the difficulties outlined above began to be resolved in papers which appeared in the early 1940s yet—because of restricted communication between scientists at that time—would enjoy wide appreciation only later. Here, just brief indications are given of how work initiated in those papers led thereafter to rather systematic developments of the 'taxonomy' both of transition and of fully developed turbulence.

Regarding transition in boundary layers, two very different 1940 papers carried knowledge of their stability to small disturbances far beyond the analysis of Tollmien–Schlichting waves for a boundary layer on a flat plate. On the one hand, for boundary layers treated as parallel flows, Schlichting himself showed [40] how, for slightly retarded flows (type (b), dV/ds small but negative, of p 820, with vorticity abstracted at the wall), the presence of a vorticity maximum in the midst of the layer implied instability to disturbances with a wide range of frequencies as in figure 10.17; moreover, a relatively low Reynolds number sufficed for the onset of this instability. On the other hand, Görtler showed [41] how boundary layers on curved walls, with the curvature concave to the

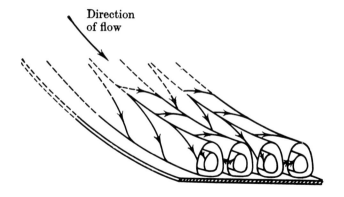

Direction
of flow

Figure 10.21.   *The thin boundary layer on a concave surface is unstable (Görtler [41]) to the onset of Taylor–Görtler vortices illustrated in this (magnified) diagram, and regarded as analogous to the vortices found by Taylor [31] when an inner cylinder rotates in a stationary outer cylinder (figure 10.18 in case $\Omega_2 = 0$).*

flow (so that velocity is greatest on the inside of the bend), are subject to unstably developing disturbances with vorticity aligned to the flow, as had been studied (figure 10.18) by Taylor [31]; such boundary-layer vortices are now called Taylor–Görtler vortices (figure 10.21).

Then, just four years after these significant further contributions to 'the first half' (p 829) of the resolution of the boundary-layer transition paradox, a paper from Russia by L D Landau [28] was to sow the seeds of a great mass of radically new theory, that would branch out far beyond studies of stability to small disturbances in contributing to 'the second half'. Although this mass of (non-linear) theory would, in its later ramifications, become formidably complex, something of its essence may be perceived already in a brief summary of Landau's 1944 paper.

Essentially, he attempted in that paper to model the distinction (p 826) between 'abrupt' and 'restrained' transition. Here, for example, 'abrupt' transition may be found in systems with a vorticity maximum at the wall, which tend to become unstable to small disturbances when R—the Reynolds number (20)—reaches a 'critical' value R_c which is relatively high; R_c being defined so that such small disturbances have an amplitude A which changes *exponentially* like

$$e^{\gamma t} \text{ with } \gamma < 0 \text{ when } R < R_c \text{ but } \gamma > 0 \text{ when } R > R_c. \qquad (30)$$

In these systems, moreover, even positive values of the growth rate γ may be only moderate. Yet any rather large disturbance can cause vorticity to be abstracted at the wall so that the vorticity maximum moves into the fluid, generating momentarily an unstable local motion with a much lower R_c and a far greater amplification of disturbances. By contrast, systems tending to exhibit—at least in early stages of transition—a regular or 'sinuous' form of disturbance, may often become unstable more readily (having much lower values of R_c) but may be subject to an effect by which, at least initially, the appearance of larger amplitudes 'restrains' the exponential growth (30) of disturbances.

Landau's highly simplified mathematical model of these differences takes the form of an equation

$$\frac{d(A^2)}{dt} = 2\gamma A^2 - \alpha A^4 \tag{31}$$

for a variable 'amplitude squared' proportional to the energy of disturbances to the mean flow. For very small A the right-hand side of (31) is dominated by its first term (evidently, one wholly consistent with the behaviour (30) for small disturbances), but Landau's second term gives a preliminary idea, on an assumption (acknowledged by him as oversimplified) that only the amplitude A of a single mode is important, of how any growth of amplitude may be either restrained ($\alpha > 0$) or enhanced ($\alpha < 0$).

Now it turns out that the substitution $y = (1/A^2) - (\alpha/2\gamma)$ throws equation (31) into the very simple form (3)—with 2γ for k and t for x— of which (4) is the general solution. This means that Landau's equation (31) has the general solution

$$A^2 = \frac{1}{(\alpha/2\gamma) + ae^{-2\gamma t}} \tag{32}$$

whose form is plotted (figure 10.22) in two cases:

(i) restrained transition ($\alpha > 0$) where the solid lines give the behaviour of disturbances when $\gamma > 0$ (implying, by (30), that $R > R_c$); and

(ii) abrupt transition ($\alpha < 0$) where disturbances with $\gamma < 0$ (with subcritical Reynolds number $R < R_c$) behave as shown by the broken lines.

(Note: all solutions assume one of the shapes shown, though with a possible horizontal shift.)

In case (i), very small disturbances begin by growing exponentially. However, this does not continue indefinitely; after a certain time, a definite level of disturbance is maintained—as in the example (figure 10.18) of Taylor vortices.

In case (ii), by contrast, an 'abrupt' transition becomes possible already at subcritical Reynolds number (with $\gamma < 0$). There is a certain threshold level of energy $A^2 = 2\gamma/\alpha$ such that, for initial disturbances below it, $a > 0$ and so the disturbance energy damps out to zero. On the other hand, for initial energies above that threshold, $a < 0$; and then there is a runaway instability—the 'abrupt' form of transition—such that,

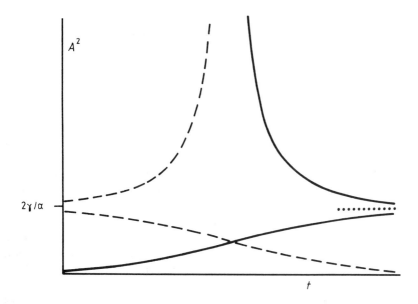

Figure 10.22. *Solutions of equation (31). Solid lines, case where $\alpha > 0$, $\gamma > 0$. Broken lines, case where $\alpha < 0$, $\gamma < 0$.*

in a finite time, the energy has already grown without limit (reaching levels, of course, where other terms besides those retained in equation (31) become important).

For the paradox concerning transition to turbulence in the boundary layer on a flat plate, such considerations opened the way to completing 'the second half' of its resolution, based on the layer's enhanced instability to disturbances above a certain level. Both halves were triumphantly confirmed soon after by Galen Schubauer and his colleagues at the National Bureau of Standards, Washington, DC (the inheritors, and further developers, of Dryden's precision techniques for aerodynamic instrumentation).

They showed on the one hand [42] that, when an enormous effort was made to suppress flow disturbances in the wind-tunnel's working section, and also any roughness elements (common sources of disturbances) on the solid surface, it became possible to excite growing Tollmien–Schlichting waves, at the predicted frequencies, by very small vibrations of a 'ribbon' flush with the surface (figure 10.23). This discovery has since been developed much further in researches detailing all those stages through which the growing waves develop three-dimensional characteristics and, ultimately, the chaotic nature of turbulence.

But the same group had next followed up Emmons's discovery [25] of randomly originating turbulent 'spots' (p 825) with a comprehensive demonstration [43] that these indeed were the typical form of transition

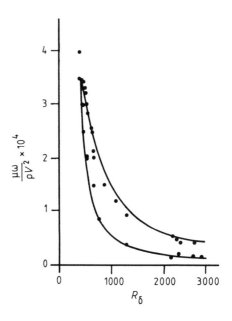

Figure 10.23. *Instability of boundary layer on a flat plate to small wave-like disturbances. These are able to grow, at given Reynolds number R_δ, for frequencies ω between the upper and lower branches of the 'neutral curve' shown. (——, predicted; ●, observed) neutral curve. (Note: here, the value of δ used to define R_δ is the 'displacement thickness' of the layer, defined so that its defect in mass flux is $V\delta$.)*

in boundary layers on flat plates (and also in pipes and channels), with each spot growing in dimensions amid the surrounding laminar flow at quite a regular rate as it is swept downstream. Locally, such a transition takes a literally 'abrupt' form in the sense that within each spot an almost fully developed turbulence is already present; thus, fluid on passing inside a spot adopts almost instantaneously a chaotic form of motion. On the other hand, some averaged quantity, like the frictional stress in the wall, may exhibit just a gradual change through a 'transition region' that extends from where the spots are first making their appearance to where the statistical assemblage of spots has grown to fill the whole layer (figure 10.15).

Subsequent work to uncover those mechanisms by which the spots of turbulence maintain sharp boundaries as they grow within the surrounding laminar flow has made rather slow progress; mainly, by using ideas (cognate to those of section 10.1.4) from the theory of non-linear wave propagation—although spots need of course to be viewed [44] as three-dimensional 'wave packets' (see section 10.3) rather than as anything similar to long-crested waves. Only in the century's concluding decade do answers to this challenging question seem at last to be emerging [45].

In the meantime, many other theoretical studies of transition at subcritical Reynolds numbers have followed somewhat more closely Landau's idea of improving the equation for development of a small-disturbance mode with a higher-order addition—but, now, one calculated from the equations of fluid mechanics [46, 11]. The most important higher-order effect is that associated with distortion of the mean-flow distribution $U(y)$ by action of a Reynolds stress (26) calculated from values of the small-disturbance velocities u and v. At a given subcritical Reynolds number, this once again yields an amplitude level above which the mode grows without limit.

For restrained transition (case (i) above) theoretical and experimental studies have combined to elucidate that 'spectral evolution' of disturbances which, at sufficiently supercritical Reynolds number, follows any initial amplitude growth at the frequency for which the system is most unstable. By the 1960s already, many papers were demonstrating those sequences of 'period-doubling bifurcations' which would later be acknowledged as standard routes to chaos. These may occur either as a process in time [47, 48] or, alternatively [49], as a gradual development with increase of R beyond R_c in a system like that of figure 10.18. In either case, the end result is fully developed turbulence.

For turbulence itself, just as for transition, a key paper offering the subject increased coherence emerged from Russia in the early 1940s. Kolmogorov's 1941 paper [50] explained how, even though ideas of turbulence 'as a single integrated phenomenon' are erroneous (section 10.2.3), and although turbulence (like transition) exists in a variegated taxonomy of types, nonetheless all those types have their small-scale features in common at very high Reynolds number.

Moreover, because one of the types is isotropic turbulence (section 10.2.3), which obeys certain special rules, including equation (29) for the spectrum $E(k)$ of turbulent energy, and because any small-scale features are associated with properties of spectra at relatively large magnitudes k of the wavenumber vector, this law of Kolmogorov implies that every type of turbulence becomes isotropic at relatively large wavenumbers and obeys those special rules. Isotropic turbulence has, then, this one 'generic' property (see the end of section 10.2.3), that all turbulence exhibits locally isotropic small-scale behaviour at very high Reynolds number.

In brief summary, Kolmogorov's analysis allows for the fact that, while $E(k)$ is the spectrum of the kinetic energy, $\frac{1}{2}(u^2 + v^2 + w^2)$ per unit mass, a corresponding spectrum of the rate

$$\varepsilon \text{ per unit mass} \tag{33}$$

of viscous dissipation of turbulent energy into heat is a spectrum of squares (and products) of gradients of u, v and w which should be proportional to $k^2 E(k)$; it turns out, for example, to be $(2\mu/\rho)k^2 E(k)$ for isotropic turbulence. Now, in a long spectrum, $k^2 E(k)$ reaches its maximum at much higher wavenumbers than $E(k)$, and this leads to the concept of energy being dissipated in 'small eddies' (that is, spectral components of motion with large k) even though the main energy-containing eddies have a much larger scale (smaller k). Also, expression (27) for the rate at which energy is fed into the turbulence includes a product of velocities u and v (not of their gradients) and its spectral peak is close to that of the turbulent energy itself.

Turbulence is a process, then, where the main energy-containing eddies, while receiving energy directly, are passing it down a 'cascade' of eddy sizes to those small eddies which dissipate it into heat. At every stage of the cascade, non-linear effects stemming from the basic non-linearity of momentum transport tend to generate 'overtones' or 'summation tones' of somewhat higher wavenumber; yet many successive stages, all involving random elements, are needed to reach large wavenumbers. After just a few stages, indeed, the wavenumber spectrum $E(k)$ becomes 'statistically decoupled' from the main energy-containing eddies, and assumes a universal 'equilibrium' form, independent of the type of turbulence in which the small eddies are embedded.

The principles of dimensional analysis (section 10.2.1) tell us that this large-wavenumber behaviour of the energy spectrum $E(k)$, dependent only on the four variables k, ρ, μ and ε (see (33) above), must really involve the non-dimensional form

$$\varepsilon^{-2/3} k^{5/3} E(k) \tag{34}$$

of the spectrum in a dependence on just a single non-dimensional variable ηk, where

$$\eta = (\mu/\rho)^{3/4} \varepsilon^{-1/4} \tag{35}$$

is known as the Kolmogorov dissipation length and characterizes the size of the energy-dissipating eddies. Extensive experimental data support this conclusion, with a dependence of expression (34) on ηk close to that shown on a log–log plot (solid line, derived from Pao's 1965 studies [51]) in figure 10.24. It is noteworthy that, within the 'equilibrium' range of wavenumbers where this dependence applies, there is a subrange $\eta k < 0.1$ (commonly called 'inertial' subrange) for which expression (34) is approximately constant, with $E(k)$ decreasing like $k^{-5/3}$; yet the spectrum of energy dissipation, proportional to $k^2 E(k)$, is still increasing (broken line) towards its peak around $\eta k = 0.3$.

In parallel with the researches on these large-wavenumber features common to different types of turbulence, studies of specific aspects of each type at the level of the energy-containing eddies made steady progress. Here, although space does not allow detailed descriptions, it

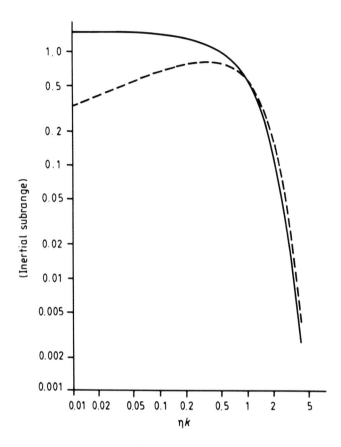

Figure 10.24. *Energy spectrum of turbulence in the equilibrium range. The solid line shows the dependence of expression (34) on ηk, while the broken line plots a quantity $(\varepsilon^{-2/3}\eta^{1/3})k^2 E(k)$, proportional to the spectrum of energy dissipation.*

may again be interesting to note an individual property which all the types have in common, as suggested first by Alan Townsend in his 1956 book [52].

Townsend amassed evidence that a role of some importance in guiding the development of turbulence was played by what he called 'big eddies'. These were substantially larger in scale than the main energy-containing eddies, and tended to fill the whole of any shear layer (boundary layer, wake, jet); in recent years they have often been described as 'coherent structures' within some largely chaotic shear layer.

The full wavenumber spectrum of turbulence is analysed, then, (figure 10.25) into 'big', 'energy-containing' and 'small' eddies—with the equilibrium range of 'small' eddies subdivided further (as the log–log plot of figure 10.24 shows best) into an inertial subrange and the main energy-dissipating eddies. Here, Townsend's graph reproduced as figure

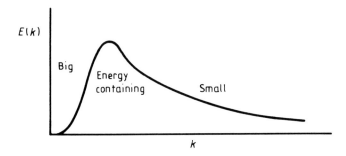

Figure 10.25. *Analysis of the energy spectrum $E(k)$, derived [53] from measurements of isotropic turbulence behind a grid of bars of mesh M in a stream of velocity U, where $\rho U M/\mu = 5300$, into 'big', 'energy containing' and 'small' eddies. (Note: linear scales for k and $E(k)$ are used here; for a continuation of the small-eddy range on a log–log scale, see figure 10.24.)*

10.25 is a spectrum derived (1951) by equation (29) from measurements of $\Phi(k)$ for a particular case of isotropic turbulence [53], but such a 'fourway split' is now seen as characteristic of turbulence in general.

For example, in any type of turbulence adjacent to a solid surface, those big eddies which exercise a guiding role in its development have the form of elongated vortices, their vorticity being aligned to the wind but with crosswind alternations of sign. These, as indicated in figure 10.26 (where the mean flow is perpendicular to the paper), have two important effects on the mean vorticity (here, directed from left to right):

(a) they produce a correlation of vortex-line stretching (p 818) with inflow, which greatly increases vorticity levels near the solid surface [11]; yet

(b) elsewhere, they push vorticity maxima out into the midst of the fluid so that, from this highly unstable configuration, there emerge 'bursts' of intense turbulence which feed the system as a whole [54].

Both effects (a) and (b) influence also the distribution of mean velocity $U(y)$ found near a solid surface (figure 10.27): associated with a given frictional stress at the surface, there is an equilibrium (and approximately logarithmic) distribution of $U(y)$, aptly named 'law of the wall' in the 1955 studies of D Coles [55]—who interpreted it a lot more convincingly than had the investigators (obsessed with 'mixing length' ideas) of the 1930s.

But turbulent flows away from any solid surface—with mean-flow distributions referred to as 'law of the wake' by Coles [56]—are influenced very differently by their big eddies; whose mixing action tends not to intensify differences in the shapes of mean-flow distributions from those characteristic of laminar flow, but to smooth them away. An extreme example of this is the turbulent jet, with its aeronautical importance (section 10.4) for exerting a thrust F equal to the total rate of

Figure 10.26. *Tendency of 'big eddies' in a turbulent boundary layer to generate both (a) correlation of vortex-line stretching with inflow and (b) convection of vorticity maxima into the midst of the fluid.*

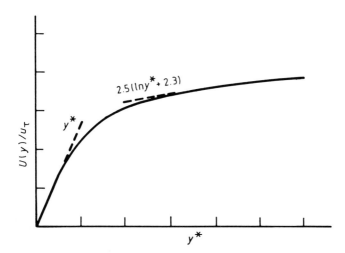

Figure 10.27. *Illustrating how, in turbulent flow near a smooth solid surface on which a frictional stress $\tau = \mu U'(0)$ is exerted, the mean velocity $U(y)$ may be related to the 'friction velocity', $u_\tau = (\tau/\rho)^{1/2}$, and to the non-dimensional distance from the wall, $y^* = \rho u_\tau y/\mu$.*

momentum transport in the jet. The quantity $(\rho F)^{1/2}$ has the dimensions of viscosity, and the fully developed jet flow has an effective Reynolds number $(\rho F)^{1/2}/\mu$ independent of distance from the orifice (indeed, its diameter increases linearly with distance, in inverse proportion to the diminution in average velocity). Actually, a laminar flow with a viscosity of about

$$0.055(\rho F)^{1/2} \tag{36}$$

would have a velocity distribution (figure 10.28) almost identical to that of the mean flow [58] in a fully developed turbulent jet!—which leads to expression (36) being often referred to as the effective 'eddy viscosity'.

10.2.5. The diffusion–convection balance for scalar quantities

Stationary solid objects in a wind (the main topic of section 10.2), besides being subject to forces, are often cooled and/or dried. In other words, it is not only vector quantities like momentum—or vorticity—that may be subject in boundary layers to the combined action of convection and

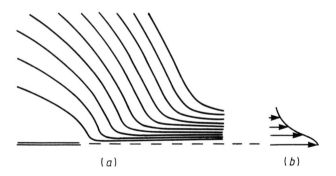

Figure 10.28. *(a) Flow lines for an air jet with momentum flux F, as calculated* [57] *for laminar flow with the viscosity (36). (Flow is symmetrical about the broken lines.) Note the entrainment of external air into the jet, which causes its mass flux (although not its momentum flux) to increase with distance from the orifice. (b) Distribution of velocity in the jet.*

diffusion, but also scalar quantities like heat or water vapour. Prandtl himself recognized early [59] how, in particular, this must overturn views on the modes (radiation, conduction, convection) of heat transfer.

It used to be thought that, apart from any radiative loss, a body in a wind would lose heat exclusively by convection. This is impossible, however, since at the solid surface itself the air is stationary (section 10.1.3). Conduction must, therefore, operate to move heat out into the boundary layer before convection can sweep it away.

Accordingly, the rate of heat transfer from unit area of solid surface at temperature T_s into a wind at temperature T_w may be written as

$$k\frac{T_s - T_w}{h} \tag{37}$$

where k is thermal conductivity and h is a certain fraction of the boundary-layer thickness δ (numerically derived for an important case in 1921 by Prandtl's colleague K Pohlhausen [60], and calculated later in many other cases). Just like viscosity, then, conduction of heat can never be disregarded.

Similar remarks apply to water vapour, which is transferred from a wet surface at a rate

$$D\frac{q_s - q_w}{h} \tag{38}$$

here, D is the diffusivity of water vapour, defined as rate of transfer of water mass across unit area divided by the gradient in its volume concentration q—while q is q_w in the wind and takes its saturated value q_s at the surface. Actually, this diffusivity D for water vapour is almost identical numerically with the diffusivity of heat (which is $k/\rho c_p$ because the volume concentration of heat has its gradient in a ratio ρc_p to that of temperature), and it follows that h is the same fraction of δ in both (37) and (38), being determined by an identical diffusion–convection balance in each case [11].

It was G I Taylor (1933) who first saw [61] how this explains why a wet body in a wind reaches the equilibrium ('wet-bulb') temperature

$$T_s = T_w - \frac{L_v D}{k}(q_s - q_w) \tag{39}$$

where L_v is the latent heat of evaporation. Equation (39) is called for by the condition that the rates of transfer of sensible heat (37) and of latent heat (expression (38) multiplied by L_v) add up to zero.

In the case of a turbulent boundary layer, the turbulence itself generates practically all the diffusion; whether of momentum, heat or water vapour (in laminar flow, by contrast, the diffusivity μ/ρ of momentum is about 30% less than that of heat or water vapour). At the extraordinarily early date of 1874, Osborne Reynolds had noticed [62] how, in any turbulent flow, this implied almost identical diffusion–convection balances for momentum and for heat (now known as the Reynolds analogy).

Thus, for a body in a wind of speed U, if τ is the frictional stress at the surface (rate of transfer of momentum to unit area) and Q is the heat transfer rate, then to a close approximation

$$\frac{Q}{\rho c_p (T_s - T_w)} = \frac{\tau}{\rho U} \tag{40}$$

in turbulent flow, because the same diffusion–convection balance governs the right-hand side (momentum transfer divided by momentum per unit volume in the wind) and the left-hand side (heat transfer divided by difference in heat content per unit volume between surface and wind). Equation (40), to which a similar result for water–vapour transfer in turbulent flow may be added, has continued to be widely applied in twentieth century fluid mechanics [29].

Although all further discussion of 'forced convection' (heat transfer from, or to, a body in a wind) is postponed to section 10.4.2, some brief comments are needed here on 'free convection', which, in the absence of any wind, is generated simply by buoyancy forces acting on air warmed by heat transfer from the body. In free convection from a vertical surface (to be described first), it was again a penetrating analysis by Pohlhausen, working in collaboration with the brilliant experimenter E Schmidt, that established a proper boundary-layer description of the phenomenon, and won excellent agreement (see figure 10.29) between theory and experiment [63].

In the boundary-layer equation (13), one term $\rho V \, dV/ds$ could be suppressed because there was no flow outside the layer, but an additional

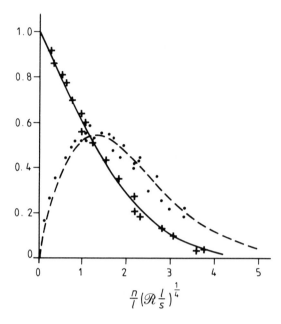

Figure 10.29. *The free-convection boundary layer on a vertical flat plate of height l maintained at a temperature T_1 exceeding the ambient air temperature T_0. Calculated and measured dependence [63] of the temperature T and the velocity v in the boundary layer at a normal distance n from the plate and at a height s above its base.*

force per unit volume—the buoyancy force $\rho g(T - T_0)/T_0$, where T_0 is the ambient air temperature—needed to be incorporated on the right-hand side

$$\rho v \frac{\mathrm{d}v}{\mathrm{d}s} = \rho g \frac{(T - T_0)}{T_0} + \mu \frac{\mathrm{d}^2 v}{\mathrm{d}n^2} \qquad (41)$$

while the diffusion–convection balance for the temperature T could be similarly expressed as

$$\rho c_p v \frac{\mathrm{d}T}{\mathrm{d}s} = k \frac{\mathrm{d}^2 T}{\mathrm{d}n^2}. \qquad (42)$$

Pohlhausen had used equation (42) on its own to solve the forced-convection problem [60], with v obtained from an independent solution of (13); but now it was necessary [64] to solve (41) and (42) as two coupled equations to obtain the results of figure 10.29.

In such free-convection problems the absence of any external wind U means that the Reynolds number (20) has to be replaced by a different non-dimensional parameter, known (for reasons indicated below) as the Rayleigh number $\mathcal{R}$. The form

$$\mathcal{R} = g \frac{T_1 - T_0}{T_0} \ell^3 \left(\frac{\rho}{\mu}\right) \left(\frac{\rho c_p}{k}\right) \qquad (43)$$

of $\mathcal{R}$, for a vertical surface of height ℓ maintained at temperature T_1, is suggested by the closely parallel importance of the diffusivities μ/ρ in (41) and $k/\rho c_p$ in (42). For the total rate of heat transfer from unit width of such a vertical surface, Pohlhausen obtained the expression

$$0.52\mathcal{R}^{1/4}k(T_1 - T_0) \tag{44}$$

proportional to the fourth root of the Rayleigh number.

Similar calculations can be made (but with g replaced by its component parallel to the surface) for a flat surface at an angle to the vertical—provided that the angle is acute! By contrast, the nature of free convection above a horizontal surface is altogether different, depending entirely on the instability of fluid heated from below.

Aspects of this problem on the huge length scales ℓ (more precisely, huge values of $\mathcal{R}$) that are relevant in meteorology involve, as described in section 10.5, a highly chaotic motion; with randomly spaced—and timed—ascents through cool air, from a hot ground, of those masses of warm air known as 'thermals'. This main section 10.2, however, is now concluded with a brief account of free-convection instability on a laboratory scale, that follows on rather naturally from earlier discussions of instability and transition.

See also p 899

Laboratory experiments on fluid heated from below are carried out either

(a) on a gas or liquid between two solid horizontal surfaces, or
(b) on a liquid between a solid surface and a free surface.

Some famous 1901 experiments of Bénard [65] had used system (b); much later (1958), however, it was recognized [66] that the instability which they demonstrated was primarily driven not by buoyancy forces, but by the temperature dependence of surface tension! (Warm liquid rising to the surface is pulled onwards by adjacent cold liquid with its greater surface tension.)

For problem (a) on the other hand, coupled equations similar in general character to (41) and (42), but without any simplifying boundary-layer approximation, may be written down and analysed in the small-disturbance limit. This was first done (1916) by Rayleigh [67], who, however, was not able to solve them with realistic boundary conditions. Later workers [68] showed in case (a) that the critical value $\mathcal{R}_c$ (minimum value of $\mathcal{R}$ for instability to small disturbances) is 1700. (In a liquid, the T_0^{-1} factor in (41) and (43) needs to be replaced by the liquid's coefficient of cubic expansion.)

The mode of disturbance to which the system is most unstable takes the form of long 'rolls'; moreover, transition is of the restrained type—corresponding to a positive α in Landau's equation (31)—so that the rolls assume a definite amplitude when $\mathcal{R}$ only moderately exceeds $\mathcal{R}_c$. The

Figure 10.30. *Free convection: motions of fluid heated from below, made visible by Koschmieder* [69] *in a cylindrical container between two horizontal circular plates of glass.*

1966 experiments of Koschmieder [69] showed that the detailed form of the rolls within a container reflects above all its side-wall geometry. Figure 10.30, indicating the true form of free-convection instability in liquid between two horizontal glass plates, is here offered as a sort of counterpoise to the historical overemphasis on 'Bénard cells'.

10.3. Non-linear effects on the generation and propagation of waves

10.3.1. Waves with hidden energy loss

Waves and flows interact [70]; in a sound wave, for example, values of the air velocity u are propagated (p 809) at a signal velocity $u + c$ (the local sound speed relative to the local air motion) which exceeds the undisturbed sound speed c_0 by an amount $1.2u$ of which five-sixths† results from convection of sound by the air motion u. This section 10.3 outlines twentieth century developments in the understanding of wave/flow interactions, beginning (section 10.3.1) with further consequences of larger values of u 'catching up' with smaller ones

† *Sound waves in water (with its different compressibility properties) behave similarly except that the coefficient 1.2 is replaced by about 4, so that only one quarter (instead of five-sixths) of the excess signal velocity results from convection in this case.*

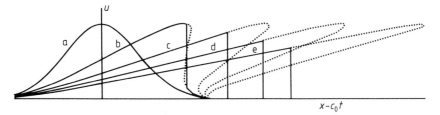

Figure 10.31. *Propagation of plane sound pulse. Curves a and b reproduce from figure 10.5 the initial wave form and an early wave form incorporating a shock wave, while curves c, d and e show—at equal time intervals—the later development of the wave form (derived from the dotted curves by Whitham's equal-area rule).*

and going on (section 10.3.2) to the effects of more complex (e.g. turbulent) flows on sound, before illustrating some very different types of interaction between flows and dispersive waves (those where the small-disturbance wave speed is not a constant c_0 as with sound but may depend e.g. on frequency).

For plane sound waves, consequences of the fundamental 1910 discovery (section 10.1.4) that 'catching up' proceeds until an extremely sharp discontinuity is formed—the shock wave, with thickness proportional to a diffusivity so that diffusion can balance convection within it whilst also dissipating wave energy into heat—were gradually developed during the next half-century. Essentially, the conclusion [70] was that sound propagation outside the shock wave continues to be governed by constant-entropy laws (section 10.1.4) because the very slight entropy inhomogeneities present in air traversed by a shock wave of changing strength have negligible effects on propagation. By contrast, the overall wave energy is progressively diminished by the accumulated effects of energy losses 'hidden' inside such an apparent discontinuity.

Not only energy, but also information, disappears in this process!—as may be illustrated by continued tracing (figure 10.31) of the non-linear propagation, with shock-wave formation, depicted in figure 10.5. At each stage the necessary discontinuity is incorporated at that location for which overall mass conservation is retained; this condition, the 'equal area' law of Whitham [71], requires that the discontinuity leaves unaltered the area under the curve (strictly, area under the corresponding curve of density ρ against distance [70]—although a closely linear relationship of density variations to u allows the law to be applied, with good approximation, to the graph of u itself).

Absolutely all information about the original shape of the compression pulse—except about the area Q under the curve—has disappeared in the later stages of this process, when the waveform has become a right-angled triangle of area Q with a hypotenuse of slope $(1.2t)^{-1}$ (the reciprocal slope $\delta x/\delta u$ increases at a rate 1.2 because a value $u + \delta u$ propagates faster than a value u by a signal-speed excess of

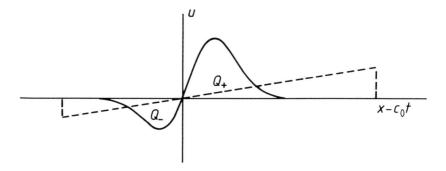

Figure 10.32. *Case of an initial waveform (solid line) with positive lobe of area Q_+ and negative lobe of area Q_-. Broken line, subsequently developing N-wave with unchanged areas for positive and negative lobes.*

$1.2\delta u$). Such a triangular waveform has a height $(5Q/3t)^{1/2}$—the strength of the shock wave—and a length $(12Qt/5)^{1/2}$. Moreover the energy of the wave, proportional to length and to amplitude squared, diminishes as $t^{-1/2}$ (what an enormous contrast to energy attenuation following an exponential law, as found for small-amplitude waves!) and the rate of attenuation, proportional to $t^{-3/2}$, is found to agree exactly with the rate of energy loss 'hidden' inside the discontinuity.

The corresponding conclusion for a sound pulse incorporating negative as well as positive values of u is that it develops asymptotically into an 'N-wave' (figure 10.32) which consists of two right-angled triangles, with areas Q_+ above the $u = 0$ axis and Q_- below it and with aligned hypotenuses of slope $(1.2t)^{-1}$. The shock strengths are $(5Q_+/3t)^{1/2}$ and $(5Q_-/3t)^{-1/2}$ and the overall length is $(12Q_+t/5)^{1/2} + (12Q_-t/5)^{-1/2}$; while, yet again, wave energy diminishes as $t^{-1/2}$ with all energy loss 'hidden' inside the two shock waves.

It is not just in plane waves that such behaviour is found; as may be illustrated (again, following Whitham [71]) by analysing the 'exploding wire' phenomenon that results from instantaneous discharge of a condenser through a very thin wire. A cylindrically expanding sound wave is generated in the surrounding air by the wire's sudden vaporization; however, an interesting difference from the plane-wave case emerges already from a simple linear analysis of the type familiar to nineteenth century physicists: cylindrical wave propagation, on such an analysis, converts the sound source (the outward-pushing vapour) into a wave with both positive and negative values of the outward velocity u. Thus, at radial distances r beyond a certain value r_1

$$u = \left(\frac{r_1}{r}\right)^{1/2} u_1(t - c_0^{-1}r) \qquad (45)$$

where the waveform $u_1(t)$ comprises a positive phase followed immediately by a negative phase of identical area Q (making $\int u_1(t)\,dt = 0$, and thus reconciling the necessary r^{-1} dependence of energy flux to the impossibility of any indefinite growth with r in the total outward displacement $2\pi r \int u\, dt$).

This means that, on non-linear theory, an N-wave (and, indeed, a 'balanced' N-wave with equal areas $Q_+ = Q_- = Q$) is necessarily formed. Whitham's analysis showed how a value $u_1 = (r/r_1)^{1/2}u$ is once more propagated at a signal speed $dr/dt = c_0 + 1.2u$ (replacing the simple c_0 value of (45)), giving to a close approximation

$$\frac{dt}{dr} = c_0^{-1} - 1.2uc_0^{-2} = c_0^{-1} - 1.2\left(\frac{r_1}{r}\right)^{1/2}u_1c_0^{-2} \qquad (46)$$

which can be integrated to give

$$t = c_0^{-1}r - 0.6(r_1r)^{1/2}u_1c_0^{-2} + \text{ constant} \qquad (47)$$

as the altered time when a given value of u_1 is found. Thus the temporal waveform, of u_1 as a function of the time t, is now sheared (backwards) in such a way that the reciprocal slope $\delta t/\delta u_1$ takes an asymptotic value

$$-0.6(r_1r)^{1/2}c_0^{-2} \qquad (48)$$

whose magnitude increases indefinitely with r.

The ultimate N-wave therefore takes the form of two right-angled triangles of area Q whose aligned hypotenuses have the reciprocal slope (48). Each triangle has a height (discontinuity in u_1) equal to

$$\left[Qc_0^2/0.3(r_1r)^{1/2}\right]^{1/2} \qquad (49)$$

which varies as $r^{-1/4}$; but, since $u = (r_1/r)^{1/2}u_1$, this yields Whitham's remarkable inverse-three-quarters-power prediction for the strengths of both shock waves—which has been accurately verified in experiments on exploding wires (figure 10.33). The time interval separating the shock waves takes a value

$$\left[1.2Q(r_1r)^{1/2}c_0^{-2}\right]^{1/2} \qquad (50)$$

which increases as the fourth root $r^{1/4}$ of the distance travelled, while the wave energy decays as $r^{-1/4}$; with all the energy dissipation hidden, once more, inside the two shock waves.

Although the small part of twentieth century research on shock-wave dynamics that is mentioned in Chapter 10 has been selected with the aim of omitting studies on explosions for warlike purposes, the brief introduction given here (and in section 10.1.4) is further amplified in section 10.4.2. There, in descriptions of aeronautically generated shock waves [72], the 'double bang' described above as an N-wave is again characteristically found.

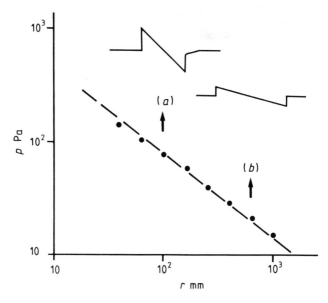

Figure 10.33. *Illustrating how an exploding wire generates pressure–time curves of N-wave shape, (a) and (b), at different radial distances r, and comparing the pressure rise at the initial shock wave with an $r^{-3/4}$ law (broken line).*

10.3.2. Sound from flows

Rayleigh's genius for recognizing when a phenomenon was simply unexplainable on the basis of existing knowledge (see section 10.1.4) was exhibited too in his 1879 paper on the aeolian harp [73], an instrument which responded with 'eerie' or 'ethereal' sounds when placed in a wind. Strouhal had accumulated data on the tones emitted by a wire of diameter ℓ in a wind of speed U, expressing their frequency f in a non-dimensional form

$$S = \frac{f\ell}{U} \tag{51}$$

which under the name Strouhal number is still used in many contexts; but he called them Reibungstöne in a false analogy with the effect of a bow on a violin string. Rayleigh preferred the expression 'aeolian tones' and showed that they could be present with or without vibrations of the wire but that such vibration when it occurred was at right angles to the wind direction. He correctly concluded [13] that production of the aeolian tone must be a consequence of some fluid-mechanical phenomenon still to be discovered.

The challenge was answered in various accounts of experiments (beginning with a—this time, flawless—paper [74] published in 1908 by H Bénard) which Theodore von Kármán would interpret in a famous 1911 study [75]. At Reynolds numbers between about 50 and about 2000, the entire flow past the wire exhibits an instability of the 'restrained' type

Figure 10.34. *The Kármán vortex street in the wake of a thin wire (observations [76] at $R = 71$).*

(section 10.2.4), involving the periodic shedding of vorticity of opposite signs from the upper and lower surfaces of the wire to form a well defined 'Kármán Vortex Street' (figure 10.34) behind it. The wake has a steady rate of growth in its momentum ($\frac{1}{2}\rho$ times the moment of the vorticity distribution) in a direction opposed to the wind, which Kármán could identify with the drag (equal and opposite to the force with which the wire acts on the air); while periodic vortex shedding leaves a periodically varying circulation of alternating sign around the cylinder, with which oscillatory lift is associated. The Strouhal number (51) for the oscillations of lift takes the same value (about 0.2, though falling away towards 0.1 at the lower end of the Reynolds number range) as for the aeolian tones.

Reciprocally, the wire exerts on the air an equal and opposite oscillatory force, such as would be supposed in nineteenth century acoustics to act as a dipole source of sound waves. Gradually, it became recognized that the waves generated are indeed the dipole field associated with this force; a 1955 paper by N Curle first demonstrated [77] that the presence of flow insignificantly affects the radiated sound field. He pointed out too that the radiated power takes a form

$$\langle \dot{L}^2 \rangle / (12\pi\rho c_0^3) \tag{52}$$

in terms of the mean square rate of change of total lift L on the whole length of wire; which at frequency f is $(2\pi f)^2 \langle L^2 \rangle$ so that, as the wind speed U increases, the proportionality of lift to U^2 and the dependence (51) of f on U imply a sixth-power law (like U^6) of dependence of sound radiation on wind speed.

Expression (52) for the sound radiated is perfectly valid whether or not the oscillatory lift causes the wire to vibrate. In the fluctuations of total lift L, on the other hand, considerable cancelling can result from fluctuations in circulation not being perfectly in phase all along the wire. Then a resonant vibration of the wire—as in one of the variously tuned strings of the aeolian harp—may produce an increase in sound intensity simply by bringing all these fluctuations into phase, 'locked on' to the phase of that vibration.

Much more musically·important are those many wind instruments which use the interaction of a narrow air jet with a sharp edge to generate

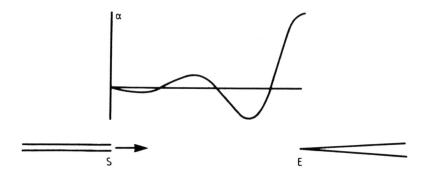

Figure 10.35. *When an air jet issues horizontally (at moderate Reynolds number) from a long slit S, variations in the extremely small angle α of upward deflexion of the jet take the form of a travelling sinusoidal wave of exponentially increasing amplitude. At the instant shown, the edge E experiences a positive angle of attack α and exerts a downward force on the fluid, causing positive rate of increase of the deflexion α at S and therefore tending to reinforce the travelling wave.*

sound—which, again, is often intensified by such 'locking on' to some resonant frequency (in an adjacent pipe). Curle identified in the jet-edge system a fundamental feedback loop which allows the air to exert on the edge an oscillatory lift force with its related dipole sound field [78].

Specifically, the dipole field produces a minute oscillation in flow direction at the jet orifice, and this disturbance as it travels downstream becomes amplified in consequence of the instability (again, of 'restrained' type) associated with the parallel flow in the jet. Feedback is effective, therefore (figure 10.35), at those frequencies for which the amplified direction change on reaching the edge has the right phase to reinforce the lift fluctuations.

Sound from flows, then, may take a dipole form, with power output (52), if a flow exercises a fluctuating force L on a solid body immersed in it; and at moderate Reynolds number, when 'restrained' instability can produce periodic fluctuations in L, the sound may be a musical note. At higher R, on the other hand, the fluctuations become chaotic, so that they generate acoustic noise; yet continued 'Strouhal scaling' (51) of typical frequencies f in the turbulent flow ensures a continued U^6 dependence of acoustic power output on the wind speed U. The 'roar' of a high wind is generated by fluctuating forces arising in its interactions with solid objects.

Various faster airflows, however, can generate significant noise even when not interacting with solid objects. The problems of estimating the noise that turbulent jets may radiate—without edges (or other objects) placed in them—began to be studied around 1950, when a fruitful method since named 'the acoustic analogy' was introduced [79].

Outside the jet, where any sound generated is just a minor by-product

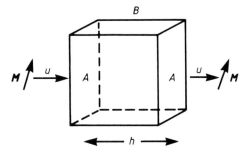

Figure 10.36. *Showing the part of the momentum transport into a box-shaped element B which occurs across one pair of opposite faces, of area A, a distance h apart.*

of the turbulence, the ordinary linear equations of sound may be applied. Essentially these equate (i) the rate of change of mass in a box-shaped volume element B to the net rate of mass flow into B, and (ii) the rate of change of momentum in B to the pressure force acting. (They are a closed system of equations because momentum per unit volume coincides with mass flux, while mass per unit volume is adiabatically related to pressure.)

But in accurate fluid mechanics the rate of change of momentum in B should include also the net rate of momentum flow into the element (figure 10.36). Momentum transport, of course, involves a product of two velocities, of the kind whose mean value appears in the Reynolds stress (26)—although in the acoustic analogy we must concern ourselves with its fluctuations. This product of two velocities is legitimately neglected outside the jet, where the linear equations of sound are applied, but cannot be ignored in the turbulence itself. There, it effectively appears in the second acoustic equation (ii) as an additional force, equal to the net rate of momentum transport into the element; thus, the ordinary equations of sound become correct if—simply, within the turbulent flow—such an additional force is incorporated within them.

These arguments demonstrate the validity of the acoustic analogy between the sound field of the turbulence and the sound field that would be generated in a linear acoustic medium by such a distribution of additional forces. This analogy is fruitful in many ways; not least, by showing how the absence of any solid body in the flow makes an essential difference to the radiated sound.

Indeed, without any resultant action on the fluid of a fluctuating external force L with its associated dipole radiation (52), the generation of sound must depend entirely on the net effect of a distribution (with zero resultant) of internal forces between elements. Now, the net radiation emitted (at an angle θ) by two equal and opposite dipoles separated by a distance h arises entirely (figure 10.37) from the difference $c_0^{-1}h\cos\theta$ in times of emission of signals received simultaneously by a far-off observer.

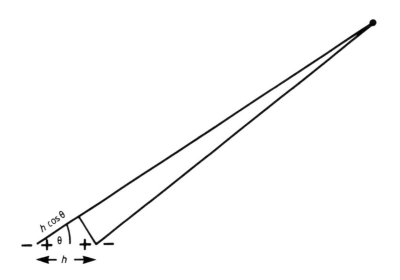

Figure 10.37. *For two equal and opposite dipoles, a distance h apart, the far field in a direction making an angle θ with the line joining the dipoles results from the difference h cos θ in distances travelled, leading to a difference $c_0^{-1} h \cos\theta$ in times of emission, of signals received simultaneously.*

So this radiation—described as that of a quadrupole, of strength equal to h times the dipole strength—takes the form

$$(c_0^{-1} h \cos\theta)\left(\frac{\mathrm{d}}{\mathrm{d}t}\right) \quad \text{(dipole radiation)}. \tag{53}$$

Here, the new differentiation with respect to time (additional to that which is already present in the radiation field of a dipole of strength L) produces an extra frequency factor in the sound field; which, with Strouhal scaling (51), and also the c_0^{-1} factor present in (53), yields an additional Mach number factor (U/c_0) in amplitude and the same factor squared in the power radiated. Essentially, this is why flows at low Mach number where fluctuating body force is present generate sound of purely dipole type, compared with which any quadrupole radiation is negligible.

In a jet, however, fluctuations in body force are absent. Then the U^6 dependence of the acoustic power which they would generate is converted by the extra $(U/c_0)^2$ factor into an eighth power dependence (like U^8) of jet noise radiation [80]. The enormous difference between this U^8 law for jet noise and the U^2 dependence of jet thrust (section 10.2.4) was to prove crucial for making possible (section 10.4.1) an environmentally acceptable development of jet engines for air-transport purposes.

854

Reconsideration of the box-shaped element B gives a simple idea of how quadrupoles arise (figure 10.36). Net transport of momentum into B across any one pair of opposing faces, each of area A, involves an input MuA at one face and $-MuA$ at the other, where u is the velocity normal to those faces and the vector M represents momentum per unit volume. Each rate of change of momentum is equivalent to a force and so to a dipole and the two dipoles together form a quadrupole of strength $MuAh$ in a volume element Ah; thus a quadrupole of strength Mu per unit volume is associated with this pair of faces (with similar associations for the other two pairs).

It is indeed rather closely correct (as suggested by such crude arguments) that the strength per unit volume of the quadrupole sound source is equal to the momentum transport. Of course, while either a force or the equivalent dipole is a vector, with a direction as well as a magnitude, momentum transport involves two directions (the momentum's direction and that in which it is transported) and therefore is a tensor—exactly like a quadrupole strength, which involves both the direction of the component dipoles and that of the displacement between them.

Identification of the effective quadrupole source strength with the momentum transport tensor proved a solid foundation for developments in the second half of the twentieth century in the science of sound from flows—which, in its aeronautical application, has become known as aeroacoustics [81]. The many other considerations that have needed to be taken into account in these developments include, for example, these:

(i) that there is phase coherence between such sources at points P and Q only when they are close enough to permit the covariance plotted in figure 10.19 to be substantial; and

(ii) that, in a jet, all fluctuations are convected by the mean velocity distribution [80].

This brief account of sound from flows may be concluded by an even briefer postscript on flows from sound. Although some aspects of such 'acoustic streaming' were well covered in Rayleigh's *Theory of Sound*, a completely new aspect emerged in the second quarter of the twentieth century with the widespread availability of piezoelectric quartz crystals as powerful ultrasonic sources. Early experimenters were puzzled by 'the quartz wind'!—that was often observed to accompany such sources.

The acoustic energy density in an ultrasonic beam takes the value $\rho\langle u^2 \rangle$ (twice the kinetic energy density because of equipartition between kinetic and potential energies). But this expression is also the mean momentum flux in the beam (transport, per unit area, of u-momentum in the u-direction). Integrating across the area of the beam, we conclude that the total momentum transport in the beam is equal to the acoustic energy per unit length; or to c_0^{-1} times the total energy transport (as indeed might be expected by analogy with light).

Now, the energy of an ultrasonic beam in the megahertz range falls away very steeply from a variety of dissipative effects; yet momentum has to be conserved. Necessarily, then, a mean flow of jet type is formed to carry that part of the beam's original momentum transport which can no longer be present in the beam after its energy losses [82]. Ultimately, the jet takes the form indicated in figure 10.28, corresponding to a total rate of transport F of momentum in the jet equal to c_0^{-1} times the power of the ultrasonic source. Between such a 'turbulent jet generated by sound' and the problem described earlier of 'sound produced by a turbulent jet' there is an intriguing reciprocity—with momentum transport as the key to both phenomena.

10.3.3. Non-linearity competing with dispersion
Waves where, even for very small amplitude, the speed of propagation may vary (e.g. with wavelength) are called dispersive, because different sinusoidal components of an initial local disturbance will travel at different speeds and so become 'dispersed' from one another. In the ocean, for example, a local storm can create an extremely complicated disturbance of the water surface; yet, after a large time t, ocean waves of approximately sinusoidal form are observed as 'swell' (see section 10.5) at distances $c_g t$ from the storm region, where the velocity c_g takes different values for waves of different length λ.

Nineteenth century physicists, who had named c_g the 'group velocity'—or speed of travel of the energy in a group of approximately sinusoidal waves—had exploded the fallacy that this speed is equal to the 'wave velocity' c with which the crests of a sinusoidal wave train move; demonstrating rather [13] that if c varies with λ all the energy in a group of waves travels, not at any such illusory speed c suggested by the movement of crests, but at the group velocity

$$c_g = c - \lambda \frac{dc}{d\lambda}. \tag{54}$$

Twentieth century oceanographers systematically verified this statement by measurements of swell at great distances (many thousands of km) from where it had been generated [83].

For these waves on 'the surface of the deep' (section 10.3.4), satisfying the classical relationship

$$c = \left(\frac{g\lambda}{2\pi}\right)^{1/2} \quad \text{so that } c_g = \tfrac{1}{2}c \tag{55}$$

both velocities vary so steeply with λ that any modest non-linear effects on signal speed, such as were discussed for sound waves in section 10.3.1, become unimportant by comparison; indeed, the substantial non-linear effects on ocean waves which have been investigated in the twentieth century are of a markedly different character. Before outlining these studies, however, we devote section 10.3.3 to the interesting topic of wave propagation where there is competition on a roughly equal footing between non-linearity and dispersion in their effects on signal speed [70].

Such propagation is typified by waves in 'shallow' water; that is, water whose mean depth h_0 (say, above a flat horizontal bottom) is very much smaller than the wavelength λ. A superficial, yet illuminating, analogy with sound waves suggests that any local value of the depth h should be propagated horizontally with wave speed $c = (gh)^{1/2}$. Very briefly, what governs the propagation in just two (horizontal) dimensions of these essentially longitudinal waves is the relationship between the vertically integrated values, ρ_v and p_v, of the density and pressure, given by equations

$$\rho_v = \rho h, \quad p_v = \tfrac{1}{2}\rho g h^2 \quad \text{so that} \quad c^2 = \frac{dp_v}{d\rho_v} = \frac{\rho g h \, dh}{\rho \, dh} = gh \qquad (56)$$

is the square of the wave speed.

Pursuing the analogy, we see that a first approximation to the increased wave speed with which values of h in excess of h_0 are propagated is

$$c = c_0 \left(1 + 0.5 \frac{h - h_0}{h_0}\right) \quad \text{where } c_0 = (g h_0)^{1/2} \qquad (57)$$

is the linear-theory wave speed. Here, of course, c is the velocity of propagation relative to any water motion; whose velocity u is given on linear theory as

$$u = c_0 \frac{h - h_0}{h_0} \qquad (58)$$

because any rate of increase of depth in a wave travelling at velocity c_0 must take the value $-c_0 dh/dx$ and be balanced by the downward flux gradient $-d(h_0 u)/dx$. These crude approximate arguments suggest augmented values of the wave speed and the signal speed

$$c = c_0 + 0.5u \quad \text{and} \quad c + u = c_0 + 1.5u \qquad (59)$$

respectively; where the factor 1.5, replacing the 1.2 in equation (17), implies that 2/3 of the signal-speed excess is due to convection of waves at the water velocity u and 1/3 to the increase in c.

Once again, a full non-linear analysis by Riemann's methods [15] of such shallow-water propagation—assumed unidirectional as well as longitudinal—proves the equations (59) to be completely accurate for waves of any amplitude; with depth changes related to changes in c by equation (56), and with values of u propagated at precisely the speed $c_0 + 1.5u$. Just as with sound, then, a waveform develops as shown in figure 10.5 (except that 1.2 is replaced by 1.5) and poses the same enigma of how the tendency to form a discontinuity is compatible with the unavoidable loss of wave energy—a loss, per unit mass of water, shown by Rayleigh in a 1914 paper [84] to be

$$\frac{g(h_2 - h_1)^3}{4h_1h_2} \qquad (60)$$

wherever the depth increases discontinuously from h_1 to h_2.

Much later, however, the resolution of this enigma was found to differ fundamentally from that which Rayleigh [6] and Taylor [7] had revealed for sound waves. Briefly, shallow-water waves have a very slightly dispersive character; which becomes more important as the gradient of the waveform in figure 10.5 is increased and, 'in competition with' non-linear effects, resolves the enigma.

Dispersion is found even in linear theory, where a sinusoidal waveform

$$h - h_0 = a \sin[k(x - c_0 t)] \quad \text{of length } \lambda = 2\pi/k \qquad (61)$$

with the water velocity (58), gives potential and kinetic energies per unit mass of water as

$$\frac{ga^2}{4h_0} \quad \text{and} \quad \frac{c_0^2 a^2}{4h_0^2} \qquad (62)$$

respectively; equipartition being assured by equation (57) for c_0. Yet not all the kinetic energy corresponds to the horizontal component u of water motion; a changing depth involves a vertical velocity varying linearly from dh/dt at the surface to zero at the bottom, which makes an additional contribution $(k^2 c_0^2 a^2/12)$ from vertical motions to the kinetic energy per unit mass. Equipartition now modifies the wave velocity from $c_0 = (gh_0)^{1/2}$ to a reduced value

$$\left[gh_0/(1 + \tfrac{1}{3}k^2h_0^2)\right]^{1/2} \simeq c_0\left(1 - \tfrac{1}{6}k^2h_0^2\right). \qquad (63)$$

With this value for c in equation (54), the group velocity c_g becomes

$$c_g = c - \lambda\frac{dc}{d\lambda} = c + k\frac{dc}{dk} = c_0\left(1 - \tfrac{1}{2}k^2h_0^2\right) \qquad (64)$$

so that energy travels just a little slower than the wave crests in these slightly dispersive waves.

It follows [70] that a travelling discontinuous increase in depth—or 'hydraulic jump'—can drag along behind it a regular wave train with

Figure 10.38. *An undular bore on the river Severn photographed by Professor D H Peregrine.*

crests that travel at the same speed; yet whose energy, moving more slowly, is being gradually carried backwards relative to the jump, thus achieving practically all of the energy loss (60). The UK's famous Severn bore, generated as the flood tide propagates up the narrowing Severn estuary and river, often takes this form (figure 10.38).

A properly balanced description of the competition between weak non-linear and weak dispersive effects is obtainable from the equation introduced in 1895 by D J Korteweg and G de Vries [85], yet applied by others only much later to shallow-water propagation as well as to other physical phenomena influenced by these competitive effects. That equation makes the rate of change of u a sum of the term $-c_0 \mathrm{d}u/\mathrm{d}x$ appropriate to constant signal velocity and terms

$$-\tfrac{1}{6}c_0 h^2 \frac{\mathrm{d}^3 u}{\mathrm{d}x^3} - 1.5u\frac{\mathrm{d}u}{\mathrm{d}x} \tag{65}$$

which, respectively, give the corrected linear dispersion relation (63) and the non-linear effect (59) on signal speed [70]. The regular waves behind a hydraulic jump are shown by this equation to be not sinusoidal but 'cnoidal'!—with shapes described by the Jacobian elliptic function cn.

The Korteweg–de Vries equation has also the famous 'soliton' solution, observable in shallow water as a propagating 'solitary wave' of locally increased depth whose effective wavelength is small enough for dispersive effects to cancel completely the steepening tendency of figure

10.5 so that the waveform remains of permanent type. Solitons, although largely a curiosity in fluid mechanics, have been shown to exhibit intriguing types of mutual interaction which make them of widespread interest to theoretical physicists in general [86].

10.3.4. *The surface of the deep*

Two great interacting components of the Earth's fluid envelope (section 10.5) are the ocean and the air above it, where winds in a turbulent boundary layer generate transfer of water vapour (section 10.2.5) from ocean to atmosphere, and transfer of heat between them (in either direction); while also transferring momentum from air to ocean, where much of it assumes the form of surface waves. As in other fields of physics, ocean waves of energy E (say, per unit horizontal area) carry momentum E/c in the direction of propagation, with c as the wave velocity [70]. However, when agencies such as wave breaking which diminish E reduce also the momentum E/c carried by the waves—without, of course, any change in total momentum—the remainder is necessarily converted into currents (compare section 10.3.2 on acoustic streaming).

In the absence of such energy dissipation, and of any renewed momentum transfer from winds, ocean waves have in the twentieth century been shown, even at amplitudes where non-linear effects are important, to satisfy conservation laws which are interestingly reminiscent of wave–particle duality principles. These laws are outlined here, though further reference to geophysical implications of knowledge on surface waves is deferred to section 10.5.

Duality relates waves to particles carrying action in specific amounts; similarly, a key concept for ocean waves is the wave action, A per unit horizontal area. For small-amplitude waves on still water, A is defined so that the energy E is $A\omega$ in terms of the frequency ω while the wave momentum is Ak in terms of the vector wavenumber k (its magnitude, therefore, being E/c as stated above).

These quantities, moreover, can still be defined for regular long-crested waves even at large amplitudes [86], when surface waves (unlike the minimally dispersive waves of section 10.3.3) acquire from non-linear effects no continual steepening of a periodic waveform of length λ but just a limited modification from the sinusoidal shape into one [87] with peakier crests and flatter troughs (figure 10.39). Then k is simply defined as a vector in the direction of propagation with magnitude $2\pi/\lambda$, and ω as 2π divided by the wave period; while Whitham's fine 1965 studies [88] proved that the water motions possess momentum Ak and kinetic energy $\frac{1}{2}A\omega$—although departures from equipartition due to non-linear effects make the associated potential energy slightly less.

Figure 10.40 shows the results of accurate energy computations made in 1975 by M S Longuet–Higgins [89] for regular long-crested waves of

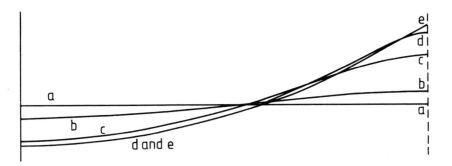

Figure 10.39. *Waveforms [87] for periodic deep-water waves of length λ and of different heights H between crest and trough. The broken line marks the plane of symmetry at each crest. (a) Undisturbed water surface. (b) Waveform for $H = 0.030\lambda$. (c) Waveform for $H = 0.100\lambda$. (d) Waveform for $H = 0.130\lambda$. (e) Waveform for the maximum wave height $H = 0.141\lambda$.*

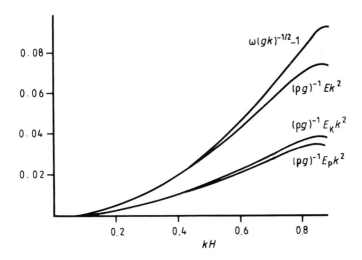

Figure 10.40. *Frequency ω, and wave energy E (divided into kinetic E_K and potential E_P) calculated [89] for deep-water waves of wavenumber k.*

wavenumber k (the slight rise in frequency ω above its small-amplitude value $(gk)^{1/2}$ is also shown). Both the kinetic and potential energies, as well as their sum the total wave energy E, increase with amplitude to a maximum value, beyond which the wave assumes a sharp-crested form (figure 10.39) such that, with further addition of energy, dissipation arises [90] through foaming at the crests (the celebrated 'white horses' phenomenon).

Conservation of wave action holds, on the other hand, when such dissipation is absent. Indeed, even where waves propagate into regions

861

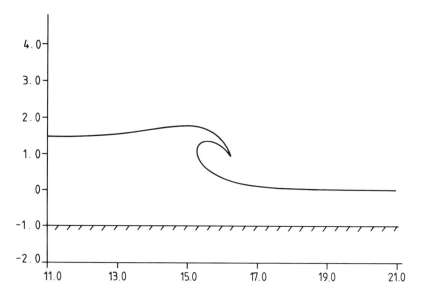

Figure 10.41. *Form of a plunging breaker computed by Professor D H Peregrine.*

of ocean current (so that wave energy need not be conserved because of exchange with the energy of currents), Whitham's action-conservation principle still applies. The kinetic energy of the waves becomes $\frac{1}{2}A(\omega - V \cdot k)$ in a current of velocity V, where $\omega - V \cdot k$ describes a modification to the effective frequency resulting from the familiar Doppler effect. Thus waves which propagate from still water into a region of opposing current (negative $V \cdot k$) may experience substantial gains in wave energy.

Space allows reference to but two other developments in non-linear ocean-wave theory. First, the need to make non-linear propagation studies still more precise (those based on action-conservation laws retain good accuracy only where values of the amplitude change per wavelength are very small) led in the able hands of K Stewartson and his colleagues to the fruitful description of a wavetrain's changing envelope by means of a non-linear Schrödinger equation, excellently adapted for convenient numerical solution [91].

Secondly, the study of 'spilling breakers' (waves in a slowly varying wavetrain which, as described above, attain the amplitude for maximum energy only to lose some of it by foam production at crests) has been valuably complemented by investigations of those 'plunging breakers' that arise when waves move into shallow water. Superb modern developments in computational fluid dynamics for unsteady motions of water with a free surface of complicated shape have permitted great insight into this alternative, and more spectacular, type of breaking to be achieved (figure 10.41).

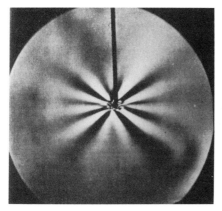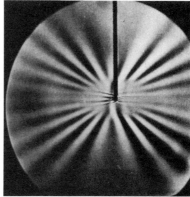

Figure 10.42. *Schlieren photographs* [92] *of waves with along-crest energy propagation (left, at time $t = 10$ s; right, at time $t = 25$ s).*

10.3.5. Along-crest energy propagation

The vector group velocity c_g, or velocity of propagation of the energy of waves in fluids, may differ not only in magnitude but also in direction from the wave velocity c of propagation of crests normal to themselves. This section is devoted to dispersive waves where the two velocities are orthogonal!—so that the wave energy is propagated along crests instead of at right angles to them [70].

Figure 10.42 shows the results [92] of experiments in a uniformly stratified salt solution (where the density reduction with height follows a constant gradient). Here, the vertical rod's function is simply to support at its lowest end an immersed horizontal cylinder which was given a brief horizontal displacement at time $t = 0$. Waves from this source have been made visible by schlieren photography at $t = 10$ s and at $t = 25$ s.

These waves are seen to have crests which—far from being concentric circles, as when energy generated at the source travels at right angles to crests—are arranged radially, because of along-crest energy propagation. A motion picture shows all crests moving normal to themselves, downwards for those above the source but upwards for those below it; yet wave energy, travelling radially outward, has reached considerably farther at $t = 25$ s than at $t = 10$ s.

Fundamental properties of disturbances to stratified fluids are geophysically important [93] because of density stratification in both the atmosphere and the ocean (section 10.5). This section, however, concentrates on the basic physics of along-crest energy propagation.

It arises [70] in any system where the wave frequency ω depends only on the direction and not on the magnitude of the wavenumber k. The simplest formula $c_g = d\omega/dk$ for the group velocity in one-dimensional propagation—which, since $\omega = ck$, agrees with an expression like

$c+kdc/dk$—can be generalized in three dimensions to a statement that the vector group velocity c_g is the gradient of the frequency in wavenumber space (on duality principles, indeed, waves move like particles, with the Hamiltonian expression for particle velocity as the gradient of energy in momentum space adapted through energy and momentum being associated with frequency and wavenumber). Where ω depends only on the direction of k it must stay constant along the vector k itself—which, therefore, is orthogonal to c_g (as the gradient of ω).

Figure 10.43 shows the nature of plane waves in a stratified liquid like that of figure 10.42, with a straight-line dependence

$$\rho = \rho_0(1 - \varepsilon z) \tag{66}$$

of density on the height z. The effective incompressibility of the liquid prevents the particle motions from having any 'longitudinal' component (a component perpendicular to crests, as in sound waves); they are forced, rather, to lie in surfaces of constant phase—where, being subject to a pressure-gradient force BC perpendicular to those surfaces and a vertical gravity force BD, they must move always up or down a direction BE of steepest descent. Now a particle displaced upwards by a small distance s at an angle θ to the vertical finds itself denser than its surroundings by an amount $\rho_0\varepsilon(s\cos\theta)$ and subject to a downward gravity force $g\rho_0\varepsilon s\cos\theta$ per unit volume, so that

$$g\rho_0\varepsilon s \cos^2 \theta \tag{67}$$

is the resultant restoring force in the direction BE. Therefore such a particle, of mass ρ_0 per unit volume, performs simple harmonic motion with frequency

$$\omega = N \cos \theta \quad \text{where } N = (g\varepsilon)^{1/2} \tag{68}$$

is known as the Väisälä–Brunt [94, 95] frequency. (Note: when such arguments are applied to atmospheric air in section 10.5.2, the only essential modification is a reduction in N that results from a change to the restoring force linked with the particle's adiabatic density drop in response to pressure loss with height.)

As foreshadowed, then, it is only on the direction of the wavenumber vector that the wave frequency ω depends. If moreover θ is slightly reduced to $\theta - \delta\theta$, the frequency increases by

$$(N \sin \theta)\delta\theta = (Nk^{-1} \sin \theta)k\delta\theta \tag{69}$$

while the wavenumber k experiences a displacement $k\delta\theta$, which points downwards or upwards (figure 10.43) according as k has an upward or downward component, respectively [70]; so the group velocity vector c_g has

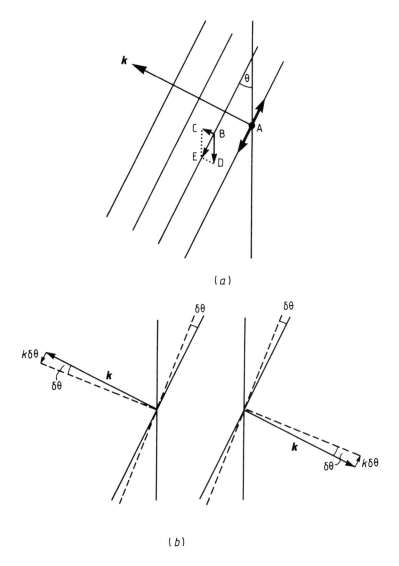

(a)

(b)

Figure 10.43. *(a) Illustrating plane waves in uniformly stratified fluid, including (at A) particle movement parallel to crests at angle θ with vertical, and (at B) resultant of pressure forces BC and gravity forces BD. (b) When θ is reduced to θ − δθ, the change (of magnitude kδθ) in the vector wavenumber **k** has vertical component of opposite sign to that of **k**.*

(i) magnitude $Nk^{-1} \sin \theta$ (see (69)) and
(ii) direction at an angle θ to the vertical, with
(iii) a vertical component opposite in sign to that of $\mathbf{k}$.

Property (iii) explains why waves with the vertical component of

865

c_g positive or negative (found above or below the source in figure 10.42) have downward or upward directions of propagation, respectively. Property (i) indicates that, in terms of the wavelength λ, the radial distance travelled by wave energy in time t is $c_g t = (N\lambda t/2\pi)\sin\theta$ so that any small angle subtended at the source by crests a distance λ apart takes the value

$$2\pi/(Nt\sin\theta) \tag{70}$$

which again agrees with figure 10.42 (the angle varies with t like t^{-1} and with θ like $(\sin\theta)^{-1}$).

These 'internal' gravity waves within a stratified fluid, besides displaying the unusual features just discussed, exhibit interesting interactions between waves and flows: the central theme of section 10.3. Such interactions may be exemplified first in a uniform flow and then in a sheared one.

It is possible [96] for a two-dimensional pattern of internal gravity waves to remain stationary in a uniform stream of velocity U—so enabling them to form part of the steady flow past a long obstacle stretched perpendicularly to it—provided that their wavenumber has magnitude $k = N/U$. Then crests remain stationary because by equation (68) the wave velocity $c = \omega/k$, with which they move normal to themselves, cancels the velocity $U\cos\theta$ which is the wind's opposing component normal to crests. The energy in such waves, if generated at the obstacle, has travelled a distance $Ut\sin\theta$ parallel to crests (by (i) above) as well as being swept downstream by the greater distance Ut (figure 10.44(a)), so simple trigonometry makes the crests normal to a radius vector from the obstacle. Thus they are circular arcs; yet, because the energy's net distance downstream of the obstacle is necessarily positive, they are not full circles but semi-circles (figure 10.44(b)).

Early theoretical studies, having demonstrated that disturbances made by an obstacle in a uniform stream of stratified fluid satisfy equations governing stationary cylindrical waves, wrongly concluded that crests would be circles. This error is underlined here as one representative example out of many hundreds of studies of waves in fluids where determination of the uniquely correct wave pattern has needed application not only of equations of motion but also of the above condition (analogous to Sommerfeld's radiation condition in electromagnetic theory) that energy flow must be directed away from the source.

Accompanying this energy flow–say, E per unit length of obstacle—is a flow

$$(E/c)\cos\theta = E/U \tag{71}$$

of the component of wave momentum in the direction opposed to the wind. This momentum is also generated at the obstacle; which must experience an equal and opposite force, E/U per unit length, known

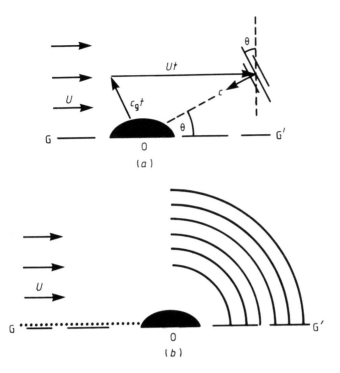

Figure 10.44. (a) In a uniform stream of speed U, internal waves of wavenumber $k = N/U$ have two remarkable properties: (i) crests remain stationary, because their propagation at wave velocity $c = \omega/k = (N\cos\theta)/k$ cancels the stream's component $U\cos\theta$ normal to the crests, while (ii) energy travels from a source (the obstacle O) at group velocity $c_g = Nk^{-1}\sin\theta = U\sin\theta$ at an angle θ to the upward vertical (along-crest propagation), besides being convected at velocity U by the stream, and so is found downstream in a direction making an angle θ with the horizontal; that is, in a direction normal to crests. (b) Consequential circular-arc form of crests. Also (see next page) the 'upstream wake' is indicated. (Note: the plane GG' may represent either a flat ground beneath the wind flow (wherein wave crests are quarter-circles) or else a plane of symmetry for a large flow (in which the wave crests with their mirror images in that plane are semi-circles.)

as wave-making resistance. The associated momentum transfer is not to wake fluid behind (and about on a level with) the obstacle as with frictional resistance; but, as first emphasized by F P Bretherton [97], to fluid vertically remote from it.

Although these conclusions have been explained in one very special case—uniform stratification, uniform stream, long obstacle of uniform cross-section—where simple results are readily derived, conclusions that are essentially similar can be worked out in an enormously wider range of cases, where any or all of the above elements may be non-uniform. In every case, stationary waves are found downstream of (yet vertically

867

remote from) the obstacle, which experiences a wave-making resistance associated with the upstream component of their momentum. Such momentum transfer to vertically remote fluid has to be taken into account in modelling atmospheric flows over mountain ranges (section 10.5.2).

Paradoxically, 'upstream wakes' are possible in all these cases!—as can be seen in the simple case treated above by allowing θ to take its limiting value $\frac{1}{2}\pi$ (horizontal crests; a possibility tacitly ignored in the earlier analysis). Then the frequency (68) vanishes so that waves of any (vertical) wavenumber k can be stationary. Yet their along-crest energy propagation (at velocity Nk^{-1}, by (i) above) may exceed the stream velocity U if $k < N/U$, so that such waves can exist upstream of any obstacle tall enough to generate significant disturbances with vertical wavenumbers $k < N/U$. (Another approach to this surprising phenomenon is to remark that the stream's kinetic energy may be inadequate to supply the gravitational potential energy needed for lower parts of it to surmount a tall obstacle.) Comprehensive non-linear accounts of upstream wakes—with which, again, a resistance is associated—were first given by R R Long [98].

There is one very special capability that along-crest energy propagation gives to internal gravity waves in a sheared wind (one of velocity V varying with altitude z): they can intensify the wind by giving up all their energy to it. Indeed, conservation of wave action (section 10.3.4) tells us that, at a 'critical' altitude $z = z_{cr}$ where the relative frequency $\omega - V \cdot k$ falls to zero, wave energy is reduced to zero through exchange with the flow.

Here, relative frequency means the frequency in a frame of reference with the air at rest; which equation (68) identifies with $N \cos \theta$, so that the angle θ tends to $\frac{1}{2}\pi$ at the critical altitude. Waves, as they propagate through the sheared wind, must maintain constant values of the absolute frequency ω and the horizontal component k_H of wavenumber, but its varying vertical component $k_H \tan \theta$ can become larger and larger, in such a way that the vertical component of group velocity, $Nk_H^{-1} \cos^2 \theta \sin \theta$, tends to zero as the square of the distance from the critical level $z = z_{cr}$ where $\omega - Vk_H = N \cos \theta$ vanishes. This allows the energy unlimited time to reach the critical level—so that mere reflection at that level, which occurs in many other physical problems, is impossible (as Booker and Bretherton [99] first showed), being replaced by total absorption. Some high-level jet-like winds are maintained by such absorption of energy from internal gravity waves.

Along-crest energy propagation occurs also in other fluid-mechanical systems; for example, in a homogeneous liquid rotating with constant angular velocity Ω. Here, plane waves with crests making an angle θ to the axis of rotation have frequency $\omega = 2\Omega \sin \theta$, which again depends only on the direction of the wavenumber vector; and various properties of internal gravity waves are repeated with N and θ replaced by 2Ω and

$\frac{1}{2}\pi - \theta$. Most spectacularly, the analogue of the upstream wake becomes a wake at $\theta = \frac{1}{2}\pi$, the 'Taylor column'; thus, when slow flow at right angles to the axis of rotation encounters a three-dimensional body, it was shown by G I Taylor [100] to become a two-dimensional flow about a column of fluid which amounts to a sort of extension of the body shape in the direction of that axis!

10.4. Transforming the human condition through aeronautics and ocean engineering

10.4.1. *Fluid mechanics of flight efficiency*

Human life on Earth was transformed in the twentieth century by various technologies, including those that made the Earth 'more of a neighbourhood' through fast and efficient movement of people and valued goods across it. This section 10.4 outlines the critical contribution of fluid mechanics to such developments, especially in the air but also in the ocean; which were achieved in close cooperation with developments (founded on other physical advances) in the worldwide transmission of information.

Aviation contributed decisively to these changes as a direct result of two great discoveries of Prandtl: that avoidance of boundary-layer separation can maintain very low drag on a non-lifting body (p 807) while the extra drag due to wing lift (p 824) can be kept very low for high wing spans. It was above all the 'Earth-shrinking' effects of ever greater range (distance travelled by an aircraft on a single flight) that responded so sensitively to these, as well as to some later, discoveries.

The elementary 'range equation' which give insight into such effects (see Küchemann [101], for example) describes the distance X travelled by a transport aircraft under those 'cruise' conditions—constituting the main part of its flight—where the lift L essentially balances the aircraft weight W while the drag D is balanced by the engine thrust T. The specific fuel consumption of the engine is defined as the ratio

$$s = \frac{\text{weight of fuel consumed per unit time } (-dW/dt)}{\text{useful work done per unit time } (UT)} \tag{72}$$

and it follows that

$$\frac{dW}{dt} = -sUT = -sUD = -sU\frac{D}{L}W \tag{73}$$

so that the cruise range X (the distance $\int U \, dt$ covered, at speed U, as the aircraft's weight W is reduced from W_0 at the start to W_1 at the end of cruise) is

$$X = \int U \, dt = \frac{L}{D}\frac{1}{s}\int_{W_0}^{W_1}\left(-\frac{dW}{W}\right) = \left(\frac{L}{D}\right)\left(\frac{1}{s}\right)\left(\ln\frac{W_0}{W_1}\right) \tag{74}$$

on the reasonable assumption that both L/D and $1/s$ are to be kept essentially constant (close, that is, to their maximum values) under cruise conditions.

Although the total range exceeds X (by the much smaller distances covered during the climb to, and descent from, cruise altitude), equation (74) is a valued first indication that achievable range may be dominated by a product of three separate factors, related to respective achievements in the three professional disciplines (aerodynamics, propulsion, structures) that make up aeronautical engineering. Evidently, the aerodynamicist and the propulsion engineer need to secure high maximum values of L/D and $1/s$, respectively. Moreover, successes in the design of a reliably safe structure which eschews unnecessary weight determine how much fuel for cruise, $W_0 - W_1$, can be carried in addition to other components of the all-up weight W_0; which include above all structure weight—along with fuel for climb and descent and for potentially necessary diversions, and of course 'payload' (used as a shorthand for all weights stemming from the aircraft's transport function).

Prandtl's discoveries [4, 5] led to a recognition of how high maximum lift-drag ratios L/D could be achieved. Briefly, the drag includes a 'frictional' element D_f, associated with the (mainly, turbulent) boundary layers on an aircraft's wings and fuselage, together with a drag due to lift, D_L, related to the work done in generating that vortex wake which produces the lift L. Equation (24) shows that the least possible value of D/L is

$$\frac{D_f}{L} + \varepsilon = \frac{D_f}{L} + \frac{L}{2\pi\rho U^2 b^2} \tag{75}$$

which is itself minimized when both terms are equal. This specifies

$$L = \left(2\pi\rho U^2 b^2 D_f\right)^{1/2} \quad \text{and Max } \frac{L}{D} = \left(\frac{\pi\rho U^2 b^2}{2D_f}\right)^{1/2}. \tag{76}$$

Here, given meticulous design to ensure that separation is avoided all over the airframe, the 'frictional' drag D_f takes a form $k_f\rho U^2 S$ linked mainly to turbulent diffusion of momentum over a total area S of wings and fuselage, with k_f as a small constant—its smallness stemming, of course, from d'Alembert's Theorem—of order 10^{-2}. Then equations (75) and (76) yield results

$$\varepsilon = \left(\frac{k_f}{2\pi}\frac{S}{b^2}\right)^{1/2} \quad \text{and Max } \frac{L}{D} = \frac{1}{2\varepsilon} = \left(\frac{\pi}{2k_f}\frac{b^2}{S}\right)^{1/2} \tag{77}$$

which demonstrate the benefits to be derived from sufficiently high wing spans b—or, more precisely, from sufficiently large values of the ratio

b^2/S—when lift–drag ratios around 20 or more (corresponding to values of $\varepsilon < 0.025$) have proved readily achievable [101].

Fluid mechanics, then, contributed greatly towards increases in the cruise range (74) through this achievement of large lift–drag ratios; while structural branches of aeronautical engineering made another big contribution through an increase in W_0/W_1, resulting from designs of reliably safe structures with reduced ratios of structural to all-up weight. Of course, different demands interact: the semi-span b of the wings may, essentially, be determined as a compromise between lift–drag benefits of increasing b and those structure-weight penalties which may result e.g. from the associated increase of bending moment at the wing 'root'.

The third great contribution to increasing the cruise range (74) arose from the evolution of power plants with massively reduced values of the specific fuel consumption s. Alongside refined thermodynamic and engine-design developments, these achievements depended also on advances in fluid mechanics which, yet again, interacted with considerations of lift–drag-ratio maximization.

Paradoxically [101], important reductions in the specific fuel consumption (72) accrue from increasing the aircraft speed U. For (say) a propeller aircraft, if an air mass Q enters the propeller disks per unit time at velocity U and is accelerated to velocity V, then that air's rates of increase of momentum and of kinetic energy are, respectively,

$$T = Q(V - U) \quad \text{and} \quad dE_K/dt = Q\left(\tfrac{1}{2}V^2 - \tfrac{1}{2}U^2\right) \tag{78}$$

so that equation (72) becomes

$$s = \frac{(-dW/dt)}{dE_K/dt} \frac{\tfrac{1}{2}V^2 - \tfrac{1}{2}U^2}{(V - U)U} = \frac{(-dW/dt)}{dE_K/dt} \frac{V + U}{2U}. \tag{79}$$

Here, the first factor is (inversely) related to the thermodynamic efficiency of conversion of the fuel's chemical energy into mechanical energy. But it is the second, more strictly fluid-mechanical, factor which so strikingly depends on the airspeed U, and which is reduced substantially as U increases (whether for given V, or for given $V - U$, or for given $\tfrac{1}{2}V^2 - \tfrac{1}{2}U^2$).

Similar equations for jet aircraft, though a little more complex because the exhaust gas is not air but an air-fuel combustion product, give a fluid-mechanical factor in s which is very similarly reduced (actually, a bit more) as U increases. With typical energy increases $\tfrac{1}{2}V^2 - \tfrac{1}{2}U^2$ derived from combustion, such considerations suggest that valuable increases in s may come from increasing the airspeed U to very large values.

Interactions of this conclusion with lift–drag-ratio maximization emerged in two principal stages. First, the cruise condition (76), with $W = L$, fixes the value of the ratio

$$W/\rho U^2 = (2\pi b^2 k_f S)^{1/2} \tag{80}$$

Figure 10.45. *A Boeing 747/400, with Rolls-Royce RB211-524G turbofan engines, and with all-up weight 395 tonnes, cruise Mach number 0.85 and maximum range 15 400 km.*

so that large airspeeds U demand low values of the air density ρ in cruise, and a further reduction in ρ as W decreases from W_0 to W_1. In other words, cruise must take place at rather high altitudes, increasing slightly during flight. (This cruise phase of flight is of course bracketed by other, briefer phases—takeoff and climb, descent and landing—when values of $W/\rho U^2$ may need to be far higher than the value (80) which maximizes lift–drag ratio.)

Secondly, the potential for increases in U becomes severely limited as U approaches the sound speed c_0. As indicated in section 10.4.2 below, energy losses in aeronautical shock waves represent a potent additional source of drag for aircraft at supersonic speeds; and, even when the Mach number (19) is somewhat less than 1, passage of the airflow over the wing can accelerate it to supersonic speeds and allow shock-wave formation. On the other hand, sweptback wings (figure 10.45) have allowed cruise Mach numbers to be increased to quite high subsonic values [102], around 0.85; essentially, because just the component of airflow perpendicular to the wing is accelerated by passage over it, while the parallel component remains unaltered.

Another fluid-mechanical limit on the early evolution of jet-propelled aircraft proved to be the intolerably high level of noise radiation from the jets themselves. This noise problem threatened to make the development of increasingly powerful jet engines environmentally incompatible with acceptable existences for communities around airports.

It is, perhaps, historically unusual that two big problems are found to have a single common solution! Yet two great obstacles to development of efficient jet aircraft—limitations on the reduction of s placed by an upper limit on cruise Mach number, and environmentally imposed restrictions on noise radiation from jets—were overcome by essentially the same development [80].

The theory of jet noise sketched on p 854 indicates that the acoustic energy radiated varies as the eighth power of the jet exit velocity, written there as U but here rewritten as V. High engine thrusts are needed at takeoff and in the early stages of climb, when the airspeed U is not large so that the thrust for a jet of given cross-section varies as V^2, as opposed to the V^8 dependence of the noise radiated to nearby communities. This contrast points to the possibility of increasing thrust, yet reducing noise, by large increases in jet cross-section accompanied by reductions in V.

The turbo-fan engine has been highly successful in achieving this objective. Although the energy acquired from air-fuel combustion would give its products a very high velocity if they emerged as a simple jet, most of that mechanical energy can first be extracted in order to operate a turbine which, by rotating a very large fan, accelerates huge masses of air that have by-passed the combustion chamber. Values of V for both this air and the combustion products can then be kept down to moderate values, so that the ratio (proportional to V^6) of noise emitted to thrust exerted is enormously reduced: the engine is both quieter and more powerful.

Simultaneously, these developments have led to yet another augmentation of the cruise range (74). Essentially, they allow the fluid-mechanical factor $(V + U)/(2U)$ in the specific fuel consumption (79) to be diminished significantly, even under strict conditions of an upper limit on the cruise Mach number U/c_0, simply because V itself takes a reduced value.

Powered flight—over distances less than 1 km—had first been achieved by the Wright brothers in 1904. Then progress in the science of fluid mechanics, that started in the very same year, made a massive contribution to those twentieth century developments which, by allowing increases in cruise range for transport aircraft to over 15 000 km, with closely associated cost-reduction benefits, would significantly transform the character of human life on Earth.

10.4.2. Aeronautical shock waves

This section 10.4.2 outlines that 'exciting change, with the prominent appearance of shock waves' (p 813) which airflow patterns undergo when the Mach number (19), defined as the ratio of wind speed U to sound speed c_0, exceeds 1. (In aeronautics, of course, this wind speed U means the speed of the air relative to a moving aircraft.) After that change, not only do 'waves and flows interact' (section 10.3.1), but the flow consists almost entirely of waves whose propagation has been annulled (that is, brought to rest) by the flow.

For example, propagation at an angle θ to the wind of very weak sound waves at the undisturbed sound speed c_0 can be annulled by the opposing component $U \cos \theta$ of airflow if

$$U \cos \theta = c_0. \tag{81}$$

Provided that $U > c_0$, an angle θ satisfying (81) exists, and such waves can be a permanent feature of the flow [19].

For waves that are not so weak, the true signal velocity (17) replaces c_0 in equation (81), so that θ is reduced for positive u (or positive excess pressure) and increased for negative u. Moreover, if a shock wave appears, its own speed of propagation replaces c_0 in (81).

An extra source of drag is represented by the 'hidden energy loss' arising (section 10.3.1) in any shock waves [103], such lost energy needing to be restored from additional work done by thrust to overcome this component of drag. So 'aeronautical shock waves' need to be kept as weak as possible.

There is a huge contrast between the low-Mach-number flow around a blade with an 'aerofoil' section appropriate to such flow (figure 10.3) and the pattern of flow around a supersonic aerofoil (figure 10.46). The latter shape must be even thinner, and must also have a sharp leading edge, so that all disturbances to the oncoming supersonic stream remain small and the shock waves generated are weak. Moreover, the whole visible flow pattern takes the form of stationary waves propagating at angles θ satisfying either equation (81) or similar equations with modified right-hand sides [103].

Each wave arises from a disturbance to the incident airflow by the aerofoil surface. The point A where such disturbance is zero ($u = 0$) emits a wave in accordance with equation (81). But the direction of the surface at B is associated, as figure 10.46(*b*) shows, with propagation of a positive value of u at an increased speed $c_0 + 1.2u$ giving a reduction in θ; while the direction at C is shown in figure 10.46(*c*) to be compatible with $u < 0$ and an increased value of θ. The excess pressures, positive at B and negative at C, produce a resultant drag, related rather precisely [103] to the hidden energy loss in the N-waves (compare figures 10.46(*d*) and 10.32) generated by such non-linear propagation. But practically no waves arise in the region behind the rear shock wave, where the incident airflow is undisturbed.

At the same time, just as thin boundary layers are shown in figure 10.3 as attached to the surface, so also (to complete the comparison) boundary layers with broadly similar distributions of the velocity v are present on a supersonic aerofoil. Yet there is an astonishing difference in the temperature distribution from that discussed in section 10.2.5. At supersonic speeds, a cold wind no longer cools a body; it heats it!

This 'aerodynamic heating' phenomenon arises [103] because the energy per unit mass which is diffused in a boundary layer is no longer just the heat energy $c_p T$ but the total energy $c_p T + \frac{1}{2}v^2$ obtaining by adding on the kinetic energy. Because that total energy has the value $c_p T_0 (1 + 0.2M^2)$ in an incident airflow of temperature T_0, the temperature T at the solid surface (where $v = 0$) tends to rise close to the value

$$T = T_0(1 + 0.2M^2). \tag{82}$$

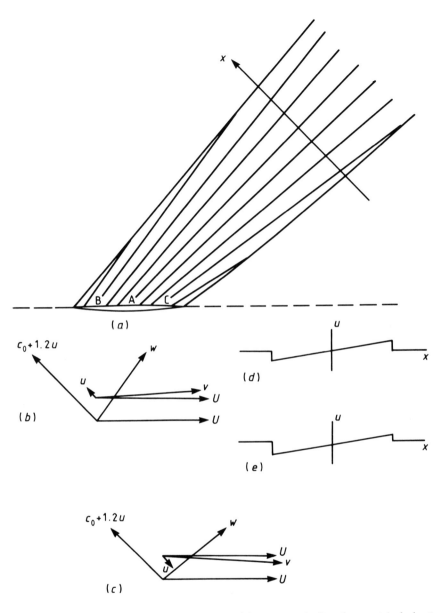

Figure 10.46. *(a) Stationary waves generated in supersonic flow (symmetrical about the broken line) around a thin aerofoil. (b) At B, the wind velocity U and a velocity u carried by the waves have a resultant v that slopes upwards (along the tangent to the aerofoil surface). The stationary wave is in the direction w (resultant of U and the signal velocity). (c) At C, the negative value of u causes the resultant v to slope downwards (again along the tangent) and the stationary wave now has a reduced angle to the wind. (d) Form of the balanced N-wave at the section shown. (e) At positive angle of attack, this becomes an unbalanced N-wave, with downward momentum.*

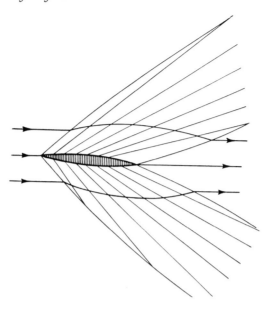

Figure 10.47. *The complete flow field* [103] *around a lifting aerofoil (actually, one twice as thick as that in figure 10.46).*

When, for example, $M = 2$ while T_0 takes a stratospheric value 210 K, this temperature (82) is 378 K, around the boiling point of water!

And one further comparison may be made: just as the symmetrical flow in figure 10.3 changes, at a positive angle of attack, into the lift-generating flow of figure 10.12(a), so also such an angle of attack converts the symmetrical flow of figure 10.46(a) into a flow that gives lift; briefly, because each value of u is decreased on the upper surface while values on the lower surface are increased, so that all pressures on the lower surface exceed the corresponding upper-surface pressures. And the aerofoil exerts once more an equal and opposite force, which confers downward momentum on the fluid. Yet this momentum appears, not in any vortex-wake region behind the aerofoil, but almost exclusively in the flow between the shock waves!—where the balanced N-waves of figure 10.46(d) become unbalanced, as figure 10.46(e) shows, and include [103] a downward component of momentum (see also figure 10.47).

Nevertheless, designers of supersonic aircraft in the twentieth century were increasingly drawn away from wings with such aerofoil sections by hard facts about aeronautical shock waves. The aim of keeping these weak so as to reduce shock-wave drag (and, often, shock-wave noise: the 'supersonic boom' heard at ground level) is achieved best, not by shapes that are thin in just one dimension (figure 10.46), but by shapes that are slender in both dimensions perpendicular to the flow [103]. Accordingly, the rather simple concepts of supersonic aerofoil theory (figure 10.46) required careful extension to involve three-dimensional wave propagation—using ideas outlined in section 10.3.1 above—if essential features of the 'slender-body aerodynamics' needed for $M > 1$ were to be usefully exploited [104, 72].

A good first suggestion of what may be called for is given by the analysis (p 848) of shock waves produced in air by an 'exploding wire'. Here, the source of sound is the appearance in the fluid of foreign matter, the outward-pushing vapour, whose cross-sectional area S increases very suddenly with time; generating, on a linear analysis, air motions (45), which are converted by non-linear effects into an N-wave.

Flight at supersonic speeds through a mass of air similarly introduces foreign matter (the aircraft), whose overall cross-section S in contact with that air mass increases very suddenly with time—though it then falls abruptly to zero. For a slender shape at supersonic speeds, this suggests the possible importance of a function $S(x)$, describing variation in the wind-direction x of the body's cross-sectional area S normal to that direction.

Slender-body aerodynamics showed how shapes where the function $S(x)$ varies rather smoothly and gradually along the aircraft's length have two advantages: (i) shock-wave strengths are kept relatively low; and (ii) they are quite well estimated [71], as for an exploding wire, in a two-stage analysis—linear, and then non-linear—with source strengths depending on the total area $S(x)$ of each cross-section rather than on its detailed shape. Of course the waves spread, not cylindrically, but conically at the angle θ defined by equation (81); as the linear part of the analysis makes clear already because equation (81) represents a 'stationary phase' condition.

This initial, linear analysis can be further improved [105] if the non-directional simple-source radiation is modified by adding a distribution of downward-pointing dipoles of strength $L(x)$, representing (section 10.3.2) forces on the fluid equal and opposite to the lift forces on aircraft cross-sections. This improvement, after the necessary non-linear adaptation, produces modest increases of N-wave strength below, and decreases above, the aircraft.

During the twentieth century, a significant first step in extending air transportation to supersonic speeds was made with Concorde [106], which offers busy passengers the benefit of journey times of only 3 hours over ranges of around 6000 km. Its cruise Mach number $M = 2$, attained at stratospheric altitudes between 15 and 16 km, limits aerodynamic heating to temperatures (82) at which aircraft materials retain safe properties; while its external shape (figure 10.48), with a sharp-edged slender delta wing, is a remarkable compromise between aerodynamic requirements in the supersonic and the low-Mach-number parts of its flight.

Thus, in spite of substantial shock-wave drag, it achieves a supersonic

Figure 10.48. *An Aérospatiale/British Aerospace Concorde with Rolls-Royce/Snecma Olympus 593 turbojet engines, and with all-up weight 185 tonnes, cruise Mach number 2.0 and maximum range 6500 km.*

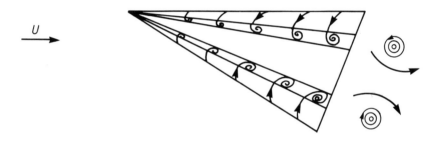

Figure 10.49. *Schematic illustration of lift generation by a sharp-edged delta wing at low Mach number. At a moderate angle of attack, a delta wing like that of Concorde sheds vorticity from the sharp leading edges—which rolls up into a vortex pair similar to that shed from the trailing edge in figure 10.13(b). Creation of downward momentum in the flow associated with this vortex pair balances the weight of the aircraft.*

lift–drag ratio around 10, and the shock waves produced below it are found, after propagation into the much denser air at ground level, to involve pressure jumps of only about a millibar. Yet the most innovative feature of the Concorde wing design is its fluid-mechanical method for achieving lift stably at low Mach number: the vortex pair essential (as figure 10.13(b) suggests) to this purpose is shed, not from the trailing edge, but from the sharp leading edge (figure 10.49), generating downward momentum which supports the take-off weight of Concorde stably at angles of attack of over 20°.

Development of such a shape needed extensive testing of models, both in ordinary low-speed wind-tunnels (figure 10.20) and also in supersonic wind-tunnels. Actually, it had been studies by Prandtl, initiated [18] in yet another fine paper of 1904, which produced insight into how wind-tunnel testing at supersonic speeds could be achieved.

Although an airflow's *convergence*—that is, reduction in cross-sectional area, giving an increase in ρV, the mass flow per unit area—produces the expected acceleration at subsonic speeds (as in figure 10.20), it has the opposite effect at supersonic speeds. Indeed, the balance (section 10.1.4) of mass-acceleration with pressure gradient

$$\rho V \frac{dV}{ds} = -\frac{dp}{ds} \left(= -c^2 \frac{d\rho}{ds} \quad \text{in adiabatic motion} \right) \tag{83}$$

makes ρV change with distance s at a rate

$$\frac{d(\rho V)}{ds} = \rho \frac{dV}{ds} + V \frac{d\rho}{ds} = \rho \left(1 - \frac{V^2}{c^2} \right) \frac{dV}{ds} \tag{84}$$

implying, as first recognized in the 1880s by the great Swedish engineer Laval, that acceleration to supersonic speeds needs a convergence followed by a divergence. What else does it need?

Prandtl suggested the answer with a diagram (figure 10.50) where the solid curves show all possible adiabatic distributions of pressure in a certain 'convergent–divergent nozzle', given the upstream pressure p_0. The fainter curves, describing wholly subsonic flow, apply for downstream pressures exceeding a certain value p_1; while the bold curve, describing continuous acceleration to supersonic flow, is achieved for downstream pressure p_2. But for downstream pressures between p_2 and p_1, where no continuous solution exists, Prandtl proposed theoretically and confirmed experimentally that the pressure follows an essentially discontinuous progression from the bold curve to one of those dotted curves which describe solutions at increased entropy. This penetrating discovery indicated how, whereas supersonic motion at a point is often determined just by upstream conditions, there may nonetheless be exceptions where the value of a downstream pressure is highly influential.

Extensions of such 'one-dimensional' analyses, including extensions to unsteady flow, allowed refined design of, and effective 'starting' procedures for, supersonic wind-tunnels; as achieved by those, beginning with A Busemann [107] in Göttingen, who would build up in the twentieth century such a deep experimental knowledge of aerodynamics at $M > 1$. (See figure 10.51 for a fine schlieren photograph [108] taken in an early supersonic tunnel at Göttingen.) But here, rather than pursuing this theme further, or extending it to that so-called 'hypersonic'

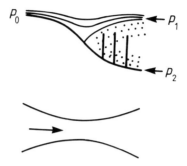

Figure 10.50. *Possible pressure distributions in a convergent–divergent nozzle. Solid lines, adiabatic distributions, of which the bold curve shows acceleration to supersonic speeds and the fainter curves describe wholly subsonic flow. Dotted lines, distributions with higher entropy, towards which the pressure rises after the discontinuities indicated.*

Figure 10.51. *Schlieren photograph of flow around a supersonic aerofoil published in 1927 by J Ackeret, another pioneer (then at Göttingen, but later in Zürich) of supersonic aerodynamics.*

aerodynamics [109] of spacecraft re-entry (with M as high as 20) in which shock-wave drag may actually be a friend—but the enemy is aerodynamic heating—this section 10.4.2 is concluded with comments on transonic flow: the transition régime between subsonic and supersonic aerodynamics [110].

Emphasis was placed in section 10.4.1 on the importance for flight efficiency of achieving high subsonic values of cruise Mach number without incurring shock-wave drag penalties. To help appreciate how shock waves may arise, the contrast between figure 10.46 at $M > 1$ and figure 10.3 at low Mach number may here be 'interpolated' with a diagram (figure 10.52) of symmetrical flow over an aerofoil at high subsonic Mach number.

Here the acceleration of air over the aerofoil surface generates a local region of supersonic flow. On the other hand, as in Prandtl's nozzle

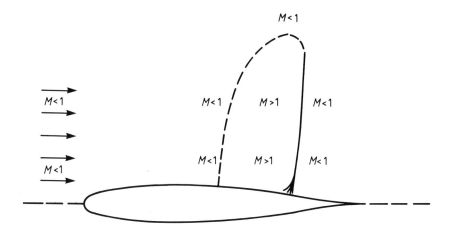

Figure 10.52. *Illustrating a flow (symmetrical about the broken line) around an aerofoil at high subsonic Mach number, when there appears in the flow a limited supersonic region terminated by a 'discontinuity' that incorporates a shock wave/boundary-layer interaction.*

example (figure 10.50), this flow can adjust to the level of downstream pressures only through a 'discontinuity'. Moreover, it is recognized now—in both cases—that this discontinuity involves not just a simple shock wave but a complex 'shock wave/boundary-layer interaction', including upstream influence [111] of the shock wave on the boundary layer (that is, an influence extending much farther upstream that might have been expected). And it is noteworthy that, here, the standard analysis (section 10.1.3) of an air flow into two parts (an exterior flow and a boundary layer) requires modification; often, by means of a 'triple deck' analysis in which the boundary layer is further subdivided, with a very thin inner part of the boundary layer near the wall responding to the external flow with particular sensitivity [112, 113, 114].

In the design of transport aircraft it is above all such shock-wave effects—being harmful both to lift–drag ratio and in other ways—which need to be avoided by ideas such as sweepback (figure 10.45). Other concepts valuable in design for transonic speeds include the celebrated 'area rule' of Whitcomb [115]. Briefly, an aircraft for which $S(x)$, the distribution of cross-sectional area defined above, varies smoothly and gradually with x will experience aerodynamic forces which, to an even greater extent at transonic than at supersonic speeds, are determined almost entirely by this distribution. (For such a determination, however, the earlier two-stage analysis—first linear and then non-linear—needs at transonic speeds to be replaced by a single-stage, fully non-linear analysis.)

But, beyond any such conceptual aids to aerodynamic design, it is above all on advanced computational fluid dynamics (CFD) that aircraft companies rely for detailed analysis of the expected performance of an aircraft configuration in complicated conditions like those characteristic of transonic flight. Modern CFD programs [116] give good representations of the steady flow outside the boundary layer around intricate aircraft shapes in three dimensions; also, where continuous flow is impossible, these programs identify the location and strength of any resulting shock wave—although without describing its interaction with the boundary layer. The data obtained, taken together with experimental knowledge of real interactions of external flows with boundary layers, are found invaluable for all aspects of design, including those directed (section 10.4.1) towards the improvement of flight efficiency.

10.4.3. Fast ships and secure offshore structures

Alongside the revolution in long-distance transport of people and of high-value goods that emerged from aeronautical engineering, some significant transformations of bulk carriage by sea, and of other fields within ocean engineering, were taking place. In section 10.4.3, selected contributions from fluid mechanics to these developments are outlined.

The surface of the deep (section 10.3.4) offers evident advantages for carriage of heavy goods, solid or liquid, in bulk-carriers or tankers whose overall weights can be in hydrostatic balance with the weight of water displaced. However, that surface's ability to propagate waves creates some counterbalancing disadvantages, including both speed and weather limitations.

At sea, relatively fast travel can once again yield benefits—above all, ships and crew complete more journeys each year—but, just as a limit on the speed of economical air transport is placed by the onset of extra drag due to shock-wave generation, so also some limitations on speeds of economical marine transport are placed by the onset of significant drag due to surface-wave generation. Here, methods for pushing back this onset to higher speeds (analogous to the use of sweepback in aeronautics) are sketched. On the other hand modern 'ship routing' methods for minimizing delays from stormy weather depend on advanced forecasting techniques and are postponed to section 10.5.2.

Away from any storms, hydrodynamic forces on ships can be estimated by linearly combining any forces [117] due to ambient ocean waves (especially, swell from far-off storms) with forces which the ship's steady motion at speed U would produce in a calm sea. But such calm-sea forces, on which we concentrate, may be influenced by the ship's own power to generate waves.

Waves so generated remain stationary relative to the ship, so that if they propagate at an angle θ to the ship's motion with wave velocity c they satisfy the equation

$$U \cos \theta = c \tag{85}$$

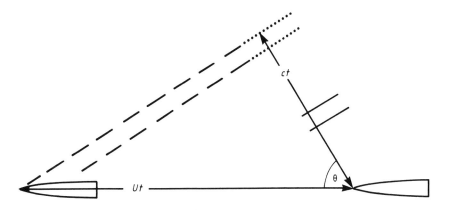

Figure 10.53. *Waves generated by a ship moving at speed U are propagated at various angles θ, but each has velocity c = U cos θ. If they travelled a distance ct during time t, while the ship moves a distance Ut, then by simple trigonometry the crests (dotted) would be on (broken) lines through the new position of the ship. Actually, the energy in the (plain-line) crests has travelled half-way towards these lines at the group velocity $c_g = \frac{1}{2}c$. These arguments, including consideration of all angles θ, lead to the Kelvin ship-wave pattern of figure 10.54 below.*

analogous to (81) for sound waves. With the linear-theory value (55) for c, equation (85) gives the wavelength as

$$\lambda = \frac{2\pi U^2}{g} \cos^2 \theta. \tag{86}$$

An enormous difference from sound waves arises, however, from the fact that the energy in surface waves travels, not at the wave velocity c with which crests move, but at the group velocity $c_g = \frac{1}{2}c$. Figure 10.53 points this contrast: during time t, while a ship moves a distance Ut, the waves generated travel a distance $c_g t$. If c_g took the value $c = U \cos \theta$ then any wave generated would lie on a line through the ship's present position (as with sound waves in figure 10.46). But waves with $c_g = \frac{1}{2}c$ have only travelled half as far. Trigonometry then shows [70] that—on these linear-theory arguments, first offered in the 1880s by Lord Kelvin [118]—the waves all lie within a 'wedge' of semi-angle

$$\sin^{-1}(1/3) = 19.5° \tag{87}$$

and take forms sketched in figure 10.54. But we shall see that (i) for a given ship at a given speed, only part of the Kelvin ship-wave pattern is observed, while (ii) refinements of the analysis (including non-linear considerations) produce some broadening of the wedge.

Because the greatest possible wavelength (86) for ship-generated waves is $2\pi U^2/g$ (attained in figure 10.54 by those longer waves which

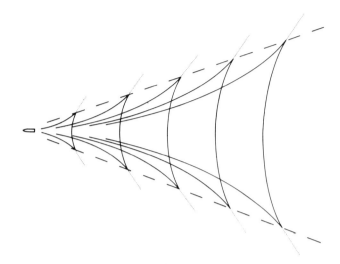

Figure 10.54. *Plain lines, ship-wave pattern suggested by Kelvin [118] from simple group-velocity arguments. Dotted lines, extensions beyond the broken-line 'caustic' derived from exact linear-theory calculations [70]. (Note: non-linear effects produce a further slight broadening of the Kelvin wedge.)*

are propagating at small angles θ), the classical approach to minimizing the energy in such waves (so that wavemaking drag is kept very small) is to ensure that the ship's length, and smoothness of shape along it, are both so great that relatively short waves, with $\lambda \leqslant 2\pi U^2/g$, simply cannot be excited. But the fine Japanese naval architect, T Inui, demonstrated in 1962 an important improvement [119] on this classical approach, which may here be briefly explained [70].

At low enough speeds U, a ship experiences only frictional drag because it generates no waves. Then the flat free surface behaves like a reflection plane for the physicist's 'method of images'; and, indeed, a common experimental approach to estimating frictional drag for a ship design of given shape is to build a 'double model'—a model of the immersed part of the hull together with its reflection in the free surface—and to test this model fully immersed (with the boundary layer made turbulent to simulate full-scale Reynolds number) in a water tunnel. Frictional drags so estimated can later be subtracted from drags measured in flows with a free surface to indicate values of wavemaking drag.

In such a double-model experiment, the half of the flow below the plane of symmetry (which, by symmetry, remains flat) represents the flow below the real ocean surface. However, it fails in just one respect to represent it with full accuracy: the distribution of pressure varies in that plane, instead of taking a constant value (the atmospheric pressure) as it should on a flat surface. Approximately, any waves generated are those

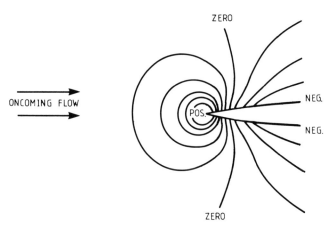

ONCOMING FLOW

Figure 10.55. *For ship hulls with conventional bows, a typical distribution of excess pressure measured (or computed) in the plane of symmetry of a 'double model' (immersed part of hull plus its reflection in that plane) is as shown, with spectral elements of highest wavenumber arising from the steep positive pressure peak just in front of the bow.*

which this distribution of pressure, moving at speed U, would produce if applied to the free surface; and the value of U at which such waves become important is that for which the highest significant wavenumbers in the spectrum of this distribution with respect to x are around g/U^2 (corresponding to $\lambda = 2\pi U^2/g$).

Both experimental and computational studies of the pressures in the plane of symmetry confirm [70] that these highest wavenumbers are associated with a steep positive pressure peak just ahead of the ship's bow (figure 10.55), related by Bernoulli's equation (9) to a sharp retardation of flow in this region. They might be eliminated, then, by an added hull element which tends to produce a cancelling peak of negative pressure in the same region.

Such an element is the subsurface 'bulbous bow' of Inui [119], projecting ahead of the ship's visible bow. In the corresponding double model, with two bulbous projections, convergence of the flow between them produces in the plane of symmetry a local velocity peak related, by equation (9), to a local negative-pressure peak which can suppress the former positive peak and postpone the former drag rise to higher values of U.

The economics of transport by modern tankers and bulk-carriers benefited, then, from the introduction of (i) longer ships with (ii) subsurface bulbous bows. And thick boundary layers near the stern of such longer ships confer an intriguing fluid-mechanical advantage, because much of the water entering the propeller disks is water from the boundary layers. Briefly, since its velocity relative to the ship takes a reduced value $v < U$, its momentum gain (contribution to thrust) per unit mass when kinetic energy E is added becomes

$$(2E + v^2)^{1/2} - v \qquad (88)$$

which is substantially augmented as v is reduced.

Like aeronautical engineers, naval architects combine the use of conceptual aids to design with model testing and progressively refined CFD techniques for analysing flows around ship hulls [120]. These are now able to apply at the free surface, not a linearized, but an accurate boundary condition (with some associated widening of the Kelvin wedge), and have led to further valuable improvements in hull design.

The second half of the twentieth century saw also two useful innovations in the fast marine transport of middle-sized loads over fairly short ranges [121]. Both of these achieve major reductions in frictional drag by lifting the main part of the hull out of the water.

The first is the hovercraft, with an annular jet used to allow creation of a cushion of air at positive pressure which supports the weight of the hull. (Any wave drag is then generated literally, rather than just conceptually as above, by such a distribution of pressure on the water surface, moving at speed U.)

The second is the hydrofoil boat, where a wing-shaped element under the water generates, after the design speed is reached, the main lifting force that holds the hull out of the water. (Then the most important resistance is the wing drag, including the drag due to lift.)

And yet another area of ocean engineering that gained special importance in the twentieth century was the engineering of secure offshore structures, mainly for oil and gas exploration and exploitation. Essentially, this has been a field where applications of fluid mechanics could no longer benefit from d'Alembert's Theorem. Thus, whereas shapes of aircraft or ships can be 'streamlined' (section 10.1.4) so that drag forces are low when the vehicle moves ahead in its designed direction of travel, an offshore structure has no such predetermined direction for the relative motions of water and structure. Essentially, it must withstand the onset of waves from all directions.

These considerations led to the widespread use of structural members with no preferred direction, including (especially) circular cylinders [122]. Some complexities in the interaction of a circular cylinder with just a steady oncoming flow were illustrated already in figure 10.1, but offshore engineering has demanded from fluid mechanics a detailed understanding of the cylinder's far more complicated interaction with very unsteady flow fields in ocean waves. Above all, hydrodynamic loads on the cylinder are associated, of course, with changes of momentum in the fluid flow as affected by such interaction.

Here, two aspects of these changes are emphasized. First, any vorticity that is shed as a result of flow separation in one phase of an oscillatory flow may be convected back over the cylinder in later phases. Deep study of the resulting vorticity distributions led to an impressive level of knowledge of how the associated momentum ($\frac{1}{2}\rho$ times the moment of the vorticity distribution) would vary, with some vital consequences for hydrodynamic forces on the cylinder [123].

But the part of the flow that is without vorticity has its own additional momentum. This is, of course, a potential flow—with a momentum that shows no change with time (corresponding to d'Alembert's Theorem that the associated force is zero) when the oncoming flow is steady. Yet when the cylinder, of radius, a, is placed in a flow whose velocity U varies with time, the potential-flow part of the interaction has a varying momentum associated with a force whose value

$$2\pi a^2 \rho \frac{\mathrm{d}U}{\mathrm{d}t} \tag{89}$$

was already well established in the nineteenth century.

In 1928, however, G I Taylor foresaw [124] the need to calculate such potential-flow forces when a body is placed in an ambient fluid motion that varies in space as well as in time. He analysed them by means of a profound application of general Hamiltonian dynamics, and also verified the results directly with a more direct, yet far more laborious, calculation of pressure distributions. For example, in the particular case of a circular cylinder, with axis in the z-direction, placed in such a variable flow field with U and V as its x- and y-components, the x-component of the potential-flow force turns out to be

$$2\pi a^2 \rho \left[\frac{\mathrm{d}U}{\mathrm{d}t} + \frac{1}{2} \frac{\mathrm{d}}{\mathrm{d}x} (U^2 + V^2) \right]. \tag{90}$$

For many decades, G I Taylor's result (90) was used only rarely. In offshore-engineering applications, moreover, there was almost universal loyalty to the classical expression (89) for the potential-flow force. By the mid-1980s, however it became recognized that the quadratic correction in equation (90) could be important [125]; e.g. for the excitation of resonant modes of motion of structures (such as tension-leg platforms) with a very low natural frequency—which quadratic forces allow to be excited as a 'difference tone' by a spectrum of waves of considerably higher frequencies.

Interactions between offshore engineering and fluid mechanics cover an enormously wide field. Here, we have touched on just a very small sample of that field, chosen above all to illustrate differences in its character from other areas of application of fluid mechanics.

10.5. Dynamics of the Earth's fluid envelope, and its forecasting applications

10.5.1. Wave-like current patterns
The Earth's fluid envelope is a mixture of air and water. In the atmosphere air predominates, yet water (transferred from the ocean)

plays a spectacular role. In oceans, rivers and lakes water predominates, yet they teem with life because of dissolved oxygen and CO_2. Not only mass exchange between the different components but also transfers of heat and of momentum are major influences on their dynamics.

Study of that dynamics, besides its scientific interest, has crucially important applications to meeting the requirements of countless industries—agriculture, shipping, aviation, energy-supply, coastal engineering, river-management, etc—for improved weather forecasting; see section 10.5.2, where some account of climate-change forecasting is also given. Practically none of the dynamical or forecasting problems involve only one component of the Earth's fluid envelope in isolation from others; nevertheless, this section 10.5.1 is devoted to certain features of ocean and river dynamics which, while influenced by atmospheric input, assume something of a life of their own because patterns of water flow propagate in a wave-like manner.

Forecasting was first achieved successfully for ocean tides [126]. Superficially, these seem to be periodic vertical motions that alter sea level; yet practically all their kinetic energy is in the horizontal motions: those powerful tidal currents which make an exciting spectacle off many coastlines. The pattern of these currents propagates in a wave-like manner, similar in all ways but one (see below) to the shallow-water propagation described in section 10.3.3.

Deep oceans masquerade, then, as shallow water for propagation of tidal currents. Indeed the ratio $\frac{1}{3}k^2h_0^2$ between kinetic-energy contributions from vertical and horizontal motions is so small that it annuls the dispersion effect (63): at frequency $\omega = kc_0$, the value (57) for c_0 gives

$$kh_0 = \frac{\omega h_0}{c_0} = \omega \left(\frac{h_0}{g}\right)^{1/2} \tag{91}$$

which for the tidal component of lowest period (half a day) is less than 0.005 even in the ocean's deepest parts, making $\frac{1}{3}k^2h_0^2$ less than 10^{-5}.

Newton had recognized that gravitational theory gave the Moon a special influence on tides: relative to the motions of the solid Earth in the gravity fields of Sun and Moon, water nearest to each body would be subject to an excess attraction while water farthest from it would experience a defect; in other words, a relative repulsion. For each body, then, forces tending to raise sea level would reach maxima (figure 10.56) at the rotating Earth's nearest and farthest points—whose positions change with a period close to half a day. But, although the Moon was far less massive than the Sun, its proximity would yield bigger values of those gradients of gravitational force with distance which produce tide-raising forces. In addition, once a fortnight at full or new Moon, the Moon's tide-raising force would be augmented by an almost collinear solar force, giving 'spring' tides (with perfect collinearity at an equinox); while neap tides, with the forces orthogonal, would occur in between.

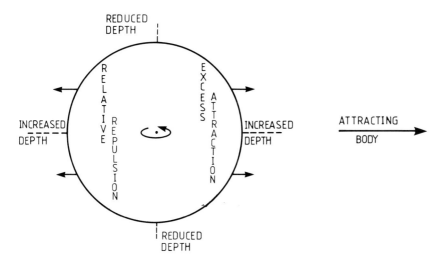

Figure 10.56. *Gravitational forces tending to raise sea level (short arrows) reach maxima at points nearest to or farthest from an attracting body. Newton argued that the resulting ocean depth distribution might be as suggested by the broken lines. (Also, see note (ii): the horizontal components of the forces play a leading role.)*

Fluid-mechanical analysis, however, in the hands of Laplace [127] and his successors, gave a picture differing from that of Newton in three important respects:

(i) it allowed for the wave-like character of the pattern of ocean-current response to tide-raising forces; of which, furthermore,

(ii) only the horizontal components could energize tidal currents (whose convergence, then, would raise sea level); while moreover

(iii) the Earth's rotation, besides determining periods for different tide-raising forces, had also an important dynamic effect.

This effect, first identified for the tides, is important too for most oceanic or atmospheric motions [93]. Vorticity was defined in section 10.2.2 so that a fluid simply rotating with angular velocity Ω has vorticity 2Ω. Accordingly, just the Earth's rotation with axial angular velocity 2π per day gives all terrestrial fluids an axial vorticity of 4π per day. But analysis of essentially horizontal tidal currents requires only a knowledge of the vorticity's vertical component, equal at latitude θ to

$$f = \frac{4\pi \sin \theta}{\text{day}} = \frac{4\pi \sin \theta}{24(3600\,\text{s})} = (1.45 \times 10^{-4} \sin \theta)\,\text{s}^{-1}. \qquad (92)$$

The sense of this 'planetary' vorticity is cyclonic: anticlockwise in the northern hemisphere ($\theta > 0$) and clockwise in the southern ($\theta < 0$).

Its importance for tides is that when these raise ocean depth from an undisturbed value h_0 to a new value h, such stretching (see figure 10.9)

of a vertical column of water stretches the vertical vorticity component
(92) to a new value fh/h_0; of which the part

$$f\left(\frac{h}{h_0} - 1\right) \tag{93}$$

must correspond to effects other than Earth's rotation. Expression (93)
represents, then, the vorticity ς of the tidal currents themselves (motions
relative to the Earth).

Simple shallow-water theory is shown in equation (56) to be a
direct analogue of the theory of sound. Two-dimensional propagation
over shallow water whose undisturbed depth h_0 varies with position
would, similarly, equate with two-dimensional acoustic propagation in
a medium with non-uniform undisturbed sound speed c_0 if, as in sound
theory, potential flow (with zero vorticity) could be assumed. However,
tidal currents—although propagating effectively over shallow water—
are not potential flows since they have a vorticity (93), where h/h_0
represents the depth increase due to convergence of the same currents.
This difference is found, very briefly, to make the waves dispersive after
all!—with a local dispersion relationship

$$\omega^2 = f^2 + (gh_0)k^2 \tag{94}$$

of frequency ω to wavenumber k for small disturbances (corresponding
to wave velocity $\omega/k = (gh_0)^{1/2}$ if f is absent).

All analyses of tides, and all interpretations of tidal observations,
start [126] from Newton's elucidation of tide-raising forces and spectrally
analyse these. The largest component is known as M_2, the Moon's 'semi-
diurnal' influence; actually, with period 12 h 25 min, greater than half
a day by 1/28 because the Moon's orbital motion with period 28 days
is in the same sense as Earth's rotation. The next largest is S_2, that part
of the solar tide-raising force which is exactly semi-diurnal (with period
12 h). There are very many other components, all with considerably
longer periods, which need to be allowed for in constructing tide tables;
however, it is noteworthy that the basic cycle of spring and neap tides
derives already from the beats between M_2 and S_2.

These, moreover, are the two components which allow equation (94)
to determine a real wavenumber k for, essentially, all latitudes θ. Thus
S_2 has frequency 4π per day, equal to a polar value of f, while M_2 has
an only slightly lower frequency, equal to the value (92) of f for $\theta = 75°$.
In both cases, the equation simply indicates that k becomes small (that
is, tidal wavelengths become large) at relatively high latitudes (figure
10.57). By contrast, all spectral components with lower frequency are
faced with a considerably lower maximum latitude for real k; which acts

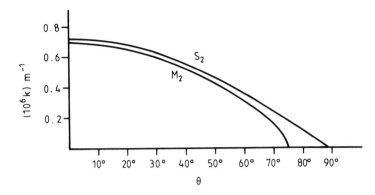

Figure 10.57. *Wavenumber k at latitude θ as indicated for the M₂ and S₂ tides by equation (94), with h₀ taken as 4 km (the oceans' mean depth).*

as a caustic bounding any waves (there is, of course, the usual tailing-off of amplitude beyond the caustic).

On any continental shelf, concentration of the tidal kinetic energy within water of much diminished depth may augment local currents, with two results: (i) dissipation of energy by bottom friction and (ii) various non-linear distortions of waveforms, some of which (section 10.3.3) may involve a 'hidden' energy loss. What is the energy source that compensates for any dissipation?

The energy source for tides is the Earth's rotation itself, which stores energy rather as a flywheel does in some machines. If the Earth rotated so slowly as to turn always the same face to the Moon, there would be no tidal movements at all but just a static elevation of water (as in figure 10.56) at points nearest to or farthest from the Moon. Tides are due to excess rotation (the day being less than the Moon's orbital period) and derive their energy from it. Thus, dissipation of that energy reduces the excess; in short, 'tidal friction lengthens the day'. By 1919 already [128], G I Taylor had successfully verified this statement through a comprehensive comparison of the rate at which Earth's rotation is becoming slower with an estimated rate of dissipation of tidal energy in shallow seas. (For the Moon's own rotation, of course, any analogous excess was long ago annulled by dissipation of tidal motions in the solid Moon itself.) Ocean tides, then, being rather lightly damped forced oscillations, take broadly the form of standing waves, with local wavenumbers given by equation (94); yet, because dissipation is localized in shallow shelf areas, nearby tidal-current patterns assume rather more of a progressive-wave character, with energy flowing towards where it is being dissipated.

Historically, tide gauges have long been abundant in those areas of relatively shallower water where tidal forecasting was needed, with a spectral analysis of their records having allowed the construction of reliable tide tables. But it was only in the second half of the twentieth century [129] that pan-oceanic pictures of tidal current patterns became achievable by numerical analysis. This uses 'shallow-water theory' (see above) with the correct distribution (93) of vorticity, with known ocean-depth distributions, and with continental-shelf boundary conditions that allow for realistic values of energy dissipation.

Figure 10.58 illustrates a pioneering computation of this type for the M_2 tide in the Atlantic Ocean [130], obtained by C L Pekeris in 1969. (It represents part of a computation for the Earth's oceans taken as a whole.) Here, wavelengths follow approximately the trend shown in figure 10.57, while the importance of vorticity is shown especially at 'amphidromic' points around which the phase of the standing wave rotates. This diagram, which is in rather good local agreement with coastal and island measurements, offers a remarkable contrast to the simple Newtonian picture of figure 10.56.

An excess rise of sea level, above that associated with the Moon's and the Sun's tide-raising forces, may occur in stormy weather; which produces an additional horizontal force on the ocean associated with momentum transfer from winds. Such an excess rise, called 'storm surge', is numerically predicted from wind forecasts (section 10.5.2) using shallow-water approximations as above [131]; with some severe coastal inundations being experienced when high tide and peak storm surge coincide.

Nevertheless, much more persistent ocean-current patterns are generated, not by sudden storms, but by the prolonged action of winds at the surface. In fact, most of the circulation of ocean waters is due to this cause [93] (the main exception being the 'thermohaline' part, driven by the sinking of water with high salt concentration generated [126] by ice formation in polar oceans).

For such forcing by slowly varying winds, another mechanism for vorticity change becomes important besides that effect of stretching of a vertical column of fluid which underlies equation (93) for the vertical vorticity ς of the current pattern itself. Even if there were no stretching, nonetheless movement of vortex lines with the fluid (figure 10.9) would imply conservation of the total vertical vorticity $f + \varsigma$ in any northward motion at velocity v; yet such motion makes the value (92) for the planetary vorticity f increase at a rate

$$\beta v \quad \text{where } \beta = (2.3 \times 10^{-11} \cos \theta)\,\text{s}^{-1}\text{m}^{-1} \tag{95}$$

implying an equal and opposite rate of change, $(-\beta v)$, for ς. In 1940 the gifted oceanographer C G Rossby [132] saw the importance of wave-like current patterns governed by this relationship $d\varsigma/dt = -\beta v$ (making the rate of change of the curl of the current pattern proportional to one component thereof).

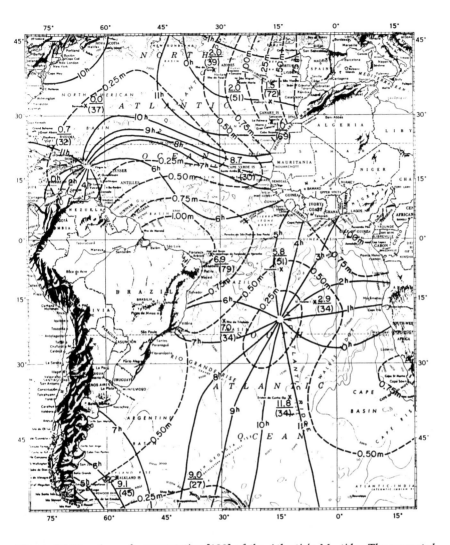

Figure 10.58. *An early computation* [130] *of the Atlantic's* M_2 *tide. The computed cotidal lines (solid) give phase lag of high tide in hours after lunar transit at Greenwich, while corange lines (broken) give tidal range in metres. Underlined figures give observed phase lags, and bracketed figures observed ranges in centimetres.*

The dispersion relationship for such 'Rossby waves' is

$$\omega k^2 = -\beta k_1 \tag{96}$$

where k_1 is the eastward component of the wavenumber vector k. If its northward component is k_2, then the curves of constant ω in a wavenumber plane with k_1 and k_2 as coordinates are circles (figure 10.59). But the group velocity (p 864) is the gradient of ω in this plane, which is directed radially inwards towards the centre of each circle [93].

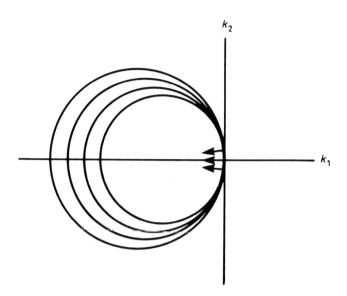

Figure 10.59. *The dispersion relationship (96) for Rossby waves describes circles in the wavenumber plane (k_1, k_2) whose radius $\beta/2\omega$ becomes less as ω increases, so that the gradient of ω points inwards towards the centre of each circle. Arrows in the direction of the group velocity refer to the case of waves with $k_2 \gg k_1$ produced by zonal forcing.*

To give just one illustrative example, the contrast between westerly mean surface winds in middle latitudes and easterly trade winds at relatively lower latitudes tends on average to generate a broad zonal pattern of oceanic vorticity that is anticyclonic (ς of opposite sign to f). Such zonal forcing has k_2 much greater than k_1, giving a westward group velocity (figure 10.59). So all the energy of the resulting current patterns travels westward (along-crest energy propagation, as in section 10.3.5), to become concentrated in a poleward current on the western boundary (the Gulf Stream in the North Atlantic, the Kuroshio current in the North Pacific, the Agulhas current in the South Indian Ocean, the East Australian current in the South Pacific, etc). Figure 10.60 shows schematically (i) the mechanism of westward travel of such a zonally distributed signal and (ii) its tendency to generate a thin boundary current [133]. This has been successfully analysed by boundary-layer methods (allowing [93] for non-linear inertial effects and effects of momentum transport across the layer) similar to those of section 10.1.3; furthermore it may separate like other boundary layers—as, for example, the Gulf Stream separates from the North American coast at Cape Hatteras, becoming a somewhat meandering jet thereafter as it makes it way across the Atlantic.

Actually, Rossby's analysis [132] took account of vorticity changes of both types. If the stretching effect (93) as well as the effect (95) of

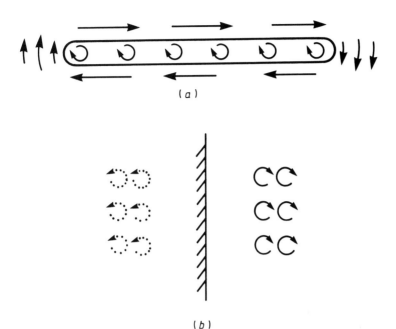

Figure 10.60. *(a) Schematic view of the mechanism for westward travel of a disturbance with nearly zonal forcing (northern-hemisphere case). Negative ς induces northward flow to the west, and so generates negative ς there by equation (95). (b) At any western boundary a westward travelling pattern of vorticity, along with its image system (broken lines), generates a northward boundary current.*

'advection' by any northward motion v is allowed for, then the total vorticity $f + \varsigma$, instead of being unchanged, must vary in proportion to the depth h. Rossby gave the quantity

$$\frac{f + \varsigma}{h} \tag{97}$$

which remains constant in any advective motion, the imaginative name 'potential vorticity', which was to prove highly influential on both oceanographers and meteorologists (section 10.5.2). The associated dispersion relationship includes the whole of (94)—except that the left-hand side (negligible at very low frequencies) is suppressed—together with (96) to give

$$\omega\left(\frac{f^2}{gh_0} + k^2\right) = -\beta k_1. \tag{98}$$

With use of (98), the geometry of figure 10.59 is a little modified, yet conclusions on western boundary currents in mid-latitude oceans remain essentially unchanged.

But an important difference between current patterns generated by tide-raising forces distributed uniformly with depth and by wind stresses concentrated at the ocean surface still needed to be addressed. Below a well-mixed layer near the ocean surface, water density shows an increase with depth that is associated with gradients of temperature and salinity (figure 10.61). The gravitational response of water with such a distribution of density can be analysed into modes [133], of which primarily the first two are important:

(i) the 'shallow-water' mode, frequently called 'barotropic', where currents distributed uniformly with depth generate sea-level changes by convergence and

(ii) the 'baroclinic' mode without sea-level changes, where a vertical column of fluid has no net momentum because currents near the surface are opposed by deeper currents, and where convergence simply tilts density contours.

Gravity acting on stratified fluid opposes such tilting; very briefly, the term gh_0 which represents the restoring force for the barotropic mode is retained for the baroclinic mode except that h_0 ceases to be the ocean depth (of many km) but is replaced by a quantity of order 1 m. Then the bracketed expression in (98) is dominated by its first term, so that the equation describes non-dispersive waves propagated to the west at the wave speed

$$\left(-\frac{\omega}{k_1}\right) = gh_0\frac{\beta}{f^2} \qquad (99)$$

which is plotted as a function of latitude in figure 10.62.

Mid-latitude oceans respond very sluggishly in the baroclinic mode. Thus figure 10.62, although merely describing small-amplitude waves, suggests that, with speeds of the order of 1 cm s^{-1}, they might need a decade to cross an ocean; and numerical models confirm that, in these latitudes, distributions of ocean current with depth take at least a decade to respond to forcing by wind—even though such surface forcing has a basic tendency to excite more baroclinic than barotropic response.

But equatorial oceans respond readily in the baroclinic mode. Figure 10.62, besides indicating this, suggests that any simple ray-theory analysis is impossible near the equator because of steep wave-speed gradients; however a detailed wave theory shows how baroclinic waves, trapped near the equator, propagate along it [134, 135]. The different trapped-wave modes, with waveforms as in the Schrödinger equation for the harmonic oscillator, include one that propagates eastwards at velocity $(gh_0)^{1/2}$, around 3 m s^{-1}, while the rest propagate westwards at somewhat lower velocities. The former contributes to that irregular pattern of tropical-meteorology changes known as the Southern Oscillation (section 10.5.2), while the latter modes play a role in the lively

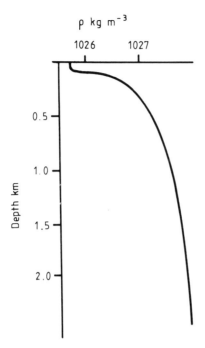

Figure 10.61. *A typical average variation of density ρ in the ocean as a function of depth.*

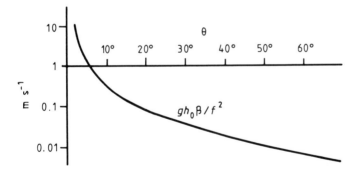

Figure 10.62. *The wave speed (99) for westward propagation of small baroclinic disturbances is plotted (on a logarithmic scale) against latitude θ for $h_0 = 1$ m.*

response of the Indian Ocean currents to onset of the Southwest Monsoon [133].

In another part of the Earth's fluid envelope—the rivers—downstream movements of flood waters generated by intense rainfall may exhibit wave-like patterns that were first elucidated [136] by a fine hydraulic engineer, J A Seddon, for the great rivers of the USA.

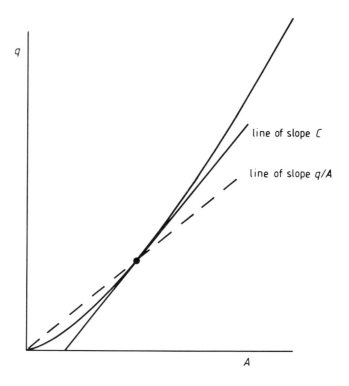

Figure 10.63. *The relationship between volume flow q and cross-sectional area A of river water is concave upwards at any location, with slope C = dq/dA exceeding the mean flow velocity q/A. Multiplying the local rate of change of A by C, we obtain the local rate of change of q in the form* $-Cdq/dx$, *so that values of q travel at velocity C.*

Anywhere on a river, there must be a relationship between the water's cross-sectional area, A, and the volume flow of river water, q. This relationship is concave upwards (figure 10.63) because, where α is the bed slope, the downstream gravity force $\rho g \alpha A$ on a section is balanced by frictional resistance (associated with turbulent flow in the river), which varies as about the square of the mean flow velocity q/A. But, besides this mean velocity, a somewhat higher velocity C defined as the slope $C = dq/dA$ of figure 10.63 plays an important part in determining how flood waters move. At any location, simple kinematics shows that the rate of change of water cross-section A (depending on inflow minus outflow) is minus the gradient of the flow q; so that, on multiplying this relationship by C, we deduce that the rate of change of q is $(-C)$ times the gradient of q itself.

This means that changes in q travel downstream at velocity C, in the form of what has been called a 'kinematic' wave. (Its wave speed C, although greater than the mean flow velocity q/A, is much less than the propagation velocity c_0 for 'dynamic' waves such as gravity waves on shallow water; which, indeed, become powerfully damped in long turbulent rivers.) Moreover, different values of q travel at different speeds C, in such a way that larger values may tend (yet again!) to catch up with smaller values [137]; until an equilibrium waveform— the 'kinematic shock wave', with thickness of order 1 km—is formed from a balance of kinematic and dynamic effects. This then is yet another flow response to atmospheric input which assumes something of 'a life of its own' in the form of a current pattern that propagates in a wave-like manner.

10.5.2. Weather and climate

An interacting system comprising atmosphere, oceans, continents and the month-by-month distribution of incident solar radiation determines the Earth's climate; that is, the average geographical distribution of winds, temperatures, cloudiness and precipitation in each month of the year. It is strongly affected by contrasts between the heat-absorbing properties of the ocean's well-mixed layer (about 50–100 m thick) and of weakly conducting continental surfaces; while evaporation of ocean water has two opposing effects on climate: the 'greenhouse' effect (interception of outgoing long-wave radiation by water vapour), and reductions of incoming solar radiation through reflection from its condensed forms (clouds, snow cover, etc).

However, the system's many unstable and indeed chaotic features bring about an enormous variability about any monthly average; and such variations on small time-scales, of course, constitute the weather. Also some of the instabilities, by generating large-scale horizontal (as well as vertical) fluxes, are influences on climate itself.

The atmosphere, no less than the ocean, is strongly affected [93] by the planetary vorticity (92), by its change (95) in north–south movements, and by a form of Rossby's 'potential vorticity' idea (97). Fundamental applications of these concepts to weather and climate are outlined in section 10.5.2 before (i) describing twentieth century successes in meeting urgent needs of agriculture, transport, civil protection, etc for improved weather forecasting, and (ii) reviewing possibilities for discerning changes of climate over periods exceeding the basic annual cycle. First of all, however, some key influences on atmospheric stratification must be described [138].

Air temperatures normally decrease with altitude, but never at a rate exceeding 1 °C per 100 m; determined as g/c_p, with the air's specific heat c_p at constant pressure around 1000 J kg^{-1} per °C. The explanation is that heat input when pressure as well as temperature changes is

$$c_p \, dT - \rho^{-1} \, dp \quad \text{which is} \quad c_p \, dT + g \, dz \qquad (100)$$

wherever pressure drop is due to an increase in altitude dz. So air rising with zero heat input (that is, adiabatically) loses temperature by 1 °C per 100 m, and would continue to rise—while falling air would continue to fall—if ever the ambient lapse rate (gradient of temperature decrease) exceeded this value. Such excessive lapse rates, in short, are rapidly eliminated by vertical mixing.

On the other hand, any rising air with a fully saturated partial pressure of water vapour (a quantity which becomes less, of course, after a temperature decrease) must as it cools experience condensation, with associated latent heat release, and a consequently more gradual temperature drop—along the 'moist-air adiabatic'—of around 0.5 °C per 100 m. It follows that, wherever the ambient air's lapse rate lies between 0.5 °C and 1 °C per 100 m, any saturated air may experience that vigorous convective mixing, involving vertical motions with associated condensation, which characterizes (say) cumulus clouds.

In the more stably stratified air just above the tops of cumulus clouds, a continued pounding from below by the more vigorously rising parcels of moist air generates internal waves (section 10.3.5). Flight through these feels much like flight through turbulence, and pilots often call them clear air turbulence. The theory of internal waves given in section 10.3.5 for a liquid with the density stratification (66) needs just one modification in the atmosphere [139]: air in an adiabatic upward displacement $s \cos \theta$ suffers a pressure drop $\rho g s \cos \theta$ and a consequent density drop $\rho g c^{-2} s \cos \theta$ where c is the sound speed (14); so it finds itself denser than its surroundings by an amount $\rho(\varepsilon - g c^{-2}) s \cos \theta$. Thus it is only when $\varepsilon > g c^{-2}$ that the stratification is stable (this, of course, is just another form of the condition on lapse rate), and then the Väisälä–Brunt frequency N takes a modified form [94, 95]

$$N = \left[g(\varepsilon - g c^{-2}) \right]^{1/2}. \tag{101}$$

However, after this change, all aspects of the theory of internal waves given in section 10.3.5 remain unaltered.

Where solar radiation strongly heats land surfaces, the powerful air movements that remove much of the heat by upward convection are described (p 845) as thermals. In each thermal, the fluid flow field is broadly of the nature of a 'vortex ring', with a large upward momentum. A thermal grows as it rises by entraining ambient air and, very gradually, slows down—although, especially in tropical regions, the upward movement may continue to altitudes of very many km.

These and many other mixing processes generate the troposphere, that fairly well mixed region of the atmosphere which extends to a height varying from 16 km over the equator to mid-latitude values around 11 km; and, in polar winters, a level as low as 8 km. Above it, those substantial temperature lapse rates that are associated with fast mixing processes disappear or are reversed, and this region with very stable stratification is known as the stratosphere.

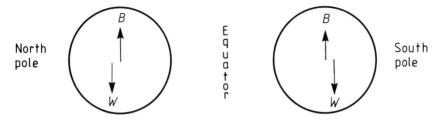

Figure 10.64. *Schematic explanation of the vorticity-generating effect of horizontal temperature gradients. The line of action of a fluid sphere's weight W is displaced to the cold side of its geometric centre through which the buoyancy force B acts.*

The general climatology of global winds is often summarized in terms of surface-wind features (e.g. westerlies in temperate zones, easterly 'trade winds' at lower latitudes), but from the standpoint of basic physics the surface winds are a second-order aspect of climate [93]. The outstanding feature of the distribution of mean winds is the enormous general increase of westerly wind with height at practically all latitudes; for example, the easterly trade winds extend only to altitude 2–3 km, above which they are overlain with increasingly strong westerlies. It is above all the large temperature difference between equator and poles (derived from the distribution of solar radiation) which makes for this general increase of westerly mean wind with height, in a way which can be interpreted [140] in terms of fundamental properties of vorticity.

Horizontal temperature gradients change the essential conclusion (figure 10.9) that, because pressure forces act through a sphere's geometrical centre, they are not altering its angular momentum. Figure 10.64 shows how a sphere, subject to an equatorward temperature gradient T' at a given pressure level, has its centre of gravity displaced polewards so that the couple formed from its weight and the Archimedes buoyancy force on the sphere generates angular momentum. The associated rate of change of vorticity in a direction perpendicular to the paper is gT'/T; and, in equilibrium, it is only the increase with height z of the westerly wind U (the component 'into the paper') which, by tilting the planetary vorticity (93) at a rate dU/dz, can generate horizontal vorticity at a balancing (equal and opposite) rate $f dU/dz$. This balance

$$f\frac{dU}{dz} = g\frac{T'}{T} \tag{102}$$

known as 'the thermal wind equation', accounts in the troposphere for typical increases of mean westerly wind by 3–4 m s^{-1} per km of altitude.

Some idea of how the planetary vorticity's change (95) in north–south motions allows flow patterns to be wave-like was outlined for the oceans in section 10.5.1. Corresponding studies for the atmosphere are a bit more complicated—because of the large vertical gradient (102) of mean velocity—so that there is space here for only a brief summary of the conclusions.

Particularly important atmospheric roles are played by two types of wave, both capable of growing to substantial amplitudes [93]. The two types are on markedly different length-scales, and they also differ in another important respect, reminiscent of the distinction (section 10.2.4) between flow instabilities of the 'abrupt' and the 'restrained' kinds. Yet they interact with each other.

One has wavelengths so large [141] that just two or three lengths embrace the globe; a small-amplitude theory of this wave shows general similarity to that of a Rossby wave, but with capabilities of amplitude growth. Yet the growth, being of a 'restrained' type, retains [142] the general wave-like form (figure 10.65); moreover, within the overall westerly wind pattern, the wave maintains a nearly stationary position which (roughly speaking) can appear 'anchored' to the topography of continents. In the wave, large maximum wind speeds are found in the famous upper-troposphere jet stream. Thus, the climatological mean wind field indicated by the thermal wind equation (102) and the equator–pole distribution of mean temperature represent smoothed-out versions of the fluctuations of an often more intense wind pattern, associated with more concentrated temperature gradients.

Another (smaller-scale) type of growing wave [143], again influenced both by the mean wind shear (102) and by the 'beta effect' (95), has something in common with the growing Tollmien–Schlichting waves found in boundary layers (p 830) and, like them, tends to generate chaotic flow patterns. These are the cyclonic disturbances which randomly appear, especially over mid-latitude oceanic regions, and are amplified—while becoming distorted in various ways—as they travel eastwards. Weather in such regions is powerfully influenced by this chaotic tendency, known to meteorologists as 'baroclinic instability', which moreover has been shown to put some definite limits (see below) on the times over which forecasting can be attempted. Also, the disturbances interact with the large-scale wave system [93]; indeed, part of their energy is capable, in a manner analogous to the absorption of internal waves at a critical level (p 868), of being fed into the jet stream.

Furthermore, climate is itself affected by all of those disturbances to pure east–west motion which have been mentioned. Very briefly, the north–south components of those disturbances produce a net northward transport of the angular momentum of westerly winds about the axis of the Earth [93]. Thus, over and above the tendency (102) for the mean westerly wind to increase with height, mid-latitude regions are gaining some additional 'westerly' angular momentum at the expense of lower latitudes. And so the observed surface wind pattern emerges as what was described earlier as a 'second-order' effect: on the average, the poleward angular momentum transport is balanced by the moments

PNA PATTERN

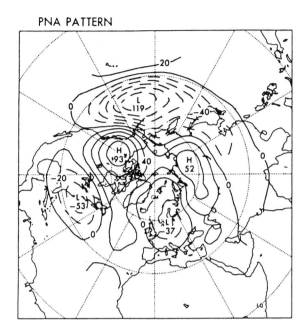

Figure 10.65. *Example of an upper-troposphere stationary-wave pattern in the northern hemisphere* [142]. *Altitude departures (in m) of the 300-millibar pressure surface from zonally averaged levels are shown for the rather common winter condition known as PNA (Pacific–North American anomaly, with a 'high' H over North America and a 'low' L over the Pacific). Tightly packed contours (in other words, unusually steep slopes of the constant-pressure surface) are associated with high-speed upper-troposphere winds.*

of the resistive forces exerted by the Earth's surface on westerly surface winds in middle latitudes and easterly 'trade winds' at lower latitudes.

Climatologically, the trade-wind zone is a relatively stable one, where a slow process of 'subsidence' (sinking of air) necessarily balances the continued rapid ascent of air (especially in 'thermals') over equatorial regions. Such slow subsidence is by no means adiabatic; instead, much of the heat gained in compression is lost by radiation, so the lapse rate is much reduced (also, the descending air is rather dry) and a highly stable stratification results [144]—except in a well-mixed layer close to the surface. Another climatologically significant feature of the trade-wind zone is that poleward transport of heat is preferentially achieved [145], not by atmospheric motions, but by such oceanic flows as western boundary currents (section 10.5.1); a good example of climate's dependence on the Earth's fluid envelope as a whole.

In the field of weather, two examples of discontinuity formation may be mentioned, one from middle latitudes and one from tropical regions. As mid-latitude cyclonic disturbances develop, horizontal gradients of temperature often become progressively sharper until there is formed, as J Bjerknes explained already in 1919, an effective discontinuity . At such a 'front' , inclined at a small angle α to the horizontal, both the temperature and the wind velocity parallel to the front make practically discontinuous changes, ΔT and ΔU, such as are permitted by equation (102) to be in balance. (Essentially, the horizontal gradient T' inside the inclined front is $\alpha dT/dz$, so that an integrated form of (102) across the front makes $f\Delta U$ equal to $g\alpha\Delta T/T$.) Fronts are striking weather features in temperate zones, with major influences on cloud formation and precipitation.

But it is a tropical weather feature which to the greatest degree can threaten disastrous effects on human populations [146]. The so-called tropical cyclones (known also as typhoons or as hurricanes) are vastly bigger and more intense than mid-latitude cyclones, exhibiting horizontal length scales of very many hundreds of km (figure 10.66) and surface wind speeds of the order of 50 m s^{-1}. They are formed when a wind pattern with substantial cyclonic vorticity comes in contact with a tropical ocean (having surface temperature 26 °C or more) provided that the atmosphere's vertical structure exhibits certain features; including, especially, a lapse rate in that range well above 0.5 °C per 100 m (see above) where fully saturated air is freely able to make vertical motions.

The discontinuity in this case is known as the 'eyewall': a vertical wall of dense convective cloud surrounding an 'eye' which, often, is nearly free of cloud (figure 10.66). The eyewall forms where air, having been continuously accelerated as it followed a long spiral path inwards over the warm ocean, has finally reached its saturation vapour pressure; then it can rise fast all the way to the top of the troposphere (at, say, 15 km altitude) where it tends to spiral outwards in a broadly anticyclonic motion. The eye, by contrast, consists of rather calm air at the same (low) pressure as the fast moving air in the eyewall.

In 1986, K A Emanuel pointed out [147] that tropical cyclones can be regarded as heat engines operating on something close to a Carnot cycle, in which the working fluid is that intimate mix of dry air with water in all its forms (vapour, droplets, ice crystals) which appears in the atmosphere. It follows the 'moist-air adiabatic' (see above) as it rises in the eyewall, after taking in energy at, essentially, the ocean surface temperature—partly as sensible heat but mainly as latent heat associated with an increase in partial pressure of water vapour. (From a physics standpoint, the energy transfer arises because the wind extracts the more energetic water molecules from the ocean surface.) Then, at a much lower upper-troposphere temperature, the working fluid loses energy by radiation and by mixing with its environment. The large difference between the heat-input and heat-output temperatures makes for an efficient heat engine.

Mechanical energy so generated is dissipated mainly by frictional resistance at the ocean surface—which, in a sense, transfers momentum

Figure 10.66. *Typical satellite photograph of a northern-hemisphere tropical cyclone with a horizontal extent of about 1000 km. Note the 'eye of the storm'.*

'back' to the ocean. Such transfer, wherever a tropical cyclone approaches a coastline, can threaten a devastating 'storm surge' (p 892) in addition to any more direct threats of disaster from extreme wind speeds—or from extreme rainfall if the storm deposits its huge moisture content over land.

Insight into the initial formation of tropical cyclones as well as of many other important atmospheric phenomena is given [146] by an extension, originally made in 1942 by H Ertel [148], of Rossby's 'potential vorticity' concept (97). Very briefly, Ertel's extension takes h as the vertical distance between neighbouring surfaces of constant entropy; so that, in adiabatic processes where such surfaces move with the fluid, the total vorticity $f + \varsigma$ changes in proportion as the distance h is stretched. Regions with anomalously high values of Ertel's potential vorticity 'have the potential' as it were to develop into large cyclonic structures. Moreover, potential vorticity was recently shown to have properties analogous to those of vorticity proper in aeronautics where (section 10.2.2) the vorticity field entirely determines the flow; similarly, meteorologists found an inversion algorithm which uniquely relates a wind field to its distribution of potential vorticity, so underlining this distribution's diagnostic value.

Weather forecasting [149] needs to be global: ships require advance warning of storms to plan their routes, winds must be forecast at all the levels where aircraft fly (section 10.4), warnings of a dangerous tropical cyclone are needed well in advance of its approach to land. Agriculture too needs forecasts several days ahead, and such forecasts must be global because over a period of a few days the influence of remote weather patterns becomes important. Predictions of both water supply and energy demand need to be based on weather forecasts, which indeed are valued in relation to most human activities.

The last quarter of the twentieth century saw the achievement [150] of truly global forecasting of increasing accuracy by means of computational fluid dynamics (CFD). The computer programs, using horizontal mesh sizes of the order of 100 km, are based on comprehensive descriptions of the dynamics of air motions at many altitude levels (around 20), with separate accounts taken of interactions with land topography and with the ocean surface, of water vapour and its condensation, of moist-air convective movements, of more general cloud physics and precipitation (and subsequent evaporation of falling rain), and of the influence of all these variables on radiative heat transfer. A global data-gathering and telecommunication network provides initial data for the programs, primarily at 00.00 and 12.00 GMT, including information at the different altitudes derived from the release of radiosonde balloons that can telemeter data on wind, temperature, humidity etc. from all levels. Some more two-dimensional, yet still very valuable data are provided from meteorological satellites and from other instrumentation, while various users of forecasts (ships, aircraft) are also communicators of data to the network. Computer analysis assimilates all the data (discarding anomalous values) and puts them into a form compatible with use as initial conditions for the CFD program. This is then used to forecast developments up to 10 days ahead, with outputs generated at very many intermediate forecast times; the output data being appropriately interpreted and communicated in various different forms to the different industries that need it as well as to the general public.

Later, all forecasts are compared with actual outcomes to determine how various measures of error increase with forecast time. Towards the end of the twentieth century, the general utility of 3-day forecasts had become very substantial, while 6-day forecasts were beginning to show a definite value.

In addition, some sort of measure of best possible 'predictability horizon' (allowing for unavoidable errors in initial data) was being estimated [150]. To this end, programs are run with just very small differences in initial data; and, as with other chaotic systems, large differences in the forecasts emerge beyond a certain predictability horizon, which seems to be around 15–20 days. An ingenious method for studying which errors produce most rapid divergences in forecasts has identified them to be errors in data at the western edge of any large mid-latitude ocean; these, of course, are the errors which are most magnified by 'baroclinic instability' effects.

In relation to such regions, efficient methods exist for reconciling the use of a mesh size of the order of 100 km with the physical tendency to form fronts—whose locations and strengths are routinely inferred from output data. In areas prone to tropical cyclones, however, problems arise from the far more drastic discontinuities at eyewalls [146].

Indeed, even the original formation of such a discontinuity remains hard to predict (although improvements may emerge now that plots of potential vorticity are included in some computer outputs); accordingly, forecasting may need to start from an 'eye' already identified by satellite and to impose a vortical structure (centred thereon) upon nearby initial data. Some CFD programs using this technique along with 'movable nested meshes' (where the finest mesh moves along with the eye region) are beginning to offer improved accuracies in forecasting the tracks of tropical cyclones, such as are urgently needed by the inhabitants of vulnerable coastlines [146].

Last of all, possibilities of systematic shifts of climate on time-scales exceeding the basic annual cycle are briefly reviewed. Work on the most important of these was pioneered in the 1920s by the Director General of the India Meteorological Department, Sir Gilbert Walker [151].

Indian agriculture depends critically on summer rainfall generated by the Southwest Monsoon. In the days before long-term storage of grain from high-yield crops, any 'bad' (low-rainfall) monsoon was a grave disaster for India. Sir Gilbert discovered long-distance correlations between monsoon behaviour over India and climatic variations stretching from East Africa to South America, and viewed such 'teleconnections' as evidence for a 'Southern Oscillation', that is, a (somewhat irregular) shift back and forth between extremes of climate over the Earth's more southerly oceans.

Later investigations of monsoon phenomena [152] suggested some of the physical basis for such an 'oscillation'. The familiar South Asian monsoons are part of a system of strong low-level winds—broadly opposed by even stronger high-level winds—on an enormous scale; for example, the Southwest Monsoon involves low-level winds beginning in the southern part of the Indian Ocean and blowing over East Africa and South Asia before moving onwards across most of the Pacific. Yet monsoons continue to be regarded as wind patterns related to large-scale ocean-land contrasts. Indeed, an expansion of either (i) global land cover or (ii) global wind patterns in spherical harmonics includes—alongside the zonally averaged components with $n = 0$—some specially important components with $n = 1$, corresponding (i) to the area of Eurasia/Africa hugely exceeding that of the Americas and (ii) to the related monsoon wind fields.

Further progress on the Southern oscillation required a prolonged collaboration of meteorologists with oceanographers. Very briefly, this irregular cycle emerges [153] from large-scale effects of sea-surface temperature distributions on winds and of wind variability on ocean current patterns. At one phase of the oscillation, for example, a marked slackening of trade winds changes the normal pattern of wind stress on the tropical Pacific and allows a 'trapped' baroclinic wave (p 896) to travel slowly eastwards across it; producing on arrival that event known as 'El Niño', which by warming Peruvian surface waters seriously harms the big local fishing industry. The complex nature of 'ENSO' (the popular name for the El Niño/Southern oscillation phenomenon) illustrates yet again the leading climatological role played by ocean/atmosphere interaction.

It is this intimate coupling which makes CFD modelling of climate change so difficult. It might be thought that, even though a CFD model of weather exhibits beyond the predictability horizon a massive divergence of solutions with only slightly different initial conditions, nonetheless the statistics of those solutions could represent climate. Some initial success with this approach (using relatively coarse-mesh models) was achieved [154] in the 1980s, but it became recognized in the 1990s that models developed primarily for weather forecasting take insufficiently detailed account of oceanic processes to be used for climate-change forecasting. Accordingly, some intricate coupled ocean/atmosphere models began to be adopted—in spite of severe difficulties in quantifying the necessary coupling [155].

A principal objective of such models was to forecast climate changes related to likely increases in atmospheric CO_2. Even though the 'greenhouse effect' (see above) is mainly associated with water vapour, carbon dioxide makes a modest additional contribution to it. Moreover a forecast doubling of atmospheric CO_2 by the middle of the twenty-first century could influence climate, not only through its direct effect on interception of outgoing radiation, but also indirectly through two positive-feedback effects: (i) increased evaporation leading to enhanced interception by water vapour and (ii) reduced snow cover raising the absorption of solar radiation. Current models, by comparing solutions with constant and with increasing CO_2 levels, have indeed estimated so-called 'global warming' as rather non-uniform—appearing preferentially over northern-hemisphere land masses [156].

The fluid mechanics of just air and water were seen in this chapter to have yielded 'a fair share' out of the many exciting developments in twentieth century physics; those mentioned having been chosen for their general importance and interest to non-specialist readers. Any specialist reader, of course, may argue that his or her specialism merited more detailed description; yet the chapter's author can legitimately counter suggestions of personal bias with a comment that his own major area of specialist expertise—fluid mechanics of the biosphere—has been omitted altogether.... Above all, he hopes that the chapter will extend the interests of many generally 'physics-minded' readers into fluid mechanics.

References

[1] Reynolds O 1886 *Phil. Trans. R. Soc.* **177** 157
[2] Basset A B 1888 *Hydrodynamics* (Cambridge: Cambridge University Press)
[3] Tietjens O 1931 *Handbuch der Experimentalphysik* vol 4, part 1, ed W Wien and F Harms (Leipzig: Akademische Verlagsgesellschaft) pp 669–703
[4] Prandtl L 1905 *Verhandlungen des III. Internationalen Mathematischen Kongresses (Heidelberg, 1904)* (Leipzig: Teubner) pp 484–91. Note: this and the other cited papers of Prandtl can be found also in his collected papers: Prandtl L 1961 *Gesammelte Abhandlungen* ed W Tollmien, H Schlichting and H Görtler (Berlin: Springer).
[5] Prandtl L 1918, 1919 Tragflügeltheorie, parts I, II *Nachr. Ges. Wiss. Göttingen, Math.-Phys. Klasse* 151, 107
[6] Rayleigh Lord 1910 *Proc. R. Soc.* A **84** 247
[7] Taylor G I 1910 *Proc. R. Soc.* A **84** 371
[8] Van Dyke M D 1975 *Perturbation Methods in Fluid Mechanics* (Palo Alto, CA: Parabolic)
[9] Lamb H 1932 *Hydrodynamics* 6th edn (Cambridge: Cambridge University Press)
[10] Zhukovski N E 1912 *Teoreticheskie osnovy aerodinamiki* (Moscow: Tekhnicheskaya uchilishcha) (French Transl. Joukowsky N 1916 *Aérodynamique* (Paris: Gauthier-Villars))
[11] Rosenhead L (ed) 1963 *Laminar Boundary Layers* (Oxford: Oxford University Press)
[12] Bernoulli D 1738 *Hydrodynamica* (Strasbourg: Argentorati)
[13] Rayleigh Lord 1878/1896 *The Theory of Sound* 1st/2nd edns (London: Macmillan)
[14] Laplace P S 1816 *Ann. Chim. Phys.* **3** 238
[15] Riemann B 1859 *Abh. Göttingen Ges. Wiss.* **8** 43
[16] Lighthill M J 1956 *Surveys in Mechanics* ed G K Batchelor and R M Davies (Cambridge: Cambridge University Press) pp 250–351
[17] Hugoniot A 1889 *J. de l'Ecole Polytech.* **58** 1
[18] Prandtl L and Pröll A 1961 *Z. Vereines Deutsch. Ing.* **48** 348
[19] Mach E 1887 *Sitzungsber. Wiss. Akad.* **95** 164
[20] Reynolds O 1883 *Phil. Trans. R. Soc.* **174** 935
[21] Prandtl L 1914 *Nachr. Ges. Wiss. Göttingen, Math.-Phys. Klasse* 177
[22] Wieselsberger C 1914 *Z. Mech.* **5** 140
[23] Lighthill J 1976 Flagellar hydrodynamics *SIAM Rev.* **18** 161–230
[24] Reynolds O 1895 *Phil. Trans. R. Soc.* **186** 123
[25] Emmons H W 1951 *J. Aeronaut. Sci.* **18** 490
[26] Rayleigh Lord 1880 *Proc. Lond. Math. Soc.* **19** 67
[27] Rayleigh Lord 1916 *Proc. R. Soc.* A **93** 148
[28] Landau L D 1944 *Dokl. Akad. Nauk. SSSR* **30** 299
[29] Goldstein S (ed) 1938 *Modern Developments in Fluid Dynamics* (2 volumes) (Oxford: Oxford University Press)

[30] Tollmien W 1935 *Nachr. Ges. Wiss. Göttingen, Math.-Phys. Klasse* 79–114
[31] Taylor G I 1923 *Phil. Trans. R. Soc.* A **223** 289
[32] Prandtl L 1921 *Z. Angew. Math. Mech.* **1** 431
[33] Tollmien W 1929 *Nachr. Ges. Wiss. Göttingen, Math.-Phys. Klasse* 21–44
[34] Schlichting H 1933 *Nachr. Ges. Wiss. Göttingen, Math.-Phys. Klasse* 181–208
[35] Taylor G I 1922 *Proc. Lond. Math. Soc.* (2) **21** 196
[36] Dryden H L 1929–36 *Nat. Adv. Comm. Aeronaut. Report* **320** (with A M Kauth) and **581** (with G B Schubauer, W C Mock and H K Skramstad)
[37] Taylor G I 1935 *Proc. R. Soc.* A **151** 429
[38] von Kármán T and Howarth L *Proc. R. Soc.* A **164** 192
[39] Batchelor G K 1953 *The Theory of Homogeneous Turbulence* (Cambridge: Cambridge University Press)
[40] Schlichting H 1940 *Jahrb. Deutsch. Luftfahrtf.* **1** 97
[41] Görtler H 1940 *Nachr. Ges. Wiss. Göttingen, Math.-Phys. Klasse* 1–26
[42] Schubauer G B and Skramstad H K 1947 *Nat. Adv. Comm. Aeronaut. Report* **909**
[43] Schubauer G B and Klebanoff P S 1955 *Nat. Adv. Comm. Aeronaut. Report* **1289**
[44] Gaster M 1968 *J. Fluid Mech.* **32** 173
[45] Smith F T 1992 *Phil. Trans. R. Soc.* A **340** 171
[46] Stuart J T 1958 *J. Fluid Mech.* **4** 1
[47] Sato H 1960 *J. Fluid Mech.* **7** 53
[48] Michalke A 1965 *J. Fluid Mech.* **23** 521
[49] Coles D 1965 *J. Fluid Mech.* **21** 385
[50] Kolmogorov A N 1941 *Dokl. Akad. Nauk. SSSR* **30** 301
[51] Pao Y H 1965 *Phys. Fluids* **8** 1063
[52] Townsend A A 1956 *The Structure of Turbulent Shear Flows* (Cambridge: Cambridge University Press)
[53] Stewart R W and Townsend A A 1951 *Phil. Trans. R. Soc.* A **243** 359
[54] Kim H T, Kline S J and Reynolds W C 1971 *J. Fluid Mech.* **50** 133
[55] Coles D 1955 *Fünfzig Jahre Grenzschichtforschung* ed H Görtler and W Tollmien (Braunschweig: Vieweg) pp 153–63
[56] Coles D 1956 *J. Fluid Mech.* **1** 191
[57] Squire H B 1951 *Q. J. Mech. Appl. Math.* **4** 321
[58] Reichardt H 1942 *Gesetzmässigkeiten der freien Turbulenz* vol 414 (Berlin: Verein Deutscher Ingenieure)
[59] Prandtl L 1910 *Phys. Z.* **11** 1072
[60] Pohlhausen E 1921 *Z. Angew. Math. Mech.* **1** 115
[61] Whipple F J W 1933 *Proc. Phys. Soc.* **45** 307 gives a full account of Taylor's (not separately published) analysis.
[62] Reynolds O 1874 *Proc. Manchester Lit. Phil. Soc.* **14** 7
[63] Schmidt E and Beckmann W 1930 *Tech. Mech. Thermod.* **1** 391
[64] Schmidt E and Beckmann W 1930 *Tech. Mech. Thermod.* **1** 391 gives a full account of Pohlhausen's (not separately published) analysis.
[65] Bénard H 1901 *Ann. Chim. (Phys.)* **23** 62
[66] Pearson J R A 1958 *J. Fluid Mech.* **4** 489
[67] Rayleigh Lord 1916 *Phil. Mag.* **32** 529
[68] Pellew A and Southwell R V 1940 *Proc. R. Soc.* A **176** 312
[69] Koschmieder E L 1966 *Beitr. Phys. Atmos.* **39** 209
[70] Lighthill J 1978 *Waves in Fluids* (Cambridge: Cambridge University Press)
[71] Whitham G B 1956 *J. Fluid Mech.* **1** 290
[72] Whitham G B 1952 *Commun. Pure Appl. Math.* **5** 301
[73] Rayleigh Lord 1879 *Phil. Mag.* **7** 161

[74] Bénard H 1908 *Comptes Rendus* **147** 839
[75] von Kármán T 1911 *Phys. Z.* **13** 49
[76] Homann F 1936 *Forsch. Geb. Ing.-Wes.* **7** 1
[77] Curle N 1955 *Proc. R. Soc.* A **231** 505
[78] Curle N 1953 *Proc. R. Soc.* A **216** 412
[79] Lighthill M J 1952 *Proc. R. Soc.* A **211** 564
[80] Lighthill M J 1963 *Am. Inst. Aeronaut. Astron. J.* **1** 1507
[81] Goldstein M E 1976 *Aeroacoustics* (New York: McGraw-Hill)
[82] Lighthill J 1978 *J. Sound. Vib.* **61** 391
[83] Munk W H, Miller G R, Snodgrass F E and Barber N F 1963 *Phil. Trans. R. Soc.* A **255** 505
[84] Rayleigh Lord 1914 *Proc. R. Soc.* A **90** 324
[85] Korteweg D J and de Vries G 1895 *Phil. Mag.* **39** 422
[86] Whitham G B 1974 *Linear and Nonlinear Waves* (New York: Wiley)
[87] Schwartz L W 1974 *J. Fluid Mech.* **62** 553
[88] Whitham G B 1965 *J. Fluid Mech.* **22** 273
[89] Longuet-Higgins M S 1975 *Proc. R. Soc.* A **342** 157
[90] Banner M L and Peregrine D H 1993 *Ann. Rev. Fluid Mech.* **25** 377
[91] Davey A and Stewartson K 1974 *Proc. R. Soc.* A **338** 101
[92] Stevenson T N 1973 *J. Fluid Mech.* **60** 759
[93] Gill A E 1982 *Atmosphere-Ocean Dynamics* (New York: Academic)
[94] Väisälä V 1925 *Soc. Sci. Fenn. Commentat. Phys.-Math.* **2** 19
[95] Brunt D 1927 *Q. J. R. Meteorol. Soc.* **53** 30.
[96] Yih C S 1980 *Stratified Flows* (New York: Academic)
[97] Bretherton F P 1969 *Q. J. R. Meteorol. Soc.* **95** 213.
[98] Long R R 1970 *Tellus* **22** 471
[99] Booker J R and Bretherton F P 1967 *J. Fluid Mech.* **27** 513
[100] Taylor G I 1923 *Proc. R. Soc.* A **104** 213.
[101] Küchemann D 1978 *The Aerodynamic Design of Aircraft* (Oxford: Pergamon)
[102] Lucas J 1988 *Boeing 747: the First Twenty Years* (London: Taylor & Francis)
[103] Sears W R (ed) 1954 *General Theory of High Speed Aerodynamics* (Princeton, NJ: Princeton University Press)
[104] Ward G N 1955 *Linearized Theory of Steady High-Speed Flow* (Cambridge: Cambridge University Press)
[105] Hayes W D 1971 *Ann. Rev. Fluid Mech.* **3** 269
[106] Morgan M B 1972 *J. R. Aeronaut. Soc.* **76** 1
[107] Busemann A 1931 Gasdynamik *Handbuch der Experimentalphysik* vol 4, ed W Wien and F Harms ch 1 pp 343–460
[108] Ackeret J 1927 Gasdynamik *Handbuch der Physik* vol 7, ed H Geiger and K Scheel (Berlin: Springer) pp 289–342
[109] Hayes W D and Probstein R F 1966 *Hypersonic Flow Theory* (New York: Academic)
[110] Zierep J and Oertel H (ed) 1988 *Symposium Transsonicum III* (Berlin: Springer)
[111] Lighthill M J 1953 *Proc. R. Soc.* A **217** 478
[112] Stewartson K 1969 *Mathematika* **16** 106
[113] Neiland V Ya 1969 *Izv. Akad. Nauk. SSSR Mekh. Zhidk. Gaz.* **4** 33
[114] Messiter A F 1970 *SIAM J. Appl. Math.* **18** 241
[115] Whitcomb R T 1952 *Nat. Adv. Comm. Aeronaut. Memorandum* RN L52 H08
[116] Jameson A, Baker T J and Weatherill N P 1986 *Am. Inst. Aeronaut. Astron. Paper* 86-0103
[117] Newman J N 1991 *Phil. Trans. R. Soc.* A **334** 213
[118] Kelvin Lord 1891 *Popular Lectures* vol 3 (London: Macmillan) pp 450-500
[119] Inui T 1962 *Trans. Soc. Nav. Arch. Mar. Eng.* **70** 282

[120] Wehausen J V and Salvesen N (ed) 1977 *Numerical Ship Hydrodynamics* (Berkeley, CA: University of California)
[121] Trillo R L (ed) 1993–94 *Jane's High-Speed Marine Craft* (London: Jane's Information Group)
[122] Chapman J C 1979 *BOSS '79 (Proc. 2nd Int. Conf. on Behaviour of Off-Shore Structures)* ed H S Stephens and S M Knight (Cranfield: BHRA Fluid Engineering) pp 59–74
[123] Bearman P W and Graham J M R 1979 *BOSS '79 (Proc. 2nd Int. Conf. on Behaviour of Off-Shore Structures)* ed H S Stephens and S M Knight, (Cranfield: BHRA Fluid Engineering) pp 309–22
[124] Taylor G I 1928 *Proc. R. Soc.* A **120** 260
[125] Rainey R C T 1989 *J. Fluid Mech.* **204** 295
[126] Defant A 1961 *Physical Oceanography* (2 volumes) (Oxford: Pergamon)
[127] Laplace P S 1775 *Mém. Acad. R. Sci. (Paris)* 75–182
[128] Taylor G I 1919 *Phil. Trans. R. Soc.* A. **220** 1
[129] Hendershott M and Munk W 1970 *Ann. Rev. Fluid Mech.* **2** 205
[130] Pekeris C L and Accad Y 1969 *Phil. Trans. R. Soc.* A. **265** 413
[131] Jelesnianski C P 1967 *Mon. Weath. Rev.* **98** 740
[132] Rossby C G 1940 *Q. J. R. Meteorol. Soc.* **66** (Supplement) 68
[133] Lighthill J 1971 *Phil. Trans. R. Soc.* A **270** 371
[134] Blandford R R 1966 *Deep-Sea Res.* **13** 941
[135] Matsuno T 1966 *J. Meteorol. Japan* **44** 25
[136] Seddon J A 1900 *Trans. Am. Soc. Civ. Eng.* **43** 179
[137] Lighthill M J and Whitham G B 1955 *Proc. R. Soc.* A **229** 281
[138] Brunt D 1939 *Physical and Dynamical Meteorology* 2nd edn (Cambridge: Cambridge University Press)
[139] Turner J S 1973 *Buoyancy Effects in Fluids* (Cambridge: Cambridge University Press)
[140] Lighthill M J 1966 *J. Fluid Mech.* **26** 411
[141] Charney J G and Eliassen A 1949 *Tellus* **1** 38
[142] Karoly D J, Plumb R A and Ting M 1989 *J. Atmos. Sci.* **46** 2802
[143] Charney J G 1947 *J. Meteorol.* **4** 135
[144] Betts A K and Ridgway W 1988 *J. Atmos. Sci.* **45** 522
[145] Vonder Haar T H and Oort A H 1973 *J. Phys. Oceanogr.* **3** 169
[146] Lighthill J, Zheng Z, Holland G and Emanuel K (ed) 1993 *Tropical Cyclone Disasters* (Beijing: Peking University Press)
[147] Emanuel K A 1986 *J. Atmos. Sci.* **43** 585
[148] Ertel H 1942 *Meteorol. Z.* **59** 271
[149] Houghton D D (ed) 1985 *Handbook of Applied Meteorology* (New York: Wiley)
[150] Manabe S 1985 *Issues in Atmospheric and Ocean Modeling. Part B. Weather Dynamics* (New York: Academic)
[151] Walker G 1928 *Q. J. R. Meteorol. Soc.* **54** 79
[152] Lighthill J and Pearce R P (ed) 1981 *Monsoon Dynamics* (Cambridge: Cambridge University Press)
[153] Philander S G 1990 *El Niño, La Niña, and the Southern Oscillation* (New York: Academic)
[154] Washington W M and Parkinson C L 1986 *An Introduction to Three-Dimensional Climate Modelling* (Mill Valley, CA: University Science Books)
[155] Houghton J T, Jenkins G J and Ephraums J J (ed) 1990 *Climate Change: the IPCC Scientific Assessment* (Geneva: World Meteorological Organisation)
[156] Carson D A 1992 *The Hadley Centre Transient Climate Change Experiment* (Bracknell: UK Meteorological Office)

Chapter 11

SUPERFLUIDS AND SUPERCONDUCTORS

A J Leggett

11.1. Introduction

11.1.1. Liquid helium: the early days

If the subject which we now know as low-temperature physics can be said to have a birthday, that day would be 10 July 1908—the date on which Heike Kamerlingh Onnes and his team at the University of Leiden first successfully cooled the element helium (^{4}He) below 4.2 K and thereby liquefied it. For the next 15 years, the only place in the world where liquid helium existed was the Leiden laboratory (now named after Onnes).

If helium, like other elements, became liquid, should it not also, like them, become solid under its own vapour pressure when cooled to sufficiently low temperatures? Onnes certainly expected this, and over the next 15 years reached lower and lower temperatures in an unsuccessful search for the freezing point. By 1922 he was speculating that helium might remain liquid even if cooled down to absolute zero. In fact, after his death, the Leiden team did succeed in inducing freezing, but only at a pressure of 30 atmospheres. Onnes' speculation was correct, the phase diagram of helium is quite different from that of any ordinary element.

In the mid- and late-1920s experimental research on liquid helium accelerated, with its production in a number of laboratories in Europe and North America, and a number of curious but apparently minor anomalies were observed, in particular at a temperature close to 2.2 K, which was eventually interpreted as a transition point between a higher- and a lower-temperature phase of the liquid. These phases were christened He I and He II, respectively. Until 1936, however, it seems that

Heike Kamerlingh Onnes

(Dutch, 1853–1926)

Onnes studied in Groningen and Heidelberg, and in 1882 was appointed to the first chair in experimental physics established in the Netherlands in Leiden. Partially with a view to testing van der Waals' hypothesis of the law of corresponding states, he embarked on a programme to cool various gases to low temperatures, and in particular to liquefy helium, a goal which he achieved in 1908. From then until his death in 1926 the Leiden laboratory was the acknowledged world leader in low-temperature physics; it was there, in 1911, that the phenomenon of superconductivity was discovered, and Onnes and his co-workers subsequently established many of its principal characteristics. He received the Nobel Prize in 1913 for his work on helium. Onnes had a reputation as a far-sighted and careful worker who, despite his considerable theoretical talents, always emphasized the primacy of experiment: his motto was *'door meten tot weten'* ('through measurement to knowledge').

See also
p 974

liquid helium was regarded as a curiosity of Nature, but mainly because it did not freeze under its own vapour pressure (a property eventually explained to most people's satisfaction in terms of the anomalously large quantum-mechanical zero-point energy). It was not suspected that its behaviour might be qualitatively different from that of any other known liquid. It is astonishing that the characteristic feature of He II which we now know as superfluidity, must have been almost an everyday occurrence in low-temperature laboratories for a quarter of a century, without ever being consciously observed!

An important development came in 1932, when Willem Keesom and his daughter A P Keesom conducted careful experiments on the thermal properties of liquid helium near the apparent transition between He I and He II at 2.2 K. They were surprised to find no latent heat at the

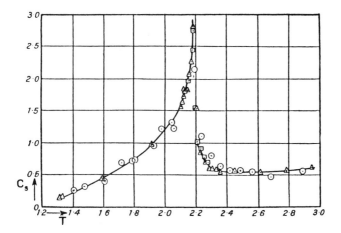

Figure 11.1. *The specific heat of liquid ^{4}He around 2 K. The shape of the curve resembles the Greek letter lambda (λ).*

transition, but a remarkable variation of specific heat (figure 11.1) with an apparent discontinuity at the transition itself. This characteristic shape led to the name lambda-transition, but the nature of the transition and the difference between He I and He II remained unclear until the late 1930s.

11.1.2. Superconductivity up to 1933

Having successfully liquefied helium, and thereby obtained the means to cool other substances to temperatures of a few degrees absolute, Onnes pressed ahead with his assistant Holst to investigate, among other things, the electrical properties of metals in this temperature range. He believed that the resistance of a pure metal should disappear on cooling to absolute zero. After some inconclusive experiments on platinum he decided to try mercury. On the basis of previous work it had been concluded that 'at very low temperatures ... the resistance would, within the limits of experimental accuracy, become zero. Experiment has confirmed the forecast. While the resistance at 13.9 K ... is still 0.034 times the resistance of solid mercury extrapolated to 0 °C, at 4.3 K it is only 0.0013; and at 3 K it falls to less than 0.0001'. Within a few weeks the true significance of what was going on had been appreciated: a second paper, entitled 'Disappearance of the electrical resistance of mercury at helium temperatures' sets an upper limit on the resistance at 3 K of 3×10^{-6} Ω, one ten-millionth of that at 0 °C. Onnes concluded that by this means 'conductors of zero resistance are obtainable': he had discovered the phenomenon of *superconductivity*.

Like liquid helium, superconductivity remained until 1923 exclusive to the Leiden laboratory. Many important features of the phenomenon were soon established. For example, it was found that the onset of the zero-resistance state was abrupt, taking place over an unmeasurably small range of temperature, that this state was destroyed by a sufficiently large current (*critical current*) and that it could also be destroyed by application of a *critical magnetic field*, of a few hundredths of a tesla. These last two observations were unified by Silsbee's conjecture that, as experiment later demonstrated, the critical current was simply the current necessary to produce the critical magnetic field at the surface of the wire. Also, superconductivity was observed in tin and lead as well as mercury (though not in gold or platinum). Finally, it was confirmed that the phenomenon of zero resistance manifested itself not only as a zero potential drop along a current-carrying superconducting wire, but far more sensitively in the failure to decay, over an observable timescale, of a current set up in a ring made entirely of superconducting metal. On observing this stability of a ring current, Onnes commented on the similarity to the molecular currents envisaged by Ampère in his explanation of ferromagnetism.

By the mid-1920s research on superconductivity was started in Toronto and in the Physikalische-Technische Reichsanstalt (PTR) in Berlin as well as Leiden, and many more materials were found to be superconducting; the element with the highest transition temperature, T_c, was niobium (8.4 K). Many alloys and chemical compounds were also found to be superconducting, including copper sulphide and gold–bismuth alloys where neither element was by itself a superconductor; this established that the phenomenon could not be associated with individual atoms.

In 1932 the head of PTR, Walter Meissner, reviewed the current state of experiment and theory in superconductivity, and formulated the most important question regarding superconductivity as follows 'Is the superconducting current just a variation (*"Abart"*) of the usual current, or are we dealing with a totally different phenomenon?'. In particular, are the electrons which carry the supercurrent the same ones that carry the current in a normal metal, or are they newly released (e.g. from atomic traps) when the metal becomes superconducting? Meissner gives arguments for each hypothesis, and recounts various experiments intended to shed light on this question, including experiments, at that time inconclusive, on the important question of whether the supercurrent is a bulk or a surface effect. As a result of this work Meissner drew the conclusion that 'the electrons which carry the supercurrent cannot move from the superconductor into a [normal] metal closely attached to it'—an unjustified conclusion from a modern point of view. Towards the end of his discussion he remarks that one of the most important experimental questions is 'whether other physical properties [besides the electrical ones] also undergo a jump on the transition to the superconducting state. In all experiments to date this happens for no other physical property, which makes superconductivity appear particularly mysterious'. The absence of a jump in the thermal conductivity at T_c seems especially to tell against the first hypothesis above (since if the electrons suddenly cease to be scattered at T_c, why does the thermal conductivity not jump up?) but perhaps 'only a small fraction of the ordinary electrons are superconducting at T_c'. Meissner's final speculation is that a proper explanation of superconductivity would require not only quantum mechanics but also quantum electrodynamics 'which is not yet fully developed'.

Actually, there were few leading theorists of the period who had not tried their hands at the problem, particularly during the years 1928–32 when the recently developed concepts of quantum mechanics held out high promise of a key. Some went so far as to publish papers on it, while others, such as Felix Bloch and Niels Bohr, developed theories which eventually failed to satisfy them and thus were never published. Of the ideas floated in these years, two in particular stand out. The first, originated by Bloch and developed by Landau and Frenkel, was that the ground state of a superconductor is characterized by the existence of spontaneous currents, which, however, flow in random directions and therefore on average cancel until ordered by an externally imposed current. This model, no doubt largely motivated by analogy with a ferromagnet whose different domains are oriented in mutually cancelling ways until aligned by an external field, has not stood the test of time (though it had useful by-products). A second idea, which was advocated by Kronig, Bohr and Frenkel and which strikes more of a resonance today, is that superconductivity must result from electron–electron interactions and consist in a *correlated* motion of all the electrons such that they cannot individually be scattered. However, this correlated state was apparently thought of as *crystalline* in nature, which is far from the modern picture. Over the whole discussion there hung a famous theorem of Bloch, which stated that *no* current-carrying state can be the ground state, so that the supercurrent (as then understood) can only be *metastable*. It is not surprising that little progress was made in this period, since a crucial piece of the puzzle was still missing.

11.1.3. The Meissner effect and other experimental developments in superconductivity up to 1945

The critical year in the history of experimental work in superconductivity is 1933. For the first 20 years it had been almost universally taken for granted that the only significant difference between a superconductor and a normal metal lay in the property of perfect electrical conductivity; consequently theoretical efforts had focused on attempts to show how the collisions which limit the conductivity in the normal state could be switched off. Now, it is easy to show that if a superconductor is no more than a metal with infinite conductivity, then the magnetic flux in the interior of a completely superconducting body cannot change. If one attempts to change it by imposing (or changing) an external magnetic field, the time variation of the magnetic field will give rise to an electric field which, although transient, will set up a circulating current that is just sufficient to cancel the externally imposed field in the interior of the body. By virtue of the infinite conductivity, this current will thereafter never decay (so long, of course, as the external magnetic field is not further changed), and the superconductor will behave as a permanent magnet. A particular and apparently trivial special case is that when the metal is cooled through T_c in a steady magnetic field, the superconductor retains the original flux.

This leads us to ask what is the equilibrium state of a superconducting sphere (for example) at a temperature below T_c and in a weak external magnetic field H ('weak' means less than the critical field at that temperature)? Now, we can reach this state by at least two different paths: in path (a) we first cool the sphere below T_c in zero field and then switch on the field, while in path (b) we first switch on the magnetic field while the temperature is above T_c and only thereafter cool through T_c.

In case (a) it is clear from the above considerations (and had been established by 1933 in numerous experiments) that the magnetic flux cannot penetrate the body, and the field lines behave as in figure 11.2(*a*). On the other hand, in case (b), when the field is switched on in the normal state, the magnetic field lines penetrate the sample as usual (figure 11.2(*b*)). When the sample is cooled below T_c, in constant external magnetic field, the above argument indicates that there should be no currents set up and no change in the flux in the interior of the body, so that the final state also should be that represented in figure 11.2(*b*). If this is correct, the final state of the sphere for the given temperature T and field H is not unique, but is a function of its history. There is nothing intrinsically absurd about this proposition—many other cases are known in physics in which, depending on the history, various 'metastable' states can be generated—but it has the surprising feature that the non-uniqueness persists for arbitrarily small values of the magnetic field, that is, the two different final states can be brought arbitrarily close together without one being able to relax into the other.

As a matter of fact, by 1932 indirect evidence had begun to accumulate, from experiments on the hysteretic behaviour of superconducting wires when the external field on them was cycled, that the above picture might be too simple. Ironically, however, the crucial experiment was actually designed to look for something else. Meissner had at the forefront of his mind the question whether the current in a superconducting wire flowed uniformly over the cross section or only in a thin layer at the surface. Max von Laue, who was a consultant to the PTR, suggested that this question could be resolved by measuring

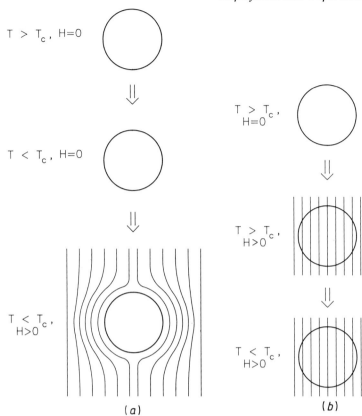

$T > T_c$, $H=0$

$T < T_c$, $H=0$

$T < T_c$, $H>0$

$T > T_c$, $H=0$

$T > T_c$, $H>0$

$T < T_c$, $H>0$

(a)   (b)

Figure 11.2. *Pre-Meissner expectations about the expulsion of magnetic flux from a superconductor.*

the magnetic field at a small distance from a pair of current-carrying superconducting cylinders. Normally, in such an experiment one would cancel the earth's magnetic field before cooling the sample through T_c (in zero current), but in some of the PTR experiments this was not done (it is still debatable whether on the first occasion this was an accident!) and, as the sample cooled through T_c, a clear jump in the measured field was seen, indicating that all or at least most of the magnetic flux was expelled from the interior of the superconductor: the final state, in this case also, corresponds to figure 11.2(a) and not figure 11.2(b)! (On the other hand, no change was detected in the flux threading a hollow cylinder.)

The fundamental significance of this effect, the Meissner (or Meissner–Ochsenfeld) effect, which was rapidly reproduced at other laboratories, was immediately recognized: superconductivity is more than simply a vanishing of the electrical resistance, it is a state with totally different thermodynamic properties from the normal state of metals. The fact that the magnetic field was actually expelled from the interior of the superconductor shows beyond reasonable doubt that the zero-flux state—that of figure 11.2(*a*)—is the *equilibrium* state of the body, not just a consequence of a failure to relax to true equilibrium. Theorists at last had something to get their teeth into, and progress was not long in coming.

By this time political events in Europe had begun to affect the course of low-temperature physics. Following the accession to power of the Nazi government in Germany, the years 1933–34 saw the exodus of many Jewish physicists, including Kurt Mendelssohn, Franz Simon, Nicholas Kurti and the London brothers, all of whom were attracted to Oxford by F A Lindemann (later Lord Cherwell) at the Clarendon Laboratory, which thereby rapidly became a major centre in experimental low-temperature physics. At the same time superconductivity at the PTR suffered a severe blow when Meissner left for Munich and von Laue was fired by the new Nazi-appointed head. In the next few years politics in the Soviet Union also had an impact. The Soviet experimentalist Pyotr Kapitza, who had almost completed his helium liquefier in Cambridge, was detained on a return trip to Moscow and spent the rest of his long working life there; and the work of the Kharkov laboratory, which despite its recent origin had produced one of the most convincing confirmations of the Meissner effect and pioneered work on superconducting alloys, came to a halt when its leader, A Shubnikov, was arrested in the Stalinist purges of 1937. He died in prison in 1945.

Apart from the Meissner effect, the most significant experimental developments in superconductivity in the years 1933–45 were probably those on superconducting alloys. It had long been appreciated that in some alloys the critical field (the magnetic field beyond which superconductivity disappears) was much larger than in pure elements. It was primarily Shubnikov's group in Kharkov who showed that in many alloys there are *two* 'critical fields' (called by Shubnikov H_{c1} and H_{c2}, a designation that has stuck). Below the 'lower critical field' H_{c1}, the field cannot penetrate (just as in pure elements). However, between H_{c1} and the much larger 'upper critical field' H_{c2} the field can penetrate partially, without destroying superconductivity, which vanishes only when H reaches H_{c2} (figure 11.3). This is the property which, much later, allowed coils made of alloy to become the routine method of producing very strong magnetic fields. The characteristic behaviour of alloys, including marked magnetic hysteresis, is known as type-II superconductivity, in contrast to the type-I behaviour of pure elements, and the region between H_{c1} and H_{c2} is known as the *mixed state*.

The historical development of our understanding of superconductivity would no doubt have been smoother and more logical—though it is not clear that it would have been any faster—had type-II superconductivity never been discovered until a microscopic theory of the type-I version was in place. As it was, these new experimental developments gave rise to a considerable amount of debate and, in retrospect, confusion. It be-

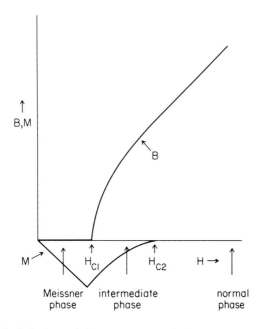

Figure 11.3. *The behaviour of the magnetization (M) and magnetic induction (B) of a 'type-II' superconductor as a function of external magnetic field H.*

came clear only slowly that the mixed state of alloy superconductors between H_{c1} and H_{c2} had essentially nothing in common with the intermediate state of elemental superconductors, in which one also gets partial flux penetration; the latter, unlike genuine type-II superconductivity, can be accounted for in terms of macroscopic energetics. It was also debated whether the characteristic features of alloy superconductivity were really a thermodynamic equilibrium effect, as claimed by Gorter and H London, or rather a consequence of hysteresis, as postulated in the *sponge model* of Mendelssohn. But by the mid-1960s the Gorter–London view was generally accepted.

Other experimental advances in this period will be mentioned later for their theoretical implications, but one discovery must be mentioned for its technical consequences. Justi and his group at the PTR discovered, in 1941, superconductivity in niobium nitride at 15 K, a record transition temperature at the time, and significant because it meant that liquid hydrogen could in principle replace liquid helium as a coolant. This was the forerunner of a whole class of superconducting materials with transition temperatures ~16–24 K and which, though not disordered, show characteristic alloy behaviour.

11.1.4.   Liquid helium: the experimental revolution
At the beginning of 1936 liquid helium was still just one more inert-gas liquid, which was peculiar only in a few respects, mainly that it

did not solidify under its own vapour pressure and that it appeared to have two different liquid phases, He I and He II, with the curious 'lambda-transition' between them at 2.17 K. Within three years the picture had changed totally, and it was clear that liquid helium had properties qualitatively different from those of any other known substance. While the Leiden and Toronto laboratories were involved in the earlier phase, the main breakthroughs came in the single year 1938, almost entirely from the more recently established low-temperature groups at Oxford and Cambridge in the UK and Moscow and Kharkov in the Soviet Union.

It had been noted by McLennan in Toronto in 1932 that when helium which had been boiling furiously just above the λ-temperature was cooled into the He II phase, the boiling abruptly stopped and the liquid became quiescent. The significance of this phenomenon became clear in 1936, when the Keesoms found that the thermal conductivity of He II was greater than that of He I by a factor of at least 3×10^6, a behaviour they named 'supra-heat-conductivity'. A year later a team in Cambridge found that the heat current in He II was not even proportional to the temperature gradient as in ordinary materials, so that a thermal conductivity in the normal sense could not even be defined.

This might have been explained as convective transport of heat, if the viscosity of He II had been low enough. But although the viscosity was indeed found to decrease rapidly below the λ-point, the reduction was not nearly enough. This, together with some observations of the strange propensity of He II to leak from otherwise sound containers, which had apparently been common knowledge among experimenters but had not previously been thought worthy of systematic investigation, suggested to Kapitza in Moscow and to Allen and Misener in Cambridge that a direct measurement of the viscosity by flow through thin capillaries might be of interest. They were indeed: the two papers, published simultaneously in 1938, showed unambiguously that the viscosity of He II when measured in this way must, if it existed at all, be less than one part in 1500 of that of He I! For this totally unexpected and unique behaviour Kapitza coined, by analogy with superconductivity, the term *superfluidity*.

This was not the only remarkable property of He II that now came to light. If the liquid was placed in an open-ended U-tube with the bottom packed with fine powder (so that it, unlike an ordinary liquid, could by virtue of its superfluidity leak through it) and one side was heated by shining a flashlight on it, it rushed from the cold to the hot side— so violently that a 'fountain' could be produced! (figure 11.4). Later, the inverse of this effect was discovered—an increase of pressure could cause a rise in temperature (*mechanocaloric effect*)†. Another curious effect which

---

† *H London applied to these phenomena a thermodynamical argument devised by Kelvin to relate thermoelectric effects, and in this way demonstrated the quantitative connection between them.*

Figure 11.4. *The fountain effect in superfluid liquid ^{4}He.*

had been briefly noted by Onnes, but apparently not taken seriously, was rediscovered: if a small beaker is placed inside a larger one, with different levels of fluid in the two, they rapidly equalize, even though the rim of the smaller beaker is well above either surface! It seemed as if a thin film of liquid helium covered the whole surface of the inner beaker, and that below the λ-temperature this film, like the helium in narrow capillaries, became superfluid. This was soon verified, and the thickness of the film estimated to be 20–30 nm—a conclusion which was confirmed directly and elegantly by optical methods a dozen years later.

By the end of 1938 it was clear that superfluid helium possessed at least as many bizarre properties as superconducting metals. The stage was set for a theoretical synthesis.

11.1.5. Theoretical developments, 1933–45
The first hint that the superconductors and liquid He II might have something in common came from a study of the thermodynamics of the two phase transitions. Prompted in part by the observation that while the specific heat appears to be discontinuous across the λ-transition there is no latent heat, i.e. no discontinuity in the entropy, Ehrenfest formulated in 1933 the general concept of a higher-order phase transition, in which the first derivatives of the free energy (entropy, volume) are

continuous but higher-order derivatives could undergo a discontinuity. In Ehrenfest's classification, second-order phase transitions (as identified by the behaviour of the specific heat) included not only the λ-transition of helium but also the transitions of metals such as iron to the ferromagnetic state and of some binary alloys to the ordered state in which the two types of atoms were arranged on regular sublattices. Shortly thereafter, the superconducting transition was added to the list when measurements at Leiden showed a specific heat jump there also. As discussed in Chapter 7, Landau developed a theory of such transitions in terms of an *order parameter* which grew from zero as the temperature was lowered through T_c. But although the nature of the order parameter was clear enough for magnetic transitions and binary alloys, it remained obscure for liquid helium and superconductivity. In the latter case, it was tentatively identified by Landau, in one of the most interesting of the pre-Meissner theoretical papers on superconductivity, as the critical current. In Gorter and Casimir's two-fluid model, the fraction of electrons which acquired superconducting properties rose from zero at T_c to 1 as the temperature decreased to zero, and served as an order parameter. Like Bohr and Kronig, they visualized the superconducting electrons as forming some kind of crystalline lattice.

See also p 549

Ehrenfest's work and the discovery of the Meissner effect gave a fillip to attempts to apply standard thermodynamic arguments to the superconducting transition. Ehrenfest's student Rutgers formulated a relation between the critical field and the specific heat of the normal and superconducting states, and this was rapidly verified experimentally. Shortly thereafter Gorter and Casimir gave a general discussion of the transition in samples of different shapes, showing in particular that under certain conditions it would be energetically advantageous for normal and superconducting regions to be interleaved—the configuration now known as the *intermediate state.*

The first decisive theoretical breakthrough in superconductivity was made in 1935 by the brothers Fritz and Heinz London. They observed that since superconductors do not obey Ohm's law (steady-state current proportional to electric field) it is most natural to suppose that it is the *acceleration* of the current which is proportional to field (Box 11A). This assumption, combined with Maxwell's electromagnetic equations, leads to the conclusion that in a superconductor in equilibrium the magnetic field consists of two parts: a part which dies off exponentially with distance from the surface over a layer (now called the London penetration depth) of thickness $\lambda_L \sim 10^{-5}$ cm† and a *frozen-in* field which is simply the field the sample possessed when cooled into the superconducting

---

† *This conclusion had actually been drawn ten years earlier by de Haas Lorentz, but the paper was in Dutch and not widely read. It was also derived in a 1933 paper by Becker, Heller and Sauter, and slightly later by Braunbek.*

state. But given Meissner's disproof of the frozen-in fields, they proposed to start instead from an equation—now known as the London equation (equation (A3) in Box 11A), which rather than relating the *acceleration* of the current to the *electric* field, relates the *equilibrium* value of the current to the *magnetic* field. This leads (in a simply connected superconductor) to a field that contains only the first of the above two components, and therefore accords with Meissner's observations. They further supposed that the n which enters the definition of λ_L (equation (A2)) is the number of superconducting electrons and is temperature-dependent; in particular the penetration depth should become infinite as T_c is approached. In a second paper the Londons also discuss the case of a *multiply connected* superconducting body (such as the rings used by Onnes), and conclude that in this case a frozen-in flux and persistent current can exist, in accordance with experiment†.

The phenomenological London equation is now generally believed to give a correct description of most of the electromagnetic properties of superconductors in thermodynamic equilibrium. The most interesting and prescient part of the Londons' work, however, is their speculations about the microscopic origins of this equation. They recognized clearly that it would follow, if one made the assumption (see Box 11A) that the wavefunctions of the electrons are unaltered by a magnetic field, somewhat as happens with the electron in a hydrogen atom. The system would, in fact, be behaving exactly like a huge atom!

But why should the wavefunctions of the electrons be unaltered by the magnetic field, quite unlike what happens in normal metals? 'Suppose the electrons to be coupled by some form of interaction, in such a way that the lowest state may be separated by a finite interval from the excited ones. Then the disturbing influence of the field on the eigenfunctions can only be considerable if it is of the same order of magnitude as the coupling forces.' This was to prove a key to the development of a microscopic theory.

The work of the Londons made an immediate impact, and a number of experiments were undertaken to verify its predictions, in particular that the penetration depth should tend to infinity as the transition temperature T_c is approached. In a classic series of experiments on colloidal mercury, Shoenberg determined the temperature variation of $1/\lambda_L^2$ (i.e. the 'number of superconducting electrons'). Some years later his result was shown to fit rather well the formula $1 - (T/T_c)^4$, precisely the temperature-dependence predicted by Gorter and Casimir's two-fluid model. In addition, experiments on (the lack of) thermoelectric effects in the superconducting state conducted on the eve of World War II (but

---

† *In their 1935 paper, the Londons come within a hair's breadth of predicting the phenomenon we now know as 'flux quantization' (F London in fact predicted it explicitly in 1948).*

BOX 11A: THE LONDON EQUATION

An electron, of mass m and charge e, not subject to collisions, acquires drift velocity v from an electric field E in accordance with the equation

$$\dot{v} = eE/m.$$

In a gas of n electrons per unit volume, the current density J is nev, whence it follows that

$$\dot{J} = ne^2E/m. \tag{A1}$$

This is the acceleration equation for current. Now in a changing magnetic field, Faraday's law of induction can be written $\mathrm{curl}E = -\mu_0\dot{H}$. Combined with (A1) this gives

$$\lambda_L^2\mathrm{curl}\dot{J} + \dot{H} = 0 \quad \text{where } \lambda_L^2 = m/\mu_0ne^2. \tag{A2}$$

The Londons proposed to drop the time-derivatives and take as the basic equation for the supercurrent

$$\lambda_L^2\mathrm{curl}J + H = 0 \tag{A3}$$

where λ_L is the London penetration depth.

In a simply connected superconductor (A3) is equivalent to the relation

$$J = -\lambda_L^{-2}\mu_0^{-1}A = -(ne^2/m)A \tag{A4}$$

where A is the electromagnetic vector potential, defined by $\mathrm{curl}A = H$. In their paper the Londons remark that this relation may be intuitively understood as follows: for a single electron the expression for the electric current may be written

$$J = -(i\hbar e/2m)(\psi\boldsymbol{\nabla}\psi^* - \psi^*\boldsymbol{\nabla}\psi) - (e^2/m)|\psi|^2A \tag{A5}$$

CONTINUED

only reported in full after the war) were interpreted as suggesting the idea of an energy gap, though no direct measurement of its value was made until much later.

Meanwhile, the question of the nature of the He II phase was also of great interest to theorists, particularly after the rash of discoveries in 1938.

BOX 11A: *CONTINUED*

In a normal metal the wavefunctions ψ deform in response to the application of a potential A in just such a way as to cancel the principal effects of the last term, so that to a first approximation the current is zero. However, for $A \to 0$ this deformation requires, according to the principles of perturbation theory, mixing in to the original $(A = 0)$ wavefunction ψ_0 states of arbitrarily low energy. If for some reason no such states exist in the superconductor, then even for finite A we may have $\psi = \psi_0$, $\nabla \psi = \nabla \psi_0 = 0$, and only the second term is left. Since $|\psi|^2$ is the single-electron probability density, summation of (A5) over all the electrons leads directly to (A4).

Once the idea that the λ-transition was an 'order–disorder' transition was established, the most obvious hypothesis was that it corresponded to some kind of short-range positional ordering of the atoms of the liquid, and such a scenario was considered by Fröhlich, Fritz London and others. However, the fact that x-ray diffraction studies seemed to show no difference between the ordering of the atoms in He I and He II was discouraging. In 1938 London came up with a radically new hypothesis: it had been pointed by Einstein in 1924 that a set of non-interacting atoms†, if cooled to a sufficiently low temperature, would undergo the phenomenon now known as *Bose* (or *Bose–Einstein*) *condensation*, in which a finite fraction of the atoms, which tends to one as the temperature falls to zero, occupies the single state with lowest energy. Possibly in part because of objections raised by Uhlenbeck, this proposal had apparently not been taken very seriously. London now resurrected it and, noting that the temperature T_c estimated for a gas of free atoms at a density equal to that of liquid helium, (3.13 K) was not so different from the observed λ-temperature (2.17 K), proposed to identify the λ-transition with the onset of Bose condensation. He supported this identification by pointing out that the specific heat of the He II phase, while not identical to that expected for a Bose-condensed gas, was qualitatively similar. Tisza quickly followed with his two-fluid model, the condensed and non-condensed atoms acting as two mutually interpenetrating fluids; the condensed atoms are responsible for the characteristic superfluid properties while the rest (*normal*) behave much like atoms in an ordinary gas. Thus 'the viscosity of the system is due entirely to the atoms in excited states'. Tisza recognized clearly that

---

† *Later it became clear that only atoms of integral total spin, such as ^{4}He but not ^{3}He, obey Bose–Einstein statistics.*

in measurements, for example, of the damping of an oscillating cylinder the normal atoms would confer viscosity, whereas only condensed atoms would flow through a superleak, without viscous drag. He also predicted the existence of a 'temperature wave' in which the two sorts of atoms oscillated relative to one another, with no net oscillation in the total density, but the idea was criticized, among other things, because there was no obvious way of inhibiting collisions between the normal and the condensed atoms.

Although the war seriously disrupted low-temperature research in Western Europe and America, rather surprisingly its impact was less in Moscow, where Kapitza continued his experiments and Landau his related theoretical work. In 1941 there appeared a classic paper in which Landau did for superfluid helium approximately what the Londons had done for superconductors. Starting from a general consideration of the quantum mechanics of a liquid of Bose particles at zero temperature, he deduced that the low-lying excited states of such a system could be described in terms of *elementary excitations*—entities which carry a definite momentum p and energy ε (the energy depends on the momentum). Provided there are not too many of them, the total momentum and energy of the system is just the sum of that carried by the elementary excitations. The ground state of the system is free of elementary excitations, but except at zero temperature a certain number will be thermally excited with an energy distribution appropriate to a Bose gas of particles which can be created or destroyed. The concept of an elementary excitation (or *quasiparticle*) is a very general and important one which has been applied to many other condensed-matter systems besides superfluids and superconductors. In one special case, that of quantized lattice vibrations or phonons in crystalline solids, the concept was already implicit in the standard description, but Landau's 1941 paper probably marks its first appearance in a more general context.

See also p 1363

Landau deduced that in a quantum liquid the excitations were of two kinds. The first are phonons, quantized sound waves for which ε and p are related by the equation $\varepsilon = cp$, c being the velocity of sound; these dominate the thermodynamics at temperatures well below the λ-point, giving a specific heat proportional to T^3. The second branch of excitations, *rotons*, corresponded to quantized rotational motion (vorticity) and have a finite energy gap Δ: in the original paper Landau assumed that the minimum energy occurred for $p = 0$, so that the spectrum is $\varepsilon(p) = \Delta + p^2/2\mu$ (where μ is a sort of 'effective mass' for the roton), but he was later forced by experiment to modify this to $\varepsilon(p) = \Delta + (p - p_0)^2/2\mu$, i.e. the lowest-lying roton has a finite momentum, p_0. In this paper he gave a famous argument to the effect that if the liquid flows through a capillary with velocity v, it will be unstable against the creation of an avalanche of elementary excitations as soon as v exceeds a *critical velocity* v_L (Landau critical velocity) given by the minimum value of $\varepsilon(p)/p$. For the assumed spectrum v_{L} is finite (it is the smaller of the quantities c and $(2\Delta/\mu)^{1/2}$) and therefore it is possible that the flow is dissipation-free (superfluid) for $v < v_{\mathrm{L}}$. Landau himself was careful to state that the criterion $v < v_{\mathrm{L}}$ was a necessary condition for superfluidity but not a sufficient one; this caveat seems not always to have been remembered by subsequent workers.

He now proceeded to derive a conceptually much more satisfactory version of the two-fluid model originally suggested by Tisza. He assumed that one could think of an unexcited part of the liquid and a 'gas' of elementary excitations, like two interpenetrating fluids obeying different laws of motion. However, he stressed that this model should not be taken too literally and was only a manner of speaking. The hydrodynamic equations obeyed by the gas of thermally excited elementary excitations—the *normal component*—were identical to those of an ordinary liquid; in particular the drift velocity v_{n} vanishes at the walls of the containing vessel. By contrast, the velocity v_{s} of the unexcited superfluid component was postulated to be *irrotational*—that is, in a simply connected bulk sample the integral of v_{s} around any closed curve should be zero (cf section 11.2.6). This condition is similar to that satisfied, in the London theory, by the electrical current in a superconductor in the absence of electric and magnetic forces. In a remarkable *tour de force*, Landau succeeded in deriving from these simple considerations a complete 'two-fluid hydrodynamics' for the description of a superfluid liquid. In this description the entropy S is associated entirely with the normal component and can be calculated if one knows the excitation spectrum $\varepsilon(p)$ of the elementary excitations: the apparent infinite thermal conductivity is simply due to convective counterflow of the two components. The fact that the superfluid carries zero entropy offers an immediate explanation of the fountain and mechanocaloric effects, since in those experiments it is only the superfluid which flows through the superleak.

From his two-fluid hydrodynamics, Landau (apparently unaware, because of war-time conditions, of Tisza's detailed work) also predicted a second type of sound wave. However, he did not identify it as a pure temperature oscillation, so that the first attempts to detect it, which used mechanical excitation, were unsuccessful. It was only after Lifshitz pointed out that temperature changes would be a far more effective excitation mechanism that Peshkov, using a heater and thermometer, detected this wave in 1944. Landau also pointed out that the moment of inertia of the helium in a rotating bucket would be proportional to the 'density of normal excitations' ρ_{n} and hence could be used to measure the latter, as was done in a famous experiment by Andronikashvili at the end of the war period.

11.2. The period 1945–70

11.2.1. *Liquid helium*

While World War II interrupted low-temperature research in the USA and Western Europe, it had some beneficial side effects. Technological spin-offs from war-time research included sophisticated electronic techniques and the development of the Collins helium liquefier, which made research at helium temperatures accessible to a large number of laboratories. On the theoretical side, Bohm's war-time work on plasma problems related to the isotopic separation of uranium led him, together with Pines and others, to investigate the problem of the electron gas in metals—thus making an essential contribution to the eventually successful theory of superconductivity.

See also
p 1361

The year 1946 not only marked the resumption of large-scale low-temperature research in the West, it is also a landmark year for the first genuinely microscopic theory of either helium or superconductivity that has stood the test of time, namely N N Bogoliubov's theory of the properties of a dilute Bose gas with weakly repulsive interactions. Starting from London's assumption that such a gas would undergo Bose–Einstein condensation he developed a systematic field-theoretic perturbation theory to show that the long-wavelength energy spectrum would have the phonon form predicted by Landau. At shorter wavelength the atoms behave essentially as if free. This work is important not only in its own right, but as the beginning of a spate of theoretical papers which treated various many-body problems by the methods of quantum field theory.

Meanwhile, continued measurements by Peshkov of the velocity of second sound had failed to give particularly good agreement with either Tisza's or Landau's theory, and in 1947 led Landau to modify his spectrum so that the rotons were no longer a different branch, but joined continuously on to the phonon spectrum (figure 11.5). Tisza continued to maintain his own theory, and the issue was not settled in favour of Landau's prediction until 1950.

The physical meaning of the Landau spectrum was much clarified by Feynman and Cohen. Using a variational wavefunction, Feynman was able to show that only the low-momentum excitations have the phonon form postulated by Landau, but that at higher momenta the motion of a particular atom would induce a backflow of the rest of the liquid. Using this idea, he obtained precisely Landau's predicted dip in the energy spectrum at high momenta (figure 11.5). A few years later $\varepsilon(p)$ was measured directly by neutron scattering, and the theoretical model confirmed. It has to be said that despite this work the detailed microscopic nature of the 'roton' excitations has remained somewhat obscure to this day. The general opinion, however, is that they have

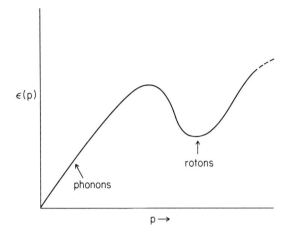

Figure 11.5. *The excitation spectrum of He II, as postulated in Landau's 1947 paper and subsequently confirmed by experiment.*

at most a distant connection with the original ideas about quantized vorticity in which their name originates.

In fact, ideas about vorticity in superfluid helium developed in a quite different direction. According to Landau, the velocity field of the superfluid should be irrotational, which means that at sufficiently low temperature, when there is essentially no 'normal component' left, liquid helium should not be able to rotate at all, and therefore in a rotating bucket would not take up a parabolic surface. It was therefore a considerable surprise when in 1950 the first experiment of this type showed that He II, even at low temperatures, behaved like any other liquid. The explanation was given by Onsager in a famous conference discussion remark, and later in a published paper by Feynman: the liquid acquires vortex lines on which the density of superfluid is zero, and around which the superfluid flows with quantized circulation, as will be described in more detail later. When the container is rotated, the vortices arrange themselves so that the macroscopically averaged velocity is the same as in an ordinary liquid, and the macroscopically averaged surface (all that could be measured in the experiment) is correspondingly identical.

See also
p 944

Evidence for Onsager's hypothetical vortices was soon forthcoming: Hall and Vinen showed that the more severe damping of second sound in a rotating container could be explained quantitatively by the presence of vortices, and in 1961 Vinen's vibrating-wire experiment showed the *quantum of circulation* to be close to the expected h/m. In recent years even more direct observations of the pattern of vortices in the rotating liquid have been made by electron microscopy: since free electrons preferentially travel along vortex lines, the pattern of their arrival at a detector placed above the liquid images the pattern of the vortices. These vortices play an important role in many experimental properties of He II; in particular, it is believed that under most conditions its inability to sustain persistent currents above a 'critical velocity' v_c is due to the fact that pre-existing vortices become free to move across the current, thereby reducing it.

Among many other experiments which probed the conceptual foundations of the theory of superfluidity, two are particularly noteworthy: in one, suggested by London and carried out by Hess and Fairbank, the liquid was cooled through the λ-temperature while rotating very slowly (so slowly that not even one vortex line could be formed) and it was observed that the *equilibrium* angular momentum decreased in proportion to the normal density. While the apparently similar experiment of Andronikashvili is the analogue of the failure of magnetic flux to penetrate an already superconducting sample, this experiment of Hess and Fairbank is the analogue, for a neutral system, of the Meissner effect. The second experiment by Kukich, Henkel and Reppy, probed the critical velocity very close to T_λ, and confirmed a theoretical prediction that the circulating-current state should not in fact be perfectly metastable: rather, it should be possible to make transitions, by a process of thermal activation, to lower-energy (and lower-current) states, leading to an observable decay of the current with time.

Despite the very considerable body of circumstantial evidence that accumulated in these years in support of the theoretical picture of superfluidity, it was always a source of unease that the phenomenon held to be the key to it, Bose condensation, had never been directly observed. Neutron scattering experiments have been interpreted as showing the existence of a finite condensate fraction (of the order of 10%), but cannot be said to have demonstrated it beyond reasonable doubt. Indeed, to this day a direct demonstration of the existence and magnitude of the condensate in ^{4}He is lacking.

11.2.2. Superconductivity—experiment and phenomenology, 1945–56
The idea that, for the wavefunction of the superconducting state to possess the rigidity required by the London equation, a minimum energy was needed to excite single electrons—the *energy gap*—was emphasized by Ginzburg in his 1946 book. It was also a central result of a microscopic theory by Heisenberg and Koppe, which, although eventually abandoned, had a big influence in the late 1940s. In their 1946 report of their experiments conducted on the eve of the outbreak of war, Daunt and Mendelssohn inferred the existence of such a gap from the absence of a thermoelectric effect (an inference which in the light of modern ideas is not compelling). More convincing evidence came in 1953–54, with experiments on the thermal conductivity and specific heat which indicated rather firmly that the 'density of superconducting

electrons' decreased *exponentially* as $T \rightarrow 0$, as would be expected if there were an energy gap. Yet more direct evidence came from measurements of electromagnetic absorption in the superconducting state, which occurs above a certain critical frequency, with a fairly abrupt threshold. (To conclude the story, in 1960, when the BCS theory was already established, a classic experiment by Giaever gave what is still the most direct evidence for the gap: when a normal and a superconducting metal are joined by a thin insulating barrier, no current flows until the voltage bias reaches a certain value—that necessary to overcome the energy gap in the superconductor.)

A very critical experiment was on the *isotope effect*—the influence of the isotopic mass of an element upon its superconducting transition temperature. Ironically, this question had been of interest (under the influence of now discarded ideas about ionic 'vibrators') in the very early days of the subject, and Onnes sought an effect in lead, but without success. The development of isotopic separation during and after the war helped, and in 1950 two groups simultaneously announced that the transition temperature of mercury decreased with increasing isotopic mass. It was immediately clear that the mechanism of superconductivity must have something to do not so much with the interaction of electrons with static lattice ions (which would be independent of isotopic mass) as with the dynamics of the lattice (i.e. the phonons). Coincidentally, Fröhlich had developed a theory in which the electron–phonon interaction played a vital role, and which predicted that T_c should be inversely proportional to the square root of the isotopic mass, as was immediately verified for isotopes of tin.

But an even more crucial role in the development of the theory in this period was played by the experiments of Pippard in Cambridge on the penetration depth. Starting from an idea of H London, that a magnetic field should affect the 'number of superconducting electrons' and therefore the London penetration depth λ_L (hereafter λ), Pippard decided to measure the dependence of λ on field (from the change of resonant frequency of a microwave cavity containing the specimen). The result was a surprise: λ varied much less with field than predicted. It appeared that whatever was responsible for the London equation was not itself confined to the London penetration depth! Thus Pippard was led to postulate that there was some kind of 'long-range order' in the superconductor which extended over perhaps 20 times the zero-temperature penetration depth λ_0, or about 10^{-4} cm. Noting that specimens much smaller than this (indeed smaller than λ_0) had been shown to be fully superconducting, Pippard commented that 'the range of order must therefore not be regarded as a minimum range necessary for the setting up of an ordered state, but rather as the range to which order will extend in the bulk material'.

An even more exciting result was found when Pippard went on to show that the penetration depth increased markedly with alloying, while the thermodynamic properties were essentially unchanged. Since in the simple London theory it is the same quantity, n_s, which determines both the thermodynamics and the penetration depth, something had to give. Part of Pippard's PhD thesis had been concerned with the anomalous skin effect in non-superconducting metals which arises when the current is confined to a surface layer thinner than the electronic mean free path. The relation between current and field then becomes *non-local* (see Chapter 17). In his thesis Pippard had noted that a parallel consideration should apply in Heisenberg and Koppe's model of superconductivity. While the London equation relates the current $J(r)$ directly to the vector potential $A(r)$ at the same point, a non-local theory would relate it to some average of A over a region around that point. He now realized the possible significance of this observation: the range over which the non-locality extends is essentially the 'range of order' (*coherence length*) introduced previously. When the mean free path is reduced to much less than this length, e.g. by alloying, only a fraction of the region which 'ought' to be contributing to the response does so, so the 'effective' number of superconducting electrons goes down and the penetration depth correspondingly increases. The microscopic meaning of Pippard's coherence length became beautifully clear with the BCS theory.

See also
p 1346

Meanwhile, the phenomenology of superconductivity was being developed in other directions. In 1950 Fritz London (now long settled at Duke University in North Carolina) published the first volume on superconductivity of his two-volume series *Superfluids*. As the name implies, he strongly emphasized the parallels between the superconductivity of metals and the superfluidity of He II, and, among other things, reiterated his recent prediction that the magnetic flux through a thick superconducting ring should be *quantized* in units of h/e (h = Planck's constant, e = electron charge). At the time, the anticipated difficulty of the experiment apparently deterred people from trying to verify this remarkable prediction; had they done so, they would have had a shock and the history of the subject might have gone a little differently!

Unknown to most people in Western Europe and America because of Cold War conditions, a momentous development was meanwhile taking place in Moscow. Ginzburg and Landau, building in part on London's ideas, attempted a specific realization for superconductors of Landau's earlier general theory of second-order phase transitions. With another of the inspired guesses which have been characteristic of the history of the theory, they identified the order parameter of the Landau theory with a *quantum-mechanical wavefunction* such as the one invoked intuitively by London—they did not say of what—and proceeded to treat its interaction with a magnetic field just as one would for a single electron. In the Ginzburg–Landau theory there is a sort of kinetic energy associated with the spatial variation of the order parameter (just as there would be for a single electron), and one can therefore define a (temperature-dependent) characteristic length $\xi(T)$, namely the length ξ over which the order parameter has to be 'bent' before the required energy exceeds the condensation energy of the superconducting state. Since the theory also incorporates London's formula for the penetration depth $\lambda(T)$, one can form a dimensionless ratio κ of these two quantities, which turns out to be temperature- independent and is therefore an intrinsic characteristic of the material. Ginzburg and Landau noted that if κ were greater than $1/\sqrt{2}$, the effective surface energy between a superconducting and a normal region of the metal would become negative, which would allow superconducting filaments to persist in the normal metal beyond the thermodynamic critical field. A very similar idea, though phrased in terms of his coherence length, had been put forward by Pippard (and indeed earlier by Gorter) and used to explain the flux penetration in superconducting alloys. However, Ginzburg and Landau surprisingly remark that 'from the experimental data, it follows that κ is always $\ll 1$' and that it was therefore unnecessary to examine what would happen in the case $\kappa > 1/\sqrt{2}$!

This state of affairs did not last long. Studying results obtained by his experimental colleague Zavaritskii on the magnetic behaviour of amorphous thin films, Landau's student, A A Abrikosov, began to wonder whether the restriction $\kappa > 1/\sqrt{2}$ was really sacrosanct, and to think about what would happen were it to be violated. He concluded that the magnetic behaviour would indeed be that characteristic of superconducting alloys (type II), and, eventually, that the way that the magnetic field would penetrate in the mixed state would be in the form of *vortices*, that is, regions with a core which is similar to a region of normal metal, allowing penetration of the field; currents flow around these regions in such a way as to screen the field out from the rest of the metal. These vortices are in fact the exact analogue, for a superconducting (electrically charged) system, of those which occur in superfluid (uncharged) ^{4}He. This theory was published in 1957, more or less simultaneously with the BCS microscopic theory. Ten years later the predicted *Abrikosov vortex lattice* was observed directly by means of magnetic decoration (a technique in which fine magnetic particles deposited on the surface of a superconductor are attracted to the positions of the vortices), finally dispelling lingering doubts about its reality.

The Ginzburg–Landau theory initially attracted little attention in the West, and not all of that favourable. It was only later, after it had been shown to be derivable under appropriate conditions from the more microscopic BCS theory, that it began to be taken seriously and its formidable potential for solving complicated problems of magnetic behaviour was recognized. Nowadays it is a standard item in textbooks on superconductivity.

11.2.3. Pre-BCS attempts at a microscopic theory

It is somewhat remarkable that while the early period of superconductivity saw a number of attempts at microscopic, first-principles theory, the dozen years following Meissner's momentous discovery brought very few—perhaps because it was now obvious that no simple consideration of scattering mechanisms could solve the problem. The late 1930s saw some qualitative work by Slater and a more quantitative attack by Welker, but neither stood the test of time. When, after the war, renewed attempts at a microscopic model were made, they rather surprisingly echoed the pre-Meissner work of Frenkel and Landau in assuming that a superconductor is characterized, even in the absence of external electric and magnetic fields, by the presence of currents flowing in random directions: the role of external fields is mainly to align these currents appropriately (the force of Bloch's theorem seems to have been less than fully realized in its application to these models). Born and Cheng hoped to avoid Bloch's theorem by taking into account the (static) periodic potential of the ions: in their model, the Coulomb interaction transfers electrons from the valence to odd corners of the conduction band, where they are responsible for spontaneous currents. The model of Heisenberg and Koppe started with electrons moving freely (no lattice potential) but, again under the influence of the Coulomb force, rearranged to produce spatially localized wave packets which then moved in a correlated way under the influence of fields. While this theory did not in the end turn out to have much relation to the true explanation of superconductivity, it had two interesting aspects which foreshadowed some features of the BCS theory. Firstly, it predicted an energy gap for one-electron excitations, and, secondly, it caused the authors to observe that the only reasonable explanation, both of the very small condensation energy of the superconducting state ($\sim 10^{-7}$ eV/atom, compared to typical Coulomb energies of a few eV/atom) and of the enormous variation between systems of the value of T_c (a factor of order 100), must lie in the transition temperature being exponentially dependent on material properties.

Fröhlich's attempt lay closer to what turned out to be the successful line of development. He started from the idea that the polarizability of the ions gives rise to an indirect interaction: as one electron passes by, it polarizes the ionic lattice (figure 11.6) and since the ions are heavy their relaxation is slow, so that a second electron passing by some time after will be attracted to the lingering positive charge density. However, while the idea that this attraction would lead to an instability of the Fermi distribution was to play a crucial role in the BCS theory, Fröhlich's specific conjecture for the solution was eventually shown not to give a Meissner effect; so too with his second theory published four years later, which turned out to be a model not of superconductivity but of the *sliding charge density-wave conduction* now known to be characteristic of some quasi-one-dimensional metals such as NbSe$_3$.

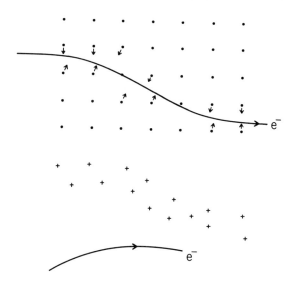

Figure 11.6. *The mechanism of electron–electron attraction in a metal, as envisaged by Fröhlich and subsequently by Bardeen, Cooper and Schrieffer.*

In 1955 Bardeen and Pines went further, taking into account both the ion–electron interaction and the direct Coulomb repulsion between the electrons to produce a more realistic interaction which had a crucial feature: provided the energy exchanged between two electrons near the Fermi surface was not too great, it could be *attractive*. (Fröhlich had already shown that the electron–phonon interaction *alone* would lead to an attraction, but this was not convincing so long as it seemed that it might be outweighed by the Coulomb repulsion.) In retrospect it can be seen that all the pieces were now in place for a definitive microscopic theory of superconductivity.

One piece of work, indeed, which already foreshadowed some of the qualitative features of the BCS theory was that of Schafroth, who in 1954 suggested, in the spirit of London, that superconductivity results from the Bose condensation of quasi-molecules formed by pairs of electrons (an idea which had actually been suggested as long ago as 1946 by R A Ogg in the context of a very specific problem in chemical physics). However, in Schafroth's formulation it was very difficult to calculate specific properties of the system, and in a subsequent paper with Blatt and Butler on the problem of formation of 'molecules' in Fermi systems he confined himself to 'speculations' about superconductivity, without claiming to have a concrete theory.

11.2.4. BCS and its aftermath
The critical breakthrough in our microscopic understanding of superconductivity (at least of the old-fashioned sort!) came in the twelve

months July 1956–July 1957, at the University of Illinois at Urbana-Champaign, at the hands of John Bardeen (professor), Leon Cooper (a postdoctoral research associate whom Bardeen had brought from Princeton) and Robert Schrieffer (a graduate student of Bardeen). The first hint came when Cooper, trying out various forms of interaction which he hoped might yield an energy gap, took up the somewhat artificial problem of a single pair of electrons, with total momentum zero, which attracted one another, not in free space but in the presence of the Fermi sea, so that states below the Fermi surface were blocked. He quickly realized that such a pair would form a bound state, that is, a state with energy less than the minimum allowed for freely moving electrons subject to the same restriction. A significant feature was that the magnitude of the binding energy of the molecule-like object so formed depended exponentially on the strength of the interaction—the very feature which Heisenberg had identified as an essential ingredient in any viable theory.

It therefore looked promising to attribute the onset of superconductivity to the formation of many such *Cooper pairs*, and to identify the energy gap, which was becoming increasingly indicated by experiment, with the binding energy of the pair. Clearly, however, the pairs could not be regarded as independent objects, even if one allowed them to undergo Bose condensation *à la* Schafroth: a simple calculation showed that the separation of the pair was so large (typically $\sim 10^{-4}$ cm) that between any two putatively paired electrons there would be millions of others, and the whole notion of identifying 'partners' was meaningless. The critical technical problem, then, was to implement the idea of Cooper pairing quantitatively, taking into account the Fermi statistics obeyed by the electrons. The first step was to guess the nature of the many-body wavefunction; this was taken, in effect, to be simply that of a collection of Cooper pairs all occupying the same state (thus somewhat resembling a Bose condensate) but with the critical proviso that the wavefunction was correctly antisymmetrized to take account of the Fermi statistics. The second, and highly non-trivial, step was to find a way of actually carrying out calculations on a system described by such a wavefunction. This problem was eventually solved with the aid of a trick similar to one used ten years earlier by Bogoliubov for liquid helium and a quantitative theory was born.

The BCS theory of superconductivity, as it is now universally known, starts from the Bardeen–Pines interaction which is attractive at low energies, but it replaces the rather complicated Bardeen–Pines form by a simpler interaction, which is attractive and constant when the electrons are close to the Fermi surface, but otherwise zero (this turned out to be a serendipitous simplification). A guess is then made: the most important interactions are likely to be those between electrons of opposite momentum and spin (as in the Cooper problem), so all others

John Bardeen

(American, 1908–91)

Bardeen studied electrical engineering as an undergraduate at the University of Wisconsin and, after a few years in industry, went on to do his doctorate at Princeton with E P Wigner. He then went back to industrial research, and it was as a member of the technical staff at Bell Telephone Laboratories that he invented, with Brattain and Shockley, the transistor, a discovery which won him the Nobel Prize in 1956. After joining the University of Illinois in 1951 he devoted himself to the problem of finding a microscopic basis for the theory of superconductivity, and in 1956–57 produced, with Leon Cooper and J Robert Schrieffer, the theory which is nowadays universally known as the BCS theory and which won him and them his second Nobel Prize in 1972.

He remained at the University of Illinois until his death in 1991. Bardeen always regarded himself as an electrical engineer as much as a physicist, and at Illinois held chairs in both departments; his theoretical work is characterized by an intimate contact with real experimental systems and a very individual brand of intuition which he sometimes found difficult to convey to his colleagues, but which paid off in many different areas of solid-state physics.

are neglected. The problem is then in essence exactly soluble, and what is found is that indeed at zero temperature the ground state does in some sense represent a Bose condensate of 'diatomic-molecule-like' objects—Cooper pairs—which all have the same wavefunction, corresponding to zero centre-of-mass momentum and, as regards the internal state, zero total spin and orbital angular momentum. To break up a pair and create an excited state takes a minimum energy equal to twice the energy gap, a quantity which turns out to be exponentially dependent on the strength of the interaction and hence on the material parameters. Because of this feature, the electronic wavefunction has indeed the kind of rigidity needed in the London approach to produce the Meissner effect and moreover to guarantee the stability of persistent currents. The Pippard coherence length, in pure metals, turns out to be nothing other than the effective radius of the Cooper pairs: when an electric field is applied to one 'side' of the pair, the effect is automatically propagated over a distance of the order of this radius. At finite temperatures the superconducting state is stable below a temperature T_c, where kT_c is of the same order as the zero-temperature energy gap and depends on the isotopic mass as $M^{-1/2}$. The superfluid density (density of superconducting electrons), and hence the London penetration depth, can be calculated as a function of temperature and turns out to have just the behaviour postulated in the phenomenological London theory.

A very exciting aspect of the BCS theory was that it could make quantitative predictions of various experimental quantities, some of which had not at the time been measured. Particularly striking were the predictions for the ultrasonic attenuation rate and nuclear spin relaxation rate. Both these quantities are in some sense a measure of the number of electrons available to exchange small amounts of energy with other excitations (phonons and nuclear spins, respectively), and because of the energy gap one would not expect the superconducting electrons to contribute; so the naive prediction is that both rates would fall off fast below T_c. However, because of some very specific features of the BCS wavefunction, it turns out that while the naive prediction is correct for ultrasonic attenuation, the nuclear spin relaxation rate is predicted to *rise* just below T_c, before falling at lower temperatures. The confirmation of this counter-intuitive prediction in the spring of 1957 by Bardeen's experimental colleagues Hebel and Slichter was a dramatic event which probably did more than anything to promote the general acceptance of the theory, despite some initial scepticism by the authors of competing theories.

The BCS theory was indeed enthusiastically received and over the next few years spawned literally thousands of theoretical and experimental papers. A particularly noteworthy development occurred in 1959, when Landau's student, Lev Gor'kov, extended the BCS formalism to give a microscopic derivation of the phenomenological Ginzburg–Landau theory (still unappreciated in the West), and identified their order parameter as nothing other than the local energy gap. One striking 'correction', however, had to be made: it was clear that the order parameter has to do with Cooper pairs, and therefore the charge to be associated with the elementary entity is $2e$. An immediate consequence (pointed out explicitly by Byers and Yang on rather more general grounds a year later) was that in a multiply connected superconductor flux should be quantized not in units of h/e—as in London's experimentally untested prediction— but rather in units of half the size, $h/2e$. This prediction was soon confirmed experimentally, and was generally regarded as another striking piece of evidence in favour of the BCS theory. It is amusing to note that Ginzburg had flirted with the idea of an effective charge e^*, but had been warned off by Landau.

The next fifteen years saw increasing refinement and generalization of the theory in different directions. With the help of a general technique for handling the electron–phonon interaction, due to Migdal and Eliashberg in the Soviet Union, detailed quantitative methods were developed to explain, and in one or two cases even predict, the transition temperatures of a large number of materials, even in circumstances where the original BCS model was too crude. This work also helped in understanding why the original model, despite its simplifications, works so well for so many systems. Detailed theories were developed to deal with the effect on superconductivity of both non-magnetic impurities (which have a very small effect, at least as regards the thermodynamics, although they can significantly change the electromagnetic properties) and magnetic impurities (which, because the two electrons of the pair, having different spins, see them differently, are very deleterious). Most importantly, the magnetic behaviour of type-II superconductors was quantitatively understood in the light of the ideas of Ginzburg and Landau and of Abrikosov, which were now recognized as firmly based in microscopic theory. A particularly important point is that type-II superconductors in the mixed state, while they show partial Meissner effect, may not always display zero resistance, because the Abrikosov vortices can move across the current flow and in so doing generate a finite voltage drop. In the development of high-field superconducting magnets—a major technological spin-off—it became essential to find ways of 'pinning' the vortices to the right sort of inhomogeneity. Nowadays superconducting magnets producing fields of up to 20 T, (2×10^5 gauss) with no generation of heat, are almost standard equipment.

11.2.5. *The Josephson effect*

One development which is of such fundamental conceptual and practical importance that it deserves a section to itself is the work of a young student working in Cambridge in 1962. Brian Josephson was struck by the idea that the 'wavefunction' which describes the centre-of-mass behaviour of the Cooper pairs has not just an amplitude but also a phase, and asked how this would affect the transmission of electrons between two bulk superconductors joined by a thin insulating barrier (such as those used by Giaever for his tunnelling experiments). Performing a calculation along the lines of the BCS theory, he came to the surprising conclusion that one would not only, like Giaever, see tunnelling of the excited single electrons (making allowance now for the existence of an energy gap on both sides) but also tunnelling of the Cooper pairs (now called Josephson tunnelling). This process is a highly coherent one in which the approximately 10^{20} Cooper pairs participate, as it were, as one, and has the remarkable feature that for no bias voltage the current

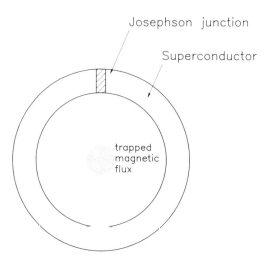

Figure 11.7. *A radio-frequency* SQUID *(superconducting quantum interference device).*

is proportional to the sine of the difference in phase of the Cooper pair wavefunctions in the two superconductors, while for a finite bias voltage V the current oscillates with the Josephson frequency $2eV/h$.

Although Josephson's own attempt to verify his prediction was unsuccessful (it turned out later that, as so often happens, others had probably 'seen' the effect but were unable to interpret it and had therefore discarded the relevant data), confirmation was not long in coming, and the Josephson effect is now the basis of an important class of electronic devices. One reason for its importance is the following: it can be shown that a Josephson supercurrent of maximum magnitude I_c is equivalent to the existence of an energy of magnitude $\hbar I_c/2e$ which depends on the relative phase of the Cooper-pair wavefunctions in the two superconductors. But this relative phase may be regarded as a macroscopic variable, and is associated with various unarguably macroscopic quantities: for example, in a so-called RF SQUID ring (a thick superconducting loop interrupted by a single Josephson junction, as in figure 11.7), there is a close association between the phase difference across the junction and the total magnetic flux trapped in the ring. Since the phase-dependent energy can be very small—of the order of the thermal energy of a single atom at room temperature—we have the extraordinary situation of a *macroscopic* variable which is controlled by a *microscopic* energy. This not only makes possible devices of unprecedented sensitivity but has in recent years allowed various tests to be made of the applicability of quantum mechanics at the macroscopic level—tests which so far the theory has passed with flying colours.

11.2.6. The modern unified picture of superconductivity and superfluidity

The understanding of superconductivity and superfluidity as macroscopic phenomena, which had evolved by the end of the 1960s (and which is still largely accepted) is firmly rooted in the ideas of the Londons and of Landau a generation earlier. In the case of superconductivity the microscopic underpinning was spectacularly provided by the work of BCS, but a microscopic understanding of superfluidity came more gradually and imperceptibly, and even today is not quantitatively as satisfactory as for superconductivity. In this section, I give a brief overview of the current picture, emphasizing the ideas which are common to the two branches of the subject. This section is necessarily somewhat more technical in nature than the rest of the chapter.

Let us start with the superfluidity of He II. The key notion is London's idea that the λ-transition corresponds to the onset of Bose condensation in this system. Generalizing this beyond the original case of thermodynamic equilibrium, we assume that below T_λ, for any state of the liquid, we can find a set of single-particle wavefunctions (in general these will depend on position r, and the choice of them may depend on time) such that exactly one of them—labelled $\chi_0(rt)$—is occupied on average by a number $N_0(t)$ of atoms which is a significant fraction of the total number N, while all the rest will have an occupation of order 1. (It is *not* assumed that N_0 need ever be equal to N, even in thermal equilibrium at zero temperature, because collisions are always scattering atoms between this 'special' state and the rest.) Given this hypothesis, we can define an order parameter (or, as it is sometimes rather misleadingly called, a 'macroscopic wavefunction' of the superfluid) $\Psi(rt)$ by the prescription

$$\Psi(rt) \equiv \sqrt{N_0(t)}\chi_0(rt). \tag{1}$$

Thus $\Psi(rt)$, like the one-particle wavefunction χ_0, is a complex quantity. We analyse it into amplitude and phase

$$\Psi(rt) \equiv A(rt)\exp i\phi(rt) \qquad (A, \phi \text{ real}). \tag{2}$$

By applying the usual expressions for the probability density and particle current of a single particle, we derive the density $\rho_c(rt)$ and mass current $j_c(rt)$ of the condensed particles

$$\rho_c(rt) = (A(rt))^2 \qquad j_c(rt) = (\hbar/m)(A(rt))^2\nabla\phi(rt). \tag{3}$$

We take the ratio of these quantities to define the superfluid velocity $v_s(rt)$

$$v_s(rt) \equiv j_c(rt)/\rho_c(rt) = (\hbar/m)\nabla\phi(rt). \tag{4}$$

Thus, the superfluid velocity is, apart from a factor, nothing but the gradient of the single-particle wavefunction $\chi_0(rt)$ into which condensation has taken place. It should be emphasized that although the quantity $\chi_0(rt)$ is indeed a perfectly legitimate single-particle wavefunction, its evolution is governed by many-body effects and therefore not described by any simple equation of the Schrödinger type.

From (4) it follows immediately that, for all points in the liquid at which $\chi_0(rt)$ is finite, $\text{curl} v_s = 0$, i.e. the superfluid flow is irrotational. This provides the basis for the two-fluid hydrodynamics of Landau, and from this point on we can essentially just follow his arguments. One thing to note is that the so-called *superfluid density* ρ_s of the liquid, that is, the ratio of the total current associated with v_s to v_s itself, is not in general identical to the condensed fraction N_0. In particular, there are rather general arguments to show that at zero temperature ρ_s is equal to the total density ρ of the liquid, even though N_0 is believed to be less than 10% of N. (In effect, the condensate particles may drag all or, at non-zero temperatures, part of the uncondensed particles along with them.)

Our microscopic identification of v_s, however, permits us to go beyond Landau by considering the case of a multiply connected geometry, e.g. a ring. We cannot now conclude that the integral of v_s around the ring must vanish (since in the space in the middle v_s is, of course, not defined). But the order parameter, like χ_0 in (1), must be single-valued, i.e. its phase $\phi(rt)$ must be well-defined within addition factors of $2n\pi$. This immediately leads to the Onsager–Feynman quantization condition

$$\oint v_s \cdot dl = nh/m \qquad (5)$$

which is nothing but the analogue of the Bohr quantization condition $l = nh$ for an electron in an atom. If it should happen that $\chi_0(rt)$ vanishes along a line within the liquid, then for any contour which encloses this line, condition (5) must be satisfied. In particular, if we consider a straight line and a circular contour of radius R, and take $n = 1$, we have

$$v_s(R) = \hbar/mR. \qquad (6)$$

This is just the classical flow pattern of a vortex line. Note that although the condensate wavefunction χ_0 must vanish at the vortex core ($R = 0$), there is nothing to say that the total density of the liquid must do so. A vortex ring (vortex line joined in a loop) is also a possible pattern.

Let us consider a particular experimental geometry, a fine tube bent into a ring of radius R, which may be rotated at will. We consider two experiments. In the first (the idealized version of the Hess–Fairbank experiment), the liquid is cooled through T_λ while the annulus is rotating with a small angular velocity ω, of the order of the quantity $\hbar/mR^2$. Above the λ-transition the helium behaves much like an ordinary liquid, and rotates with the container. When the liquid is cooled below T_λ,

See also p 932

however, a finite fraction of the atoms have to condense into a single one-particle state, with wavefunction χ_0. Because this is governed by the Onsager–Feynman quantization condition, only discrete choices are possible and the system will choose that single-particle state χ_0 which most nearly corresponds to rotation with the container; in particular, for $\omega < \hbar/2MR^2$ it will come to rest in the laboratory frame. Thus a fraction $\rho_s(T)/\rho$ of the liquid will appear to have ceased to rotate, while the rest will behave much like a normal liquid. While the ring geometry is particularly simple, similar results can be obtained for a simple bucket provided we remember the possibility of vortices.

A second experiment, which at first sight looks similar but is actually conceptually very different, involves starting with the liquid above T_λ in a container rotating rather fast ($\omega \gg \hbar/MR^2$). Once again, the helium comes into rotation with the container. We now cool through T_λ with the container still rotating; when we do so, the state of the liquid does not change perceptibly. This is easy to understand in the light of the above analysis—the liquid had velocity v such that $\oint v \cdot dl$ was already large compared with h/m, so that the Onsager–Feynman condition could be satisfied by a v_s which was imperceptibly different from v. Finally, we stop the rotation of the container, still holding the liquid below T_λ. What we observe experimentally, except when the temperature is extremely close to T_λ, is that the circulating current is persistent and will never die away while the liquid remains below T_λ (though changing the temperature can change its magnitude as more or less of the liquid is found in the condensed state).

It should be emphasized that whereas the behaviour of the liquid in a Hess–Fairbank type of experiment is exactly analogous to that which would be expected (if we could do an analogous experiment!) for a single electron in a hydrogen atom, its behaviour in this second ('persistent-current') experiment is very different. It is clear that the persistent-current state cannot be the state of thermodynamic equilibrium, and in the atomic analogue (the electron excited to a state of high angular momentum) the excited state would rapidly decay to the ground state (s-state). Why does superfluid helium behave so differently? The reason is as follows: if we consider either the wavefunction of the atomic electron, or, in the analogous quantity, the order parameter, for helium, then in both cases for the ground state the 'winding number' n which occurs in the Onsager–Feynman quantization condition, i.e. the number of 'turns' of 2π made by the phase in going around the ring, is non-zero in the initial state but zero in the final state (s-state). It is impossible to go from one state to the other without the wavefunction (order parameter) passing through zero at some point on the ring at some stage. For the atomic electron this can and does happen without difficulty. However, in the case of superfluid helium the strong interactions between the atoms resist the creation of such a zero in the order parameter, and the non-zero n state may remain metastable. One way of changing n without a prohibitive expenditure of energy is to move a pre-existing vortex line— a region where $|\Psi|$ indeed vanishes—across the ring; it can be shown that this can reduce n by 1. Also, extremely close to the λ-transition where the superfluid density and the associated energy are very small, it is possible to create a vortex ring from scratch, by a thermal fluctuation, and then expand it until it fills the whole tube, with the same effect; this is thought to be the mechanism operating in the experiments of Kukich, Henkel and Reppy.

Thus, consideration of Bose condensation plus the effects of interatomic interactions can explain both the Hess–Fairbank result (a thermodynamic equilibrium phenomenon) and the existence of persistent currents (a metastable effect). It is amusing (and rarely noted in the literature of the subject) that the classic experiment on superleak flow as carried out by Kapitza and by Allen and Misener (section 11.1) may correspond, depending on the 'de Broglie wavelength' $\lambda \equiv h/m v_s$ realized in the experiment, to either or both of these effects: if λ is small compared with the length of the superleak, 'superfluidity' must essentially be understood as a metastable phenomenon, while if the opposite condition is realized it is a thermodynamically stable one!

See also
p 922

We turn now to superconductivity. The principal complications are (i) that the electrons in metals obey Fermi rather than Bose statistics, so that there is no question of 'Bose condensation' in the usual sense, and (ii) that unlike the atoms of ^{4}He, they are charged. As to the first, BCS showed that the many-body wavefunction of a superconducting system corresponds to a state in which a finite fraction of the electrons (which in their model is unity at zero temperature) pair up to form 'di-electronic molecules' (Cooper pairs) which all have the same two-particle wavefunction ($\chi(r_1\sigma_1 : r_2\sigma_2)$ (where σ_i is the spin projection of the ith electron). Although this is not exactly Bose condensation, it is clear that there is a qualitative similarity. Indeed, for our purposes, as in (1), we may take the 'order parameter' of the superconducting state to be just the 'molecular' wavefunction $\chi_0(r_1\sigma_1 : r_2\sigma_2 : t)$ times the square root of the number of electrons condensed into it

$$\Psi_2(r_1\sigma_1 : r_2\sigma_2 : t) = \sqrt{N_0(t)}\chi_0(r_1\sigma_1 : r_2\sigma_2 : t). \tag{7}$$

Unlike the order parameter of a Bose system, this is a two-particle quantity and, as we shall see in the case of ^{3}He, can have quite a complicated structure. However, it is believed that for superconductors (at least of the pre-1970 sort), the spin wavefunction is just a singlet and moreover the dependence on the relative coordinate ρ is isotropic, with its dependence on $|\rho|$ fixed by the energetics. Thus, if we denote the centre-of-mass coordinate of the pair, i.e. the quantity $(r_1 + r_2)/2$, by r, an adequate approximation to (7) for our purposes is

$$\Psi_2(r_1\sigma_1 : r_2\sigma_2 : t) = \Psi(r, t) \cdot f(|\rho|) \cdot \frac{1}{\sqrt{2}}(\uparrow\downarrow - \downarrow\uparrow) \qquad (8)$$

where the last factor is a symbolic representation of the spin-singlet state. It is the centre-of-mass wavefunction $\Psi(r, t)$ which in thermodynamic equilibrium becomes the order parameter of Ginzburg and Landau (and more generally the analogue of the superfluid-helium order parameter defined by (1)). The other terms are constant and may be ignored for present purposes (although, of course, a knowledge of $f(|\rho|)$ is important when we want to discuss the internal properties of the pairs such as the Pippard coherence length.)

We can now proceed exactly as in the helium case, with one important proviso: since electrons, unlike He atoms, are charged, the relation between the superfluid velocity v_s and the phase of the wavefunction χ_0 must be modified to include a possible electromagnetic vector potential $A(rt)$. For a single particle the correct replacement would be $\nabla\phi \rightarrow \nabla\phi - eA/\hbar c$; however, since the 'phase' in question is that of the centre-of-mass wavefunction of a *pair*, e should be replaced by $2e$ (and m by $2m$) and we therefore get as the generalization of (4)

$$v_s(r, t) = \frac{\hbar}{2m}\left(\nabla\phi(rt) - \frac{2e}{\hbar c}A(rt)\right). \qquad (9)$$

The electric current j associated with the pairs is taken to be proportional to v_s, with a constant of proportionality $\rho_s(T)$ which is roughly speaking the fraction of electrons bound into Cooper pairs (though, as with helium, it is not identical to the N_0 introduced above). Hence, on taking the curl of (9) we find, for any region in which the order parameter is non-vanishing, the London equation

$$\mathrm{curl}\, j \propto \mathrm{curl}\, A \equiv H \qquad (10)$$

from which the whole of the London phenomenology follows, in particular that in a bulk sample the magnetic field is screened out in a penetration depth λ_L which is proportional to $\rho_s^{-1/2}$. Moreover, if we apply (9) to a thick ring, and consider a path which lies well inside the material, where no field can penetrate, we may take $v_s = 0$ and hence $\nabla\phi = 2eA/\hbar c$ (figure 11.8). Integrating this relation around the loop and using the fact that ϕ must be single-valued up to additive terms of $2n\pi$, we obtain the famous quantization condition for the total flux Φ threading the ring

$$\Phi \equiv \oint A \cdot dl = n\phi_0 \qquad \phi_0 \equiv hc/2e \qquad (11)$$

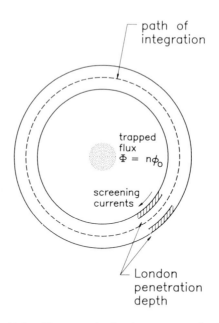

Figure 11.8. *Flux quantization in a superconducting ring.*

as originally predicted by London (with, however, ϕ_0 given by an expression twice as large) and subsequently by Byers and Yang.

Thus, if we compare a charged system (superconductor) with a neutral one such as superfluid helium, we see that the Hess–Fairbank 'no-rotation' effect and the Meissner effect, are analogues, as are the Onsager–Feynman quantization of circulation (5) and the flux quantization (11). We recall that the Ginzburg–Landau parameter κ, which determines the magnetic behaviour of a superconductor, is defined as λ_L/ξ: since, as we let the electric charge tend to zero, λ_L tends to infinity we may say that liquid helium (for which λ_L is of course not strictly defined) corresponds to the limit $\kappa \to \infty$, i.e. an 'extreme type-II' superconductor. Thus, at not too high angular velocities it is always energetically advantageous for rotating helium to admit the penetration of vortices (type-II behaviour): by contrast, as we have seen, superconductors in a magnetic field may show either type-I or type-II behaviour depending on the value of κ. Alloying tends to reduce the superfluid density ρ_s, and hence to increase λ_L and κ and promote type-II behaviour.

The explanation of the metastability of persistent currents in a ring proceeds along the same lines as outlined above for He. In the present case the finiteness everywhere of the order parameter Ψrt, which guarantees the topological stability of the current-carrying states, is due to the attractive interaction which binds the Cooper pairs. To break up a pair so as to reduce Ψ requires an energy of at least twice the energy gap. As in the helium case, it is a delicate question whether 'superconductivity' in the original sense of finite current flow under zero voltage drop is better thought of as essentially a manifestation of an equilibrium phenomenon (Meissner effect) or of a non-equilibrium one (metastability of supercurrent flow). In type-II superconductors, just as in helium, the motion of existing vortex lines across the current can lead to its decay. This a serious problem in the design of type-II superconductors for high-field magnets, but one which has been overcome by, for example, cold-working, to provide defects which will pin the vortices.

Finally, it should be noted that the Josephson and related effects arise because the order parameter is essentially a single-particle wavefunction (in helium) or the centre-of-mass wavefunction of a pair of particles (in a superconductor). These effects, discovered in superconductors in 1962, had to wait until the 1980s for their confirmation in superfluid helium.

11.3. New developments

11.3.1. The superfluid phases of ^{3}He

Superfluid ^{3}He is unusual among the topics of this chapter in that its existence, if not most of its detailed properties, was predicted theoretically before its experimental discovery. The light (mass-3) isotope of helium is extremely rare in nature (about 1 part in 10^7 of naturally occurring helium), and experiments on it only became feasible with the development of nuclear reactors, which produce ^{3}He via the decay of tritium (^{3}H). By the early 1950s it was known that this isotope liquefied at a temperature only a little below that of the common mass-4 isotope, and a few properties of the liquid had been measured. Like ^{4}He, it appeared that it would not solidify under its own vapour pressure.

Since the ^{3}He atom contains an odd number of spin $-1/2$ elementary particles (2 protons, 1 neutron, 2 electrons) it should differ from ^{4}He in obeying Fermi–Dirac statistics; liquid ^{3}He was the first system of this kind which could be cooled without freezing to the point of *degeneracy* where the statistics might play an important role. It was recognized that there was a close parallel with electrons in (normal) metals, which could be understood qualitatively as a system of non-interacting Fermi particles, and in 1956 Landau provided a firmer theoretical base for this notion with his theory of a Fermi liquid, i.e. a strongly interacting Fermi system which is liquid in the degeneracy regime. As in his earlier work on the Bose liquid ^{4}He, Landau introduced the idea of an elementary excitation: only in this case the elementary excitations of the strongly interacting system were in one-to-one correspondence with those of the free gas, and in particular, like them, obeyed Fermi statistics. As a result, the properties of the liquid are mostly qualitatively similar to those of the free gas, the only major exception being the modification, in the strongly degenerate regime, of the ordinary sound waves into a peculiar collective excitation called by Landau *zero sound*. Over the next decade or so measurements down to about 3 mK showed that below 100 mK ^{3}He is indeed very well described by Fermi-liquid theory.

The atoms of liquid ^{3}He in this regime, then, behave qualitatively much like the electrons in a normal metal. But a metal which is normal at temperatures of more than a few degrees may in the very low-temperature regime become superconducting, and once the microscopic (Cooper-pairing) mechanism for this was understood in the work of BCS it became an obvious question whether something similar might not happen in liquid ^{3}He at sufficiently low temperatures. The very earliest speculations assumed that if this happened the Cooper pairs would form in an s-wave, spin-singlet state as in superconductors, i.e. their wavefunction would be of the general form of equation (8), but it was very soon realized that the very strong repulsion between the 'hard cores' of the helium atoms would make this form of pairing energetically unfavourable, and attention shifted to the possibility of pairing with non-zero orbital angular momentum l (which would reduce the effect of the core repulsion because the centrifugal force automatically keeps the atoms from getting too close). Because of the Pauli principle, even-l pairs must be in a spin-singlet state (with the spins of the two atoms anti-parallel) while odd-l pairs must be spin triplets (spins parallel). The majority (though not universal) opinion throughout the 1960s was that the state with $l = 2$ (hence spin-singlet) was likely to be the most stable, and the properties of this state were explored in some detail, in particular in an influential 1962 paper by Anderson and Morel. Since the internal wavefunction of the Cooper pairs in this case does not possess spherical symmetry, neither do the physical properties: the liquid should be anisotropic. While a system of electrically neutral atoms could not be superconducting, it was predicted to display the analogous property of superfluidity. Thus this as yet hypothetical phase of matter became known as an *anisotropic superfluid*. Throughout the 1960s and early 1970s repeated attempts were made by theorists to predict the transition temperature of liquid ^{3}He into the anisotropic superfluid phase, with results ranging from temperatures just below the lowest currently achieved in the liquid all the way down to 10^{-17} K.

In the autumn of 1971, a graduate student at Cornell University, Doug Osheroff, was measuring pressure changes during the cooling of a mixture of liquid and solid ^{3}He. Between 3 and 1.5 mK, the record of pressure against time showed two small but reproducible anomalies: a change in the slope of the curve at 2.6 mK (the *A feature*), and at a rather lower temperature a tiny discontinuity in trace (*B feature*). The first paper in which Osheroff, with his supervisor D M Lee and R C Richardson, reported these results was entitled 'Evidence for a new phase of solid ^{3}He'; however, within weeks magnetic resonance experiments showed

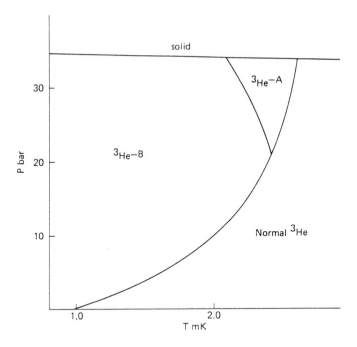

Figure 11.9. *Phase diagram of liquid ^{3}He below 3 mK.*

unambiguously that whatever was going on, at both the A and B features, was going on in the liquid rather than the solid. Was this the long-anticipated anisotropic superfluid phase?

The outcome of the feverish period of experimentation and theoretical analysis which followed could hardly have been anticipated: below 3 mK liquid ^{3}He has not one but three new phases, now called A, B and A_1, each of which is in some sense an 'anisotropic superfluid'. In zero magnetic field the phase diagram is approximately as shown in figure 11.9 (but the A phase can be supercooled considerably below the equilibrium 'AB line'). As the field increases, the A phase fills a larger and larger region of the diagram, suppressing the B phase entirely by about 6 kG, while the second-order A-to-normal transition splits into two, with the A_1 phase a thin sliver between them.

The origin of this surprising result is the fact that the primary mechanism of attraction which binds the Cooper pairs does not arise, as in superconductors, from the polarization of an ionic background (there is none!) but rather through a conceptually similar mechanism involving polarization of the spins of the 'background' liquid. This feature had indeed been anticipated in the 1960s, and it had been appreciated that such a mechanism, if dominant, would favour pairing in a spin triplet (figure 11.10) and hence odd-l state rather than the generally favoured $l = 2$ state. What had not been realized, however, is the richness

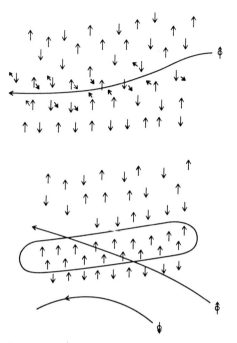

Figure 11.10. *Mechanism of indirect attraction in liquid ³He. An 'up' spin atom creates a region of predominantly 'up' polarization near its path; this region then attracts other up-spin atoms but repels down-spin ones.*

of behaviour which this kind of state permits. Suppose that, as we now believe, the favoured pairing configuration is a p-wave ($l = 1$) triplet, then there are three Zeeman spin sub-states and three orbital ones, and this gives nine possible 'components' of the wavefunction, each of which can enter multiplied by a complex coefficient. Now, if we just follow through the BCS argument, there is precisely one combination which is the most stable over the whole of the region of existence of the superfluid phase: this corresponds to the so-called Balian–Werthamer (BW) state which is now almost universally identified with the experimentally observed B phase of ³He. This state had been described, and its properties discussed in some detail, in an important paper by Balian and Werthamer in 1963. (It had actually been introduced earlier by Vdovin in the Soviet Union, but the paper was published in an obscure conference proceedings and was virtually unknown in the West.) In this phase, which in the terminology of atomic physics would be called ³P₀, the Cooper pairs all have orbital angular momentum $l = 1$ and total spin $S = 1$ (spins 'parallel'), but the vectors l and S are oppositely directed so as to give a total angular momentum J equal to zero (but see below). The energy gap and hence the thermodynamics of such a phase are indistinguishable from those of an ordinary (s-wave) superconductor.

Within this scheme, however, one cannot understand the occurrence of the A phase. In a seminal paper written within a few months of the experimental discovery, Anderson and Brinkman pointed out a crucial difference between the lattice-polarization *phonon exchange* mechanism of attraction between electrons in a superconductor and the spin-polarization *spin-fluctuation-exchange* mechanism believed to operate in liquid ^{3}He. In the former the polarized medium is the ionic lattice whose behaviour is hardly affected by the onset of Cooper pairing of the electrons. By contrast, in ^{3}He the polarized medium is the liquid itself, i.e. the same system as forms the Cooper pairs: pair formation can therefore affect the polarizability, and hence the strength of the interaction which binds the pairs. This puts a premium on those states which retain the full normal-state polarizability, which is possible only if two of the three Zeeman sub-states (corresponding to $S_z = \pm 1$ along some axis, not also $S_z = 0$) are formed. Less obviously, it favours a particular one of this class of states, which had been discussed in the early 1960s by Anderson and Morel and is nowadays known as the Anderson–Brinkman–Morel state (ABM) state. The principal characteristic of this state is that, irrespective of their spin states ($S_z = \pm 1$), all Cooper pairs have an apparent angular momentum $\hbar$ directed along the same axis. Moreover, unlike the BW state or the s-wave pairing state of superconductors, this state has an energy gap which vanishes for certain directions of motion of the quasiparticles. Further implementation of these ideas has given a fairly quantitative account of the relative stability of the A and B phases.

For completeness, it should be added that in the 'sliver' of A_1 phase it is believed that only a single Zeeman sub-state (e.g. $S_z = +1$) is occupied, the two spin directions being no longer equivalent in a magnetic field, so that one may have a slightly higher transition temperature to the Cooper-paired state than the other.

What is especially exciting about the new phases of liquid ^{3}He is that the order parameter (Cooper-pair wavefunction) in each has, as well as the usual centre-of-mass degree of freedom which is possessed also by the electron pairs in superconductors, one or more internal degrees of freedom which are not fixed by the energetics of the pair-forming process (sometimes called 'broken symmetries'). For example, in the A phase the pairs are characterized by two axes: one (conventionally denoted d) is perpendicular to the axis (which can actually be chosen to be any axis in a plane) along which pairs with $S_z = \pm 1$ are formed, while the other (denoted l) is the axis along which they have relative orbital angular momentum $\hbar$. As far as the 'gross' energies which are responsible for the pair-formation process are concerned, these axes are quite arbitrary. For the B phase the 'broken symmetry' is a bit more subtle; it corresponds to the possibility of starting from the '$^3P_0$' state and performing an arbitrary rotation of the spin coordinates relative to the orbital ones. The resulting state still has an isotropic energy gap, and indeed those of its properties which depend only on the spin coordinates or only on the orbital axes are unaffected. However, the effect on these properties which involve *correlations* between the spins and orbital coordinates, is profound.

The crucial point is that the formation of the superfluid state (the 'Bose condensation' of the Cooper pairs, if one likes to think of it that way) requires that all pairs have the same wavefunction not only with respect to their centre-of-mass coordinate but also with respect to their *internal* degrees of freedom. This feature was, of course, already present for the electrons in superconductors, but without striking consequences, because the relative wavefunction was uniquely fixed by the energy considerations which formed the pairs in the first place. By contrast, for superfluid ^{3}He the effects are spectacular—effects which would be totally negligible in a single molecule can be amplified by the 'Bose condensation' to the point where they dominate the behaviour. The best-known example concerns the interaction of the nuclear magnetic moments of the ^{3}He nuclei. If we think of these nuclei as tiny classical magnets, the usual theory of magnetism tells us that, if constrained to be parallel, they would prefer to lie end to end rather than side by side. When rotating around one another, while pointing in a constant direction, they achieve a favourable configuration for about half the time if the magnets lie in the plane of rotation, rather than normal to it. However, the energy E_d which favours the parallel orientation over the perpendicular is so tiny that even if two ^{3}He atoms in, say, the gas phase were to form a diatomic molecule (which in fact they do not), thermal disorder would completely swamp it, and the molecules of the gas would rotate with random orientations. In the superfluid phase, however, all N of the Cooper pairs must rotate in the *same* way. The energy difference between the two configurations (in-plane and perpendicular) is now not E_d, but NE_d, which is always much larger than the thermal energy kT. Thus the in-plane configuration is always overwhelmingly favoured and, except in special circumstances, is always realized. Furthermore, when this configuration is disturbed, for example by applying an RF magnetic field to the spins, the existence of this coherent dipole energy has a spectacular effect on the resonance behaviour. Historically it was in fact these effects which permitted the earliest convincing identification of the microscopic nature of the three superfluid phases. Various other ultra-weak effects may be similarly amplified (it has even been proposed that it might be possible in this way to manifest the effects on a macroscopic scale of the parity-violating 'weak neutral currents' introduced in the 'Standard Model' of the electroweak interaction in particle physics, on a macroscopic scale).

See also p 695

The above discussion assumes implicitly that the internal degrees of freedom are constant in space. However, spectacular consequences follow also when we consider the possibility of their spatial variation.

954

Perhaps the most intriguing of all relates to the question of the (meta)stability of supercurrents in an annulus. We saw that in a BCS superconductor the reason for this stability was essentially topological in origin: the number of twists of 2π the phase ϕ undergoes as we go once around the annulus (*winding number*) cannot be changed without at some stage depressing the magnitude $|\Psi|$ of the order parameter to zero. The situation is essentially the same in the B phase of superfluid ^{3}He, but in the A (and A_1) phase there is a crucial difference. Suppose we start with the 'orbital vector' l constant in direction throughout (which allows us to define a phase $\phi(r)$ unambiguously), it is then possible to go from winding number n to $n-2$ (though not to $n-1$) via a process in which the magnitude of the order parameter never changes from its original value, provided that at intermediate stages we allow the direction of l to vary in different places. Thus, in an ideal situation where there is no energy to 'pin' l, 'superfluid' ^{3}He-A would actually not be superfluid, at least in the sense of showing metastability of supercurrents. (It would still be expected to show a Hess–Fairbank effect.) That in real life the A phase appears to behave much like ^{4}He, with the usual metastability, is attributed to the combined effects of the nuclear dipole force and the walls in 'pinning' the vector l. This circumstance shows spectacularly the necessity, emphasized earlier, of distinguishing between the 'stable' and 'metastable' manifestations of superfluidity.

11.3.2. *Miscellaneous recent developments*

In this section, I briefly review a few systems, other than traditional superconductors and the two pure isotopes of helium, in which superconductivity, superfluidity or related phenomena either exist or it is believed that they might occur under conditions not yet experimentally realized. The most spectacular of these, the high-temperature superconductors, are treated in the next section.

The major surprise of the 1970s in the area of superconductivity was its discovery in some members of a class of materials known as the *heavy-fermion* systems. As already mentioned, a good description of many systems of strongly interacting fermions, such as the electrons in ordinary metals in their normal state is given by Landau's Fermi-liquid theory. In this theory the elementary excitations, which behave qualitatively much like free electrons, are characterized by an *effective mass* which for most ordinary metals is of the order of the original electron mass: the electronic specific heat at low temperatures is proportional to this mass. However, some metal compounds containing rare-earth or actinide elements, such as $CeAl_3$, $CeCu_2Si_2$ or UPt_3, have very anomalous properties at low temperatures, in particular specific heats so huge that if the Landau theory applies the effective mass of the electrons must be hundreds or even thousands of times the real mass (hence the name heavy-fermion). Indeed, at first it was suspected that this enormous specific heat might arise from localized electrons to which the concept of effective mass does not apply, but the picture changed dramatically in 1979, when a superconducting transition was discovered in UBe_{13} and subsequently in various other members of the class. The fact that the specific heat drops rapidly below T_c, as in a typical BCS superconductor, is generally taken to indicate that the electrons contributing to it must be those that form Cooper pairs and thus must be mobile. Both the normal and the superconducting properties of the heavy-fermion systems are the subject of considerable ongoing research, and it is already clear that the normal state cannot be described by any simple variant of the so-called Bloch–Sommerfeld picture which has been so successful for simpler metals. While the behaviour in the superconducting state resembles that of a typical BCS superconductor, there are indications that in UPt_3, at least, and possibly in others, the Cooper pairs form in an 'anisotropic' state similar to that which occurs in superfluid ^3He. However, the rather complicated effects of the crystal structure, and the absence of any simple diagnostic probe analogous to nuclear magnetic resonance in ^3He, has prevented this from being firmly established so far.

A second class of novel superconducting materials is the alkali-doped fullerene crystals. The fullerenes are crystals of hollow polyhedra of carbon—C_{60} is the most studied—and were first synthesized in bulk in 1990. Pure C_{60} is an insulator, but when it is doped with alkali atoms these contribute conduction electrons and the crystal becomes metallic. Moreover, for certain levels of doping it also becomes superconducting, with a transition temperature as high as ~ 33 K—which, had it been discovered seven years earlier, would at that time have been the world record! While all the evidence is that superconductivity in these materials is due to formation of s-wave Cooper pairs as in the 'old-fashioned' superconductors, it is unclear whether the mechanism which binds the pairs is the usual phonon exchange or (as it is thought to be in the heavy-fermion superconductors) a sophisticated effect of the inter-electron interactions.

It is worth noting briefly that there are several areas outside condensed matter physics where the ideas developed in the theory of superconductivity and superfluidity have made an impact. The idea of Cooper pairing has been applied with considerable success to explain some of the properties of heavier nuclei. Since in this case we are talking about a system of at most a few hundred particles, macroscopic concepts such as superfluidity have no real meaning, but the occurrence of pairing is reflected in the reduced moment of inertia of the nucleus (compare the Hess–Fairbank effect). A more direct, if more speculative, application of BCS theory is to neutron stars in which, depending on the pressure, etc, various superfluid phases, analogous to those of ^3He, are predicted to occur. Finally, ideas analogous to those of the BCS theory have found major application in particle physics, in the *dynamical symmetry breaking* scenario of Nambu and Jona-Lasinio and the mechanism, originally recognized by Anderson and elaborated by Higgs, by which the intermediate vector bosons of the unified electroweak theory acquire mass; a concept inspired by the Meissner effect.

Finally, there are systems in which superfluidity might occur but has not yet been demonstrated unambiguously, for example, the dilute ($\lesssim$ 8%) fermion system of ^{3}He atoms which is stable in (itself superfluid) ^{4}He at low temperatures, spin-polarized atomic hydrogen (a Bose gas), and the metastable Bose system formed by excitons in some insulators such as Cu_2O (where indirect manifestations of 'superfluid' behaviour may have been seen in recent experiments). The observation of superfluidity in spin-polarized atomic hydrogen ($H_\uparrow$) would be particularly exciting because, if the mechanism is indeed Bose condensation as is almost universally assumed both here and in He II (see section 11.2.6), then this latter phenomenon, which has to date resisted direct observation in He II, should be detectable almost with the naked eye; a sharp change is predicted to occur, in $H_\uparrow$ under the relevant experimental conditions, in the spatial density profile. In this way we should be able to plug, in an experimental sense, the last loophole in the general argument of section 11.2.6.

11.3.3. High-temperature superconductivity

Over the three-quarters of a century following Onnes' 1911 discovery, the maximum temperature achieved for the transition to the superconducting state crept up gradually from the 4.2 K of mercury to an apparent limit in the region $\sim$20–25 K (figure 11.11). True, claims had been made of superconductivity at higher temperatures, for example 60 K for CuCl, but these had all been later discredited. Moreover, many theorists had convinced themselves that the phonon mechanism could never produce a T_c much higher than those observed; and while some, such as Little in the US and Ginzburg in the Soviet Union, speculated that alternative mechanisms might produce a superconducting state at much higher temperatures, perhaps even room temperature, specific attempts to build such a 'high-temperature superconductor' had always ended in failure. Thus, in the summer of 1986 most physicists familiar with the subject would probably have bet 100 to 1 that T_c would never go above 30 K. In the event, within a year virtually every laboratory in the world was almost routinely observing superconductivity above not 30 but 90 K, well above the boiling point of nitrogen; the record today stands at over 150 K, half way to room temperature. Many physicists would regard this as the most significant development in solid-state physics for half a century.

The story begins with a paper from IBM Zurich in November 1986 by Alex Müller and Georg Bednorz, reporting the onset of a partial Meissner effect around 30 K in a compound first made a few years

See also p 490

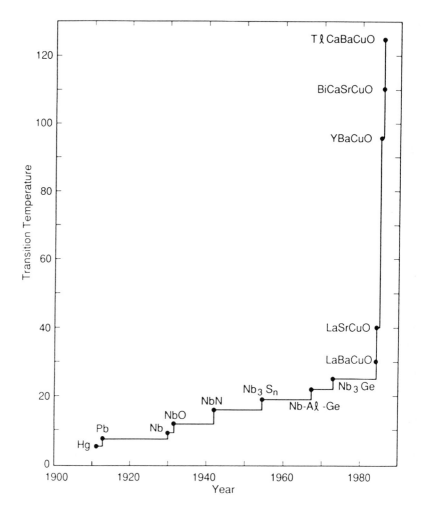

Figure 11.11. *The history of the maximum known temperature for superconductivity to occur (up to 1991).*

earlier, La_2CuO_4 doped with a small amount of barium. They tentatively (and in view of earlier false alarms, bravely) identified this as indicating a superconducting transition. (In a rare example of the award of the Nobel Prize satisfying the original condition that the work recognized should have been done in the preceding year, Müller and Bednorz received the prize in 1987.) The significance of this work was rapidly appreciated, and over the next few months many laboratories frantically explored the chemical 'neighbours' of this material to try to increase the transition temperature. A notable step, in early 1987, was the discovery by a Houston–Alabama collaboration led by Chu and Wu of superconductivity in $YBa_2Cu_3O_7$ (usually nowadays known as YBCO or '123') at 90 K, and a year or so later the maximum transition temperature was raised to 125 K in a Tl–Ba–Ca–Cu–O compound ('2223'). At the time of writing, about 100 different compounds are known which have temperatures above the 1986 dream of 30 K (not counting the fullerenes); the current record, about 160 K, is held by a Hg–Ba–Ca–Cu–O compound under pressure.

Although the high-temperature superconductors (HTS) belong to several different classes, they share a number of features[†]. First, they are all chemically complicated (few HTS are known which contain less than four different elements) and have fairly complicated crystal structures with large unit cells, but invariably containing well-separated planes (or pairs or triples thereof) of copper and oxygen atoms (CuO_2 planes) (figure 11.12). For this reason they are often referred to as *copper oxide superconductors*. It seems overwhelmingly probable that the mechanism of superconductivity is to be sought in those CuO_2 planes, and that the role of the off-plane atoms is mainly to act both as donors of electrons (or, more usually, of holes) to the planes and as spacers between them; this partially explains the enormous number of chemical compounds which show high-temperature superconducting behaviour. Secondly, in most cases superconductivity occurs in a region of the phase diagram close to an antiferromagnetic phase (figure 11.13). Thirdly, the properties of these materials in the normal state are highly anomalous. In particular, the electrical resistivity is proportional to temperature from 1000 K down to the transition temperature. A resistivity proportional to temperature is of course a feature of ordinary metals at temperatures above the Debye temperature θ_D, and one might think that the copper oxide superconductors simply have rather low values of θ_D. However, this idea seems implausible when one finds that $Bi_2Sr_2CuO_6$ behaves like the HTS in all respects except that its superconducting transitions are much lower (9 K); but resistivity is still linear all the way down to T_c, far below any remotely plausible value of θ_D (which in any case can be measured directly and is not specially low for the HTS). The Hall coefficient and NMR (nuclear magnetic resonance) behaviour are also highly unusual in the normal state of the HTS.

While the HTS are indeed superconductors in that they show, under appropriate conditions, both a Meissner effect and resistance-free flow of current, the difference of their electromagnetic behaviour from that of the old-fashioned superconductors is so marked as to be almost qualitative. In the first place, they show type-II magnetic behaviour in an extreme form: while the lower critical field, H_{c1}, is probably typically of the order

---

† *A slight complication is caused by the discovery, since 1986, of superconductivity above 30 K in $BaKBiO_3$. This has none of the characteristics described and is usually thought to be a simple BCS superconductor with an anomalously high T_c.*

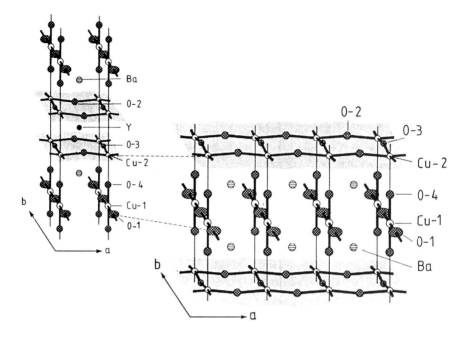

Figure 11.12. *The crystal structure of* $YBa_2 Cu_3 O_7$.

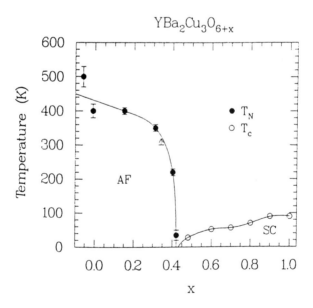

Figure 11.13. *The phase diagram of* $YBa_2 Cu_3 O_{6+x}$, *as a function of temperature T and oxygen concentration x.* T_N *is the temperature of transition to the* AF *(antiferromagnetic) state and* T_c *that to the* SC *(superconducting) state.*

of 100 gauss, the upper one, H_{c2}, is so enormous that for temperatures only just below T_c it cannot actually be reached with existing magnets. Secondly, the *transition to the zero-resistance state* is much less sharp than for conventional superconductors, particularly in a magnetic field (figure 11.14). It is also directionally dependent: recent experiments show that there is sometimes a region of temperature below T_c in which the resistance parallel to the planes drops as expected, but perpendicular to the planes actually increases! Indeed, it is currently a matter of debate whether, in certain regions of the phase diagram, the HTS should really be described as superconducting, since the vortices move around so much more readily than in a traditional superconductor. No doubt this is partly because the order parameter is large only on the CuO_2 planes, so that rather than a continuous line one should think of a vortex as like a loosely connected string of beads, which can 'wobble' or even be broken without too large an energy cost.

Turning from the macroscopic electromagnetic behaviour to the more 'microscopic' properties of the HTS we find a similarity to BCS superconductors which is the more surprising given the qualitatively anomalous behaviour in the normal state. In fact, nearly every feature of the BCS model shows up at least qualitatively, for example, both the spin susceptibility and the electronic specific heat drop off sharply below T_c, and there appears to be a relatively well-defined energy gap Δ, in the sense that most of the electronic excitations lie above a threshold energy which it is natural to call the 'gap'. However, there are two significant differences from the traditional superconductors. First, nearly every method of measuring Δ gives a value at zero temperature whose ratio with T_c is considerably greater than the BCS ratio of 1.75, and moreover in some experiments it is not clear that Δ tends to zero as T approaches T_c. Secondly and perhaps more significantly, a considerable number of experiments (tunnelling, specific heat, NMR, penetration depth) indicate that there are some electronic excitations which have energies small compared to Δ. Apart from these two features, the similarity to the BCS predictions is remarkable.

The question of the mechanism by which superconductivity is induced in the HTS is hotly debated, and any present attempt to review it is likely to be out of date within a few years, if not months. Among the few principles which would probably be generally agreed are: (i) that the mechanism of high-temperature superconductivity is 'universal'—there is not one type of mechanism for the original lanthanum compound of Bednorz and Müller, a different one for YBCO, and so on; (ii) that the dominant role is played by the CuO_2 planes, with the off-plane atoms essentially acting as 'spectators', and the differences in T_c are due in large part to the different numbers of electrons or holes which these atoms donate to the planes; (iii) that the mechanism can reasonably be investigated theoretically by considering only a single

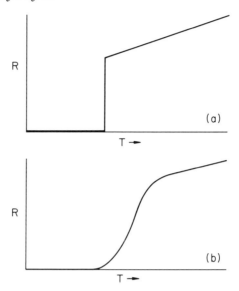

R

T →

(a)

R

T →

(b)

Figure 11.14. *Resistance as a function of temperature for (a) Al in zero magnetic field and (b) $Bi_2 Sr_2 CaCu_2 O_8$ in a field of a few tesla (schematic).*

plane (although one very recent theoretical proposal strongly disputes this), the role of inter-plane interactions being primarily to provide a Josephson-type coupling between the order parameter on neighbouring planes, which prevents the fluctuations becoming so large as to make the situation qualitatively different from that in a typical three-dimensional superconductor, and (iv) (commanding widespread though not universal acceptance) that the proximity to an antiferromagnetic instability cannot be an accident. Beyond this there is little or no agreement.

We may take the question in the following three stages.

(1) Is superconductivity due to the formation of Cooper pairs, as in the old-fashioned superconductors, or is some completely different mechanism at work?

(2) If it is Cooper pairs, what is the attraction responsible for binding them?

(3) What is the symmetry of the pair state?

Regarding (1), since the discovery of HTS a number of alternatives to the Cooper-pair mechanism have been proposed. One, which for a time looked particularly attractive, was the so-called *anyon* mechanism. Anyons are a peculiar type of excitation, with statistics intermediate between those of fermions and bosons, which can exist only in a strictly two-dimensional system, a feature which in view of the strongly 'two-dimensional' character of the CuO_2 planes in the HTS makes their attraction obvious. Moreover, it is often claimed (though the validity of this claim is not obvious to me) that at zero temperature a system of anyons is superconducting even without the need for interactions. Unfortunately, it has proved very difficult to do concrete calculations of experimental properties on the basis of the anyon model, and the few distinguishing features which have been predicted have not yet been reliably observed, so that interest in the idea, at least in the context of HTS, seems to be ebbing. Various other proposals, for example involving the Bose condensation of a fictitious 'gauge boson' associated with the strong hard-core repulsions of the electrons in the CuO_2 plane, have been made. A very strong constraint on any such 'exotic' mechanism (including the anyon one) is that it must be consistent with the observation that the unit of flux quantization in the HTS, as in the BCS sort, is $h/2e$ (and not, for example, h/e): it is not clear whether or not any theory possessing such a property must in the last resort just be another language for describing the process of Cooper-pair formation.

A variant which is similar, but not quite identical, to the Cooper-pair idea is that bound pairs of electrons are already formed in the normal state. Such pairs would be bosons, and the observed transition temperature would then simply correspond to the point of Bose condensation, as in ^4He. Some of the experimental observations, for example that in some (but not all) HTS the spin susceptibility begins decreasing at temperatures much higher than T_c, would find a natural explanation in this model, which as yet has not been developed far.

Probably the majority opinion is that the mechanism of superconductivity in the HTS is essentially Cooper pairing. If this is correct, we can tentatively apply the prescriptions of standard BCS theory to find out something of their characteristics. One feature immediately stands out if we do this: whereas in traditional superconductors the ratio of the Pippard coherence length (Cooper-pair radius) to the mean distance between electrons is thousands or even tens of thousands, in the HTS it is less than ten. This would explain why the magnetic behaviour of these materials is so extremely type-II, and would also mean that many of the approximations made in the standard BCS theory need to be re-examined, so that it would not be surprising if the quantitative aspects of its predictions are modified.

Turning to question (2), it is fairly generally agreed (by those who believe in the Cooper-pair mechanism) that the attraction which binds the pairs must be mediated by the exchange of some kind of low-energy excitation obeying Bose statistics; the question is whether this is a phonon as in traditional superconductors or some kind of collective excitation of the electron system, and if the latter, what kind. Arguments against the phonon mechanism include the absence of any appreciable isotope effect in the higher-temperature HTS, and the fact that if, as seems natural, it is collisions with this same boson which are responsible for the normal-state resistance then, as already mentioned, the linearity of the latter in temperatures down to 9 K is incompatible with any reasonable Debye temperature. However, these arguments are not incontrovertible. If the boson is a collective excitation of the electron system, what kind of excitation? Here one finds a variety of answers. One of the most popular scenarios identifies the boson as an antiferromagnetic spin fluctuation, the 'relic' of the nearby antiferromagnetic state, and this makes a striking prediction concerning question (3), the symmetry of the Cooper-pair wavefunction, to which we now turn.

Are the Cooper pairs formed in a simple s-wave state, as in traditional BCS superconductors, or with 'exotic' symmetry, as in superfluid ^{3}He and possibly some of the heavy-fermion superconductors? This question is particularly interesting because whereas, for example, a mechanism of attraction due to phonons usually favours an s-wave state, the scenario based on exchange of antiferromagnetic spin fluctuations indicates rather unambiguously that the pair state should be of the so-called $d_{x^2-y^2}$ type; this state has nodes (points on the Fermi surface where the energy gap is zero) and therefore even at low energies there should exist an appreciable number of electronic excitations, contrary to the simple BCS case where there are exponentially few. Thus, the apparent evidence for such excitations in a variety of experiments (see above) is often taken as supporting the $d_{x^2-y^2}$ hypothesis, as is the direct observation in recent photoemission experiments, of a gap which appears to be large along the x and y axis in the CuO_2 plane but small and possibly zero along the 45° axis—precisely the behaviour of the gap in the $d_{x^2-y^2}$ state. However, this argument is not totally foolproof. To see this, in figure 11.15 we compare schematic representations of the $d_{x^2-y^2}$ state and a strongly deformed s-state—a quite plausible candidate in some alternative scenarios. We see that, except very close to the 45° axis, the magnitude of the gap is almost identical in the two cases (and hence, except at very low temperatures, the number of excitations will be very similar); the principal difference between the two states is in the relative *sign* of the wavefunction in the various lobes. It would be highly desirable to measure this sign directly, and the first measurements of this type (using the Josephson effect) have very recently been made; at the time of writing the situation is confused, and it is not clear that there is *any* symmetry assignment which is compatible, given current theoretical understanding, with the totality of the data.

At present, the technically important business of finding HTS with ever higher transition temperatures is very much a matter of trial and error. Needless to say, a major motivation for trying to understand the mechanism is the hope of identifying those chemical, structural and other features which tend to raise T_c, so as to orient the search among the billions of currently unexplored compounds. While it seems unlikely that the current record will stay unchallenged for long, it is anyone's guess whether genuine 'room-temperature' superconductivity will be achieved in our lifetime (or ever); and whether, if so, the technical difficulties in

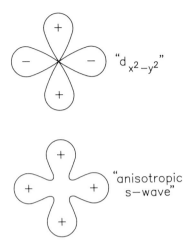

Figure 11.15. *Comparison of the angular dependence of the energy gap for (a) the $d_{x^2-y^2}$ and (b) a very anisotropic s-wave state. The + signs indicate the sign of the pair wavefunction in each state.*

the practical application of the HTS (connected, for example, with the annoyingly high mobility of the vortices) will be overcome to the extent that they may have a revolutionary effect on the technology of everyday life.

11.3.4. Further reading
In writing this chapter I have drawn heavily on references [1, 2]. Other books [3–9] also provide well-referenced accounts of related material.

Acknowledgments

This work was supported by the National Science Foundation under grant No NSF-DMR-92-14236. I thank Gordon Baym and Brian Pippard for many useful comments and suggestions.

References

[1] Hoddeson L, Braun E, Teichmann J and Weart S (ed) 1992 *Out of the Crystal Maze* (New York: Oxford University Press)
[2] Dahl P F 1992 *Superconductivity* (New York: American Institute of Physics)
[3] Keesom W H 1942 *Helium* (Amsterdam: Elsevier)
[4] Shoenberg D 1952 *Superconductivity* (Cambridge: Cambridge University Press)
[5] London F 1950 *Superfluids I Macroscopic Theory of Superconductivity* (New York: Wiley); 1954 *Superfluids II Macroscopic Theory of Superfluid Helium* (New York: Wiley)

[6]   Gorter C J and Brewer D F (ed) 1955 *Progress in Low Temperature Physics* (in 13 volumes) (Amsterdam: North-Holland)
      Note particularly:
      Feynman R P 1955 *Applications of Quantum Mechanics to Liquid Helium (Progress in Low Temperature Physics 1)* (Amsterdam: North-Holland) p 17
      Bardeen J and Schrieffer J R 1961 *Recent Developments in Superconductivity (Progress in Low Temperature Physics 3)* (Amsterdam: North-Holland) p 170
[7]   Wilks J 1967 *The Properties of Liquid and Solid Helium* (Oxford: Clarendon)
[8]   Vollhardt D and P Wölfle 1990 *The Superfluid Phases of Helium 3* (London: Taylor and Francis)
[9]   Ginsberg D M (ed) 1994 *Physical Properties of High-temperature Superconductors* (in 4 volumes) (Singapore: World Scientific)

Chapter 12

VIBRATIONS AND SPIN
WAVES IN CRYSTALS

R A Cowley and Sir Brian Pippard

12.1. The beginnings of lattice dynamics

The elastic solid, considered as a structureless continuum, claimed the attention of many applied mathematicians in the nineteenth century. It was a matter of controversy whether the most general formulation (stress being assumed proportional to strain) demanded 21 independent elastic constants as Green rightly believed, or only 15 as followed from Cauchy's model of material points interacting through central forces. The majority, however, were more concerned with applications of the theory, generally in simplified form, to real problems of wave propagation, or arising from engineering structures. The scale of their achievements may be judged from Love's [1] formidable treatise, which opens with a valuable historical survey before settling into its dense mathematical exposition.

The obvious weaknesses of Cauchy's model—it required Poisson's ratio to be $\frac{1}{4}$ while something around $\frac{1}{3}$ was much more usual—probably discouraged widespread speculation on the atomic structure of solids until x-ray crystallography gave the necessary impetus. Before that, however, Einstein had initiated a different line of enquiry with his theory of the specific heat of solids. The specific heat is so central a topic of crystal lattice dynamics that we must begin with the events that led up to Einstein's seminal paper.

12.1.1. Specific heats

The very term, specific heat, implies something different from the modern approved term, thermal capacity. Just as specific gravity refers the weight of a body to the weight of an equal volume of water, so specific heat refers its thermal capacity to that of an equal mass of water. This was a natural way of interpreting the standard procedure for measuring specific heat by the method of mixtures—dunking the hot body in a water-filled calorimeter—which was good enough to allow Dulong and Petit to formulate their law in 1819. We need not rehearse the early history of this law; by 1900 it was accepted that the atomic heat of most solid elements is about 6 calories per degree. If the chemists, who found it a valuable starting-point for estimating the atomic weight of a new element, wondered what was responsible for this remarkable regularity they seem to have been wise enough not to record their thoughts. They were aware of exceptions, notably diamond, whose specific heat is much smaller than it should be. Only in 1898 did Behn's [2] measurements begin to reveal that other substances showed at lower temperatures the same discrepancy as diamond at room temperature.

On first reading, Behn's technique seems very primitive, in that it still employed the method of mixtures. A block of metal would be cooled in liquid air (−186 °C) or in a mixture of solid CO_2 and alcohol (−79 °C), and transferred to a water calorimeter. In this way he deduced the mean specific heat between −79° and 18° and between −186° and −79°; for aluminium he found 5.3 and 4.2 cal deg^{-1}, the room-temperature value being 6.0. The demonstration was convincing enough, and he allowed himself to speculate that the specific heat might vanish at absolute zero. Ten years later, thanks to Nernst and Einstein, this idea was on the way to being generally accepted, and it was these two who were principally responsible for initiating a radically new understanding of the whole field.

Although Einstein's contribution came first, in 1907, it made little impact until the experiments of Nernst's school provided massive empirical support. It was his assistant, Eucken [3], who built the first electrically heated calorimeter (figure 12.1), which was elaborated and improved by Nernst and successive members of his team, including Lindemann and Simon [4]. At first the lowest temperature (Nernst mentions −210 °C) was attained by pumping on liquid air, but later liquid hydrogen and liquid helium brought temperatures as low as 1 K into the available range. It should be remembered that in a specific heat measurement a temperature difference, often quite small, must be measured, and this places considerable demands on the calibration of the thermometer against the absolute scale represented by a perfect gas thermometer. If there is a calibration error t which varies with temperature T, it is not the relative error t/T that matters, but the differential dt/dT. Any departure from smoothness in the calibration curve can be serious, and the establishing of a reliable scale, especially at extremes of temperature, has been a major occupation of standards laboratories throughout the century. Compared with this, other technical problems in calorimetry are quite minor.

Nernst's principal interest in calorimetry stemmed from his Heat Theorem of 1906, now known as the Third Law of Thermodynamics,

See also
p 1249

See also
p 535

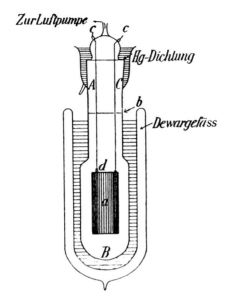

Figure 12.1. *Eucken's vacuum calorimeter. The hollow copper block a, into which samples can be sealed, hangs in an evacuated glass vessel, suspended from the leads to the platinum winding which serves both as heater and as thermometer. The whole is immersed in liquid air. The ground glass stopper is surrounded by mercury to make it airtight. Radiation from outside is minimized by silvering the inside of the vessel, and by the tin-foil-covered mica plate b.*

an essential ingredient in the technically very important process of determining the thermodynamical parameters of chemical reactions from specific heat measurements. Thus much of the emphasis in his school centred on acquiring data of immediate interest to the chemical industry, and the study of specific heats of elements for their own sake was hardly more than a by-product. Nevertheless, a picture soon emerged of the temperature variation of specific heats, remarkably similar for most elementary solids, and largely confirming the ideas put forward by Einstein.

The development of Einstein's thought on quanta has been described in detail by Kuhn [5]. Being convinced that the radiation field was quantized, he found himself forced to accept that the oscillations of charged particles, in dynamic equilibrium with the radiation, were also quantized, and hence that all oscillations, whether of charged or uncharged particles, were quantized. In his 1907 paper [6] he supposed that every atom of a solid vibrates at the same frequency, which he identified with the *reststrahlen* frequencies (in the far infrared) at which Rubens and Nichols had found almost perfect reflection, as will be discussed later. Owing to Einstein's omission of references it has sometimes been supposed that he introduced the concept of $3N$ atomic

See also p 152

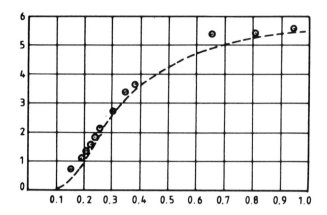

Figure 12.2. *Einstein's comparison of his specific heat formula with the available data for diamond. The abscissa is (in modern terms) $\hbar\omega_E/kT$, where ω_E is a characteristic frequency chosen to give a satisfactory fit to theory. The ordinate is specific heat in calories/gram-atom.*

oscillators—N atoms vibrating in three dimensions—but in fact it was mentioned by Boltzmann [7] in a footnote with which Einstein most certainly would have been familiar.

Boltzmann notes that the specific heat of a solid should be twice that of a perfect gas of the same atoms (Dulong and Petit's law) but Einstein, applying Planck's formula, shows how it should fall as the temperature is lowered, and cites the room-temperature values for three light elements (B, C, Si) in support of this view. Better still, he takes Weber's measurements of the specific heat of diamond between 222 and 1258 K and shows they lie reasonably close to his theoretical curve (figure 12.2) if he assumes the vibration frequency to be that of infrared radiation of 11 μm wavelength. 'Thus' he says, 'according to the theory it is to be expected that diamond shows an absorption maximum at $\lambda = 11\mu$'; but here he had over-reached himself, as he soon realized, for the atomic oscillators need not be electrically charged. So a correction [8] soon appeared: 'diamond either shows an absorption maximum at $\lambda = 11\mu$ or it has no optically demonstrable proper frequency whatsoever'. The latter is now known to be a good approximation to the truth.

By about 1911 Nernst had shown that Einstein's formula was roughly correct, but gave far too small a specific heat at the lowest temperatures. His first attempt, with Lindemann [9], to patch it up involved supposing there were two natural frequencies, one twice the other, and each accounting for half the total number of degrees of freedom. This had no theoretical justification, nor had Nernst's [10] next attempt which was to add a term proportional to $T^{5/2}$. Meanwhile Einstein [11] himself had realized the basic weakness of his own model—a single vibrating atom rapidly radiates an acoustic wave into the surrounding medium, and is

Peter Debye

(Dutch, 1884–1966)

Born and schooled in Maastricht, Peter Debye found it convenient to cross the border into Germany and enter college at Aachen, where Wien and Sommerfeld were professors. Until 1940 he spent most of his time in Germany, but when he found himself under Nazi pressure to adopt German citizenship he moved to Cornell where he spent the last 26 years of his life. He was an astonishingly versatile physicist and physical chemist who laid the foundations of many areas of study—the specific heat of solids, the theory of dipole moments, strong electrolytes, x-ray diffraction from random polycrystals, the Compton effect, adiabatic demagnetization and many others that are described in Mansel Davies' biographical memoir (Davies M 1970 *Biographical Memoirs of Fellows of the Royal Society* (London: Royal Society) p 175). Over a period of 60 years he published nearly 250 papers, despite Heisenberg's view that he 'had a certain tendency to take things easy ... I could frequently see him walking around in his garden and watering the roses even during duty hours of the institute. But the centre of his interest was undoubtedly his science'; and according to Henri Seck 'he will live in our memory as a brilliant scientist, a great teacher, a fatherly and helpful adviser and, above all, as a happy man.'.

in fact a heavily damped oscillator to which the application of Planck's formula must be considered very dubious. In a last-minute addendum to his paper of 1911 he remarked that if the atoms are capable of vibrating with a range of frequencies, each separate frequency contributes a Planck term to the thermal energy. He did not, however, suggest any way to calculate the frequency spectrum.

This was the problem that Debye on the one hand, Born and von Kármán (BvK) on the other, tackled independently. First to publish was

Debye [12] with a characteristically brilliant approximation that could be worked out completely, followed by BvK [13] with an attempt at a complete solution whose realization was quite beyond their powers. They agreed, however, that the vibrations to be quantized were not vibrations of individual atoms, but of the solid as a whole. The enumeration of the number of independent modes of vibration in a chosen range of wavelengths had been given by Rayleigh [14] in his classical treatment of black-body radiation, but Debye was apparently unaware of this and, having acquired from his earlier work on diffraction theory a taste for Bessel functions, he worked it out with considerable labour for a spherical body (Rayleigh had chosen a cube and written down the answer in two lines). BvK introduced the idea of periodic boundary conditions and also found the answer easily. In 1911 Weyl [15] proved, what every physicist takes for granted (though it is in fact a subtle matter), that the shape of the body is immaterial. The number of independent modes with wavenumber between q, defined as $2\pi/\lambda$, and $q + dq$ is proportional to $q^2 \, dq$, and if the waves are non-dispersive, so that frequency ω is proportional to q, the density of states $g(\omega)$, being the number in unit frequency range, is proportional to ω^2. This was Debye's assumption, not because he believed it true but because he argued that it would be true at very low temperatures, when only long waves were excited, and that he could satisfy the Dulong–Petit law at high temperatures by cutting off the spectrum at a certain maximum frequency ω_0, so that the total number of modes was three times the number of atoms. If the curve fitted the facts at low and high temperatures, it could not be too bad in between, and in any case he did not see his way to calculating anything more sophisticated. BvK attempted a theory of a real atomic lattice, and laid down the mathematical framework for a solution but could make little progress without severe approximations.

See also p 982

We shall return to them later; meanwhile Debye holds the stage. Like everyone else, he assigned the Planck energy, $\hbar\omega/(e^{\hbar\omega/kT} - 1)$, to each mode. There is no need to enter into details which can be found in many textbooks. The principal features of his theory are that the specific heat C is a function with one parameter only, the Debye temperature $\Theta_D = \hbar\omega_0/k$, and that at low temperatures $C \propto T^3$. It is striking that the Nernst–Lindemann formula fits Debye's curve within 2% all the way down to $\Theta_D/12$; only below this does the T^3 law show the superiority of Debye's theory.

Nernst and his school espoused Debye's theory enthusiastically even before Eucken and Schwers [16] confirmed the T^3 law in fluorspar. As more substances were investigated the observed agreement seems to have conferred on the theory a status approaching that of Holy Writ. It soon became customary to express the measurements of specific heat C in terms of the variation of Θ_D with temperature; that is, the Debye function $C = D(T/\Theta_D)$ is inverted to $T/\Theta_D = D^{-1}(C)$, or $\Theta_D = T/D^{-1}(C)$. Since for many substances the Debye formula is well obeyed, $\Theta_D(T)$ does not vary over a wide range and a compact representation of the results is achieved, without any implied interpretation. If every substance had shown the same pattern of $\Theta_D(T)$ it would have been clear that this represented the effects of Debye's approximations. When, however, different patterns were found there was a tendency (very reasonably discussed in the Ruhemanns' book [4]) to dissect C into a Debye-like part and an anomalous extra for which some cause ought to be discovered. On the whole this rather arid exercise was kept within bounds, real and fictitious anomalies being clearly discerned, and perhaps the chief consequence of Debye's great success was that generations of students grew up in the belief that it was the last word on the subject. As we shall see later, it would be truer to regard it as almost the first and in itself something of a dead end, since it was the atomic model adumbrated by BvK that developed into the elaborate modern theories.

12.1.2. *Zero-point motion*

Between 1910 and 1912, as Kuhn [5] has described in detail, Planck turned again to the quantum theory of black-body radiation, with the intention of resolving a deep enigma—how could a material oscillator absorb energy only in discrete quanta? If the radiation was exceedingly weak, it would take a long time to excite an extra quantum of oscillation, and what would happen if the radiation were turned off half way through? We need not follow the steps by which he concluded, entirely contrary to Einstein's view, that the process of absorption was continuous while emission was discontinuously quantified. The end-product of his thoughts was that his original formula for the mean energy of an oscillator, $\hbar\omega/(e^{\hbar\omega/kT} - 1)$, must be replaced by $\frac{1}{2}\hbar\omega/(e^{\hbar\omega/kT} + 1)/(e^{\hbar\omega/kT} - 1)$. All he had done was to add $\frac{1}{2}\hbar\omega$ to the mean energy, but it implies that even at absolute zero an oscillator retains some energy. In due course Schrödinger's theory of the harmonic oscillator produced the same result automatically, and by that time there was enough experimental evidence to make the concept of zero-point energy generally acceptable. It appeared at first to be an elusive effect, having no influence on the specific heat unless, as Einstein and Stern remarked, the oscillator frequency changes with temperature; but convincing examples were lacking. Later Bennewitz and Simon [17] developed an argument that was interesting in itself and had even more interesting consequences. They noted that Trouton's empirical rule broke down in a systematic way with light elements. Trouton's rule states that for all liquids the latent heat of evaporation per gram-molecule bears a constant relation to the boiling point; i.e. the entropy change on evaporation is the same for all. This is an extension of van der Waals' law of corresponding states, and would be exactly true if the law of force between molecules took the same form for all, apart from scale factors defining the strength of the potential, ε, and its characteristic range, σ†, but with argon and hydrogen, and especially helium, the latent heat is markedly smaller than Trouton's rule predicts. If, however, the zero-point energy (which vanishes in the gas) is more or less the same, $\frac{9}{8}k\Theta_D$, in the liquid as in the solid, being particularly large for light atoms, it very satisfactorily accounts for the deficit in latent heat. A striking example of the failure of the law of corresponding states, because of zero-point energy, is provided by the heat of sublimation of solid hydrogen and deuterium. If ever two substances should confirm the law it is these, since the mass of the molecule should be irrelevant; yet sublimation of deuterium takes 50% more energy than hydrogen.

Further support for the existence of zero-point motion came from x-ray diffraction, but this had to wait until wave mechanics had been applied by Hartree to determine the electron density distribution in an atom. James and Brindley [18] were then able to compute absolutely the scattering amplitudes of the potassium and chlorine atoms, and hence the strength of various reflections from a crystal of KCl at liquid-air temperature. They found the measured amplitudes less than expected on the assumption that the atoms were at rest, but in agreement with the assumption of zero-point motion.

To return to Bennewitz and Simon: noting how large was the effect of zero-point motion in liquid helium, on account of the lightness of the atom and the weakness of the cohesive force, they suggested that it acted as an internal pressure, expanding the liquid to a much lower density than it would achieve otherwise—so much lower that the atoms could not hold a rigid structure. This would explain why, even at absolute zero, a pressure of 25 atmospheres was needed to reduce its volume to the point of solidification. Helium has a liquid–vapour critical temperature (5.2 K) but no triple point. The argument was generally accepted and when, in about 1948, the possibility arose of extracting enough of the light isotope, ^{3}He, from the atmosphere, its properties were the subject of lively, mainly informal, discussion. There were those who believed that the even greater zero-point motion would prevent condensation taking place at all, except under pressure, so that the liquid–vapour critical point would also be absent. In the middle of these speculations, de Boer [19] resolved the issue with a dramatic coup. He first restated the rationale of the law of corresponding states—if only two scale factors, ε and σ, are needed to specify the forces between molecules, the equation of state must be the same for all materials when written in terms of dimensionless *reduced* variables; $P^* = P\sigma^3/\varepsilon$, $V^* = V/N\sigma^3$ and $T^* = kT/\varepsilon$, N being

---

† *Another example is Lindemann's theory of fusion, according to which a solid melts when the amplitude of atomic oscillations reaches a certain fraction of the spacing between atoms. As Einstein and Kamerlingh Onnes both pointed out, this is a dimensional consequence, like the law of corresponding states, of the assumption of a universal law of force between molecules. It tells one nothing of the mechanism of melting.*

the number of molecules in the sample. As was well known, the molecular mass m cannot be incorporated in these expressions and is therefore irrelevant to the equation of state. This is no longer true when quantum effects enter, as de Boer pointed out, for a further dimensionless parameter may be constructed, $\Lambda^* = \hbar/\sigma\sqrt{m\varepsilon}$; apart from a numerical factor, Λ^* is the de Broglie wavelength of a molecule having energy ε, expressed in terms of the range σ of the force field. The reduced Debye temperature, $\Theta^* = \Theta_D/T_c$, and Λ^* involve the same combination of parameters, if zero-point motion plays no part. We therefore expect Θ^* to be a unique function of Λ^*, the two being proportional for heavy atoms. For each elementary gas, ε and σ can be found from its high-pressure behaviour, and de Boer was thus able to prove his point by the curve in figure 12.3(a).

In a second paper he and Lunbeck [20] plotted the reduced critical temperature T_c^* against Λ^* for a number of light elements, and again found a smooth curve (figure 12.3(b)). If ^{3}He differs from ^{4}He only in its mass, ε and σ being the same, Λ^* is greater in ^{3}He by a factor $\sqrt{(4/3)}$. On extrapolating their curve they deduced T_c^* for ^{3}He, and hence the actual critical temperature $T_c = 3.3 \pm 0.2$ K. In the event the limits of error were overcautious, since the measured T_c is 3.32 K.

This example is perhaps not strictly relevant to lattice dynamics, but it is so elegant an illustration of the power of dimensional analysis that it deserves to be remembered. We must now return to an earlier period.

12.1.3. Thermal expansion

Grüneisen [21], working at the Reichsanstalt, pointed out in 1908 that α, the linear thermal expansion coefficient of a solid, varies with temperature in the same way as its specific heat—$\alpha/C = $ constant for a given material. He illustrated the relationship with data from a number of metals over a wide range of temperature, and noted that a similar idea had occurred to Slotte some years earlier†. Soon afterwards he derived this result for solids governed by classical dynamics, making use of Clausius' virial theorem which had served van der Waals well in his theory of gases. Three years later [22] he provided a more straightforward and more general thermodynamical explanation, starting from Nernst's heat theorem. Because of it he could assume the entropy S to vanish at zero temperature, and therefore derive $S(T)$ as $\int C \, dT/T$. The volume expansion coefficient, 3α, follows by use of Maxwell's thermodynamics relation, $3\alpha V = (\partial V/\partial T)_P = -(\partial S/\partial P)_T$. It is necessary to know how the specific heat reacts to the volume changes due to P, and

† *Slotte had worked in Helsingfors since before 1893 on a kinetic theory of solids, using much-simplified models and elementary analysis to derive a picture of thermal expansion. Einstein was certainly aware of his work, since he wrote an abstract of one of Slotte's papers in 1904, three years before his own specific heat paper; but it is unlikely that Slotte was an effective stimulus to Einstein's thought.*

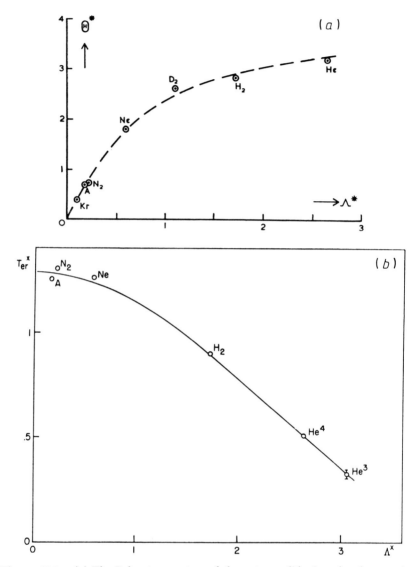

Figure 12.3. *(a) The Debye temperature of elementary solids, in reduced terms, is a smooth function of the reduced de Broglie wavelength. (b) The critical temperature is also a smooth function which can be extrapolated to estimate the critical temperature of* ³He.

Grüneisen made use of the Einstein and Nernst–Lindemann expressions, in both of which C is a function only of v/T, so that the sole variable involved is the response of the atomic frequency v to volume changes. A few lines of calculation sufficed to show that $\alpha/C = \frac{1}{3}\gamma K$, in which C is the specific heat per unit volume, K is the compressibility and γ

(Grüneisen's constant) is the relative change of atomic frequency with volume, $\gamma = -(V/v)\, dv/dV$.

Shortly after, Debye [23] tackled the same problem from the standpoint of his theory of specific heats. His approach was from first principles, and looks altogether more elaborate in starting from the statistical mechanical partition function, but the end result differs only from Grüneisen's in replacing the single atomic frequency v by its Debye equivalent, Θ_D.

Grüneisen noted that γ was approximately 2 for several materials (the law of corresponding states would have it the same for all). If the force law between molecules was strictly harmonic (force $\propto$ displacement) there would be no frequency change ($\gamma = 0$) on compression or expansion, and Grüneisen commented that the actual value of γ showed the repulsive force to rise very rapidly at short distances, as indeed van der Waals, and later Mie, had supposed. The vanishing of γ for a strictly harmonic lattice illustrates how anharmonicity and thermal expansion are intimately connected, but the importance of anharmonicity only became apparent when the problem of thermal conductivity in solids came under serious scrutiny.

12.1.4. *Thermal conductivity* [24]

It is obvious to the touch that metals conduct heat much better than non-metals at room temperature. The explanation of the Wiedemann–Franz law connecting electrical and thermal conductivity in metals was the one great success of the earliest electron theories of metals, as is discussed in Chapter 17. Here we concentrate on non-metals, especially crystals, many of which conduct so poorly that great care is needed, when measuring the conductivity, to avoid serious error from extraneous heat leaks. The once-famous Lees' disc apparatus was more successful than most, and not as crude as perhaps it appears to modern observers who, especially if they are skilled in vacuum and low-temperature techniques, may forget how difficult it was in 1900 to obtain a really hard vacuum. A good overview of the experimental position around 1920 is to be found in the *Dictionary of Applied Physics* [25], even if it has much to say about technical matters such as the heat-insulating properties of powders and fibres, and rather little about measurements down to liquid-air temperature and beyond, such as were pioneered by Eucken [26]. He found the thermal conductivity of a number of salts to vary inversely with temperature over a factor of 4.5 between 373 K and 83 K, and this regularity—about the only one known at the time—was in Debye's mind when he added a paragraph on thermal conductivity to his paper [23] on the expansion coefficient.

Up to this time little interest had been shown in the theory of thermal conduction in non-metals, perhaps through scarcity of information but also because Einstein's model (like earlier atomic models) provided no

See also p 1285

BOX 12A: INTERACTION OF SOUND WAVES

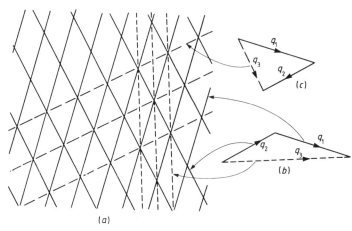

(a)

The diagram (*a*) represents two compressional waves moving from left to right and crossing in a solid. The full lines are wavefronts on which the bonds between atoms are stretched to a maximum. Halfway between these lines the atoms are pushed together. At the intersections the forces causing elongation are added; at the centre of each lozenge the compressive forces are added. Because it is harder to push two atoms together than to pull them apart by the same amount, the elongations exceed the compressions. As a result, along the two sets of broken lines the mean elongation is not zero, as it would be in a harmonic lattice, but positive; and greater than the mean elongation between the broken lines. These broken lines therefore represent a wave-like disturbance in the medium which might serve as a source for a new wave if an essential condition is satisfied: as the original waves progress, the new wavefronts progress with them, and if it happens that they move with the speed of sound waves they can excite a real progressive wave.

CONTINUED

mechanism for energy transport so long as each atom was conceived to vibrate independently. Einstein did indeed, as already noted, recognize that the energy of an oscillator would be rapidly dissipated, but he took the argument no further. Debye found himself in a quandary precisely opposite to Einstein's—so long as the acoustic waves were independent of each other, there was nothing in a perfect crystal to limit the conductivity. He seems (his full account [27] of the work is hard to

---

BOX 12A: *CONTINUED*

The vector diagrams on the right describe the situation in (*a*). The full lines, drawn normal to the wavefronts, represent the wavevectors of the original waves, while vector addition (*b*) and subtraction (*c*) give the wavevectors of the waves that might be produced by the interaction. The frequencies of the new wave motions are $\omega_1 + \omega_2$ in (*b*) and $\omega_1 - \omega_2$ in (*c*), where ω_1 and ω_2 are the frequencies of the original waves. As Pauli remarked, this is the analogue for waves of the production of combination tones when two musical notes interact in a non-linear medium (e.g. the ear). The process is possible if the new wavevector q_3 and frequency ω_3 are such that ω_3/q_3 is the velocity of a sound wave of frequency ω_3.

The condition for wave interaction can be expressed in a familiar form in terms of phonons, the particle-like quanta of wave motion which have momentum $\hbar q$ and energy $\hbar \omega$. When two phonons combine to give a single third phonon, their energy and momentum must be conserved—$\hbar \omega_3 = \hbar \omega_1 + \hbar \omega_2$ and $\hbar q_3 = \hbar q_1 + \hbar q_2$ in (*b*); in (*c*) a process of stimulated emission occurs, and plus signs are replaced by minus signs. There is, or course, no need to include $\hbar$ in these equations, which then express conservation of frequency and wavevector.

If the waves are non-dispersive, ω/q is constant, and two waves can only interact successfully if they are moving in the same direction. This does not alter the flow of energy and therefore does nothing to limit the transport of heat.

find) to have hoped at first that spontaneous density fluctuations would serve to scatter the waves; but since these fluctuations are themselves no other than thermally generated sound waves they cannot interact with other waves in a truly harmonic medium. He was forced to fall back on anharmonicity as the scattering mechanism (see Box 12A), unaware that this also is ineffective, as Pauli pointed out in 1925. We can jump straight to this moment in the story, since the intervening years are barren of interesting ideas on the subject.

Pauli's account [28] is only an extended abstract of a talk to the German Physical Society and leaves much to the imagination. According to his student Peierls [29] it is also unique among Pauli's works in that the analysis is wrong. What is not wrong is the leading idea that on an atomic lattice, as distinct from Debye's continuum, two waves may interact to produce a third. On Pauli's suggestion, Peierls studied this

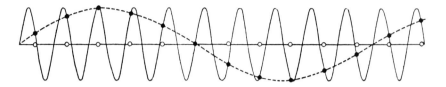

Figure 12.4. *The dark circles represent the displacement of the evenly spaced atoms (open circles) when a transverse wave passes. The full and broken lines are equally valid interpretations of the waveform responsible.*

See also
p 1304

problem and arrived at the concept of Umklapp processes. There is no need to repeat the discussion given in Chapter 17, but it should be remarked that it was in connection with the present topic, heat transport in non-metals, that the concept emerged. The essential point is that in a discrete atomic lattice there is no unique specification of wavevectors (figure 12.4); if $e^{iq\cdot r}$ describes the displacement of every atom, r being a typical position vector for an atom, then $e^{i(q+g)\cdot r}$ does equally well if g is any vector of the reciprocal lattice, having the property that $e^{ig\cdot r}$ is the same for every atom. A typical process that conserves wavevectors is expressed by $q_3 = q_1 + q_2 + g$, and this loosening of the conservation law allows waves, when they interact, to conserve energy but not flux of energy, so that a heat current can be dissipated. Momentum in the wave system is not conserved, and the balance is taken up by the lattice as a whole, in units of $\hbar g$.

Peierls' analysis is long and detailed, as befits the first substantial piece of work of a research student, but the principal results can be stated briefly, if anachronistically, in terms of phonons. At high temperatures $(T > \Theta_D)$, where Dulong and Petit's law holds, the mean free path of a given phonon, before it is destroyed in collision with another, is inversely proportional to the numbers of other phonons present, and therefore varies as $1/T$. Since the thermal conductivity κ is proportional to the lattice specific heat and to the phonon free path, and since C is constant, $\kappa \propto 1/T$ as found by Eucken and explained by Debye, though the latter's demonstration was wrong in principle. The Umklapp process is the responsible agent, but at lower temperatures it becomes less effective; when all phonons present have values of q significantly less than the smallest g, no Umklapp can satisfy the conservation conditions, and the mean free path rises steeply. In a large perfect crystal there would be nothing to limit the rise as T falls to zero, but Peierls noted that imperfections would ultimately take control by scattering the phonons; and if they did not, the free path would exceed the lateral dimensions of the sample, so that the phonons would ricochet from wall to wall in random directions. The free path would then, in a cylindrical rod, become equal to the diameter of the rod and κ would fall to zero as T^3, according to Debye's specific heat law.

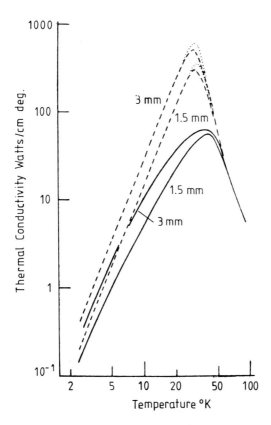

Figure 12.5. *Thermal conductivity of single-crystal sapphire rods (logarithmic scales). The abscissa is temperature and the ordinate conductivity in W cm⁻¹ deg⁻¹. As the surface of the rod is roughened the conductivity at the lower temperatures is reduced, since the phonons are scattered by the surface instead of being reflected specularly. Above 50 K scattering within the rod limits the phonon free path, and surface conditions do not matter.*

The first unequivocal verification of Peierls' predictions had to wait until 1951, when Berman began obtaining results such as that shown in figure 12.5. The conductivity at the peak is 40 times that of copper at room temperature. Further work has extended and refined both experiment and theory, without significantly affecting the foundations laid by Peierls. A review by Berman [24], and Ziman's monograph [30], bring the story up to 1960, by which time interest in the topic was declining. Ziman also covers the case of lattice conduction of heat in metals, where usually the electrons dominate, as well as very effectively limiting the phonon free path. This topic is of more interest for what it can reveal of electronic properties than of lattice dynamics, and will not be considered here.

The long free paths of phonons at low temperatures was

Figure 12.6. *Northrop and Wolfe's demonstration of the preferred directions (light image) for ballistic motion of phonons through a germanium crystal.*

demonstrated directly by von Gutfeld and Nethercot [31] in 1964. They evaporated thin films on either face of a synthetic sapphire (Al_2O_3) crystal, one to serve as heater and the other as thermometer. A very short injected heat pulse arrived at the thermometer as two well-resolved pulses, with such time delays as identified them as longitudinal and transverse waves which had travelled ballistically without intervening collisions. On account of the elastic anisotropy a wave-group does not in general travel along the direction of the normal to its wavefront, and in certain directions there is a strong concentration of radiated power. Northrop and Wolfe [32] exhibited this in a striking way by attaching a small heat sensor to one face of a germanium crystal, and scanning a focused laser beam across the opposite face in a television raster. The response of the sensor (figure 12.6) shows immediately the directions of channelling of power, the details of which can be confirmed as consistent with the known elastic constants of germanium. In itself, the experiment tells one nothing about lattice dynamics, since it can be fully explained by continuum elastic theory, but the use of ballistic phonons as a research tool has proved valuable in a variety of other contexts.

12.1.5. Lattice theories of specific heat
The huge triumph of Debye's specific heat theory, and the difficulty of handling three-dimensional lattices, caused the BvK approach to the problem almost to disappear from view. Not entirely, however, and it can now be seen providing the basis of all later developments. As such, it deserves more than the passing mention which is all we have so far accorded it. To appreciate its origins we may start with the systematic studies of the far-infrared spectrum that were pursued by Paschen, Rubens and others [33] at the Reichsanstalt as a consequence of their interest in black-body radiation.

It was found that the dispersion (variation of refractive index with wavelength) of transparent crystals could be explained by the presence of strong absorption lines in both the ultraviolet and the infrared. If such an absorption follows the standard resonance curve, the crystal will reflect very strongly—almost like a metal—wavelengths near the absorption maximum. Rubens and Nichols mounted an experiment in which heat radiation was reflected many (~6) times from crystal faces and then analysed with a reflection-grating spectrograph. They found that only narrow lines (reststrahlen or residual rays) survived the process and determined, for example, that quartz had three absorption lines at 8.50, 9.02 and 20.74 μm. In the course of time many crystals were studied, some showing reststrahlen as far out as 100 μm. As already remarked, diamond is transparent right into the far infrared.

The explanation given by Drude [34] was that, unlike the case in metals whose optical properties he rightly ascribed to itinerant electrons, the electrons in insulators are attached to atoms and give rise to two characteristic types of vibration: at high ultraviolet frequencies the electrons vibrate with respect to their own atom, while vibrations of whole atoms relative to one another take place at lower, infrared, frequencies. Reststrahlen bands therefore reveal the frequencies of the latter type of vibration—an idea taken over by Einstein in his specific heat theory. Two years later Madelung [35] greatly refined this idea, and his work lies at the root of the BvK approach. He recognized that a typical vibration involved not one atom alone (as in Einstein's theory) but the whole crystal, and that there is a natural highest frequency characterized by neighbouring atoms vibrating in antiphase. Madelung was not concerned with specific heats, but with finding a connection between the elastic constants of a crystal and its reststrahlen frequency. He assumed a model, which he thought might suit NaCl, in which the atoms are arranged in a simple cubic lattice, alternately Na and Cl, charged oppositely like the ions in salt solution, according to the ionic hypothesis that by this time was generally accepted. The ions are imagined as connected to all their surrounding neighbours by springs whose strength can be determined from the elastic constants. From the spring constants and the atomic masses the frequency of any chosen vibrational mode can in principle be found. The mode chosen by Madelung as responsible for the infrared absorption was that in which all Na ions on planes normal to a cube diagonal vibrate in phase with one another, but in antiphase with the Cl ions on planes sandwiched between them; he found good agreement between calculated and measured frequencies. In view of the fact that his model is exactly that which the Braggs determined six years later by x-ray diffraction, one might have expected them to mention Madelung, but it is likely enough that they were unaware of his work. They say there is strong evidence for the model they (and Madelung) choose, but are probably referring to the extensive studies of the packing

See also
p 428

983

of spheres by Barlow and Pope. What seems to be new in Madelung's approach is the assumption that the units are ions, not atoms. These conceptions, dating from 1906 to 1909, deserve to be remembered by text-book writers who imply that crystal chemistry started with von Laue and the Braggs.

To progress beyond Madelung's chosen vibrational mode to the general case of a standing wave of arbitrary wavevector q was the essential core of the BvK approach. The difficulties were that too little was known about the interatomic forces, and that the numerical work needed for detailed calculations was too demanding until the brute force of computers could be brought to bear on the problem. A one-dimensional chain of atoms coupled by harmonic elastic springs can be analytically solved, and BvK give the solution, presumably not knowing, as Debye did, that Rayleigh had already discussed the problem in detail. Instead of the angular frequency, ω, of a normal mode being related to its wavenumber q by a constant velocity u, $\omega = uq$, they found $\omega = \omega_0 \sin(aq/2) = \omega_0 \sin(\pi a/\lambda)$, where a is the spacing of the atoms and ω_0 depends on the mass and spring constant; ω_0 is the highest vibrational frequency, attained when $\lambda = 2a$, and alternate atoms vibrate in antiphase. Over the band of wavenumbers q around π/a, the frequency varies only slightly from ω_0, and since the wavenumbers of the modes are evenly spaced in q, there are a large number of normal modes with frequencies close to ω_0. This feature differs from the Debye approximation, and gives rise to the temperature dependence of the Debye temperature deduced from the specific heat, as discussed above.

BvK simplified the three-dimensional problem by assuming that the lattice is isotropic with the same dispersion curve (dependence of ω upon q) for the normal modes in every direction of q. It was then straightforward to obtain the energy of the thermal vibrations, and hence the specific heat. As Blackman comments in a useful review of the early work [36], BvK's expression is more complicated than Debye's, but fits the experimental results no better. At nearly the same time, 1914, Born [37] applied the BvK theory to the specific case of diamond, whose crystal structure had been only recently determined by the Braggs. The theory required a knowledge of the interatomic force constants in diamond, and to obtain these Born developed the idea that the long-wavelength modes had to be consistent with the macroscopic elastic waves which could propagate in the material. As a result of this Born showed that the number of elastic constants obtained from BvK theory of non-Bravais lattices, with forces which depended on the angles between the atomic bonds, agreed with macroscopic theory and not with the predictions of Cauchy's oversimplified Bravais model. These results were then in agreement with the known results for the elastic constants of diamond.

After this the development of the BvK approach rested, apart from some minor improvements, until Born and his group began work again

Max Born

(German, 1892–1970)

Max Born was born in Breslau in December 1892. His father was an embryologist who held a Chair at the University. He went to school in Breslau and then to university where he attended lectures on physics, chemistry, zoology, philosophy, logic and mathematics. He moved to Göttingen and completed his doctorate in 1907 where he studied and worked with Carathéodory, Courant, Hilbert, Minkowski, Schwarzschild and Voigt.

After his marriage in 1913 to Hedwig Ehrenberg, he moved to a Chair in Berlin in 1914 and then, in 1919, exchanged Chairs with Max von Laue at Frankfurt. In 1921 he became Director of the Physical Institute in Göttingen in succession to Debye where his assistants included Pauli, Heisenberg and Jordan, and the Institute was in the forefront of the development of quantum theory. In 1933 Hitler came to power and Max Born was deprived of his Chair and he accepted an invitation to move to Cambridge.

In 1936 he was appointed to the Tait Chair of Natural Philosophy in Edinburgh where he established a research school and was elected a Fellow of the Royal Society. He retired in 1953 and returned to Bad Pyrmont in Germany and became active in the cause of the social responsibility of scientists. He was awarded the Nobel Prize in 1954 for his work on quantum theory.

Throughout his life he was a keen and good musician and he died in Göttingen in 1970 leaving his widow, one son and two daughters.

in the thirties. For von Kármán it had been a brief intermission from his life's work in aerodynamics, while Born, reminiscing in 1965, admitted that he lost interest in the field after his work with von Kármán, but that it was a valuable source of problems for research students.

Blackman was one of these students and he tackled the problem

See also p 851

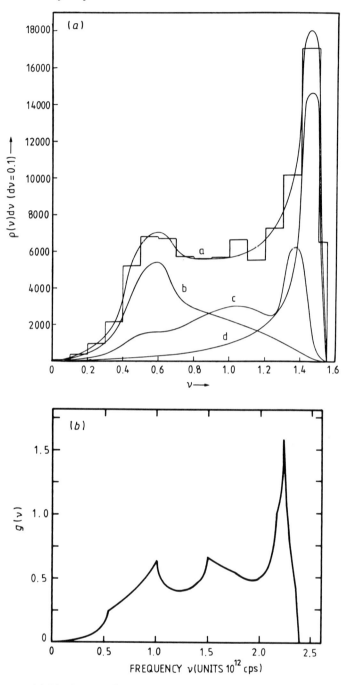

Figure 12.7. *(a) The density of states for a simple cubic lattice with nearest-neighbour interactions. Curve 'a' gives the total density of states* [36]. *(b) The frequency distribution function for potassium at 9 K calculated from 60 466 176 normal modes in the Brillouin zone* [38].

of a simple cubic solid with identical atoms connected by elastic springs between the neighbouring atoms along the cube edges and face diagonals. The model was not very realistic, but computing the specific heat [36] with even this simple model was a formidable problem when only hand-operated calculators were available. The formalism of BvK immediately reveals where the hard work arises: the dynamical equations for the displacements of the atoms are, for each wavevector q in three dimensions, a set of three coupled equations. The normal mode frequencies are the eigenvalues of the 3×3 matrix, or equivalently the equation for the frequencies of these modes is cubic and gives three independent normal modes, which in symmetry directions are a longitudinal and two transverse modes. The calculation of the specific heat requires repeating the procedure (which involves numerical solution of the cubic equations) for a large number of different wavevectors throughout the Brillouin zone. In figure 12.7(a) we show the frequency distribution calculated by Blackman with 8000 points in the Brillouin zone, and it is compared with a more recent calculation, figure 12.7(b), which included interatomic forces up to fifth-nearest neighbours and many more points in the Brillouin zone. The calculation of Θ_D from Blackman's density of states then gave qualitatively the features observed in many solids—a minimum of Θ_D at intermediate temperatures which reflects the peak in the density of states from the transverse acoustic modes at the Brillouin zone boundary.

Because of the difficulty of making these calculations, considerable effort was then directed by Montroll [39] and others towards developing analytic methods for a variety of simple two- and three-dimensional atomic models. The results have now largely been superseded by detailed computer calculations, apart from the work of van Hove [40] on the nature of the singularities in the density of states, arising from the maxima, minima and saddle points of the dispersion curves. The singularities are mostly ones in which the derivative of the density of states with respect to frequency is discontinuous at the frequency of the singularity as illustrated in figure 12.7(b).

After his move to Edinburgh in 1936, Born's group continued its work on the theory of lattice dynamics. Of particular note is the work of Kellermann [41] on the alkali halide, NaCl. Kellermann used Madelung's model for alkali halides, of charged ions interacting with electrostatic forces and a short-range repulsive interaction, and computed the vibrational frequencies for 48 different wavevectors in the Brillouin zone. The new difficulty which was overcome in performing this calculation was the evaluation of the electrostatic forces. The displacement of the ions gives rise to electrostatic dipoles which interact with other dipoles, through a dipole–dipole interaction which decreases with distance as $1/r^3$. The summation over all the dipole–dipole interactions is of very long range, and indeed if wavevector q is zero is only conditionally convergent and dependent on the macroscopic shape of the crystal. These problems were solved by Kellermann by the use of the Ewald transformation which converts the long-range dipole–dipole sum into two rapidly convergent sums. The same procedure is still used today with computers, but is much more easily implemented than it was for Kellermann working with a hand calculator.

The dispersion curves for various high-symmetry directions are shown in figure 12.8. The results show longitudinal and transverse modes of both an acoustic and optical nature. The lower-frequency acoustic modes have the Na and Cl ions moving largely in phase, and in the higher-frequency optic modes they are largely out of phase. Of particular interest as the wavevector, q, tends to zero is that the limiting frequency of the transverse and longitudinal optic modes is different. This is a direct consequence of the long-range forces being only conditionally convergent at long wavelengths, and is characteristic of all infrared-active optic modes in ionic crystals.

Kellermann's calculations and similar calculations by Iona [42] for KCl gave good agreement with specific heat measurements. Later Huang [43] extended the work of Kellermann, using Maxwell's electromagnetic equations instead of Coulomb's electrostatic law to calculate the forces between the dipoles. The differences in the results are only important where the wavevector is so small (10^{-4} of the Brillouin zone dimension) that the optic modes and light waves have similar frequencies. For these wavevectors the light waves, photons, interact with the lattice dynamical waves, phonons, to produce coupled waves as illustrated in figure 12.9. Later these results were less elegantly rediscovered, and named *polaritons*, and later again Raman scattering measurements were made which showed the correctness of this theory.

The development of lattice dynamics up until 1954 is very completely summarized in the monograph by Born and Huang [45], which still for many remains the definitive account of the formal theory. In addition to the topics discussed above, the work of Born and his students had laid down the framework of the theory for the optical properties— infrared absorption and Raman scattering—and for the elastic and thermal properties in terms of the interatomic forces. The difficulties at this time were that although the formal theory had been developed, very little was known about the interatomic forces, and the calculations needed to give quantitative results to compare with measurements of, say, the specific heat were excessively lengthy and tedious. Consequently the subject as a whole was somewhat arid and did not attract a great deal of attention. It needed more precise experimental data against which to compare the theory, and more information about the interatomic forces. One approach, to refine the measurements of the specific heat and to deduce from them more information about the frequency distribution, proved unrewarding—thermal expansion and anharmonic effects cause

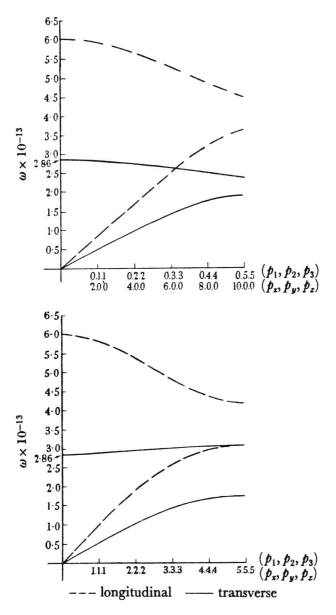

Figure 12.8. *The dispersion curves calculated by Kellermann [41] for NaCl. The wavevector $p = 10\,aq/2\pi$, and the figure shows the dispersion curves in the (100) and (111) directions.*

small changes in the specific heat, which are difficult to estimate and can cause substantial errors in the resulting frequency distributions. Clearly the information required was a direct measurement of the frequencies ω for particular wavevectors, q, and preferably the dispersion curves $\omega(q)$

989

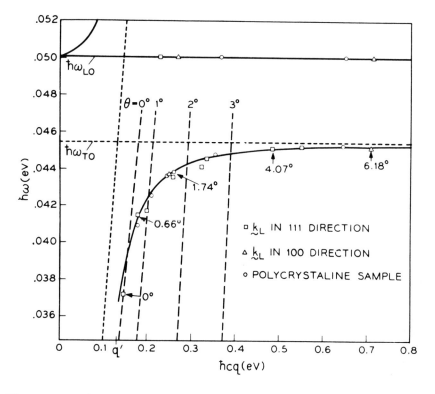

Figure 12.9. *The measured dispersion curves for polaritons, the coupled longitudinal optic and photon excitations. The measurements were made by Raman scattering at small angles and the material was GaP [44].*

as illustrated in figure 12.8.

The first direct measurements of the dispersion curves were made using x-ray scattering techniques. The long and involved history of this topic is recounted at least in part in Chapter 6, and in more detail in a review by Lonsdale [46]. X-ray scattering from a crystal can arise not only from the Bragg reflections, but also when phonons are produced or absorbed. The wavevector transfer in the scattering, Q, is then $q + g$ where g is a reciprocal lattice vector (Bragg reflection) and q is the phonon wavevector. This gives rise to diffuse haloes around each Bragg reflection, due to scattering from the acoustic modes, which were investigated by Laval [47] and more quantitatively by Curien [48]. The detailed theory shows that the intensity of the scattering by the phonon is inversely proportional to the square of the frequency, and dependent on the pattern of the displacements in the normal mode. By making careful quantitative measurements, Walker [49] managed to deduce the dispersion curves for the phonons in aluminium, as shown in figure 12.11. The method is, however, very difficult to apply generally because corrections are needed for Compton scattering of the x-rays and for multiple phonon scattering. As a result the methods cannot give such accurate measurements as one in which the frequencies are directly measured rather than inferred from intensity measurements.

It was at the end of the 1950s that the technique of neutron scattering became available which enabled detailed measurements of dispersion curves to be made, and these have had a great influence on subsequent developments. At about the same time light scattering methods became far more efficient because of the development of intense monochromatic laser sources, and these two techniques have greatly increased the amount and usefulness of the experimental information available about phonons and interatomic forces.

12.2. New experimental techniques

12.2.1. *Neutron scattering*

The neutron was discovered in 1932, and soon afterwards it was shown that neutrons were diffracted by crystalline materials. Before progress could be made, however, a more intense source was needed and this was provided by the nuclear reactors built during and after World War II. For these reactors to operate, they needed neutrons that were thermalized in a moderator, and these thermal or moderated neutrons then had a Maxwell–Boltzmann distribution of energies centred around the temperature of the moderator multiplied by Boltzmann's constant. The wavelength of the de Broglie waves for neutrons from a room-temperature moderator is 0.18 nm, and very similar to the wavelength of x-rays and the interatomic spacing in crystals. Diffraction of neutrons by single crystals is then similar to that of x-rays but, as discussed in Chapter 6, neutrons have some advantages over x-rays in that the scattering is related to the nuclear properties rather than the atomic number and so they can readily be used for light atoms and to distinguish atoms with similar atomic numbers. The neutron, having a magnetic moment, can also be used to measure magnetic structures, and is now the accepted tool for the determination of the details of these structures.

See also p 453

After the development of the first nuclear reactors, experimenters at the Oak Ridge Laboratory began using neutrons to study structures. Nearly monochromatic (i.e. mono-energetic) beams of neutrons were obtained by Bragg reflecting the beams from the reactor with a large single crystal, and these monochromatic beams were then used together with conventional crystallographic techniques to determine a wide variety of crystallographic and magnetic structures. The neutron had, however, more to offer solid-state physics than the determination of structures. The energy of these thermal neutrons is very similar to the quantized energy associated with a normal mode of vibration, $\hbar\omega(q)$. When a neutron is scattered by a single phonon, with energy transfer

991

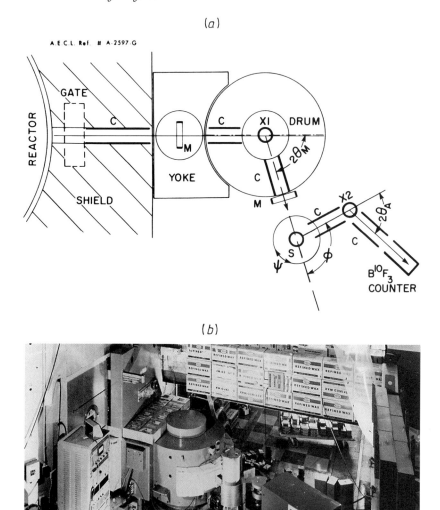

Figure 12.10. *(a) A schematic diagram of the original triple-axis crystal spectrometer built by Brockhouse* [50]. *(b) A photograph of the spectrometer showing the monochromator drum, the specimen cryostat and the analysing spectrometer.*

$\hbar\omega(q)$, and wavevector transfer $Q = g + q$, both $\omega(q)$ and Q are in principle measurable because the changes are large in comparison with

Bertram N Brockhouse

(Canadian, b 1918)

Bertram Brockhouse was born in 1918 and served in the Canadian Navy from 1939 to 1945. He obtained his BA from the University of British Columbia in 1947 and then moved to the University of Toronto to study for his PhD.

He married Doris Miller in 1948 and in 1949 became a lecturer at the University of Toronto until he completed his PhD in 1950. After his PhD he joined Atomic Energy of Canada Ltd as a research officer and worked initially with Donald Hurst on neutron diffraction.

Apart from an extended visit to Brookhaven National Laboratory when the NRX reactor at Chalk River was out of use, he remained at Chalk River until 1962. Most of his important experiments were performed between 1956 and 1962 and he was the Branch Head of the Neutron Physics Branch from 1960 to 1962. In 1962 he became Professor of Physics at the McMaster University from which post he retired in 1984. He has been awarded many honours for his work including the Nobel Prize in 1994, Fellowship of the Royal Society in 1965 and the Buckley Prize of the American Institute of Physics in 1962.

He is a Canadian patriot and has four sons and one daughter.

the original energy and wavevector of the neutron. This is different from the situation with x-rays where Q is easily measured, but the change in energy, one part in 10^5, is very small.

In the 1940s the theory of thermal neutron scattering by materials was of importance in the design of the moderators for reactors, and little of the early work was directed to basic solid-state physics. By 1950 it was realized in several laboratories that neutron scattering, in principle, provided a tool for measuring phonon dispersion relations, as was explicitly spelt out by Placzek and van Hove [51] in 1954. This possibility attracted several groups, and the most successful experiments were performed at Atomic Energy of Canada by Brockhouse. He realized that the experiments would be difficult and the intensities very low, but had the advantage of the reactor at Chalk River which had the highest flux of any in the world for this type of experiment. He therefore began developing the necessary techniques to measure the energy of the scattered neutrons. He improved the neutron detectors and the shielding around the instruments and, possibly most importantly, the size and reflectivity of the monochromating crystals. These developments enabled him to build the first triple-axis spectrometer (figure 12.10), in which the scattered energy is measured by a second monochromator crystal. The first experiments [52] used a fixed incident neutron energy, and enabled measurements to be made of the phonon dispersion curves in aluminium (figure 12.11). These were the first direct observations of phonons in metals, but the triple-axis crystal spectrometer became a much more useful tool when it was realized that it could be controlled to measure the frequencies at certain predetermined wavevectors, the so-called constant-Q technique. This was implemented in 1958, and in 1960 the spectrometer was redesigned so that it was controlled by computer-generated punched paper tape. A large number of very important measurements were made with this spectrometer, and it is still used in much the same form as when it was developed, though with further improvements in the efficiency of the monochromators and in the direct computer control which makes it far more adaptable and convenient.

In the 1950s most other laboratories chose a different approach to measure the energy transferred by the neutrons. A chopper was used to produce a pulsed monochromatic beam and the time-of-flight between chopper and detector measured to determine the scattered energy. This type of instrument has proved to be not nearly so powerful for single crystals as the triple-axis crystal spectrometer, but for isotropic systems (liquids and amorphous materials) it has proved invaluable, especially with modern computers. The first successful experiment performed with the time-of-flight technique was the measurement of the Landau phonon–roton dispersion relation in superfluid liquid helium by groups at Atomic Energy of Canada, Stockholm and Los Alamos in 1957, following the suggestion by Cohen and Feynman [53].

See also
p 931

The first neutron scattering experiments were all performed on reactors built largely for the production of isotopes and material testing. The importance and success of the early experiments led to a rapid increase in the number of groups performing these experiments and to the construction of reactors at Brookhaven and Grenoble solely for the purpose of providing intense beams for neutron scattering. The reactor at Grenoble has hot and cold sources to adjust the temperature of the moderated neutrons, and makes extensive use of guide tubes to transport the neutrons away from the reactor. The instruments have become more complex and refined, but the basic principles have remained similar to those of the early experiments.

The neutron flux of these reactors is only 5–10 times that of the reactor

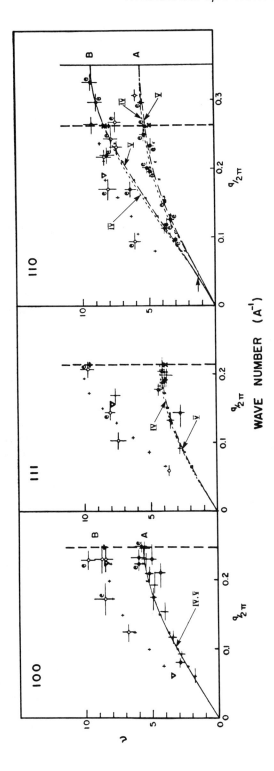

Figure 12.11. *The phonon dispersion curves of aluminium along the three principal symmetry directions; ν is in THz, the lines are various models, the + and × are from x-ray measurements [49] and the ● and ○ from neutron measurements [52].*

used by Brockhouse. The increase in flux is small because it is limited by heat transfer problems in the reactor. As a result new sources of neutrons using proton accelerators and the spallation process are being developed at the Rutherford Appleton Laboratory and other laboratories around the world. Meanwhile the range of problems in condensed-matter science to which neutrons are making a crucial contribution has spread from physics into chemistry, biology and materials science.

12.2.2. *Raman scattering*

See also
p 1397

Raman scattering, the inelastic scattering of light by phonons, was first observed in solids by Landsberg and Mandelstam [54] in 1928, at about the same time as Raman and Krishnan [55] observed it in liquids and gases. Landsberg and Mandelstam observed that sharp spectral lines developed sharp satellites, resulting from what was later interpreted as scattering from single optical phonons of long wavelength; in contrast to this Fermi and Rasetti [56] observed in 1931 a continuous spectrum which arises from the simultaneous scattering by two phonons. The Raman effect is very weak, and because the frequency changes in the scattering are typically about 10^{-3} of the incident frequency, the incident light must be highly monochromatic and the scattered light detected by a high-resolution spectrometer. These were difficult conditions when a white-light mercury lamp was all that was available for a source. As a result few experimental groups continued with the study of Raman scattering. The field was completely changed in the early 1960s by the development of laser sources to provide intense monochromatic and highly collimated beams of light, and the technique was rapidly taken up in many laboratories. The high resolution, and relative ease, of performing experiments has made Raman scattering widely available, and a powerful tool for studying the frequency and lifetime of phonons.

The theory of the first-order Raman effect was first discussed classically by Mandelstam in 1930 [57], and in the same year quantum mechanically by Tamm. After Placzek's 1934 review [58], there was little progress until Born and Bradburn [59] discussed the two-phonon spectrum of alkali halides in 1947, and Smith [60] did comparable calculations for diamond. The theory is summarized very clearly in the text by Born and Huang [45]. Full understanding of the Raman effect involves on the one hand a knowledge of the frequencies and lifetimes of the phonon modes which couple to the light, and on the other hand the nature of the coupling between the light and the phonons. The former problem is the basic problem of lattice dynamics discussed in other parts of this chapter. There have been several valiant attempts to develop a theory of the coupling mechanism by Placzek and Born, expanding the polarizability in terms of the displacements of the atoms, and by Loudon [61] using third-order perturbation theory to describe the virtual electronic transitions. Nevertheless these are difficult to apply

996

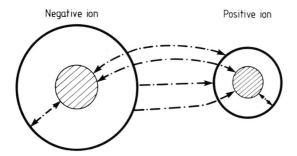

Negative ion   Positive ion

Figure 12.12. *The shell model for ions in crystals (of an alkali halide), showing the different short-range forces and shell–core coupling.*

quantitatively, and this limits the application of an otherwise valuable technique.

12.3. The development of lattice dynamics

12.3.1. Ionic crystals

Born's theory of ionic crystals is unsatisfactory because it neglects the polarizability of the ions, which is why the optical refractive index of crystals is different from that of free space—ions become polarized in the electric field of the light. The early phases of this story have been described in Chapter 6, and we take it up again with the proposal by Dick and Overhauser [62] of a shell model of the ion (figure 12.12). The outer electrons are described by a massless shell which can be displaced from the core to produce a polarization dipole. The shells interact with other cores and shells both by short-range overlap forces and by electrostatic forces, and are attached by a spring to their own cores. Woods, Cochran and Brockhouse [63] extended this model to explain the recently measured dispersion curves for NaI and KBr, and obtained very reasonable agreement (figure 12.13). There was no doubt that the shell model was successful in capturing much of the essential physics, and since then it, and various equivalent models, have been widely used to describe the interatomic forces of ionic crystals.

Despite this success, a careful comparison of the results of the shell model with the experimental results showed that there were still difficulties. In particular the frequencies of some modes could only be correctly obtained when some of the parameters had unphysical values, such as giving the shell of electrons a positive charge! This difficulty was traced to the neglect of another type of electronic distortion. Schröder [64] pointed out that it was necessary to include distortions of the ions in the form of an isotropic radial breathing motion of the electronic distributions. This does not produce a charged dipole and is

997

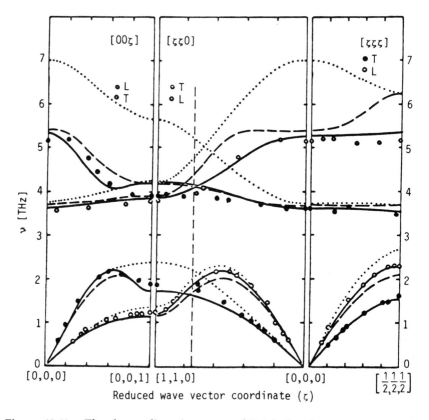

Figure 12.13. *The phonon dispersion curves of NaI [63]. The points are measured frequencies, the dotted lines are a rigid-ion model, the dashed lines a shell model and the solid lines a breathing-shell model.*

not produced by long-range dipole forces, but solely by the short-range forces.

Clearly one must ask whether it is necessary to introduce yet further distortions of the ions. Fischer *et al* [65] have suggested that quadrupolar distortions of the Ag ion are needed to describe the forces in AgCl. In most ionic materials, however, the simple mechanical shell model does provide a very satisfactory account of the interatomic forces with short-range overlap forces only between a few near neighbours. It may seem surprising that a complex electronic system can be so well described by such a model, but the electronic distortions of the spherical ions are presumably related to the nature of the excited electronic states, and for closed shells these have usually a character no more complicated than s or p waves [66].

12.3.2. *Metals*

Initially the existence of well-defined phonons was less certain in metals than in ionic crystals, because of the possibility that the free electrons might scatter the phonons; or, to express the problem differently, it was unclear that the energy of the crystal could be expressed in terms of a potential function depending only on the atomic coordinates. This is the adiabatic approximation introduced by Born and Oppenheimer [67], but they were unable to show that it was a good approximation for metals. This did not prevent solid-state physicists from taking the existence of phonons for granted, and eventually the situation was clarified by the experiments of Brockhouse and Stewart [52], who showed that they really did exist in aluminium, and by the theoretical work of Migdal [68].

There followed more detailed measurements of the phonon dispersion curves of metals, particularly sodium and lead. The dispersion curves of sodium [69] could be explained by a BvK model including forces (classical springs) between near neighbours. In detail, forces up to fifth neighbours were needed to give agreement with the experimental data. This model showed that the forces must arise through the conduction electrons, as a direct overlap between the atomic cores could not extend as far as the fifth-nearest neighbours; the model was also unsatisfactory in that it involved 13 parameters. At about the same time as these measurements were made, new light was thrown on the coupling between the conduction electrons and the atomic cores through the development of the pseudopotential method [70], which enabled the interatomic forces to be calculated in terms of the electron–phonon interaction. Movement of one atom changes the conduction electron distribution, and thus leads to a force on another atom. This theory of the origin of the interatomic forces goes back at least as far as Bardeen's important paper [71] in 1937, but was first applied in detail by Toya [72] in 1957 to calculate the dispersion curves in sodium, although he did not explicitly use pseudopotential theory. Successful application by Cochran, to sodium, of the formulation developed particularly by Sham [73] demonstrated the power of the pseudopotential approach.

See also p 1353

Sodium is a particularly simple metal, with a near-spherical Fermi surface, and the potential of the ion cores has little effect on the conduction electrons. Lead is a more complex metal with four conduction electrons per atom and with a stronger electron–phonon interaction, for which Brockhouse and his collaborators [74] found that interatomic forces of much longer range were required to explain the measured dispersion curves. This is a consequence of an effect pointed out by Kohn [75] in 1959. For wavevectors q of magnitude greater than the diameter of the Fermi surface, $2k_F$, the screening differs from that with q less than $2k_F$ in that only in the latter case is it possible for phonons to excite electron–hole pairs with both the electron and the hole at the Fermi surface. At the critical value, $q = 2k_F$, the dispersion curve is predicted to show a kink, such as is shown in figure 12.14. In free-electron metals like sodium the electron–phonon coupling is too weak to allow a Kohn

999

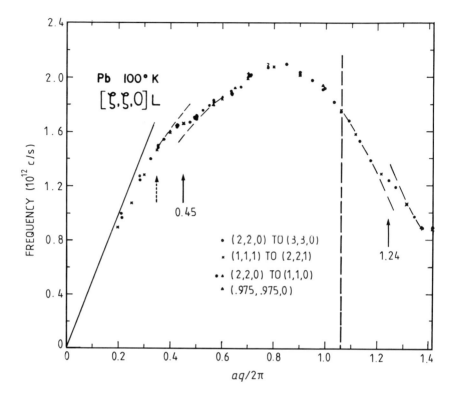

Figure 12.14. *The phonon dispersion curve of lead [74] along the (110) direction. Note the abrupt anomalies for $aq/2\pi = 0.45$ and 1.24.*

anomaly to be observed. As well as the electron–phonon coupling, the strength of the anomaly depends on the density of states at the Fermi surface, which becomes particularly large if the material is quasi-one-dimensional, owing its metallic behaviour to linear chains of conducting material. As first predicted by Peierls [76] in 1928, the lattice is then unstable against a distortion which opens up a gap in the electronic energies at the Fermi surface; in terms of the Kohn anomaly there is an infinite singularity in the dispersion curves at $q = 2k_F$. Although the precise behaviour of this strongly interacting system is still unclear, the phonon dispersion curve [77] of the one-dimensional metal KCP, whose chemical formula is given in figure 12.15, provides good evidence for a very strong Kohn anomaly, such as might be expected in a quasi-one-dimensional material, and which gives rise to a distorted crystal structure.

Since 1960, a large body of experimental information has been built up about the phonon dispersion curves in most simple metals and ordered alloy structures. Theoretically, however, progress has been less satisfactory in that pseudopotential theory has not been extended with

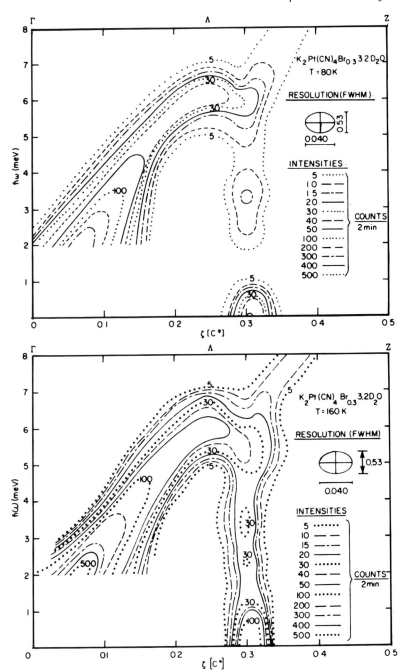

Figure 12.15. *The scattering observed [77] for wavevectors along the chain direction c, q = ζc* in KCP at 80 K and 160 K. The column of scattering for ζ = 0.3 arises from the Kohn anomaly associated with the nearly one-dimensional Fermi surface and at 80 K there is a Peierls instability and elastic Bragg peak.*

any great success to the more complex metals. For these the problem of calculating the screening by the conduction electrons is far from being solved, and there are further calculational difficulties in including the real band structure and Fermi surface.

12.3.3. Semiconductors

The nature of the interatomic forces and phonon dispersion curves in semiconductors is an important question, because they are intermediate between ionic crystals and metals. The initial calculations of H M J Smith used nearest-neighbour forces and the BvK theory to calculate the dispersion curves but, as with ionic crystals and metals, better experiments showed the limitations of this approach. After Brockhouse and Iyengar [78] measured the dispersion curves of germanium in 1959, Herman [79] showed that if one was going to use BvK theory long-range forces, at least up to fifth-nearest neighbours, would be needed.

Mashkevich and Tolpȳgo [80] suggested that induced dipolar forces were important in covalent crystals, and Cochran [81] used the shell model to describe germanium. A nearest- and next-nearest-neighbour shell model gave a surprisingly satisfactory account of the experimental results, but this model is not as satisfactory for semiconductors as it is for ionic crystals. Firstly, the detailed fits to the dispersion relation are less good unless a large number of parameters are introduced, and secondly the covalent binding in semiconductors is directional and it is unclear how this can be modelled by spherical shells. As a result a large number of different models have been introduced, with charges placed at the centre of the bonds between the atoms to describe the bonding electrons. With sufficient adjustable parameters all these models provide a description of some of the experimental results, but none are completely convincing [82].

An alternative approach is to calculate the interatomic forces or phonon dispersion curves from a full microscopic model. Hohenberg, Kohn and Sham [83] used their local density functional formulation to calculate the frequencies of a few phonons by comparing the energy of distorted and undistorted crystals. This procedure has given reasonable agreement for a few phonons in simple semiconductors but is not easily generalized to give the full phonon dispersion curves.

12.3.4. Anharmonic effects

The interatomic potentials in crystals are not harmonic but also include non-linear (anharmonic) terms which, as well as being needed to explain the thermal conductivity and thermal expansion, are also responsible for relatively small corrections ($\sim$10%) to the specific heat, the frequencies of the normal modes and the elastic constants. Since the effects are relatively small, many can be treated by perturbation theory, as was used by Born and Blackman [84] to account for the temperature dependence

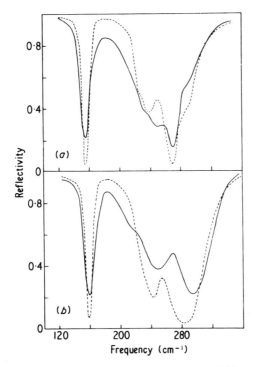

Figure 12.16. *The infrared reflectivity [86] of 2.5 μm NaCl films observed at (a) 45°
and (b) 65°. Note the complex shape of the observed peaks (full line) and theory (dotted)
due to the frequency dependence of the optic mode self-energy.*

of the linewidths of the normal modes observed in the far-infrared
spectrum of ionic crystals. The development of many-body techniques
in the late 1950s led to a more elegant recasting of the theory, and its
extension to systems where anharmonic effects are large. This has proved
particularly fruitful in the theory of the infrared spectrum. The model
developed originally by Madelung for alkali halide crystals predicts
an uncomplicated resonant absorption curve for each normal mode
which can be excited by the infrared, but detailed measurements show
considerable fine structure in the wings of the main resonance. The
many-body theory describes this structure as, in some sense, an artefact
of the method of observation. Since the lifetime [85] of the infrared mode
is determined by its decay into pairs of other phonons, it is dependent
on the density of two-phonon states, and this varies with the frequency
which is used to study the normal mode. In an anharmonic crystal it
is not then possible to specify the frequency and lifetime of a mode,
without also specifying the frequency used to study the mode. This fine
structure was first calculated in 1963 [85], and the results obtained by
Hisano *et al* [86] in 1973 are shown in figure 12.16.

The method of long waves developed by Born and his collaborators was based on the realization that the long-wavelength acoustic modes must be equivalent to the waves obtained from microscopic elasticity theory. In an anharmonic crystal the situation is more complex. At very long wavelengths the period of the wave is much greater than the lifetime of most of the phonons in the material; there are many phonon collisions during each period of the wave and the material is at every point in a state of thermodynamic equilibrium as defined by the local temperature and strain. In contrast, at higher frequencies there is insufficient time for all the phonons to reach thermal equilibrium in each period of the wave, and the acoustic modes propagate in a collisionless mode. In the former case the waves are known as *first sound*, and can be described by thermodynamics in a similar way to the propagation of sound in a gas. In the collisionless regime the waves are known as *zero sound*, and though heavily overdamped in most liquids and gases, they are well defined in a crystal and can be studied by neutron scattering. The frequencies thus measured yield different elastic constants from those determined at low frequencies by, for example, the velocity of ultrasonic waves [87].

Anharmonic effects are particularly large in quantum crystals such as solid helium. The zero-point motion is large, and expands the interatomic distances well beyond the position of the minimum in the interatomic potential; so far, indeed, that the second derivative of the interatomic potential is negative, and according to conventional theory the crystal is unstable. The difficulty was overcome by Koehler and Nosanow [88] who used the self-consistent renormalized harmonic approximation, first suggested by Born [89] and by Horton, in which the second derivatives of the potential are not calculated at the interatomic distance, but are averaged over the distances sampled as the atoms move. The average is then positive, so the crystal is after all stable, and calculated normal mode frequencies are in reasonable accord with the measurements. After further improvement, by including the residual anharmonic effects, the results give very good agreement with experiment for the frequencies, lifetimes and the details of the measured neutron scattering [90].

12.3.5. *Phonons in crystals containing defects and in glasses*

The changes in the normal modes of vibration of systems caused by the introduction of defects was studied by Rayleigh, who proved a number of important theorems. The subject then lay almost dormant until the work of I M Lifshitz, which began in 1943, and of Montroll and his collaborators in 1955 [91]. Much of this early work was theoretical, and concerned with the mathematical properties of very idealized, usually one-dimensional, models. The basic changes in the phonon spectrum, when a few defects are introduced, are the development of either localized or resonant modes associated with the defect. Localized modes occur when the characteristic vibrational frequency of the defect is different from that at which any normal mode can propagate through the host crystal; this tends to occur when the defect has a smaller mass than the host. The motions in these modes are then confined to the immediate neighbourhood of the defect. If, however, the frequency of the defect lies within a band of normal mode frequencies of the host crystal, as when a massive atom is introduced, the dispersion curves are perturbed around the frequency of the defect in a characteristic resonant way. The defects also break the translational symmetry of the crystal so both effects may be observed with optical techniques. Localized modes were first observed with infrared techniques by Schaefer [92] who studied colour centres in alkali halides, and resonant perturbations caused by Ag in NaCl and KCl. Subsequently there have been a large number of very detailed optical experiments, and detailed calculations using the shell models described above [93]. The theory has also been tested for metals by using neutron scattering techniques; localized modes were observed when defects (e.g. chromium) were introduced into a tungsten crystal, and resonant perturbation was seen with gold in copper [94].

A different type of defect, but one for which the theory is not too different from that of isolated defects, is the crystal surface. The first treatment of waves on the surface of an isotropic continuous medium was by Rayleigh in 1885, and his work and that of Love in 1911 form the basis of our understanding of seismic and other surface waves. Solid-state physics is, however, concerned with processes and effects at the atomic level, and the extension of surface wave theory to atomic lattices was initiated by Lifshitz and co-workers from 1948 onwards [91]. They used similar techniques to those they had developed for isolated defects, and the results are qualitatively similar. If the surface is perpendicular to z, and periodic in x and y, a surface mode is characterized by a wavevector of only two components, q_x and q_y. Perpendicular to the surface, the amplitude of a surface mode falls off rapidly with distance into the bulk of the crystal. The surface modes are therefore described by dispersion curves in q_x and q_y, and there are a series of branches, depending on whether the modes are polarized in the plane or perpendicular to it, and whether they have optical or acoustic character. In the same way as for isolated defects there are localized or resonant modes depending on whether the mode frequency is outside, or within, the band of frequencies of the host, so now the behaviour of the surface mode with z depends on whether the frequency is within the band of host frequencies with the same q_x and q_y.

The surface modes can influence many of the properties of crystals, particularly if the crystal size is very small. Montroll [95] in 1950 discussed the specific heat of small crystals, and predicted that the surface modes would give rise to a term proportional to the surface area and to T^2. Detailed measurements of the dispersion curves for surface waves had, however, to await the development of suitable experimental techniques. A probe is needed which can transfer the appropriate energy and wavevector, yet interacts only with the surface. A suitable probe is helium atom scattering and the experiment is very similar to that of neutron scattering, except that the source is a beam of helium atoms. The first successful experiments performed with this type of equipment were by Toennies *et al* [96] in 1983.

The developments just described led to an understanding of the properties of crystals containing a few defects, although there are still problems associated with the determination of the force constants coupling defects to the lattice and the atoms at surfaces. The formal theory of isolated defects is not easily applied when the number of defects becomes large in mixed crystals. One can start by treating the mixed crystals with the *coherent potential approximation* [97] in which the phonons are treated as being scattered by defects arising from the deviations away from the average lattice. The theory has had some success but fails to explain how a sufficient concentration of defects can lead the localized modes to cooperate in forming a band of propagating phonons. Experimentally, optical measurements on mixed ionic and semiconducting crystals have shown two types of behaviour. In 1928 Kruger *et al* [98] showed that the frequency of the transverse optic mode in a typical mixed alkali halide, $KCl_x Br_{1-x}$ varied continuously from that of KCl to that of KBr. In contrast, in semiconductors such as $InP_x As_{1-x}$, there are two frequencies of optical absorption, one close to that of InP and the other to that of InAs [99]. There is no general theory that encompasses both types of behaviour.

One of the unexpected features of very disordered systems came to light in 1971 when Zeller and Pohl [100] measured the specific heat and thermal conductivity of glasses. At low temperatures (<1 K) they found that the specific heats of vitreous silica, vitreous GeO_2 and selenium were proportional to temperature and much larger than for crystalline materials, while the low-temperature thermal conductivities were proportional to T^2. Other properties of glasses such as the velocity of sound, ultrasonic attenuation and dielectric response have also been found to differ in several respects from those observed in crystalline materials, and the anomalies were later found in almost all non-crystalline materials. An explanation for these results was put forward independently in 1972 by Anderson *et al* [101], and by Phillips [102], who suggested that in glasses there are two-level systems corresponding to parts of the glass structure flipping between two different configurations. The density of states of these levels was further assumed to be independent of energy to explain the specific heat measurements. The theory was further developed by Black and Halperin [103] in 1977, to account for the ultrasonic properties, but there is still no satisfactory understanding of the microscopic nature of these two-level systems.

On a large scale, disordered solids behave like a homogeneous three-dimensional material, in the sense that the number of particles is proportional to the cube of the linear dimension of the sample. More generally, $N \propto l^d$, where l is the size of the system and d is the dimensionality. However, on length scales less than some characteristic L, some disordered systems have $N \propto l^D$ where D is the fractal dimension, less than d. Examples of fractal systems are percolating clusters, polymers, rubbers and gels. The phonon density of states, $g(\omega)$, for modes with a wavelength greater than L has the form $g(\omega) \propto \omega^{d-1}$, but for modes with wavelengths less than L, $g(\omega) \propto \omega^{\bar{d}-1}$, where $\bar{d}$ is known as the fracton dimensionality; the modes in this frequency range were called *fractons* by Alexander and Orbach [104] in 1982, who also suggested that $\bar{d} = 4/3$. Boukenter *et al* [105] measured the Raman scattering from silica gel and found that over a limited but considerable frequency range, the scattering varied with a power of the frequency transfer very different from that observed in non-fractal glasses, such as As_2S_3. They deduced that $\bar{d} = 1.27 \pm 0.16$ consistent with Alexander and Orbach's conjecture.

See also
p 574

12.4. Structural phase transitions

It is very common for the crystal structure of a solid to change as the temperature or pressure is altered, and the methods of x-ray or neutron diffraction enable us to determine what has changed in the arrangement of the atoms, as described in Chapter 6. In a first-order transition there is a latent heat, and the two phases are distinct and may coexist in the same way as water and ice can coexist. More commonly, however, the transition is continuous or nearly continuous, and there is no question of distinct coexisting phases. Indeed, in the neighbourhood of the transition there are small-scale fluctuations of one phase in the other. A full theory of the phase transition must take account of these fluctuations, though they were not included in the very important phenomenological theory put forward by Landau [106] in 1937 in a general form, and rediscovered and applied to the ferroelectric transition in $BaTiO_3$ by Devonshire [107] in 1949. The theory makes use of symmetry to write down the free energy of the crystal in terms of the displacement of the atoms from their positions in the high-temperature phase, and assumes the simplest possible temperature dependence. Close to the transition temperature the theory is equivalent to the molecular field theory of magnetism which also neglects fluctuations, and despite its shortcomings has been very successful in describing a wide range of different phenomena. The Bragg–Williams theory of the order–disorder transitions in alloys was a similar theory, and much thought was put into ways of including the short-range order or fluctuations, as described in Chapter 7.

See also
p 549

The understanding of structural phase transitions developed differently, and the next step was the soft-mode theory. Although earlier

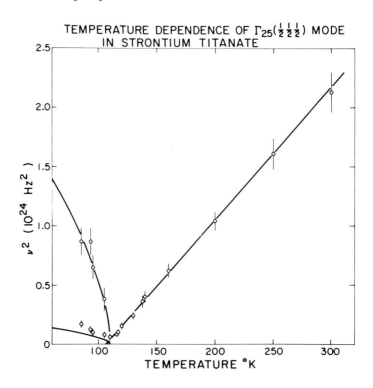

Figure 12.17. *The frequencies of the soft phonon modes with $q = (2\pi/a)$ $(\frac{1}{2}, \frac{1}{2}, \frac{1}{2})$ in SrTiO$_3$ [109]. Below the phase transition at 110 K, the distorted structure splits the degeneracy of the mode into two.*

proposed by Raman and by Anderson, it was the work of Cochran [108] in 1960 which led to direct tests of the suggestion that the phase transition resulted from a particular type of instability of the crystal. As the transition is approached the frequency of one of the normal vibrational modes decreases as $(T - T_c)^{1/2}$; at T_c the restoring force associated with this mode of distortion vanishes, and the crystal spontaneously distorts to a new structure. Infrared measurements and neutron scattering measurements showed evidence of a strongly temperature-dependent mode in SrTiO$_3$; although it does not in fact undergo a ferroelectric transition, the measurements lent considerable support to the soft-mode hypothesis. This was further strengthened by the later discovery of the antiferrodistortive transition in SrTiO$_3$ and detailed studies of the associated soft modes by Raman and neutron scattering techniques [109], as shown in figure 12.17.

These measurements showed that at least some structural phase transitions were associated with a soft mode, but there was still no explanation for why this particular normal mode was so strongly temperature dependent. It was shown that the theory of anharmonic crystals could qualitatively account for the temperature dependence assumed for the mode and for the Landau expansion of the free energy, provided that the self-consistent phonon theory was used to remove inconsistencies associated with unstable modes. From detailed calculations of the anharmonic effects it appeared that they could account in a consistent manner for a large number of different properties of $SrTiO_3$, but subsequently it has been questioned whether the detailed model used for the anharmonic interactions is correct. The difficulty lies in our lack of knowledge about the anharmonic part of the interatomic potentials [110].

While this work was progressing, soft modes were being found in many different crystals, but studies of the statistical mechanics of other phase transitions were progressing differently. The distortions occurring below the transition vary, according to Landau theory, as $(T_c - T)^{1/2}$, whereas exact solutions of the two-dimensional Ising model and other phase transitions give $(T_c - T)^\beta$, with an exponent β much less than 0.5.

The discrepancy springs from Landau's neglect of fluctuations. The first experimental evidence of their importance for structural phase transitions came from a detailed study of the order–disorder transition in β-brass by Als-Nielsen and Dietrich [111] who showed that the exponents were clearly different from those given by Landau theory. Thus fluctuations must play a crucial role at least for order–disorder transitions, in which slow diffusion is the mechanism by which phase transformation is achieved. The situation is very different in structural phase transitions like that in $SrTiO_3$, but here also the Landau theory fails in matters of detail, as was found by Mueller and Berlinger [112] who showed that the rotation of the oxygen octahedra below T_c was proportional to $(T_c - T)^{1/3}$. This made clear that critical fluctuations were important at structural phase transitions involving soft modes, as well as those involving the much slower diffusive motions in alloys.

The development of renormalization group techniques in the early 1970s clarified the role of critical fluctuations at phase transitions, and the concepts were applied to the diversity of structural phase transitions. In particular, the results showed that transitions involving elastic distortions as a primary order parameter were well described by Landau theory; the critical fluctuations did not cause changes in the behaviour. Uniaxial ferroelectrics were also described by Landau theory, albeit with logarithmic corrections to the temperature dependencies. We have reached the point where further development would involve the detailed description of particular phase transitions and technical arguments. Instead of following this road we refer the reader to some recent reviews [110].

The modern theory of continuous phase transitions depends on the concept of scaling, which implies that the fluctuations are controlled by a single length scale which diverges as the transition is approached.

Similarly the dynamics of the fluctuations is determined by a single time-scale which also diverges as the transition is approached. Structural phase transitions have provided some evidence that these concepts may need modification. The first evidence came from the experiments performed by Riste *et al* [113] in 1971 which showed that the dynamic response of $SrTiO_3$ above T_c consisted of two components; a quasi-elastic component and the oscillating soft mode. Similar results have now been found for many other transitions. These unexpected results do show that there are two time-scales involved in the fluctuations and there have been many attempts to explain these effects. Some of the explanations [110] concentrate on developing more complex treatments of the critical fluctuations, but these theories have not, as yet, reproduced the large difference between the time-scales. Halperin and Varma [114] suggested that the short time-scale arose from the phonons, while the long time-scale arose from interactions of the phonons with defects. The difficulty is that the appropriate defects have not been identified.

More recently, high-resolution x-ray measurements have identified two length scales [115] for the critical fluctuations. The experimental results are also inconsistent with existing theories of phase transitions, and furthermore the role, if any, of defects has not been clarified. These two results, showing that there are more than one time-scale and length scale for the critical fluctuations, pose severe problems for the understanding of structural phase transitions in terms of the conventional scaling theory of phase transitions. Further work is needed to understand these effects, and to decide whether the results show that defects nearly always play a crucial role at structural phase transitions, or whether the scaling theory of phase transitions is inadequate.

12.5. Spin waves

The spin waves play a similar role for magnetic materials to the one that the normal modes of vibration play for crystal structures: they describe the deviations of the atomic magnetic moments away from the ideal magnetic ordering. They were first introduced for ferromagnetic materials by Bloch [116] who calculated the quantized energy of the excitations as a function of the wavevector, q. He showed that if the magnetic interactions were of Heisenberg form

$$H = -J \sum S_i \cdot S_j$$

then the energies were given for small wavevectors by

$$\hbar\omega(q) = Dq^2$$

where the constant D depends on the exchange constant, J, and the crystal structure.

Since the spin waves describe spin deviations, the thermal excitation of spin waves decreases the ordered magnetic moment and Bloch showed that, for bulk ferromagnetic materials at low temperatures, the magnetization varied with temperature as

$$M(T) = M(0) - CT^{3/2}$$

where C is a constant which depends on J. This result has now been accurately tested experimentally.

A similar analysis to that of Debye for the specific heat shows that the magnetic contribution to the specific heat is proportional to $T^{3/2}$. The difference between the T^3 law of Debye and the $T^{3/2}$ arises from the linear dispersion curve of the normal modes and the quadratic dispersion curve of the spin waves. The difficulty of separating the magnetic $T^{3/2}$ term from the electronic T and normal mode T^3 term seems to have inhibited clear tests of this prediction.

Further theoretical work then concentrated on obtaining higher-order corrections to the magnetization by considering corrections to the low-k behaviour of the dispersion relation, and spin-wave interactions. Dyson [117] in 1956 concluded that linear spin-wave theory with non-interacting spin waves is good enough for all practical purposes.

At long wavelengths the long-range magnetic dipole interactions become important and modify the dispersion curves in a way that depends on the geometry of the sample. The results are essentially similar to those described above for the dielectric properties of crystals but, in practice, were worked out again without awareness of the dielectric results [118]. The theories gave good results in comparison with ferromagnetic resonance measurements.

The spin waves in antiferromagnetic structures were not studied theoretically until the work of Kubo *et al* [119] was done in the early 1950s. The difficulty is that the simple antiferromagnetic structure is not an eigenstate of the Hamiltonian and so spin waves might destroy the long-range order. They showed that, at least in three dimensions, this was not the case and that the antiferromagnetic spin waves had the dispersion curve $\omega = cq$, as for the normal modes of vibration. The spin waves therefore make a similar contribution to the specific heat.

As with the development of the understanding of the normal modes of vibration, further progress was inhibited by the lack of any experimental technique to study the spin waves in detail throughout the Brillouin zone. The solution to this was provided by the development of neutron scattering and Raman scattering techniques. The neutron has a magnetic moment and this interacts with the magnetic fields in a magnetic material giving rise to scattering of the neutron. For what appear to be completely fortuitous reasons, the strength of the scattering from a typical magnetic system is very similar to that of the nuclear scattering. This fortunately enables both types of scattering to be determined. Because the magnetic dipolar interaction is more complex than the Fermi pseudopotential describing the neutron–nucleus interaction, the form of the scattering cross-section is more complex, but the basic concepts of conservation of energy and crystal momentum are the same for spin waves as for normal modes of vibration.

Brockhouse [120] was the first to use neutron scattering to determine the spin-wave energies in the ferrimagnet magnetite and the ferromagnet cobalt [121]. He identified the scattering as arising from spin waves by applying a magnetic field in different directions and ensuring that the cross-section varied as predicted for the magnetic scattering. These measurements showed that short-wavelength spin waves do exist and that exchange constants could be determined experimentally. Shortly after Brockhouse's work, measurements were made on a number of different materials and one of the most complete was on the simple antiferromagnet, MnF_2, for which the dispersion curves are shown in figure 12.18 [122]. These results showed that spin waves did occur in antiferromagnets, and had the linear dispersion curve at small wavevectors as predicted by the theory, apart from the effects of anisotropy.

Since these first measurements, the spin waves have now been measured in a large number of different materials enabling the exchange constants to be determined in many materials. Measurements on simple insulators have shown the essential correctness of the theory of super-exchange, while the longer-range interactions found in rare-earth metals have confirmed the correctness of the indirect-exchange-coupling theory developed by Ruderman *et al* [123].

A different type of development has been the theory and experimental study of systems in which the electronic/magnetic energy levels are strongly influenced by the local crystalline environment, spin–orbit coupling and orbital magnetic moments. Much of our understanding of these systems comes from spin resonance experiments on the dilute salts. This work has now been extended to concentrated systems, the detailed study of the spin-wave/exciton dispersion relations and the interaction between the excitons on different atoms.

The nature of the spin waves in itinerant metals is also clarified. The theory developed by Stoner [124] suggested that in a magnetic metal the different spin bands had different energies resulting in a different filling of these bands and hence a spontaneous magnetization. Herring and Kittel [125] then showed that there were two types of excitation in these metals: collective spin waves with $\omega \propto Dq^2$, and single electron–hole excitations corresponding to transitions of the electrons from one band to the other. The details depend on the band structure but both types of excitation have been observed in magnetic metals such as the weak ferromagnetic MnSi.

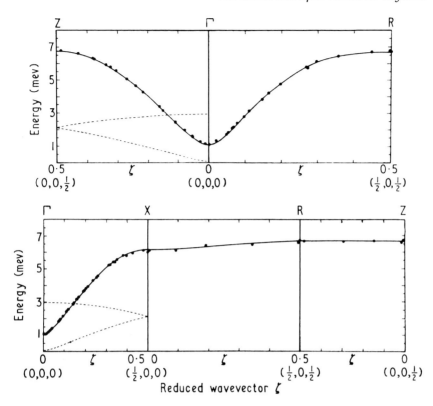

Figure 12.18. *The spin-wave dispersion curves in the antiferromagnet* MnF$_2$ [122]. *The solid line is a fit with dipolar forces and nearest- and next-nearest-neighbour exchange interactions.*

Raman scattering [126] has also proved to be a useful probe of spin waves, particularly in insulators. The one-magnon scattering gives information about the zone centre magnon frequency. The two-magnon scattering gives information about the two-spin-wave density of states. Since, however, the two spin waves are created on neighbouring sites, the effects of spin-wave—spin-wave interactions give rise to a marked change in the peak and shape of the spectrum.

The effect of the interactions between spin waves is particularly strong in materials that have magnetic chains, or other one-dimensional arrangements of the spins. These interactions are strong for small values of the spin for which the quantum fluctuations are large. In one-dimensional systems long-range order can be destroyed by the presence of a single spin-wave excitation. Since, for temperature $T > 0$, there is always an excitation somewhere on an infinite chain, no one-dimensional systems can have long-range magnetic order except when $T = 0$. In addition, Bethe [127] showed in 1931 that the quantum fluctuations destroy the long-range order of the $s = 1/2$ Heisenberg antiferromagnetic chain even at $T = 0$. The effect of these large fluctuations on the excitations has been studied now for several systems. For large spin, $s = 5/2$, the experiments [128] on tetramethyl manganese ammonium chloride (TMMC) showed fairly well-defined spin waves at low temperatures. The behaviour of the low-spin systems is, however, more complicated because of the quantum effects.

When a neutron, for example, creates a spin excitation at a particular site in a chain, the magnetic bonds must be broken on both sides of the site. These broken bonds are similar to domain walls and, in the case of Heisenberg systems, are known as spinons, and are necessarily created in pairs in a neutron scattering experiment. The excitations in the $s = 1/2$ Heisenberg antiferromagnet were studied by des Cloizeaux and Pearson [129] in 1962; they found the lowest-energy excitation for each wavevector, q. For many years this was interpreted as the energy of a spin-wave dispersion curve but, after computer simulations, further theoretical work and finally experiments [130], it is now appreciated that well-defined spin waves do not occur, despite their calculation in numerous tutorial and examination questions. The excitations are a continuum spectrum resulting from the excitation of pairs of spinons with wavevectors q_1 and $q - q_1$ which, for $s = 1/2$, can propagate largely independently.

The results for $s = 1$ Heisenberg chains are different. Haldane [131] showed that in this case the spinons are bound and so a finite energy is needed to break up this binding. The excitation spectrum then consists of a gap rather than the linear spin-wave excitation predicted by spin-wave theory for Heisenberg systems. This gap has now been observed [132]. Clearly the ability to produce materials containing magnetic chains has enabled the detailed study of quantum fluctuations in low-dimensional systems to be undertaken.

12.6. Magnetic phase transitions

The neutron scattering experiments described above have also enabled detailed measurements to be made of the fluctuations occurring close to magnetic phase transitions. This work has been recently reviewed [133], and so will not be described in full detail. The important feature is that the wavevector transfers and frequency transfers available with neutron scattering techniques enable the spatial extent and time dependence of the critical fluctuations to be studied. The simplest theories of phase transitions, the Landau theory described in section 12.4 or the well-known molecular field theory of magnetism, fail because they neglect the effect of the fluctuations. Most of the recent work on phase transitions has been in developing ways of including the effects of the fluctuations and, since neutron scattering enables measurements to be made of the fluctuations, they have played a crucial role in the development of our understanding of phase transitions.

Table 12.1. *Exponents of d = 3 Ising systems.*

System	β	ν	γ	Reference
β-brass	0.305(5)	0.65(2)	1.25(2)	[111]
MnF$_2$		0.634(40)	1.27(4)	[136]
CoF$_2$	0.305(30)	0.61(2)	1.21(6)	[137]
Theory	0.312	0.64	1.25	
Mean field	0.5	0.5	1.0	

Table 12.2. *Exponents of d = 3 Heisenberg systems.*

System	β	ν	γ	Reference
RbMnF$_3$	0.32(2)	0.701(11)	1.366(24)	[138]
Theory	0.38	0.702	1.375	
Mean field	0.5	0.5	1.0	

Continuous phase transitions, such as the transition from paramagnet to ferromagnet, are characterized by a change of symmetry and the development of an order parameter (ferromagnetic moment) in the low-symmetry phase. If the order parameter is the magnetization, M, then theory predicts that

$$M = M_0(T_c - T)^\beta$$

where M_0 is a constant and β is known as a critical exponent. The critical fluctuations are described by a length scale, ξ, and ξ is described by another exponent ν such that

$$\xi = \xi_0(T - T_c)^{-\nu}$$

and the susceptibility by χ where

$$\chi = \chi_0(T - T_c)^{-\gamma}.$$

Mean field theory or Landau theory give the same values for the exponents for every type of transition: $\beta = 1/2$, $\nu = 1/2$ and $\gamma = 1$. Exact solutions of some simple models, series approximations and experiments showed by the early 1970s that these values were often incorrect, and that better theories were needed. The improved theories made use of the scaling hypothesis and the renormalization group as described in Chapter 7. The scaling hypothesis states that the critical fluctuations are similar at all temperatures close to T_c provided that all lengths are scaled by the correlation length, ξ. Renormalization group theory provided an understanding of scaling theory and why phase transitions show universal behaviour in the sense that the exponents and critical properties are the same for systems having the same symmetry, dimensionality and range of forces.

See also p 562

Scattering experiments directly measure the temperature dependence of the order parameter, while the critical scattering, particularly above T_c, gives the susceptibility from the intensity and the correlation range from the wavevector spread of the scattering. The experiments then yield information about the exponents as well as giving detailed information about the form of the critical scattering.

By now many measurements have been performed and so only a brief review can be given of the results. The first measurement of the critical fluctuations was by Latham and Cassels [134] in 1952, who found that the total cross-section of iron increased at T_c and correctly interpreted this as arising from scattering by the magnetic fluctuations, as is well known from the critical opalescence of light at the critical point of materials. A major step forward [135] was the development of the magnetic correlation theory by van Hove with the use of mean field theory. In our view the first critical experiment that clearly showed the failure of mean field theory and demonstrated the power of neutron scattering techniques was the study of the order–disorder transition in β-brass [111] by Als-Nielsen and Dietrich in 1967. They measured the three critical exponents β, γ and ν and showed that they all differed from mean field theory and were in agreement with the exponents calculated from series expansions.

Soon after this result was obtained, similar experiments were performed on three-dimensional antiferromagnets and particularly careful measurements were made on MnF_2 [136] and CoF_2 [137]. These are anisotropic antiferromagnets and so belong to the $d = 3$ Ising universality class like β-brass: in table 12.1 we compare the measured exponents. The satisfactory agreement between the experiments gave good experimental support for universality. Another very careful experiment was performed [138] on a cubic antiferromagnet, $RbMnF_3$, which is an example from the isotropic Heisenberg $d = 3$ universality class. The exponents for this material are listed in table 12.2. There is reasonable agreement between theory and experiment and both give results that are different from those for the Ising system shown in table 12.1.

One of the most important developments was the growth of materials in which the magnetic ions are in sheets or chains, because they enabled two- and one-dimensional systems to be studied. Some aspects of the one-dimensional systems have been described in section 12.5 and so, in this section, we discuss only two-dimensional systems. Very important experiments were performed in 1973 by Samuelson [139] and by Ikeda and Hirakawa [140] on K_2CoF_4. In this material the square $Co–F_2$ planes are separated from one another by two K–F planes. The magnetic interactions are then strong in the planes but much weaker, by a factor of about 10^5, between the planes. Because of the crystal field effects on the Co ions, the effective interactions between the Co ions are anisotropic and

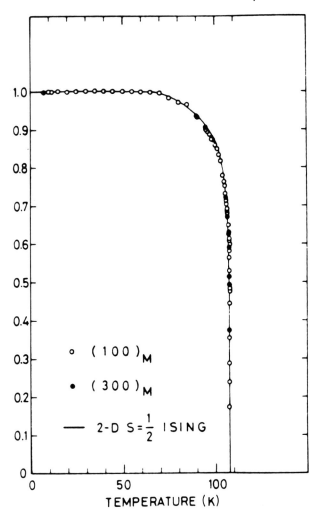

Figure 12.19. *The intensity (square of sublattice magnetization) in K_2CoF_4 which is a model two-dimensional system. The solid line shows Onsager's exact solution* [140].

See also p 554

Table 12.3. *Exponents of $d = 2$ Ising systems.*

System	β	ν	γ	Reference
K_2CoF_4	0.128(4)	0.99(3)	1.77(6)	[140]
Theory	0.125	1.0	1.75	[141]

the spins align antiferromagnetically, perpendicular to the c axis. They are therefore close to model systems for studying the two-dimensional Ising model.

This model is nearly unique in statistical physics because it was solved exactly, by Onsager [141] in 1944. In figure 12.19 we show a comparison of the measured order parameter with Onsager's solution and the agreement is excellent. In table 12.3, we list the measured values for the exponents as well as Onsager's results; the agreement is clearly very good and the values are very different from those of mean field theory.

Since these experiments were performed, the subject has widened to study phase transitions in many different areas. The reviews [133] give details of the way in which the experiments have shown that defects can change the universality class, and that percolation is dominated by the one-dimensional weak links in the backbone of the percolating cluster. Phase diagrams have been measured, and the way disorder and competing interactions in spin glasses and random field problems lead to non-ergodic behaviour and other unexpected features has been studied.

References

[1]   Love A E H 1927 *The Mathematical Theory of Elasticity* 4th edn (Cambridge: Cambridge University Press)
[2]   Behn U 1898 *Ann. Phys., Lpz.* **66** 237
[3]   Eucken A 1909 *Phys. Z.* **10** 586
[4]   Ruhemann M and Ruhemann B 1937 *Low Temperature Physics, part 2* (Cambridge: Cambridge University Press) ch 2
[5]   Kuhn T S 1978 *Black-body Theory and the Quantum Discontinuity 1894–1912* (Oxford : Clarendon)
[6]   Einstein A 1907 *Ann. Phys., Lpz.* **22** 180
[7]   Boltzmann L 1964 *Lectures on Gas Theory* (Engl. Transl. S G Brush) (Berkeley, CA: University of California) p 330
[8]   Einstein A 1907 *Ann. Phys., Lpz.* **22** 800
[9]   Nernst W and Lindemann F 1911 *Z. Elektrochim.* **17** 817
[10] Nernst W 1911 *Ann. Phys., Lpz.* **36** 395
[11] Einstein A 1911 *Ann. Phys., Lpz.* **35** 679
[12] Debye P 1912 *Ann. Phys., Lpz.* **39** 789
[13] Born M and von Kármán T 1912 *Phys. Z.* **13** 297
[14] Lord Rayleigh 1900 *Phil. Mag.* **49** 539
[15] Weyl H 1911 *Math. Ann.* **71** 441
[16] Eucken A and Schwers F 1913 *Verh. Deutsch. Phys. Ges.* **15** 578
[17] Bennewitz K and Simon F 1923 *Z. Phys.* **16** 183
[18] James R W and Brindley G W 1928 *Proc. R. Soc.* A **121** 155
[19] de Boer J 1948 *Physica* **14** 139
[20] de Boer J and Lunbeck R J 1948 *Physica* **14** 510
[21] Grüneisen E 1908 *Ann. Phys., Lpz.* **26** 211
[22] Grüneisen E 1911 *Ber. Deutsch. Phys. Ges.* **13** 426, 491
[23] Debye P 1913 *Phys. Z.* **14** 259
[24] Berman R 1953 *Adv. Phys.* **2** 103
[25] Glazebrook R (ed) 1922 *A Dictionary of Applied Physics 1* (London: Macmillan) p 428
[26] Eucken A 1911 *Ann. Phys., Lpz.* **34** 185

[27] Debye P 1914 *Vorträge über die Kinetische Theorie der Materie und die Elektrizität* (Leipzig: Teubner)
[28] Pauli W 1925 *Verh. Deutsch. Phys. Ges.* **6** 10
[29] Peierls R E 1980 *Proc. R. Soc.* A **371** 28
[30] Ziman J M 1960 *Electrons and Phonons* (Oxford: Clarendon)
[31] von Gutfeld R J and Nethercot A H 1964 *Phys. Rev. Lett.* **12** 641
[32] Northrop G A and Wolfe J P 1979 *Phys. Rev. Lett.* **43** 1424
[33] Paschen F 1895 *Ann. Phys., Lpz.* **54** 668
Rubens H and Nichols E F 1897 *Ann. Phys., Lpz.* **60** 418
[34] Drude P 1904 *Ann. Phys., Lpz.* **14** 677
[35] Madelung E 1909 *Ges. Wiss. Göttingen Nach. Math.-Phys. Klasse* **1** 100
[36] Blackman M 1935 *Proc. R. Soc.* A **148** 365
[37] Born M 1914 *Ann. Phys., Lpz.* **44** 605
[38] Cowley R A, Woods A D B and Dolling G 1966 *Phys. Rev.* **150** 487
[39] Montroll E W 1947 *J. Chem. Phys.* **15** 575
[40] van Hove L 1953 *Phys. Rev.* **89** 1189
[41] Kellermann E W 1940 *Phil. Trans. R. Soc.* **238** 513; 1941 *Proc. R. Soc.* A **178** 17
[42] Iona M 1941 *Phys. Rev.* **60** 822
[43] Huang K 1951 *Proc. R. Soc.* A **208** 352
[44] Henry C H and Hopfield J J 1965 *Phys. Rev. Lett.* **15** 964
[45] Born M and Huang K 1954 *Dynamical Theory of Crystal Lattices* (Oxford: Oxford University Press)
[46] Lonsdale K 1943 *Rep. Prog. Phys.* **9** 256
[47] Laval J 1954 *J. Phys. Radium* **15** 545
[48] Curien H 1952 *Acta Crystallogr.* **5** 392
Jacobsen E H 1955 *Phys. Rev.* **97** 654
[49] Walker C B 1956 *Phys. Rev.* **103** 547
[50] Brockhouse B N 1961 *Inelastic Scattering of Neutrons in Solids and Liquids* (Vienna: International Atomic Energy Agency) p 113
[51] Placzek G and van Hove L 1954 *Phys. Rev.* **93** 1207
[52] Brockhouse B N and Stewart A T 1955 *Phys. Rev.* **100** 756
[53] Cohen M and Feynman R P 1957 *Phys. Rev.* **107** 13
[54] Landsberg G and Mandelstam L 1928 *Naturwissenshaft* **16** 557
[55] Raman C V and Krishnan K S 1928 *Nature* **121** 501
[56] Fermi E and Rasetti F 1931 *Z. Phys.* **71** 689
[57] Mandelstam L, Landsberg G and Leontowitsch M 1930 *Z. Phys.* **60** 334
[58] Placzek G *Handbuch der Radiologie VI* ed E Marx (Leipzig: Akademische Verlagsgellschaft) p 205
[59] Born M and Bradburn M 1947 *Proc. R. Soc.* A **188** 161
[60] Smith H 1948 *Phil. Trans. R. Soc.* A **241** 105
[61] Loudon R 1964 *Adv. Phys.* **13** 423
[62] Dick B J and Overhauser A W 1958 *Phys. Rev.* **112** 90
[63] Woods A D B, Cochran W and Brockhouse B N 1960 *Phys. Rev.* **119** 980
[64] Schröder U 1966 *Solid State Commun.* **4** 347
[65] Fischer K, Bilz H, Haberhorn R and Weber W 1972 *Phys. Status Solidi* b **54** 285
[66] Further references and details of many materials are given in Cochran W 1973 *The Dynamics of Atoms in Crystals* (London: Arnold) and Bilz H and Kress W 1979 *Phonon Dispersion Relations in Insulators (Springer Series in Solid State Sciences 10)* (Berlin: Springer).
[67] Born M and Oppenheimer R 1927 *Ann. Phys., Lpz.* **84** 457
[68] Migdal A B 1957 *Sov. Phys.–JETP* **5** 333

[69]   Woods A D B, Brockhouse B N, March R H, Stewart A T and Bowers R
        1962 *Phys. Rev.* **128** 1112
[70]   Harrison W 1965 *Pseudopotential in the Theory of Metals* (New York:
        Benjamin)
[71]   Bardeen J 1937 *Phys. Rev.* **52** 688
[72]   Toya T 1958 *J. Res. Inst. Catalysis, Hokhaido Univ.* **6** 183; 1959 *J. Res. Inst.
        Catalysis, Hokhaido Univ.* **7** 60
[73]   Sham L 1965 *Proc. R. Soc.* A **283** 33
[74]   Brockhouse B N, Arase T, Caglioti G, Rao K R and Woods A D B 1962
        Phys. Rev. **128** 1099
[75]   Kohn W 1959 *Phys. Rev. Lett.* **2** 393
[76]   See Peierls R E 1964 *Quantum Theory of Solids* (Oxford: Oxford University
        Press) p 108
[77]   Carneiro K, Shirane G, Werner S A and Kaiser S 1976 *Phys. Rev.* B **13** 4258
[78]   Brockhouse B N and Iyengar P K 1958 *Phys. Rev.* **111** 747
[79]   Herman F 1959 *J. Phys. Chem. Solids* **8** 405
[80]   Mashkevich V S and Tolpygo K B 1957 *Sov. Phys.–JETP* **32** 520
[81]   Cochran W 1959 *Proc. R. Soc.* A **253** 260
[82]   For a review see Bilz H, Strauch D and Wehner R K 1984 *Handbuch der
        Physik XXV/2d* (Berlin: Springer)
[83]   Hohenberg P and Kohn W 1964 *Phys. Rev.* B **136** 864
        Sham L J and Kohn W 1966 *Phys. Rev.* **145** 561
[84]   Born M and Blackman M 1933 *Z. Phys.* **82** 551
[85]   Cowley R A 1963 *Adv. Phys.* **12** 421
[86]   Hisano K, Placido F, Bruce A D and Holah G D 1972 *J. Phys. C: Solid State
        Phys.* **5** 2511
[87]   Cowley R A 1967 *Proc. Phys. Soc.* **90** 1127
[88]   Koehler T R 1966 *Phys. Rev. Lett.* **17** 89
        Nosanow L H 1966 *Phys. Rev.* **146** 120
[89]   Born M 1951 *Festschrift der Akademie der Wissenschaften Göttingen*
        Choquard P 1967 *The Anharmonic Crystal* (New York: Benjamin)
[90]   Horner H 1972 *Phys. Rev. Lett.* **29** 556
        Minkiewicz V J, Kitchens T A, Shirane G and Osgood E B 1973 *Phys. Rev.*
        A **8** 1513
[91]   A review is given by Maradudin A A, Montroll E W and Weiss G H 1963
        Theory of Lattice Dynamics in the Harmonic Approximation (New York:
        Academic)
[92]   Schaefer G 1960 *J. Phys. Chem. Solids* **12** 233
[93]   Moller H B and Mackintosh A R 1965 *Phys. Rev. Lett.* **15** 623
[94]   Svensson E C and Brockhouse B N 1967 *Phys. Rev. Lett.* **18** 858
[95]   Montroll E W 1950 *J. Chem. Phys.* **78** 183
[96]   Brusdeylias G, Doak R B and Toennies J P 1983 *Phys. Rev.* B **27** 3662; 1983
        Phys. Rev. B **28** 2104
[97]   Elliott R J and Taylor D W 1967 *Proc. R. Soc.* A **296** 201
[98]   Kruger R, Reinkober O and Koch-Holm E 1928 *Ann. Phys., Lpz.* **85** 110
[99]   Oswald F 1959 *Z. Naturf.* a **14** 374
[100]   Zeller R C and Pohl R O 1971 *Phys. Rev.* B **4** 2029
[101]   Anderson P W, Halperin B I and Varma C M 1972 *Phil. Mag.* **25** 1
[102]   Phillips W A 1972 *J. Low Temp. Phys.* **7** 351
[103]   Black J C and Halperin B I 1977 *Phys. Rev.* B **16** 2879
[104]   Alexander S and Orbach R 1982 *J. Physique* **43** L625
[105]   Boukenter A, Champagnon B, Duval E, Durnas J, Quinson J F and
        Serughetti J 1986 *Phys. Rev. Lett.* **57** 2391
[106]   Landau L D 1937 *Phys. Z. Sov.* **11** 26

[107] Devonshire A F 1949 *Phil. Mag.* **40** 1040
[108] Cochran W 1960 *Adv. Phys.* **9** 389
[109] Fleury P A, Scott J F and Worlock J M 1968 *Phys. Rev. Lett.* **21** 16
Cowley R A, Buyers W J L and Dolling G 1969 *Solid State Commun.* **7** 181
Shirane G and Yamada Y 1969 *Phys. Rev.* **177** 858
[110] A review is given by Bruce A D and Cowley R A 1980 *Structural Phase Transitions* (London: Taylor and Francis).
[111] Als-Nielsen J and Dietrich O W 1967 *Phys. Rev.* **153** 706, 711, 717
[112] Mueller K A and Berlinger W 1971 *Phys. Rev. Lett.* **25** 734
[113] Riste T, Samuelson E J and Otnes K 1971 *Structural Phase Transitions and Soft Modes* (Oslo: Oslo Universitetsforlaget) pp 395–408
[114] Halperin B I and Varma C M 1976 *Phys. Rev.* B **14** 4030
[115] Andrews S R 1986 *J. Phys. C: Solid State Phys.* **19** 3721
McMorrow D F, Hamaya N, Shimonura S, Fujii Y, Kishimoto S and Iwasaki H 1990 *Solid State Commun.* **76** 443
[116] Bloch F 1932 *Z. Phys.* **74** 295
[117] Dyson F J 1956 *Phys. Rev.* **102** 1217, 1230
[118] For a review see van Kranendonk J and Van Vleck J H 1958 *Rev. Mod. Phys.* **30** 1
[119] Kittel C 1951 *Phys. Rev.* **82** 565
Anderson P W 1952 *Phys. Rev.* **86** 694
Kubo R 1952 *Phys. Rev.* **87** 568
[120] Brockhouse B N 1957 *Phys. Rev.* **106** 859
[121] Sinclair R N and Brockhouse B N 1960 *Phys. Rev.* **120** 1638
[122] Nikotin O, Lindgard P A and Dietrich O W 1969 *J. Phys. C: Solid State Phys.* **2** 1168
[123] Reviews are given in Stirling W G and McEwan K A 1987 *Methods of Experimental Physics 23 C* (New York: Academic) and Jensen J and Mackintosh A R 1991 *Rare Earth Magnetism* (Oxford: Oxford University Press).
[124] Stoner E C 1938 *Proc. R. Soc.* A **165** 372
[125] Herring C and Kittel C 1951 *Phys. Rev.* **81** 869
Mattis D C 1965 *The Theory of Magnetism* (New York: Harper and Row)
[126] Cottam M G and Lockwood D J 1986 *Light Scattering in Magnetic Solids* (New York: Wiley)
[127] Bethe H A 1931 *Z. Phys.* **71** 205
[128] Hutchings M T, Shirane G, Birgeneau R J and Holt S L 1972 *Phys. Rev.* B **5** 1999
[129] des Cloizeaux J and Pearson J J 1962 *Phys. Rev.* **128** 2131
[130] Tennant D A, Perring T G, Cowley R A and Nagler S E 1993 *Phys. Rev. Lett.* **70** 4003
[131] Haldane F D M 1983 *Phys. Rev. Lett.* **50** 1153
[132] Buyers W J L, Morra R M, Armstrong R L, Hogan M J, Gerlach P and Hirakawa K 1986 *Phys. Rev. Lett.* **56** 371
[133] Collins M F 1989 *Magnetic Critical Scattering* (Oxford: Oxford University Press)
Cowley R A 1987 *Neutron Scattering Part C* ed K Sköld and D L Price (New York: Academic)
[134] Latham R and Cassels J M 1952 *Proc. Phys. Soc.* A **65** 241
[135] van Hove L 1954 *Phys. Rev.* **95** 249, 1374
[136] Schulhof M P, Nathans R, Heller P and Linz N 1970 *Phys. Rev.* B **1** 2034
[137] Cowley R A and Carneiro K 1980 *J. Phys. C: Solid State Phys.* **13** 3281
[138] Tucciarone A, Lau H Y, Corliss L M, Depalme A and Hastings J M 1971 *Phys. Rev.* B **4** 3206

[139]   Samuelson E J 1973 *Phys. Rev. Lett.* **31** 936
[140]   Ikeda H and Hirakawa K 1974 *Solid State Commun.* **14** 529
[141]   Onsager L 1944 *Phys. Rev.* **65** 117

Chapter 13

ATOMIC AND MOLECULAR PHYSICS

Ugo Fano

13.1. Introduction

As recounted in Chapter 3, studies of atomic and molecular phenomena played a central role in the development of quantum mechanics, as well as in its verification, throughout the early 1930s. Mainstream physics then directed its focus to nuclear and other pursuits; yet early instances of novel atomic–molecular phenomena kept emerging. These will be reported in the next section on 'atomic and molecular physics at mid-century', which should serve as a primer for this chapter's subject and is accordingly extensive.

This chapter deals mainly with subsequent developments, powered by the steady injection of new energies (personnel and funding) and stimulated in part by the role of atomic phenomena in astrophysical and laboratory plasmas. Exploration of new areas and collection of data played comparable roles in this process, but the mechanisms of phenomena will be emphasized here, since their understanding proves essential when extrapolating knowledge of atomic processes to new substances and to new environments. It is, of course, trivial that the behaviour of atoms and molecules underlies not only chemistry but the properties of all matter, gaseous or condensed.

Sections 13.3–13.6 will accordingly deal with rather general subjects, namely: the gross response of atomic systems to the whole spectrum of electromagnetic radiation, the more specific excitation channels and resonances identified by this response and by the effects of electron impact, the main features of collisions between atoms, molecules and their ions, and key developments on aggregates (molecules and clusters). Section 13.7 will deal with inner-shell phenomena.

Sir Harrie Massey

(Australian, 1908–83)

Born near Melbourne, Harrie Massey was a student of Rutherford, Lecturer at The Queen's University of Belfast (1933–38) and Professor at University College London for the rest of his life. He was a major figure of atomic physics, providing leadership throughout the broad context of its relevance to astrophysical and upper atmospherical plasmas. His early co-authorship in 1933 of the fundamental *Theory of Atomic Collisions* (Oxford: Clarendon) was followed by the development of increasingly ample editions over several decades. His research, his personal and organizational leadership and his recruitment and training of new leaders, among them the late Sir David Bates, Michael J Seaton, Phillip G Burke and Alex Dalgarno, established the eminence and style of a British School, in the mid-century period, that spread a lasting influence throughout the world. Under his inspiration, the wealth and diversity of atomic and molecular phenomena in the cosmos provided guidance, models and targets to the development of atomic theory and of its chemical applications. (Photograph by D H Rooks.)

The last three sections will report on activities that are more focused but prove nevertheless broadly relevant: spectroscopy's role in collecting and analysing masses of data on energy levels, which serve primarily as diagnostic tools; the invention and development of novel methods and instruments of measurement, including very accurate control of atomic samples, which underlie not only advances to be reported here but also the whole area of modern metrology involving previously undreamed of precision; the optical pumping and precision handling of atoms. A comprehensive view will be finally outlined.

A few remarks may complement this introduction. The development of our subject has been understandably influenced by historical circumstances. On the experimental side, the progress of techniques and instrumentation naturally had a controlling influence, to be outlined in each relevant context, while on the theoretical side quirks and fads have had curious roles, as sketched in the following paragraphs.

The flush of quantum mechanics' success led influential figures to view the remaining tasks of atomic theory as confined to applications of analytical or numerical procedures for solving Schrödinger's equation, much as Newtonian mechanics had seemed to reduce astrophysics to constructing celestial trajectories. This view found particularly fertile ground in Britain where theory is cast as applied mathematics. Indeed Massey's school pioneered the study of collision physics for decades, combining quantum and classical procedures.

The theory of discrete stationary states centred instead on utilizing their symmetries through group theory concepts leading to algebraic procedures, complemented by numerical integrations over radial coordinates. Atomic physics seemed thus to split into two sub-fields, dealing with discrete and continuous spectra, respectively. That the transition between these spectra at ionization and dissociation thresholds is actually smooth, making their separation irrelevant, was soon realized but sank into the general consciousness imperfectly and only after a delay.

A rather poorly appreciated aspect of quantum theory was that its seemingly novel procedures, in terms of operators, eigenvalues, eigenfunctions, etc, are in fact straightforward applications of the unified view of mathematical physics developed in the 1900s by Hilbert's school; only the code name 'Hilbert's space' became popular. In fact, Hilbert had cast his procedures as extensions of Fourier's familiar analysis. Unfamiliarity with this aspect still underlies many theoreticians' reluctance to deal with eigenfunctions with continuous spectra, whose systematics had been developed in the 1911 thesis by Hermann Weyl, a student of Hilbert. This reluctance has contributed to the artificial separation of 'spectra' and 'collisions'.

A diverging trend has emerged from recent studies of atomic states with large quantum numbers, which appear *prima facie* amenable to semiclassical treatments. This interest has been fanned by the lingering (resurging?) preference of many theorists for the study of trajectories, and more generally for exploring the boundary between quantum and classical phenomena. Thereby it is possible to overlook the fact that quantum phenomena, in contrast to trajectories, inherently spreads over all three dimensions of physical space. Studies of chaotic aspects of semiclassical mechanics thus forfeit perceiving manifestations of quantum effects localized in limited portions of the accessible space.

13.2. Atomic–molecular physics at mid-century

The physics of atoms and molecules developed early in this century through many novel experiments, whose interpretation proved altogether puzzling as described in Chapter 2. Rutherford's discovery of the concentration of atomic masses and positive charges in nuclei 10 000 times smaller than atoms raised the most profound questions: which circumstances set the size of atoms? and why don't the negatively charged electrons drop on nuclei, generating currents that radiate energy away?

Analysis of such questions and of their underlying phenomena, by a new brand of physicists guided largely by N Bohr, led eventually to a scientific milestone, namely, the development of quantum mechanics in the mid-1920s. This development lies outside our scope; we shall instead introduce here the new framework of knowledge and concepts that set atomic–molecular physics on its way through the rest of the century.

13.2.1. *Atomic scale and structure*

Quantum mechanics has interpreted the size of atoms by discovering that confinement of any particle within a volume of radius a boosts its kinetic energy by $\sim \hbar^2/2ma^2$, where $\hbar$ indicates Planck's constant and m the particle's mass. This boost opposes the electrons' attraction towards atomic nuclei, which is represented by the potential energy $-Ze^2/a$ for an electron at distance a from a nuclear charge Ze. For a single electron about this charge, boost and attraction balance out at the Bohr radius $a = \hbar^2/mZe^2 = 0.053/Z$ nm, where the potential energy's magnitude amounts to twice the electron's kinetic energy (Virial theorem). This radius optimizes the electron's bond to the nucleus, represented by the 'Rydberg energy'

$$E = Ze^2/a - \hbar^2/2ma^2 = Z^2e^4m/2\hbar^2 = 13.6Z^2 \text{ eV.} \qquad (1)$$

Expression (1) is conveniently resolved into factors

$$E = \tfrac{1}{2}Z^2(e^2/\hbar c)^2mc^2 = \tfrac{1}{2}(Z\alpha)^2mc^2 \qquad (1')$$

whose dimensional element mc^2 represents the electron's (relativistic) 'rest energy'. The numerical coefficient $\alpha = 1/137.036$ represents instead the ratio of the electron's average velocity in the ground state of atomic H to the velocity of light c. Since the magnetic effects of the electron's motion are proportional to its speed, α (or $Z\alpha$) represents the relative roles of magnetic and electric forces in atomic phenomena. The small value of α implies small magnetic effects on energy levels; for this reason α is called the *fine structure* (or *relativistic*) constant. It is widely believed that the value of α is set by fundamental physics and there is a consensus that this piece of theory is beyond our present understanding. Solution of this puzzle might be viewed as determining the magnitude of the unit charge e.

Combinations of the ratio α and of the radius a are also important: the Compton wavelength αa serves as a unit for the wavelength increase experienced by x-rays recoiling from an electron. The product $\alpha^2 a =$ 2.8 fm represents the electron's classical radius, i.e. the radius of a spherical charge e whose electrostatic field has energy mc^2.

Regarding the 'shell structure' of atoms, probably familiar to readers, we note that the ratio a/Z represents the confinement radius of the innermost ('K') electron pair for an element with atomic number Z. This radius shrinks a little further for heavy elements whose values of $Z\alpha$, the ratio of electron and light velocities, are no longer $\ll 1$, because approach to the velocity of light boosts the electron's inertia and thereby the effective value of its mass m in the definition $a = \hbar^2/mZe^2$.

The confinement radii of other shells result from variants of equation (1), modified by the addition of centrifugal potentials. The latter include the centrifugal repulsion between electrons with parallel 'spin' angular momenta, a phenomenon usually called 'Pauli exclusion' or 'exchange effect' and outlined in Box 13A [1].

The radius of the outermost (valence) electron shell increases slowly with the atomic number, roughly proportional to $Z^{1/3}$. For all neutral atoms, save those of noble gases, the electron motions in this shell have a net non-zero angular momentum. These electrons, free from the constraint of cancelling one another's momentum, are thus prone to combine with electrons of other atoms to form chemical bonds and are accordingly said to be 'chemically unsaturated'—in other words, reactive—as we shall see in section 13.2.5.

'Atoms', without further qualification are implied to be electrically neutral, that is, to include a number of (negative) electrons equal to the atomic number Z which represents the multiple of unit positive charges on their nuclei. Atoms stripped of some of their electrons are called positive ions. Atoms that include supernumerary electrons—generally a single one—are called negative ions.

The structure of the compact form (ground state) of generic atomic species, considered thus far, is complemented by the occurrence of 'excited stationary states' which occupy a larger volume of space with one (or more) electrons less strongly bound than in the ground state. The term 'stationary' means that these states remain unchanged in time, in the absence of external actions. Their stability is ensured by optimization of parameters analogous to the radius a in equation (1). Their diverse properties and classifications will be introduced in section 13.2.3.

A still broader multitude of 'non-stationary' states occurs, whose time-dependence is usually represented by a Fourier-like series of sinusoidal oscillations. Each oscillation frequency equals the energy difference of a pair of stationary states, divided by $\hbar$, much as the oscillation frequency of a light beam's polarization, along its path through an anisotropic medium, reflects differences between alternative refractive indices of that material.

BOX 13A:
SYMMETRY, PARITY AND EXCLUSION

The structure of atoms and molecules frequently displays symmetry elements, typically invariance under reflection through a point, an axis or a plane. This symmetry emerges in the statistical distribution of atomic constituents for stationary states, which exhibit a maximum, minimum (often a zero) or a saddle point at the symmetry point, axis or plane. Quantum mechanics represents such *non-negative* distributions as the squared moduli of 'probability amplitudes' or 'wavefunctions', essentially square roots of probabilities, which may simply reverse their sign at a symmetry element. The absence or presence of such a sign reversal ('even' or 'odd' character) is a parameter of stationary states called *parity*, with reference to a specific symmetry.

The parity of pairs of *identical particles* under reflection through their centre of mass, an operation that leaves the pair *unchanged*, has a major relevance for atomic structure because it leaves its wavefunction unchanged, i.e. with even parity. (The wavefunction's parity for a *single* electron under reflection through a symmetry element may instead be odd.) The wavefunction of the *electron pair* consists of two factors, one pertaining to its spin and one to its position distribution; its even parity requires both factors to have *equal* parity, whether even or odd. The spin factor is even for an antiparallel spin-pair whose zero angular momentum is independent of orientation, but is odd for the parallel spin function which reverses its sign, as a vector does, under a rotation by 180° about the centre of mass and which leaves the pair unchanged. Parallel spins thus imply an odd distribution function in space, which vanishes at the centre of mass, representing an effect of centrifugal repulsion. A striking effect of this symmetry on molecular rotations will be outlined in Box 13E.

CONTINUED

13.2.2. Radiation

'Radiation' will indicate here electromagnetic oscillations of all frequencies; their exchanges of energy with atoms and molecules have provided the main evidence for atomic structure and mechanics. These exchanges proceed by discrete units (photons) proportional to the radiation frequency ω (rad s^{-1}), $\Delta E = \hbar\omega$ [2].

The principal interaction between radiation and an isolated atom or

---

BOX 13A: *CONTINUED*

Symmetry thus prevents electrons with parallel spins from approaching closely. The usual description of this 'exclusion' disregards electron identity initially by attributing labels 1 and 2 to individual electrons, it then requires their combined wavefunction to be odd under permutation of these labels. (The spin-pair's wavefunction, disjoined from position, remains even under this permutation.) Reliance on a pair's symmetry about its centre of mass, stressed by Feynman, is viewed here as preferable conceptually, but electron labelling followed by enforcing odd symmetry under permutation of each pair proves more convenient generally.

---

molecule is represented by a term of their combined energy, namely, the product of a radiation's electric field and of the electric dipole moment of an atomic *non-stationary* state. (Single atoms display no oscillating dipole while remaining in a stationary state, but are forced into non-stationary states by their coupling to radiation.) Energy transfers between radiation and atoms are particularly intense when their oscillation frequencies coincide, i.e. at *resonance*. Stationary normal modes of coupled oscillations of radiation and atoms, in which energy is shared between these partners, occur when both partners are confined in a restricted volume of space.

The most instructive processes occur, however, when the radiation spreads over volumes V far larger than atoms. In this case the unit energy, $\hbar\omega$, of each independent radiation mode is highly diluted with density, $\hbar\omega/V$, and so is its field strength acting on any single atom. This circumstance allows the evaluation of the radiation action as a 'perturbation', *linear* in the field strength A and in the electron charge, i.e. in the parameter $\alpha^{1/2}$ of equation (1'). The density of radiation modes with frequencies capable of resonating with an atomic dipole's becomes, by the same token, very high—indeed proportional to V—thus yielding a significant total contribution.

13.2.2.1. Elementary phenomena. Three alternative 'elementary' processes result from the atom–radiation interaction: (a) the resonant *absorption* of one photon by an atom, whose energy is thereby raised to a higher level, which may be bound or ionized; (b) the reciprocal 'spontaneous' emission of one photon by an atom (resonance being readily afforded by the dense spectrum of radiation modes in a large volume V); and (c) the *scattering* of a photon from one radiation mode to another,

William F Meggers

(American, 1888–1966)

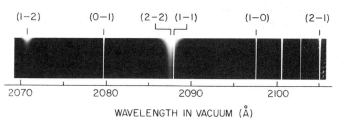

WAVELENGTH IN VACUUM (Å)

William Meggers, of the US National Bureau of Standards (NBS), was a pillar of spectral analysis through the mid-century period focusing on instrumental precision and data analysis. Under his guidance and initiative a team of spectroscopists observed, measured and otherwise collected thousands upon thousands of precise spectral line wavelengths, serving as standards throughout the world. Their lists were then patiently and persistently inspected over decades in the search of patterns that would reveal regularities of atomic mechanics.

CONTINUED

CONTINUED

Cumulative experience developed a flair that enlivened the search. Mechanization of the search started about mid-century, initially by IBM card-sorting devices but eventually through electronic computers scanning lists of spectral frequencies for pairs with equal differences; the results of these scans are embodied in the basic NBS *Tables of Atomic Energy Levels*. The clarity of interference rings beside his picture underlies the accuracy of a Hg 198 wavelength standard. The spectral lines displayed here belong to transitions between levels of two Zn multiplets with *j*-values indicated above each line. Note the differences among line widths reflecting the varying rates of decay of excited states through alternative channels. (Increasing widths near the upper edge reflect an instrumental artefact.)

---

mediated by an atom through absorption + emission (not necessarily at resonance). Photon scattering may be elastic (Rayleigh) or inelastic (Raman), the latter leaving a fraction of the absorbed photon in the atomic system.

High photon energies afford the 'Compton' variant of inelastic scattering, in which an electron recoils clear out of its atomic system. Another process, manifest mostly at high photon energies, is the emission of a continuous spectrum of radiation (bremsstrahlung) by fast electrons deflected, mainly by nuclear attraction, when traversing atomic fields.

Quantitative treatments of these processes were key elements in the development of quantum mechanics, reviewed in 1932 in Fermi's lectures [3] and more comprehensively in Heitler's book of 1936 [4].

The interaction of electrons with radiations whose photon energy approaches the electron rest energy mc^2 (= 511 keV) requires relativistic treatments that intermix radiation and electrostatic fields. A major result in this range lay in the 1932 discovery of the combination of MeV-range photons with the Coulomb field of high-Z nuclei yielding a 'pair-production', namely, the generation of electron pairs with charges of opposite sign. This process leads, in turn, to the surprising secondary process of 'vacuum polarization' in the space surrounding nuclei or other charges, a high-energy analogue of the usual electric polarization of atoms and atomic aggregates by electrostatic fields.

Electric polarization itself amounts to the non-resonant generation in matter—and also in vacuum—of the same 'dipole' displacements of electric charges as absorb or emit radiation at resonant frequencies.

Further atomic polarization effects, corresponding to very high-frequency resonances, were discovered in the late 1940s: 'Lamb shifts' of atomic energy levels and $\sim 0.1\%$ corrections to the magnetic moments of electron spins. A major problem arose in these developments from an inherent divergence of the high-frequency contribution to polarizations in the limit $\omega \to \infty$. Procedures were developed by 1950 to remove these singularities by appropriate circumscribing of the problem, thus affording very successful calculations by expansion into powers of the electron charge parameter α. However, in the view of the author of this chapter, the basic problem persists to this day.

13.2.3. *Atomic spectra*

Spectroscopes are instruments that resolve a beam of radiation into components of different frequencies, displaying the intensity of the various components on a screen or on a detector's surface. Early displays of the analysis for the radiation emitted or absorbed by monatomic vapours revealed their frequencies to lie at or about discrete values (spectral lines) in, or near, the visible range ($10^{15} < \omega < 10^{16}$ rad s^{-1}). The multitude of lines observed for all elements was soon reduced somewhat by representing them as differences between a smaller set of spectral terms, later identified as equivalent to energies of stationary states of each atom. Extracting the terms from rich spectra tested the skills and devotion of spectroscopists, until the process was computerized.

Observed lines correspond only to pairs of states generating transitions with an electric dipole moment, i.e. pairs with opposite parity (under reflection through the atom's centre) and angular momenta j differing by no more than one unit of $\hbar$. These *selection rules* afford a partial classification of spectral terms. (The 'quantum number' j represents the magnitude of the vector angular momentum $\boldsymbol{j}$, namely, the value of $|\boldsymbol{j}|^2 = j(j+1)\hbar^2$.)

Terms of free atoms are independent of their orientation in space. This 'degeneracy' is broken in the presence of external fields, electric (Stark) or magnetic (Zeeman), which split each term with $j \neq 0$ into a multiplet of levels with different quantum numbers m. Beams of atoms, or molecules, with $j \neq 0$ are resolved into components with different m values by transmission through 'Stern–Gerlach magnets'. The uniform separation of the components measures their magnetic moment parallel to the field (figure 13.1). Elastic scattering of radiation or electrons, the latter to be outlined in section 13.2.4, can form an image of the shape of an atom in a stationary state (figure 13.2).

13.2.3.1. *The simplest spectra.*

The simplest spectra occur for atoms of hydrogen and of monovalent metals, whose single valence electron moves in a field with central symmetry allowing separation of its rotational and radial motions. This electron's vector angular momentum

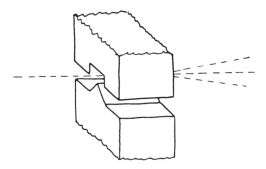

Figure 13.1. *Diagram of molecular beam analysis by a Stern–Gerlach inhomogeneous magnet. The incident beam is resolved here into three components.*

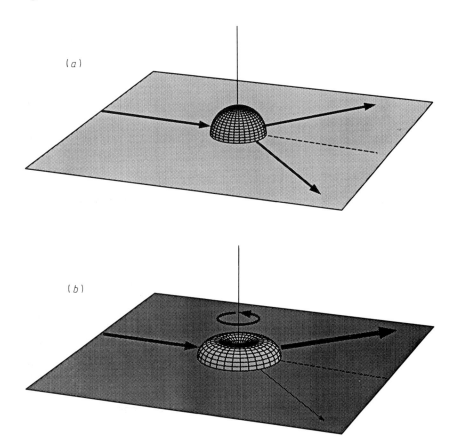

Figure 13.2. *Diagram of electron scattering by an atom. Analysis of the probability of electron deflections by different angles determines the shape of an atom. Asymmetries between deflections to the right and left reflect intra-atomic currents.*

j is accordingly a constant of the motion and might have been expected to vanish in some states, yielding a statistically uniform distribution of electron positions about the atom. The latter feature is indeed verified by certain observations on atomic ground states, but magnetic analysis of monovalent atomic beams showed them to be resolved into two components labelled by $m = \pm\frac{1}{2}$, implying that $j \neq 0$ and that the corresponding quantum number $j = \frac{1}{2}$, thus demonstrating the existence of the electron's spin [5].

The separation of rotational and radial motions is also a factor affecting the density of electron distributions. For the H ground state this density is simply proportional to $\exp(-2r/a)$ with $a = 0.053$ nm as in equation (1). For the sequence of H excited states with spherical symmetry the corresponding density is represented by the product of a similar factor $\exp(-2r/na)$ and of a squared polynomial of degree $n - 1$ (i.e. with $n - 1$ roots) in the ratio r/na, with $n = 2,3, \ldots$. The electron binding energy for this sequence of states, easily obtained from the observed H spectrum (see Chapter 2, section 2.6), is also simply related to (1) by the Rydberg formula $E_n = 13.6$ eV$/n^2$. The basic relation displayed here between the parameter na of an exponential factor of the radial density and the electron's binding energy, with an appropriate value of a, is *common to all atoms*.

Rotational motions of an electron generate directional patterns of its density distribution. Quantum mechanics represents these patterns through a set of mathematical expressions, $|Y_{lm}(\theta, \varphi)|^2$, described in Box 13B. This distribution is complemented by a radial factor $|f_{nl}(r)|^2$, analogous to that described above for the spherical distribution, but with l fewer roots in its polynomials. The energies of excited stationary states with $l \neq 0$ coincide approximately with those of equal n and $l = 0$, but depart from them through a 'fine structure' splitting, described in Box 13C.

The spectra of monovalent metals (alkaline and noble), resulting from excitation of their valence electron, resemble the H spectrum closely because this electron moves in a Coulomb field when outside the atom's (spherical) ionic core. However, this electron's penetration into the core modifies the radial density functions $f_{nl}(r)$ mainly *within* the core. Outside the core, $f_{nl}(r)$ keeps its hydrogen character to within parameter shifts in its exponential and in the roots of its polynomial factor; these are represented effectively by subtracting from the principal quantum number n a new semi-empirical quantum defect parameter μ_l, $n \rightarrow n-\mu_l$. The expression 13.6 eV$/n^2$ for the binding energy of atomic hydrogen levels is thus extended into the generic formula for levels of Rydberg series $E_{nl} = 13.6$ eV$/(n - \mu_l)^2$. The l-dependence of μ_l removes the approximate coincidence of the energies of stationary states with equal n and different l (except for l values at which μ_l is negligible).

BOX 13B: PATTERNS OF
DIRECTIONAL DISTRIBUTION

Directions of physical space are generally indicated (by analogy with coordinates on the Earth's surface) with reference to a polar axis and a 0° meridian, by a pair of angles (θ, φ). The 'longitude' φ runs from 0° to 360° (2π radians) around the axis and the 'co-latitude' θ runs from one pole to the other (0° to 180°).

Distributions in longitude are represented through a Fourier series of trigonometric functions $\{\cos m\varphi, \sin m\varphi\}$ (m = any integer) or of their equivalent complex combination $\exp(im\varphi) = \cos m\varphi + i\sin m\varphi$. The rotational motion of an electron, or other system, about the polar axis, with a constant angular momentum component $m\hbar$ along this axis, is represented by a distribution function with the single trigonometric component $\exp(im\varphi)$, whose real and imaginary parts are subdivided into m *meridian* lobes.

Distributions in co-latitude are represented through a series of polynomials of degree $l - m$ in the variable $\cos\theta$, with an additional factor $\sin m\theta$, (associated Legendre polynomials, $P_{lm}(\cos\theta)$). The distribution $|P_{lm}(\theta)|^2$ has $l - m$ lobes (e.g. $P_{00} = 1$, $P_{10} \propto \cos\theta$, $P_{11} \propto \sin\theta$, $P_{20} \propto 3\cos^2\theta - 1$). A combined distribution function $P_{lm}(\theta)\exp(im\varphi)$ is indicated in the text by the spherical harmonic function $Y_{lm}(\theta, \varphi)$. A single function Y_{lm} represents a rotational motion with squared angular momentum $|l|^2 = l(l+1)\hbar^2$ and with a component $m\hbar$ of l along the polar axis. The index l thus represents the degree of directional confinement of rotational motions.

13.2.3.2. Multi-electron spectra. Working out the energy levels of multi-electron atoms looks, at first sight, to be a daunting task. Monovalent atoms have been viewed in section 13.2.3.1 as consisting of a single electron and of a spherical ionic core, without describing how the core-electron mechanics determines the quantum defects μ_l. Other ionic cores are asymmetric (except for monovalent cores like Mg^+), so their own structure needs study.

The task is reduced considerably by the algebraic treatment of directional interactions among electrons as outlined in Box 13D, but interactions among radial motions also require attention. These interactions are themselves restricting the electrons of filled shells within ionic cores, because shell filling exhausts the angular degrees of freedom accessible to any single electron. The still current procedure,

---

BOX 13C: FINE AND HYPERFINE STRUCTURES

The fine structure of atomic spectra stems primarily from the interactions between the magnetic moments of electron spins and those resulting from intra-atomic rotational motions of electrons. The magnetic field experienced by the spin of each electron, owing to that electron's rotational motion, stems actually from the apparent rotation of *other* charges, mainly of the nucleus, about the spin itself.

The nucleus' contribution is proportional to its charge Ze and to the inverse cube of the electron's distance from it. The resulting energy of spin-orbit coupling is accordingly proportional to the integral over $|f_{nl}(r)|^2/r^3$ and to the product of the orbital and spin angular momenta $l \cdot s$. Its net effect on the spectra of monovalent metals is represented by a shift of the quantum defect μ_l proportional to $s \cdot l$ and increasing rapidly with Z, as $Z\langle|f_{nl}(r)|^2/r^3\rangle$. (An early apparent discrepancy between experiment and theory of this phenomenon was quickly removed in 1926 by L H Thomas's accurate evaluation of the field experienced by an orbiting electron.) Contributions of relativistic effects proportional to $l \cdot s$, are disregarded here.

The product $l \cdot s$ depends on the squared magnitude of the vector sum

$$|l + s|^2 = |l|^2 + 2l \cdot s + |s|^2 = |j|^2 = j(j+1)\hbar^2 \qquad (2)$$

that is, on the quantum number $j = l \pm \frac{1}{2}$, in addition to the values of l and $s = \frac{1}{2}$. The resulting energy levels, $-13.6 \text{ eV}/(n - \mu_{lj})^2$, turned out to be lower for $j = l - \frac{1}{2}$ (antiparallel mutual orientation) than for $j = l + \frac{1}{2}$. This result accounted for the earlier discovery of 'doublets', i.e. of closely spaced pairs of spectral lines. An adequate quantitative understanding of the 'simple spectra' involving a single valence electron was thus achieved by the mid-1920s.

CONTINUED

---

introduced by Hartree in 1928, exploits this angular confinement by requiring a single radial distribution function $f_{nl}(r)$ to represent the radial distribution of all electrons with the same quantum numbers. This approach, motivated initially by disregarding all electron correlations, determines the $f_{nl}(r)$ values of closed shells 'self-consistently' so as

---

BOX 13C: *CONTINUED*

Concurrently a still smaller, 'hyperfine', structure of spectral lines was resolved, analysed and evaluated, stemming from the interaction of electronic and nuclear distributions of charges and currents. We do not describe this interaction beyond stating that its energy depends on the product of electronic and nuclear angular momenta, $j \cdot I$, as reflected in the squared combined momentum $|j + I|^2 = |F|^2 = F(F + 1)\hbar^2$.

The ground state of the H atom ($j = \frac{1}{2}$, $I = \frac{1}{2}$) resolves into two components with $F = 0$ and 1. Short-wave radio emission at 1420 MHz by transitions between these levels throughout the cosmos has been of great value in assaying the distribution of atomic hydrogen. (See also section 13.2.5.3.)

The combination of orbital and spin momenta that yields energy levels with alternative values of j alters the directional asymmetry of the electron's density distribution illustrated in figure 13.2. This subject belongs to a broader class of studies whose outline and scope are introduced in Box 13D.

---

to minimize the energy of an entire closed shell core. Its Hartree–Fock variant also accounts for the exchange energy, a side-effect of the centrifugal repulsion of electron pairs with parallel spins mentioned in section 13.2.1 and Box 13A.

Each electron of an incomplete shell, whether in the ground state or excited, is also assigned quantum numbers nl (or nlj) and a radial distribution $f_{nl}(r)$, whose determination is even less straightforward. The electronic structure of ground or excited states is then represented by a 'configuration' formula, $\Pi_i(n_i l_i)^{N_i}$, where N_i indicates the number of electrons assigned to the ith subshell. The distributions $f_{n_i l_i}(r)$ are then determined so as to stabilize the total energy of the stationary state, depending on both radial and angular distributions. Bringing the energy into adequate agreement with experiment generally requires an additional step of 'configuration mixing'.

The state of this art in the mid-1930s was summarized in a treatise by E U Condon and G Shortley [10]. This state was fairly adequate to deal with excited 's' or 'p' electrons (i.e. with $l = 0$ or 1). Full unravelling of the very rich spectra of partially filled 'd' shells ($l = 2$), which accommodate up to $2(2l + 1) = 10$ electrons, began with Racah's work in the 1940s [8]. The f-shell ($l = 3$) spectra of the rare-earths defied analysis until later decades.

BOX 13D: ANGULAR MOMENTUM ALGEBRA

The evaluation of the spin-orbit coefficient $l \cdot s$ through the identity $|l + s|^2 = |j|^2$, equation (2), is a prototype of very extensive procedures central to spectral theory and to other subjects. Problems analogous to this prototype occur in evaluating the repulsion energy between two (or more) electrons with density distributions represented through spherical harmonics $Y_{lm}, Y_{l'm'}$. Electron interactions represented by the Coulomb potential $e^2/|r - r'|$, to be averaged over their combined density distributions, can be expanded into spherical harmonics. We deal here with generalizations of the trigonometry formula $\cos ax \cos bx = [\cos(a + b)x + \cos(a - b)x]/2$ which resolves a product into a sum of terms involving sums and differences of the indices (a, b). The manifold generalizations of this formula rest on the transformations of harmonic functions induced by changes of their coordinates, transformations that form a group and whose expressions in terms of indices $\{l, l' \ldots\}$ are provided by group theory.

The relevant formulations for quantum systems by algebraic functions of angular momentum quantum numbers $\{l, s, j; m, m'\}$ were provided around 1930 in parallel books by E Wigner and L Van der Waerden [6, 7]. Thereby *all averagings* over directional distributions are reduced to standard formulae. Further progress was achieved by G Racah [8], and independently by Wigner [9], relating alternative combinations of three or more functions through algebraic functions of the indices $\{l, s, j, \ldots\}$ only, which are independent of coordinate systems.

13.2.3.3. Inner-shell spectra. Inner-shell electrons are readily ejected from atoms by incident projectiles with sufficient energy. The resulting vacancy is then quickly filled by an electron with higher energy, $B_o > B_i$; the transition between these shells releases the energy difference $B_o - B_i$, a process involving a dipole oscillating current with frequency $\omega = (B_o - B_i)/\hbar$ that may, but need not, result in the emission of a photon $\hbar\omega$. Incidentally, this photon emission was anticipated by Bohr in 1913, upon his first development of the 'planetary model'; equation (1) predicted that ejection of a K-shell electron would generate x-rays with $\sqrt{\omega} \propto Z$. This prediction was soon verified by Moseley's discovery of characteristic x-rays, so called because of the contrast between their line spectra and Röntgen's earlier continuous x-ray emission. This verification provided, in turn, early evidence of the assignment of successive values of Z to the

See also
p 158

nuclear charge Ze for successive elements of Mendeleev's table.

The whole spectrum of characteristic x-rays consists of several series, in different frequency ranges, labelled K, L, M, ... series in correspondence to initial vacancies in inner shells with quantum number $n = 1, 2, 3, \ldots$. The lines of each series are labelled $\alpha, \beta, \gamma \ldots$ depending on the n value of the B_0 shell.

The, generally more frequent, alternative to x-ray emission arises by transfer of the energy $B_0 - B_i$ to an electron of the shell o, or external to it, evidenced by the detection of monoenergetic (Auger) electrons with characteristic spectra.

13.2.4. Collisions

Collisions of atomic particles have been a major tool of modern physics since Rutherford's α-particle experiments. Spectroscopy, which has provided more details about atoms, also depends largely on collisions by electrons or ions as sources of excited atoms.

'Long-range' collisions transfer energy to atoms through the Coulomb field of fly-by charged particles that remain—often far—outside a target atom or molecule. (A rapid pulse of this field acts much like a continuous spectrum of radiation.) 'Close', short-range collisions lead to formations of a *complex*, usually short-lived, in which projectile and target coalesce to varying degrees. Intermediate 'grazing' collisions, characteristic of most atomic projectiles involve only a small interpenetration of projectile and target.

The angular momentum of the complex (projectile + target combination), about its centre of mass, remains constant throughout a collision. One can take advantage of this conservation by 'partial wave analysis' of the incident beam into components with different orbital momenta. The probability of collision for each component is proportional to $(2l + 1)/v^2$ ($v =$ velocity), thus increasing with l but with a coefficient that generally drops with increasing v. Application of this approach has been mostly confined thus far to a subset of problems, despite its basic significance.

The alternative classification of collisions by 'impact parameter', the least distance of the incident projectile's trajectory from the target, is generally appropriate to all long-range collisions and to most grazing and close collisions by projectiles heavier than electrons.

The first treatise of *The Theory of Atomic Collisions* by N F Mott and H S W Massey appeared as early as 1933 [11]. It centres on a wave-mechanical formulation illustrated by experimental results.

13.2.4.1. The action of fast charged particles.

The action of fast charged particles has been a relevant ingredient of many experiments since Rutherford's, its main elements being assessed by Bohr in 1913. He stressed the contrast between the *impulsive* character, prevailing when the collision lasts less than the longest period $1/\omega$ of atomic oscillations $b\omega/v \ll 1$ (b = impact parameter, v = particle velocity) and the reversible, i.e. *elastic*, character when $b\omega/v \gg 1$. The energy dissipated by the projectile in impulsive collisions amounts to the recoil energy of all atomic electrons within the impulsive range, *as though* these electrons were recoiling freely [12].

A wave-mechanical treatment of this phenomenon, by H A Bethe (1930), added a new feature and a correction to Bohr's ideas. It viewed the Coulomb interaction between a fast particle with charge ze at r and an atomic electron at r_i as 'weak', representing it by a Fourier integral over exponentials

$$\frac{ze^2}{|r - r_i|} = \frac{ze^2}{2\pi^2} \int \frac{\mathrm{d}k}{k^2} \mathrm{e}^{\mathrm{i}k \cdot (r_i - r)}. \tag{3}$$

Each element of this integral represents the probability amplitude of transfer of momentum $\hbar k$ by the projectile to the ith electron. The squared amplitudes $|\exp(-\mathrm{i}k \cdot r)|^2$ for momentum loss by the projectile, i.e. the probabilities of its deflection, add incoherently because the deflection is observable. The corresponding amplitudes $\exp(\mathrm{i}k \cdot r_i)$ for momentum gain by individual electrons add instead coherently because only the probability of net excitation of an atomic target to its nth level is observable (unless an electron recoils clear out of the target). This probability is proportional to a function $|F_{n0}(\Sigma_i \mathrm{e}^{\mathrm{i}k \cdot r_i})|^2$ called the 'generalized form factor' of the target. Bethe verified that the resulting average energy absorption, $\Sigma_n E_n|F_{n0}(\Sigma_i \mathrm{e}^{\mathrm{i}k \cdot r_i})|^2$, equals the sum of the recoil energies of independent electrons, as anticipated by Bohr.

Bethe also noted that the minimum k-value contributing to $\Sigma_n E_n|F_{n0}|^2$ is set by the energy–momentum balance of the projectile, a limit somewhat tighter for fast electrons than Bohr's $b\omega/v \sim 1$. Mott noted that Bethe's procedure applies for massive (ion) projectiles also when $b\omega/v \gg 1$. In 1924 Fermi exploited the correspondence between the pulsed field of fast projectiles and radiation, replacing $\Sigma_n E_n|F_{n0}|^2$ by an equivalent macroscopic dielectric property of the target material.

13.2.4.2. Close collisions of slow electrons. Bethe's weak-interaction (actually impulsive) treatment [13] applies also to close collisions by fast electrons but gives unrealistically high results at incident velocities comparable to those of atomic electrons. Variants intended to remove this discrepancy did not appear desirable.

Incorporation of the incident electron into the target to form a complex with constant angular momentum might have appeared most appropriate to the present problem. However, only its application to noble gas targets with spherical symmetry proved successful in the early days (because of the complications of generic multi-electron complexes, whose treatment started only after mid-century). Within this framework, the density distribution of the colliding electron within the complex, with squared angular momentum $l(l+1)\hbar^2$, is represented by $|f_{kl}(r)Y_{lm}(\theta,\phi)|^2$ as for the valence electron of a monovalent atom in section 13.2.3.1. Here the index k, representing the momentum $\hbar k$ of the incident electron, replaces the quantum number n of the metal's valence electron. The quantum defect μ_l of the incident electron turns here into an equivalent phase shift $\delta_l(k) = \pi\mu_l$.

Experiments on elastic scattering of electrons by noble gas atoms, with unit incident flux and unit solid angle of deflection $d\Omega$, are represented in this case by the cross section

$$d\sigma(\theta) = \frac{4\pi}{k^2}|\Sigma_l(2l+1)^{1/2}\exp(i\delta_l)\sin\delta_l(k)Y_{l0}(\theta,\varphi)|^2\,d\Omega. \qquad (4)$$

(The function $f_{kl}(r)$ is in this case, outside the atomic core, a spherical Bessel function $j_l(kr + \delta_l(k))$.) Great surprise was caused by Ramsauer's early discovery that this cross section vanishes, as though atoms of Ar, Kr, Xe become *fully transparent*, at an energy of impact (< 1 eV) characteristic for each atom. It was understood later that attraction by the polarized target prevails at near zero velocity, yielding $\delta_l(k) > 0$, but is overcome at larger energies by the net repulsion of the incident electron by the 'closed-shell' atom. The phase shifts $\delta_l(k)$ vanish at the transition between these energy regimes.

13.2.4.3. *Close and grazing collisions by ions and atoms.* Atoms and molecules, whether in ionic or neutral states, can be slower than all target electrons and still pack kinetic energies in the keV range. The momenta of their nuclei are similarly large allowing the projectile to bore through targets on a near-straight line (except when passing close to a target nucleus). Their influence on target electrons is mostly elastic, but the aggregate number of excitations and ionizations along their paths is non-negligible. (The speed of thermal atoms is orders of magnitude lower than those of atomic electrons, hence thermal collisions are elastic, barring chemical reactions.) Electron transfers between heavy projectiles and targets have been the object of much study.

A curious aspect of electron transfers, specifically of 'electron capture' by an ion colliding with a neutral target, was pointed out by L H Thomas in the mid-1920s. Head-on collision of the ion with a target electron would push the electron along the ion's path, thus facilitating its capture, but the *resulting* speed of the electron, double the ion's, would prevent its capture. Velocity matching of the electron and ion requires instead electron ejection at 60° from the ion's track. A second, elastic, collision by the electron with a nucleus at rest (i.e. a 'two-shot' process) is required to turn the electron parallel to the projectile's track, thus facilitating its capture. This phenomenon and its variants were observed half-a-century later.

Close and grazing collisions between slow atoms and/or molecules, whether ionic or neutrals, belong properly to the physics of the resulting molecular complex. 'Slow' means, in this context, generally 'slower than valence electrons'. These collisions depart from ordinary molecular mechanics primarily by their high energy of internuclear motion. Their experimental and theoretical study was delayed beyond mid-century by the resulting generally multiple fragmentation, whose study requires elaborate detection and analysis techniques.

The case of grazing collisions presents an easier opportunity, especially in the elementary case when the complex has a single valence electron, whose independent motion can be studied in the field of a pair of slow moving atomic or molecular cores. A theoretical method of 'perturbed stationary states', introduced early to treat this and related phenomena [11] and widely applied later, rests on a basic molecular procedure to be described in section 13.2.5.1.

For collisions at speeds faster than those of valence electrons, a key qualitative rule was formulated early by N Bohr [14], namely, that all electrons slower than the collisions's speed are likely to be swept out of their initial states.

13.2.5. *Molecular bonds and behaviour*

Pairs of atoms and/or molecules experience a generic weak attraction, a function proportional to $1/r^6$ of their distance r, a force whose existence had been inferred from gas properties in the 1800s and which bears the name of its discoverer, van der Waals. Quantum mechanics has traced this force to mutual, synchronously oscillating, electric polarizations of the two particles. This force draws particles together until stronger forces, attractive or repulsive, become dominant upon sufficient overlap of the partners' electron distributions. It even suffices to bind pairs of rare gas atoms at temperatures of the order of 1 K, against their short-range repulsion.

The origin of the basic 'covalent' molecular bond was first established by quantum mechanics, through analysis of the prototype approach of two H atoms by W Heitler and F London (1928). The short-range force between these atoms depends critically on the mutual orientation of their respective electron spins. Parallel orientation leads to repulsion, as noted in Box 13A; opposite orientation favours instead the pair's distribution function $f(r_1, r_2)$ to centre between the two nuclei, thus optimizing the attraction of the electron pair by both nuclei and offsetting the repulsions within each pair of equal charges (four attractions, two repulsions). (The original study, based on separate atomic distributions $f(r_1, r_2) = f_{10}(r_1 - R_1)f_{10}(r_2 - R_2)$ led to an optimum bond energy $2E(H) - E(H_2) \sim 3.7$ eV, but subsequent release of $f(r_1, r_2)$ allowed the calculated energy to approach the experimental value of the H_2 bond, 4.65 eV; the corresponding average internuclear distance also approached its value of 0.074 nm.)

Mechanisms substantially equivalent to the H_2 bond were soon found to underlie the formation of most chemical bonds. Bonds between atoms with different affinity for additional electrons are asymmetric, shifting the joint distribution $f(r_1, r_2)$ closer to one than to the other atom; typically the O–H bond in H_2O draws the bonding pair closer to the O atom. In extreme cases, such as in the NaCl molecule, the bond is called ionic and is indicated by Na^+Cl^-.

Divalent atoms form bonds with two atoms, on a straight line for group II (e.g. Mg) and at an angle for group VI (e.g. O). Trivalent atoms form three bonds, at 120° on a plane for group III (e.g. B) and in trihedral form for group V (e.g. N). Tetravalent atoms similarly form four bonds in tetrahedral directions, typically in CH_4 but also in the crystalline network of tetrahedrally bonded carbon, namely, diamond.

Carbon, and other atoms to a lesser extent, develop an alternative mixed type of bonding which underlies their far richer chemistry. The graphite form of carbon utilizes three electrons per atom to form a plane 120° network of bonds and the remaining one electron per atom to run through this network in quasi-metallic form. 'Aromatic' molecules share these features as do innumerable compounds with intermingling double, or even triple, bonds. A basic treatment of these phenomena was developed by Hückel in the early 1930s.

The ground states of simple, stable molecules often consist of closed shells, with electron spins coupled to cancel each other; a notable exception occurs in the oxygen molecule with net spin $S = 1$ in its ground state and low-energy metastable excitations with $S = 0$. Low excitations with $S = 0$ within the ground-state shell occur in many molecules. Excitations of one electron to higher shells are analogous to those of single atoms, with angular momenta of that electron variously coupled to those of the molecular core.

Important conceptual progress in about 1930 traced the origin of rubber elasticity, and of analogous phenomena, to the thermal behaviour of long-chain molecules (polymers). These molecules consist of sequences of N atomic groups, e.g. CH_2, linked by bonds that are free to rotate with respect to one another. The distance of the chain's end points is proportional to N when the chain is stretched, but only to $\sqrt{N}$ for random orientation of successive bonds. Mechanical action, such as a pull by an external force, can thus stretch a length of rubber but thermal agitation tends to restore random orientations with individual molecules wrapped onto themselves.

13.2.5.1. The Born–Oppenheimer approximation. The Born–Oppenheimer approximation, basic to molecular physics, rests on the electrons moving much faster than nuclei, owing to their smaller mass. Accordingly electron distribution functions $f(r_1, r_2, \ldots)$ are evaluated assuming fixed positions of the nuclei $\{R_1, R_2, \ldots\}$ and indicated by $F(\{R_1, R_2, \ldots\}; r_1, r_2, \ldots)$. The energy $E_n(\{R_1, R_2, \ldots\})$ of a stationary electron state with distribution F_n is then viewed as the potential energy governing the slower nuclear motion. 'Non-adiabatic' corrections to the initial results thus obtained are functions of derivatives $\partial F_n / \partial R_\alpha$.

Figure 13.3 illustrates the dependence of a succession of energy levels E_n of the H_2 molecule on the internuclear distance $R = |R_1 - R_2|$. Most of the $E_n(R)$ curves display a deep (near parabolic) minimum which holds the internuclear distance within a range $\sim 10^{-2}$ nm. The *vibrational* motion of the nuclear distance, thus confined, has a sequence of its own stationary states with energies E_{nv}, whose spacings $E_{n,v+1} - E_{nv}$ are roughly uniform and of the order of 0.5 eV for H_2 (but much smaller for heavier molecules). One of the curves $E_n(R)$ has, however, no minimum; which implies that its electron state pushes the nuclei apart, indeed towards 'dissociation', corresponding to an electron pair with parallel spins, whose centrifugal effect keeps the electrons apart thus preventing the bond's formation.

'Potential' surfaces of appropriate dimensions are the analogues of figure 13.3 for polyatomic molecules. Their calculations require elaborate numerical procedures, but have been expanding into a sort of industry with the development of increasingly powerful computers.

An additional element of molecular mechanics consists of the *rotational* motion of each whole molecule about its centre of mass. Its role in the spectra, and the role of vibrations, dominate the molecular spectra to be described in section 13.2.5.2. The energy scales of rotational, vibrational and electronic energy levels differ by 1–2 orders of magnitude, depending on circumstances, thus affording the corresponding motions to proceed quite independently.

The basic items thus outlined were soon developed, by about 1930, into an extensive theoretical treatment of molecular mechanics which interpreted the extensive spectral observations then available. Chief leaders in this process were F Hund, R Mulliken [15] and G Herzberg [16].

Two important concepts emerged in this early period:

(i) The set of potential curves in figure 13.3 constitutes a *correlation diagram* that connects the 'united atom' states of the H_2 molecule at $R = 0$ to its 'separate atom' states at $R = \infty$. The former arise when $R_2 \to R_1$, forming for H_2 a doubly-charged H_2^{++} core whereby the electron stationary states coincide with those of He; the latter represent pairs of separate H atoms in various states.

(ii) Excitation of an electron, for example due to photoabsorption by H_2's electronic ground state $n = 1$ with vibrational state v, changes the initial density distribution function $F_{1v}(R; r_1, r_2)$ into a different

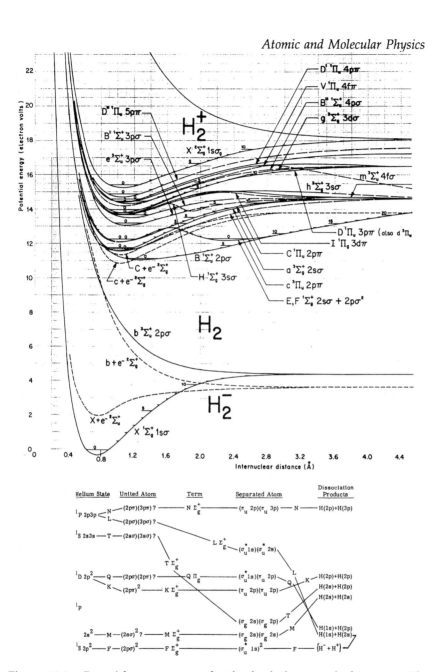

Figure 13.3. *Potential-energy curves of molecular hydrogen and of its ions. The ordinate represents the electron energy level of various stationary states at each value of the internuclear distance R plotted as the abscissa. The minimum of each curve represents the equilibrium value of that distance, the curvature at that point determines the vibration frequency; the energy values at large abscissas represent dissociation thresholds. The zero value of the ordinates corresponds to the ground-state energy of the H_2 molecule, inclusive of its vibrational energy; the curve labelled b^2 represents the repulsion of a pair of H atoms with parallel spins. Dashed curves represent energies of the H_2^- ion. Below: partial diagram of correlations between 'united' and 'separate' atom limits.*

one $F'(R; r_1, r_2)$ that is *not* stationary because the radiation of interest resonates with the electron oscillations only, owing to the larger nuclear mass. The F' distribution resolves, in time, into a number of stationary components F_{nv}. This circumstance, called the *Franck–Condon* rule and extensively verified by experiments, is being utilized currently to manoeuvre the internuclear distance R by repeated appropriate photoabsorptions so as to recast molecular structures into novel, if unstable, shapes of interest.

13.2.5.2. Molecular spectra. The aspect of molecular spectra contrasts with that of atomic spectra through the contributions of vibrational and rotational motions. Each electronic transition, which would give rise to a single spectral line in an atom, is generally accompanied by manifold transitions between alternative pairs of vibrational and rotational energy levels. Only a small subset of these transitions generates oscillating electric dipoles and is thus readily apparent in the spectra, but this subset is nevertheless large.

Typically an oscillating current arises in transitions by 'optically active' molecules in which the centre of mass of the electrons does not coincide with the centre of charge of the nuclei. An example is the water molecule, H_2O, whose electrons lean onto the O atom leaving each H atom with a net positive charge. Further restrictions (selection rules) restrict the effective intensity of dipole oscillation to transitions between successive levels of rotational or vibrational energy in each spectral sequence.

The energy separation of transitions between different pairs of vibrational levels is of the order of 10^{-2} eV and between pairs of rotational levels is still smaller by 1–2 orders of magnitude. Accordingly each vibrational transition generates a band of rotational lines, whose own spacings are linear or quadratic functions of the relevant angular momentum numbers J. The vanishing spacing as $J \to 0$, or at other 'band heads', is prominent (figure 13.4). The art of interpreting these complex manifolds of lines was well developed in the mid-1920s, at least for the diatomic molecules whose stationary states are classified by the constant angular momentum component $\lambda \hbar$ along the internuclear axis. Symmetries under inversion at the centre of mass and under interchange of identical nuclei have also been noted and interpreted.

By 1950 the spectra of numerous molecules, in the visible and near visible range of frequencies, had been studied as reported in G Herzberg's classic texts [16]. His 1950 table reported basic parameters of ~500 individual diatomic molecules: bond energies, internuclear distances, vibrational frequencies, moments of inertia, electron excitation energies, etc. Direct observations of rotational spectra by microwave spectroscopy were just beginning at that time; they will be discussed

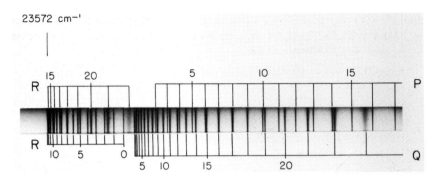

23572 cm⁻¹

Figure 13.4. *Band structure in the spectrum of electronic and vibrational excitation to states of AlH about 2.5 eV above its ground state. (Abscissa scale covers an energy range of nearly 0.05 eV, increasing towards the left.) The index N of each absorption line labels the initial (ground state) rotational energy level. The Q branch represents transitions in which the N value remained constant but the rotational energy was reduced by the increased moment of inertia. The P branch pertains to transitions N → N − 1 which reduce the rotational energy even further. The R branch pertains to transitions N → N + 1; here the increase of N prevails initially over the increased inertia but is then overwhelmed by it.*

in later sections. A remarkable rotational property of H_2 and other molecules is described in Box 13E.

Excitation of inner-shell electrons gives rise to emission of x-rays or of Auger electrons, much as outlined for atoms in section 13.2.3.3. Note, however, that temporary vacancies generated in inner shells are generally localized in one individual atom within a molecule thus disturbing the equilibrium of valence bonds and enriching the spectra.

Fine and hyperfine structures, analogous to those described for atoms in Box 13C, are also present in molecular spectra. Their initial observation, in the mid-1930s, by I I Rabi's group on beams of molecules—or atoms—in their ground states, subjected simultaneously to a constant magnetic field and to an oscillating field orthogonal to it, ushered in the new field of *radiofrequency spectroscopy*.

Molecules with a specified angular momentum component $j_z = m\hbar$, parallel to a uniform magnetic field $B\hat{z}$, were first sorted out by a Stern–Gerlach magnet, as outlined in section 13.2.3. Their magnetic energy levels in this field are represented in terms of a gyromagnetic (or magnetomechanical) ratio γ by $E_m = -\gamma m B\hbar$. Exposure of the beam to an oscillating field tuned to the precise frequency γB may cause transitions to $m \pm 1$, in which case molecules are rejected by a further analyser. The sensitivity of this selection allowed measurements of parameters γ of electrons and nuclei to unprecedented accuracy; Rabi hailed these results as the 'utilization of single molecules as nuclear laboratories'. This technique even led to the discovery of a charge asymmetry (quadrupole moment) in the nucleus of 'heavy' hydrogen ^2H.

BOX 13E: PARA- AND ORTHO-HYDROGEN

Molecular hydrogen was discovered by K F Bonhoeffer in 1929 to consist of two components that convert into one another only very slowly (in the absence of catalysts), a property shared by many homonuclear diatomics as predicted by D M Dennison. Hydrogen at very low temperature turns entirely into its *para* form.

This observation reflects the exclusion phenomenon (Box 13A) pertaining to its pair of nuclei, identical particles with spin $\frac{1}{2}$ and equivalent to a pair of electrons in this respect. The distribution function in space of hydrogen nuclei with parallel spins must change its sign under rotation of the molecule by 180°, as it does for the analogous electron pair. For the electron pair this change of sign implies a centrifugal repulsion, as it does for the nuclei of H_2, but here the centrifugal effect implies a more obvious manifestation, namely, non-zero rotational energy (more precisely, molecular rotation with an odd value of its angular momentum index J). Accordingly, the *ortho* species of H_2 disappears at low kelvin temperatures, whose minimal thermal agitation does not support even the lowest non-zero rotational level. Rising temperatures allow the resumption of H_2 molecular rotation but the spin arrangement of the nuclei is well insulated from external thermal actions, whereby the onset of odd-J rotation hinges largely on external magnetic actions.

Incidentally the nomenclature 'ortho' and 'para' for parallel and antiparallel spins originates from the empirical systematics of the atomic spectrum of helium, which also appeared to consist of two substances with different spectra, until Heisenberg traced their difference to the mutual electron spin orientation in 1927. Analogous classifications arise for electron and rotational spectra of all systems whose identical particles have alternative spin states.

13.2.5.3. Reactive collisions. Reactive collisions involve a transfer of particles between projectile and target, including the restructuring of a pair of molecules into new entities, the essential feature of chemical reactions. The large mass of molecules, as compared with electrons, makes them candidates for at least partially semi-classical treatments. We outline here a sketch of the widely studied prototype approach to such a treatment, dating from the 1930s and to be expanded in section 13.6.4.

Consider the simple reaction

$$AB + C \rightarrow A + BC \tag{5}$$

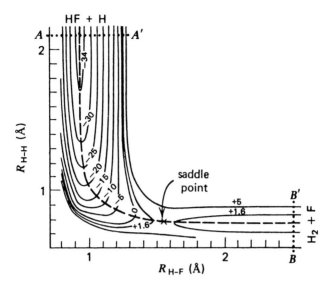

Figure 13.5. *'Contour map' of the potential energy surface for the linear molecular complex F-H-H, corresponding to equation (5) in the text (F ≡ A; H ≡ B, C). Energy contours are labelled in kilocalories per mole (1 kcal mol⁻¹ = 0.043 eV). The zero of energy corresponds to the limit of separate H_2+F. The dashed line indicates the reaction path.*

where {A, B, C} indicate atoms or molecular groups, assuming its elements to be restricted to a straight line. This model involves two parameters only, namely the non-negative distances r_{AB} and r_{BC} shown in figure 13.5. Initially r_{AB} oscillates in a potential well analogous to those of figure 13.3, whereas r_{BC} is very large; at the end r_{AB} is large and r_{BC} confined. The rearrangement takes place at $r_{AB} \sim r_{BC}$, that is, astride the potential barrier that separates the entrance and exit valleys. The ABC complex evolving in this region is said to be in its *transition state*. The relevance of this schematic model to the actual three-dimensional phenomenon does not appear to have been discussed at that time.

13.3. Completing the spectrum of radiation actions

Two major gaps remained at mid-century in the study of the radiation action, an action introduced in section 13.2.2 and elaborated in sections 13.2.3 and 13.2.5.2 within a limited context. The information provided in the latter sections dealt primarily with the action of visible light extending into the near infrared on the lower frequency side and somewhat further into the ultraviolet at higher frequencies.

Additional studies had dealt with radiofrequencies generated by electric processes and with the x-rays generated mostly by bremsstrahlung (section 13.2.2.1) and also by nuclei or cosmic rays. Technical difficulties had prevented detailed exploration of the action of radiation with frequencies in the gaps between these three ranges, each gap covering 2–3 decades of the spectrum. These difficulties concerned the generation of radiation with any desired frequency in the gaps, as well as the handling and measurements of these radiations. Typically, radiations between the far ultraviolet and near x-ray ranges are absorbed strongly by all kinds of matter.

Newer technical developments, mainly in the 1960s, have since afforded practically unrestricted exploration throughout both gaps. Their nature and results are reported separately for each gap in subsections 13.3.1 and 13.3.2. Main features of radiation action throughout its spectrum and on elements throughout the periodic system are summarized in sections 13.3.3 and 13.3.4

13.3.1. From ultraviolet to x-rays—synchrotron light

Transmission of ultraviolet light through optical systems is practically limited to photon energies below 10 eV, which is the threshold of absorption by lithium fluoride. Beyond this limit continuous spectra are produced by discharges in noble gases; their extraction into vacuum through orifices hinges, however, on the use of powerful pumps which became practical in the 1960s [17].

On the x-ray side, efficient production of bremsstrahlung was limited to photon energies in the keV range. This limit was overcome in St Petersburg by Lukirskii's clever instrumentation, reaching down to about 50 eV photons in the early 1960s [18] (figure 13.6) which, although a notable achievement, was soon overshadowed by the onset of alternative developments.

Within the gap lay the extensive emission spectra of atomic inner-shell transitions (section 13.2.3.5), whose discrete photon energies afford fragmentary glimpses of evidence on radiation action. Weissler's 1956 review of photoionization up to $\sim$50 eV reflects the modest progress achieved through such sources [19]. Evidence about photoabsorption spectra from inner shells of atoms seemed at the time compatible with continuation at lower photon energies of the saw-tooth spectra characteristic of the x-ray range (figure 13.7).

R L Platzman stressed in that period that adequate exploration of the 10–1000 eV range would hinge on extensive use of the continuous spectra from 'synchrotron light' sources. This radiation is emitted by electrons travelling on closed paths at speeds approaching the velocity of light c. Slower electrons travelling on closed paths act as radio antennas emitting a single frequency v reciprocal to their period of circulation, in directions above and below their path. At speeds approaching c, however, this emission consists mainly of extremely high multiples of that frequency, focused tangentially to the electron path and forming a practically continuous spectrum of uniform intensity. This spectrum extends to the frequency $(E/mc^2)^3 v$, where E indicates the electron's

(a)

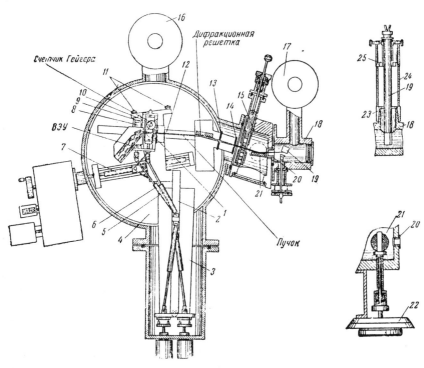

(b)

Figure 13.6. *Diagram and photograph (with T M Zimkina) of the x-ray apparatus designed by A P Lukirskii in about 1960, which extended the range of x-ray absorption measurements down to ~50 eV.*

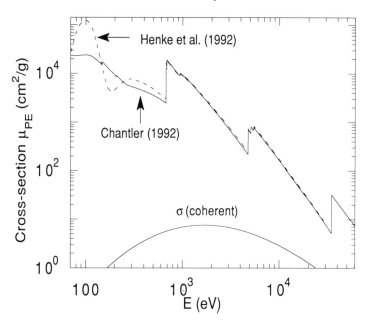

Xenon, Z=54

Figure 13.7. *Absorption coefficient of Xe gas as a function of photon energy, showing the sharp onsets of the K, L, M shell contributions (in order from the right-hand side). The units of ordinates are reciprocal to the gas layer thickness expressed in g cm^{-2}. The partial contribution of Rayleigh scattering is indicated by σ.*

kinetic energy and m its mass. (The ratio $(E/mc^2)^3$ amounts to $\sim10^{12}$ for $E = 5$ GeV.) This remarkable emission, long anticipated by theorists, was first detected in the late 1940s, and applied in 1953 to detect the photoabsorption by K electrons of beryllium at and beyond 110 eV.

The process of perceiving the full impact of synchrotron light spectroscopy and providing for its conversion into a basic tool occupied a whole decade. The striking detection by this method of a novel Rydberg series of He absorption lines near 60 eV in 1963 (figure 13.8) was, however, quickly followed by very extensive experimentation, mapping out the main features of photoabsorption spectra throughout the former gap, as reported comprehensively in a 1968 review [20]. These accomplishments required considerable technical developments for handling and measuring radiation in the new range of frequencies, such as improvements in the reflectance of mirrors and gratings and in the adaptation of photographic emulsions.

The photoabsorption and photoionization spectra of atoms and molecules thus obtained in the photon range of 10–1000 eV depart from the simple hydrogen-like pattern displayed in figure 13.7 in several

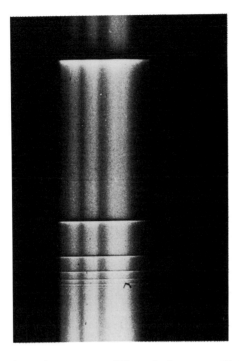

Figure 13.8. *Photoabsorption spectrum of He gas in the range ~60–65 eV (190–210 Å) showing a Rydberg series of resonances, generated by excitation to doubly excited $sp^1 P$ states.*

respects, as illustrated in figure 13.9:

(i) The rise of photoionization at the threshold for ejection from any one shell is often delayed and smoothed out by centrifugal barriers hindering the escape of slow photoelectrons.

(ii) The peaks of photoabsorption above each threshold are often broadened by the mutual repulsions between the numerous electrons of p, d or f subshells.

(iii) Characteristic deep intensity minima occur generally in the photoabsorption by electrons of incomplete shells. (This phenomenon, observed in alkali spectra since 1928, had been viewed as accidental; an indication of it in the Ar spectrum of reference [18] accordingly surprised its discoverers.)

(iv) Ubiquitous peaks often rise above the background of photoionization spectra as in figure 13.9.

13.3.2. From the near infrared to the microwave range

Extension to the infrared of the usual spectroscopy techniques meets difficulties. The lower-energy photons fail to activate usual types of

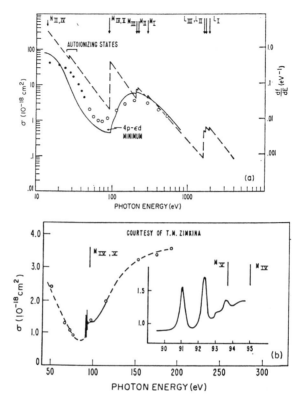

Figure 13.9. *Photoabsorption spectrum of Kr gas near the onset of the $M_{IV,V}$ (i.e. 3d) showing subshell contributions. (a) $\circ$ and $\bullet$, experimental values; $---$, hydrogenic approximation; ——, one electron model. (b) Details near the $M_{IV,V}$ threshold.*

photographic or photoelectric detectors; detection relied instead on observing the small temperature rise of 'bolometers', small black objects that absorb the incident radiation energy. Energy emission by intra-atomic currents acting as antennas is reduced in proportion to the cube of their oscillation frequency; its detection thus requires instruments with unprecedented sensitivity. The longer wavelength of infrared radiation boosts its penetration into mirrors or diffraction gratings, thus increasing energy dissipation.

The generation, manipulation and measurement of far infrared radiations of specified frequency and observation of their action hinged therefore on developing novel techniques. Three principal innovations have served this purpose progressively since the early 1960s, namely, laser sources of radiation (Chapter 19), analysis by Fourier-transform spectroscopy, and photoconductor detectors with thresholds in the far infrared (up to ~30 μm wavelengths).

Laser sources of infrared radiation were initially available at specific frequencies. The desired frequency selection was then attained indirectly through pairs of tunable dye lasers, whose beat (i.e. difference) frequency can scan the whole infrared range. An alternative consists of 'beating' the precise frequency of CO_2 or CO infrared lasers with the precise frequencies of electrically generated microwaves ($\simeq 10^{12}$ Hz). More recently, tunable lasers have been introduced, covering most of the infrared range.

Fourier-transform spectroscopy served initially to convert into frequencies, using computers without loss of accuracy, measurements of infrared wavelengths that were then more readily accessible by high-precision interferometry [21]. Direct and precise measurements of frequencies have become available more recently, including the generation of very high harmonics of microwave frequency standards.

Application of the principles outlined above have increased both the sensitivity and the resolution of measurements by several orders of magnitude. Thereby they afforded discoveries of molecular spectroscopy as will be reported in sections 13.6 and 13.8. Note that infrared radiation is hardly absorbed by electrons of single atoms in their ground state or of most substances other than metals. Its basic action consists of exciting vibrational and rotational motions of free molecules (and of their analogues in condensed matter). These excitations are restricted by the selection rules indicated in section 13.2.5.2 for single molecules. The restrictions are, however, bypassed in dense gases, through the effect of very frequent collisions that perturb the internal mechanics of each molecule. For example, single H_2 molecules have no independent electric dipole moment on which radiation may act, but absorb and emit radiation very appreciably when impacted by other molecules at high density or when polarized by external fields.

An additional related aspect of the infrared radiation's action in dense gases results from the enhanced ability of each molecule to transfer its excitation energy to surrounding molecules very rapidly. Retention of excitation by any molecule for only a brief interval Δt implies a reciprocal broadening $\Delta \omega \sim 1/\Delta t$ of its characteristic (resonant) absorption frequency ω. This broadening progresses, with increasing gas density, to the point of changing the aspect of absorption spectra radically, as illustrated in figure 13.10. It also lowers progressively the mean $\langle \omega \rangle$ of the frequencies thus absorbed.

13.3.3. *The spectrum of radiation actions*

The macroscopic action of radiation, with frequency ω and wavenumber k, on bulk matter is represented by the dielectric response function $\epsilon(\omega, k)$, often simply $\epsilon(\omega)$. A gaseous assembly of atoms or molecules is, of course, an example of bulk matter. The dielectric response $\epsilon(\omega, k)$ is a function of complex variables: its real part represents the electric susceptibility χ, i.e. the ratio of the matter's electric polarization to the inducing field strength E; its imaginary part represents the conductivity

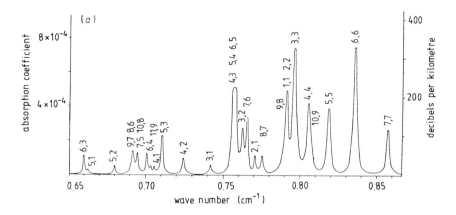

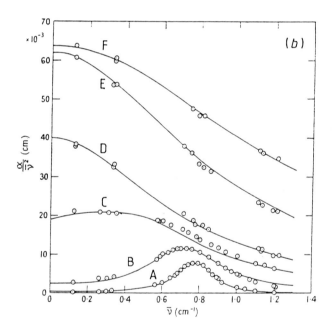

Figure 13.10. *Photoabsorption spectrum generated by excitation of the 'inversion' vibration of the ammonia (NH_3) molecule, in which the nitrogen atom oscillates back and forth through the triangular structure of the hydrogen atoms, with a frequency of 24 GHz (photon energy 10^{-4} eV). (a) In the gas at low density the oscillation frequency depends on the rotational state of the molecule, giving rise to manifold absorption peaks. (b) at increasing pressure the spectral lines are broadened by intermolecular collisions that interrupt the orderly vibrations. The entire spectrum shown in (a) has coalesced into the single broad absorption of curve A; at pressures increasing to 6 atmospheres the spectrum evolves into the curve F. The entire evolution of this spectrum has since been represented by a three-parameter formula.*

σ, i.e. the ratio of the current density to the inducing E. The dependencies of χ and of σ on (ω, k) are interrelated. Either of them has a peak slope where the other has a peak value.

The energy dissipated by radiation is proportional to $\sigma(\omega, k)$. The study of this parameter, especially in the range of high photon energies $\hbar\omega$, received great support, about and after mid-century, motivated by the need to shield workers, or even whole populations, from nuclear radiations. Here we survey briefly the spectral dependence of $\epsilon(\omega)$.

Radiofrequency energy is dissipated in conducting materials, i.e. metals, electrolytes and ionized gases (plasmas), mostly through collisions of free particles oscillating in the field. Similarly the rotation of polar molecules (i.e. molecules with electric dipole moments) is excited by radiation and then dissipated through collisions [22]. This process extends to non-polar molecules at high densities where polarity is induced by collisions, as noted in section 13.3.2.

Oscillating currents induced by radiation in non-resonant atoms or molecules act as antennas emitting in all directions (Rayleigh scattering), but these emissions combine to yield a net effect only when their density is non-uniform within a radiation wavelength. Non-uniform density fluctuations in the upper atmosphere thus combine to scatter blue light from the Sun, a process that redistributes radiation energy without absorbing it.

Resonant absorption of radiation energy by atoms and molecules of various species sets in progressively in the infrared, visible or ultraviolet frequency ranges, as described earlier and illustrated in figure 13.11 for H. Departures of this familiar process from the simple pattern of H astride its absorption threshold have been described in section 13.3.1 and are illustrated in figure 13.9. These departures subside, however, at higher frequencies ω; on the other hand the intensity of photoabsorption drops rapidly as ω increases because this process conserves both energy and momentum only within smaller and smaller volumes about the nucleus.

Radiation actions other than photoabsorption become appreciable at photon energies in the x-ray (multi-keV) range whose wavelengths are comparable to single-atom diameters. Here the electron density varies rapidly within a wavelength and non-resonant electron oscillations re-distribute radiation energy in a process of elastic (coherent) scattering. This process gradually overshadows photoabsorption as $\hbar\omega$ and the atomic number increase, being increasingly accompanied by the 'incoherent' Compton scattering, also introduced in section 13.2.2.1. Finally, the pair-production process—also introduced in section 13.2.2.1—emerges in the MeV photon range, becoming eventually the main agent of radiation-energy dissipation. The comparative contributions to energy dissipation in different materials and at different frequencies by these alternative processes have been the object of much study, as noted above. Current information is provided in a computer program [23].

13.3.4. *Spectral connection at ionization thresholds—sum rules*

The onset of resonant photoabsorption by atoms, starting from lower frequencies, occurs when the photon energy $\hbar\omega$ matches the difference between the atom's ground-state binding energy E_g and that of the lowest excited level, that is, at

$$\hbar\omega = E_g - E_n \qquad (n > 1). \qquad (6)$$

The value of E_n has been indicated in section 13.2.3.1 as 13.6 eV/n^2 for the H atom, and 13.6 eV/$(n - \mu_l)^2$ for alkali atoms. In either example the values of E_n converge rapidly to zero as n increases, indeed the photoabsorption spectra of atoms consist generally of a Rydberg series of lines converging to the limit, as $n \to \infty$, $\hbar\omega_n \to E_g$, beyond which the excited electron escapes from the ionic residue with kinetic energy $\hbar\omega - E_g$. Photoabsorption extends beyond this ionization threshold declining rapidly with increasing frequency as described above (section 13.3.3).

The qualitative difference between the discrete line spectrum below the ionization threshold and the absorption continuum above it contrasts with the basic continuity of the photoabsorption through the threshold illustrated in figure 13.11. This continuity, demonstrated mathematically by theorists in about 1960, is implied by the representation of photoabsorption through the imaginary part of the single function $\epsilon(\omega, k)$.

The magnitudes of photoabsorption in the discrete and continuum ranges, related to parameters of the $\epsilon(\omega)$ singularities, have been represented by numerical indices called 'oscillator strengths' since the earliest days of atomic physics. The oscillator strength of the nth line of the discrete spectrum, f_n, equals the ratio of its photoabsorption to that of an electron held near to equilibrium by a spring with force constant $k = m\omega_n^2$. The corresponding ratio applies to the oscillator strength of absorption in the continuum frequency range $\delta\omega$ about ω, $(df/d\omega)\delta\omega$.

The photoabsorption (at $k \sim 0$) integrated over the whole spectrum of any atom or molecule, expressed in the scale of oscillator strengths, equals its number of electrons

$$\Sigma_n f_n + \int_{E_g/\hbar}^{\infty} d\omega \, df/d\omega = Z. \qquad (7)$$

This famous 'Thomas, Reiche, Kuhn' sum rule guided Heisenberg's development of quantum mechanics. Its significance emerges by considering that the integral over the spectrum represents the action of an

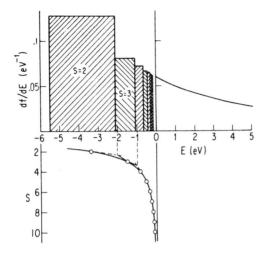

Figure 13.11. *Contributions to the photoabsorption by atomic H by its excitation to the series of discrete energy levels represented by the histogram and to the continuous spectrum of ionized states. The lower part of the diagram relates to the construction of the histogram.*

intense pulse of negligible duration, in which case each electron absorbs energy independently of all others. Useful sum rules analogous to (7) represent the spectral averages of various powers of the frequency [20].

Verification of equation (7) by experimental values of f_n for each material places a stiff requirement on the accuracy and completeness of photoabsorption measurements. An initial attempt to verify (7) for Al metal, in about 1967, fell short of the mark; critical analysis of the experimental data obtained with synchrotron light identified an instrumental defect. Successive data achieved agreement. Verification for other materials is now acknowledged as an important check of photoabsorption measurements, applied with success to elements with Z values up to ~30. Application to higher Z substances is complicated by difficulties in separating photoabsorption adequately from other processes, and by breakdown of the decay law $\omega^{-1-(7/2)}$ for photoabsorption anticipated qualitatively in section 13.3.3.

13.4. Excitation channels and resonance effects

The atomic excitation spectra described in section 13.2.3.1 pertain to unusually simple circumstances, in which a single electron moves about a spherically symmetric ion. Excitations of most atoms involve electron motions about ions with lower (or no) symmetry, often also joint excitations of two or more electrons in which the ion itself is also excited. In the case of molecules, the ionic core is apt to rotate and vibrate thus exchanging energy and angular momentum with external electrons.

The analysis of excitation spectra thus involves additional concepts and parameters that constitute the subject of the present section.

The common element of atomic or molecular excitations lies in energy that allows electrons, or other constituents, of a system to separate into fragments. The excitation energy may or may not suffice for the fragments to separate completely; full separation amounts to an ionization or a dissociation of the initial system. Alternative modes of fragmentation are called 'channels', following a custom of nuclear physics. Different stationary states (energy levels) of each channel differ in their total energy but are otherwise characterized by single values of the total angular momentum and total parity and by specified allotments of energy, angular momentum and parity among the fragments.

This allotment is, however, no constant of the motion, since interactions between fragments afford exchanges of energy, angular momentum and parity, especially when fragment pairs approach one another in the course of their motion. Thus, for example, a molecular electron may exchange energy with the ion's rotation, one of many possible effects of 'channel coupling'. The identity of each channel is thus manifest in its stages of large fragment separation, but each channel is *not* independent of other channels. The resulting shuffle of single-channel spectra obscures the simple pattern of Rydberg series described in section 13.2.3.1. However, much of this shuffle has been unravelled since the 1960s, through a major development of atomic theory.

This development started with the qualitative analysis of simple phenomena, to be outlined in section 13.4.1, observed in detail by novel instrumentation. Soon thereafter, in the mid-1960s, M J Seaton developed an analytic treatment of exchanges of angular momentum and of modest amounts of energy between a single electron and an atomic ion core. This treatment was intended for numerical *ab initio* evaluation of its parameters under simple circumstances. Its structure, on the other hand, soon proved applicable to an expanding range of spectra, through empirical fitting of parameters. Its main features will be described in the following subsections. The current need for its further extension will be discussed in section 13.10.

13.4.1. *Prototype phenomena*

A prototype of channel classification, shown in figure 13.12, rests on the energy levels of an ionic residue, He^+, omitting specification of angular momentum and parity allotment between this ion and an electron. Analysis of channel couplings centres thus on further identification and evaluation of relevant parameters. The spectrum of photoabsorption leading to excitation of He from its ground state into the cross-hatched block of the second row in figure 13.12 has been displayed in figure 13.8. The succession of intensity peaks in that spectrum corresponds to a

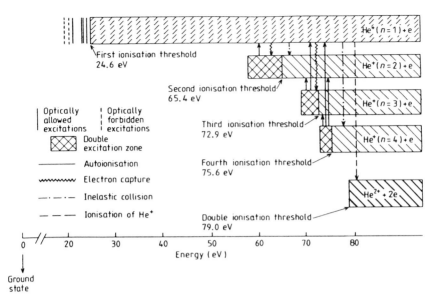

Figure 13.12. *Diagram showing an increasing number of channels in the spectrum of helium and their interconnections by auto-ionization and other processes.*

succession of doubly excited levels classed as $^1P^o$ (singlet pairing of electron spins, total orbital momentum $L = 1$, odd parity). The width and parity of each peak reflect the strength and mode of its coupling to the top channel of figure 13.12, namely, to separate fragments $He^+(1s)$ + free electron with orbital momentum $l = 1$. The relevant parameters are described in the following subsections. Roughly, the spectral width $\Delta\omega$ of each peak, in frequency units, is reciprocal to the time Δt spent by the atom in a doubly excited state of the second row before decaying into an ionized state with equal energy of the first row in figure 13.12.

The inverse process of this decay starts with the collision between an electron with kinetic energy of 35–40 eV and $l = 1$, and a helium ion He^+ in its ground state. The energy range specified here corresponds, in the first row of figure 13.12, to the cross-hatched portion of the second row. If the electron energy happens to match 1 of the energies of the peaks of figure 13.8, within its peak's width, the first-row ionized state is said to 'resonate' with the doubly excited state of the second row and can transform into it.

Such resonance phenomena were known previously at lower energies but their abundance, spread and sharpness had been widely underestimated. The first glimmer of the actual occurrence of resonances emerged two months prior to the observation of the spectrum shown in figure 13.8, in a measurement by G J Schulz of the elastic electron collisions with He atoms in their ground state, $e(\sim20 \text{ eV})$ + He [24]. The

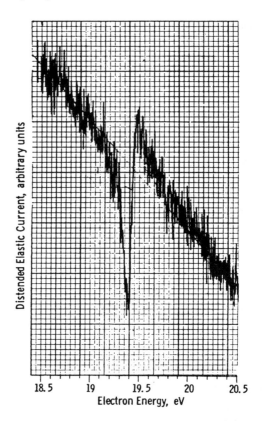

Figure 13.13. *Resonance at 19.37 eV in the spectrum of electrons deflected by 72° in elastic collisions with He.*

initial state of this collision belongs to the first row of the analogue of figure 13.12 for negative ions He⁻. The sharp oscillation in the number of 19.37 eV electrons deflected by 72°, displayed in figure 13.13, implies the temporary ($\Delta t \sim 3 \times 10^{-12}$ s) formation of a discrete resonant level of He⁻, a level classified as $1s2s^2\ {}^2S$ whose existence had been anticipated theoretically.

Schulz's observation, with its newly achieved resolution of the energy scale, was soon enlarged by J A Simpson with a more versatile instrument of equivalent resolution that detected *doublet* resonances in negative ions of all other noble gases. A significant aspect of this discovery lay in the coincidence of the doublet spacing of each *negative* ion resonance with the spacing of the ground-state doublet of the *positive* ion of the same element, implying that the resonant negative ion consists of a positive ion combined with a spinless pair of excited electrons. The versatility of Simpson's instrument yielded a steady stream of new resonance detections, in molecules as well as atoms, throughout 1963 and 1964.

A large amount of knowledge on atomic processes emerged from extensions of these pioneering observations of resonances. In particular, the, then novel, subject of negative ion resonances has recently been reviewed [25]. We shall return to this broader scope after introducing relevant theory.

13.4.2. *Quantum-mechanical parameters*

Our initial description of atomic electron states, in section 13.2.3.1, was cast in terms of radial and angular density distributions, represented as squared moduli of angular $[Y_{lm}(\theta, \varphi)]$ and radial $[f_{nl}(r)]$ 'density functions'. Such functions, generally complex valued, serve to represent electron densities and other observable quantities, much as complex amplitudes of light waves serve to represent their alternative (linear, circular or elliptical) polarizations. Light wave amplitudes are measurable, at least in principle, in macroscopic experiments; the quantum density functions are instead mathematical symbols, called 'probability amplitudes'—or more often wavefunctions—because their squared moduli serve only to represent probabilities of alternative observations of atomic phenomena, as described in Box 13F.

The number of excitation channels of any multi-particle atom or molecule relates to the number of its degrees of freedom and may thus be very large, even though finite. In practice, however, this number has been modest in most of the situations studied thus far, where many parameters remain 'frozen', barring very large energy inputs such as required, for example to destroy a whole atomic structure. Nevertheless the following developments should be viewed as being circumscribed by practical, if often implied, schematizations.

13.4.2.1. *Long- and short-range interactions—threshold effects.*

The interactions that underlie the transitions of fragments between channels extend over a range of inter-fragment distances, even though they vanish at large separations. This distinction is particularly obvious for the simple case of single-electron fragments, where the spherically symmetric Coulomb field of an ionic core predominates at large radial distances r over the dipole, quadrupole, and higher order 2^n-pole field components generated by core asymmetries, whose potentials decay as $r^{-(n+1)}$. Channel coupling stems entirely from multi-pole components. This section will deal primarily with a single-electron fragment and more briefly, in separate subsections, with multi-electron fragments and molecular dissociations.

Even though channel coupling receives contributions from interactions over a range of radial distances, it was found conceptually convenient to parametrize its effect by 'short-range transition matrices', acting at or near core surfaces, in contrast to matrices analogous to equation (8) that represent relationships at the limit of large radial distances. We have

---

BOX 13F: TRANSITION AMPLITUDES

As a prototype demonstration of probability amplitude manipulation, we describe here the amplitude that forms the core of the cross section (4) for scattering of an electron incident along a z axis into a direction (θ, φ), namely, $\Sigma_l (2l+1)^{1/2} \exp(i\delta_l) \sin \delta_l(k) Y_{l0}(\theta, \varphi)$. Its first factor $(2l + 1)^{1/2}$ represents (to within irrelevant factors) the probability amplitude for the incident electron to enter a state with orbital index l, an amplitude indicated in standard notation by $(i|l)$. The second and third factors, $\exp(i\delta_l) \sin \delta_l(k)$, represent a 'transition element', i.e. the probability amplitude $T_l(k)$ for the electron to rebound from a collision with parameter l into a direction other than its initial one. The last factor $Y_{l0}(\theta, \varphi)$, represents the probability amplitude of the electron rebounding in a specified final direction (f) forming an angle θ with the incidence. Thus one writes the probability amplitude formula for the specified collision as

$$(i, \theta_i = 0|T|f, \theta_f) = \Sigma_l (i|l) T_l(k)(l|f, \theta_f)$$
$$\propto \Sigma_l (2l + 1)^{1/2} \exp(i\delta_l) \sin \delta_l(k) Y_{l0}(\theta, \varphi). \qquad (8)$$

The set of amplitudes (8) for all pairs of initial and final directions (i, f) forms a square array, of infinite dimension, familiar in mathematics as a matrix, specifically in this case a *transition matrix*. It will serve us as a prototype for representing the connections among all pairs (i, f) of excitation channels.

The number of excitation channels to be considered in the next sections is generally finite, in contrast to the number of channels i and f of the amplitude (8). In practice, however, this amplitude also involves only a finite number of independent channels because the phase shifts $\delta_l(k)$ vanish generally for larger (e.g. two-digit) values of the orbital index l. Expansion of matrices $(i|T|f)$, connecting alternative excitation channels, into sums of terms with indices analogous to the orbital l of equation (8) is generally possible, but the physical significance of such indices is not apparent at this point of our story. Incidentally, however, we shall deal here with sets of channels having a common value of the total angular momentum index J, analogous to the index l summed over in equation (8).

previously used this approach implicitly in section 13.2.3.1 where the dependence of outer electron levels upon the structure of a multi-electron spherical core was incorporated into a single quantum defect parameter μ_l. A key feature of short-range parameters lies in their dependence on energies at a $\sim$10 eV scale, contrasting with long-range interactions whose effects are sensitive to energies near ionization thresholds at a sub-1 eV scale. Short-range parameters depend instead on energy only on the scale of potential and kinetic energies prevailing at short ranges. This concept, as well as most of its following elaboration and extensions, stems from the work of M J Seaton, often known as 'multi-channel quantum defect theory' [26, 27], partly outlined in Box 13G.

13.4.2.2. Multi-electron fragmentation. This subject was pioneered by G Wannier in a 1953 extension [28] of Wigner's threshold laws, dealing with the threshold for ionization of atoms by electron collision. The final state of this process has a pair of electrons emerging from an ion's field with near-zero energy; Wannier recognized that their joint escape requires the electrons to share their combined energy E above threshold evenly by escaping with nearly equal velocities on opposite sides of the atom. This requirement restricts the probability of joint escape severely; its analysis by classical mechanics identified a tiny bundle of suitable trajectories, accessible with probability $E^{1.127}$, the exponent being the root of a quadratic equation.

This famous result met with widespread resistance by theorists, even after its quantum-mechanical reformulation in 1970, until verified in 1973 by the brilliant experiment of Read and Cvejanovic [29]. A single electron emerging from the collision was detected, with energy $\epsilon < 0.02$ eV, as a function of the incident energy $E+24.58$ eV, 24.58 eV being the ionization threshold of the He target. Thereby only a fraction $0.02/E$ of double ionizations could be detected, with a net probability $E^{0.127}$ far easier to identify than the departure of double ionizations from a linear law. Figure 13.14(a) displays the results, extending to incident energies $E < 0$. The dip of the scored electrons astride $E = 0$ reflects the low probability of the two emerging electrons sharing their energy nearly evenly. The best fit shows the detection to be proportional to $E^{0.131}$.

The metastability of the approach to double escape identified by Wannier seems paradoxical: whereas the configuration of the two electrons on opposite sides of the ion is stabilized by their mutual repulsion, their maintenance of nearly equal velocities (and equal distances from the nucleus) during the escape seems wholly unstable. Indeed the pair's potential *peaks* at equal radial distances, $r_1 = r_2$, for a constant value of $R^2 = r_1^2 + r_2^2$. Stability is actually ensured by steady motion along R, as implied by Wannier's treatment and confirmed by later developments, but is nevertheless less than obvious.

This stability was then verified by another experiment in Read's laboratory [30]. This experiment measured the production of metastable excited states of He, at $\sim$20 eV above the ground state, by incident

BOX 13G: LONG-RANGE AND THRESHOLD EFFECTS

Seaton's analysis deals with the radial density function (wavefunction) of a single electron in an ion's Coulomb field, initially in the energy range where the fragmentation proceeds to infinite distance. We indicate this function here by $f_{kl}(r)$ as in section 13.2.4.2. The electron's energy in excess of the ionization threshold amounts to $13.6k^2$ eV if the wavenumber k, corresponding to the wavelength $2\pi/k$ of $f_{kl}(r)$'s oscillations at $r \to \infty$, is expressed in atomic units reciprocal to the radius a of equation (1). Specifically the leading term of $f_{kl}(r)$ at large r includes a main factor

$$\sin[kr + (\ln r)/k + \eta_l(k) + \delta_l(k)] \tag{9}$$

whose phase element η_l reflects a long-range influence whereas $\delta_l(k) = \pi\mu_l(k)$ reflects short-range influences, as it did in section 13.2.4.2.

The centrepiece of our parametrization connects the sine factor of $f_{kl}(r)$ with the corresponding factor of the functions $f_{kl}(r)$ pertaining to energy levels *below* the ionization threshold, as indicated in section 13.3.4. These levels, negative when measured from the threshold, were called 'binding energies' with positive values $13.6/(n - \mu_l)^2$ eV in section 13.2.3.1. A negative value of the energy, $13.6k^2$ eV implies an imaginary value of the wavenumber k. This shift to imaginary k affects, in equation (9), the short-range parameter $\delta_l(k)$ hardly at all, but changes the trigonometric dependence $\sin kr$ into a hyperbolic dependence $\sinh |k| r$. Since $\sin kr$ can be represented as the difference of two complex exponential functions, $\exp(\pm ikr)$, each of them becomes real, $\exp(\pm r/\nu)$, if we set $k \to i/\nu$. Expression (9) thus becomes

$$be^{r/\nu}r^{-\nu}\sin \pi(\nu - l + \mu_l) - b^{-1}e^{-r/\nu}r^{\nu}\cos \pi(\nu - l + \mu_l) \tag{9'}$$

with coefficients b smoothly dependent on ν. The obvious condition, that the probability amplitude $f_{nl}(r)$ remains finite as $r \to \infty$, implies that the value of $\nu - l + \mu_l$ equals an integer, as anticipated by the binding energy formula 13.6 eV$/(n - \mu_l)^2$.

Physically, the radial probability distribution $|f_{nl}(r)|^2$ of an electron bound to an ion consists of an *integer* number of lumps separated by 'nodes' of the wavefunction $f_{nl}(r)$.

CONTINUED

BOX 13G: *CONTINUED*

Smoothness of phenomena across fragmentation thresholds generally applies only to the photoionization of a single electron, and to the analogous particle–ion collisions. Extension to fragmentations with alternative interactions by C H Greene *et al* (cited in references [26, 27]) displays a variety of critical behaviours.

---

electrons of 20–25 eV. The yield of these excitations, plotted against the incident energy in figure 13.14(*b*), shows a series of peaks diagnosed as doubly excited states of He$^-$ with the pair of excited electrons at comparable distances from the nucleus ($r_1 \sim r_2$). This region of high potential energy has long been called the 'potential ridge' of two-electron systems.

The study of two, or more, electron excitations, of threshold phenomena, and of stability on the potential ridge has drawn extensive effort ever since the mid-1960s. Section 13.11 will return to this subject in a broader context.

13.4.2.3. Molecular dissociation. The dissociation of molecules, into neutral or charged fragments, is analogous to ionization in that it occurs at the limit of a succession of excited states, vibrational excitations in this case. This process has been treated in the framework of quantum defect theory by R Colle.

The succession of vibrational levels terminating at the dissociation threshold is finite, at variance with the infinite Rydberg series of electronic spectra, reflecting the difference of the relevant potentials. The vibrational spectrum has been studied in detail up to threshold in the 1980s mainly for H_2 and HD by A Carrington; photoexcitations to the continuum are beginning to be studied by multi-photon absorption.

The Born–Oppenheimer approximation that governs the vibrational motion of molecular nuclei becomes insecure at the very threshold of dissociation, because the electrons no longer move faster than nuclei in the tails of their distribution functions which are separating out. We will return to this subject in sections 13.6 and 13.11.

13.4.3. Multi-channel formulation—resonances

The theoretical elements introduced in section 13.4.2 can be used, in principle, to treat all multi-channel phenomena. In practice *ab initio* applications have been confined to phenomena with no more than two electrons outside an ionic core. Empirical determination of short-range

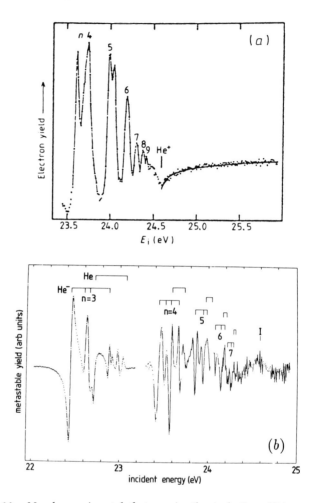

Figure 13.14. *Novel experimental features in the inelastic collision spectrum of 20–25 eV electrons with helium: (a) Yield of very slow electrons emerging from the collision. (b) Yield of metastable excitations of He.*

parameters, to be introduced in section 13.4.3.2, has afforded broader applications.

The main limitation to *ab initio* studies lies in the current want of proficiency in dealing with more than two electrons over radial distances far exceeding those of ionic cores. The problem lies in calculating the relevant transition matrices, $(i|T|f)$, or their equivalent 'R matrices' [31], for larger sets of open channels; these parameters must encompass relevant effects of electron interactions outside the core. Seaton's compendium [27] calculates R matrices within the framework of collision theory, i.e. by generalizing section 13.2.4.2 to include excitations of a single target electron to alternative discrete energy levels. C H Greene and collaborators have refined more recently effective variational calculations of R matrices in diagonal form [32], that is, with eigenvalues $\tan \delta_{\alpha J}$ (or $\tan \pi \mu_{\alpha J}$) analogous to the phase shifts δ_l of equation (8), and eigenvectors $(i|\alpha J)$ that connect the channels αJ to alternative specified channels i. The calculation utilizes sets of independent-electron orbitals in the core range, i.e. within a sphere of radius r_0, treating electrons of closed shells as forming a spherical potential $v(r)$, as was also done by Seaton. A limitation lies in formulating appropriate boundary conditions for electrons reaching the radius r_0. Analogous, but far more extensive, R-matrix calculations have been performed by P G Burke and a large number of collaborators over the last 20 years [33].

An essential aspect of channel coupling effects lies in the occurrence of 'closed' ionization or dissociation channels $|f)$, whose threshold energy for fragmentation exceeds the initial $E + E_g$. The diverging term of the factor $(9')$ of $|f)$'s density functions, namely, the term with the factor $\exp(r/v)$ should, of course, be removed as was done in that single channel example by requiring $v - l + \mu_l$ to take an integer value. At variance with removal of the analogous term $\exp(-ik_f r)$ for $E_{ft} < E + E_g$, mentioned in Box 13H, the elimination of each $\exp(r/v_f)$ requires a specific concurrence of all terms of the $\Sigma_{\alpha J}$ in equation (10). The appropriate procedure, to be introduced in section 13.4.3.1, constitutes the source of all spectral *resonances* and thereby a central element of the multi-channel quantum defect theory.

13.4.3.1. Photoabsorption by molecular hydrogen near its ionization threshold. A '2-3 channel' example. The workings of multi-channel theory came into sharp focus in 1969 through the successful analysis of a single experiment, even though the channel and resonance concepts had been previously familiar and Seaton's formalism was complete. G Herzberg aimed at observing a line spectrum converging to the ionization threshold of H_2, in order to establish the threshold's position very accurately. To this end he minimized the influence of molecular rotations and vibrations by studying samples of H_2 in its state of lowest energy, boiling off a low-temperature liquid [34]. His goal was, however, frustrated by the irregularity of the spectrum actually observed (figure 13.15(*b*)).

The origin of the irregularity, soon apparent to Herzberg, is indicated in the level diagram next to the spectrum: photoabsorption actually excites two alternative short-range (αJ) channels of H_2 with $J = 1$, each of them coupled to two different long-range channels, with ionization thresholds (E_{t0}, E_{t2}). The short-range channels correspond to excitations polarized along (σ) and across (π) the molecular axis; this distinction persists with increasing radial distance as long as the electron's motion about the core, with orbital index $l = 1$ (p state), remains anchored to

BOX 13H: *R*-MATRIX APPLICATIONS

One obvious application of R matrices consists of extending the elastic cross section (4) to include the inelastic channels of interest. For any given energy E and direction of an incident electron i, one determines its relevant admixture of channels $|\alpha)$ with angular momenta J (including spins) of the complex formed by the electron + target combination $(i|\alpha J)$. These are then combined with the R matrices to form the transition probability amplitudes

$$(i|T|f) = \Sigma_{\alpha J}(i|\alpha J)\exp(i\delta_{\alpha J})\sin\delta_{\alpha J}(\alpha J|f) \qquad (10)$$

to alternative final channels $|f)$. The set $\{|f)\}$ should include all channels that have at least one energy level lower than the sum of the target E_g and the incident E.

Here, as in equation (4), it is understood that the density functions of incident channels $(i|$ include only the 'ingoing' complex portion $\exp -i(k_i r + \ldots)$ of the sine factor (9), and those of $|f)$ only the 'outgoing' portion $\exp i(k_f r + \ldots)$. These conditions are fulfilled automatically by constructing the coefficients $(i|\alpha J)$ and by using the transition element in equation (10).

A second basic application of R matrices deals with the photoabsorption of radiation with frequency ω by a target with ground-state energy E_g and with multi-channel excitations. The photoabsorption event proper takes place within the volume of radius r_0 included in the R-matrix calculation. The probability amplitude for excitation into an eigenchannel $(\alpha J|$ of the R matrix with energy $E_g + \hbar\omega$ if often indicated by $D_{\alpha J}$. (The value of J is the vector sum of the ground-state momentum J_g and of the angular momentum, generally unity, of the absorbed photon.) The probability amplitude for excitation into the channel f with energy $E_g + \hbar\omega$ is then represented by a formula akin to (10), namely

$$(i|D|f) = \Sigma_{\alpha J} D_{i\alpha}\exp(i\delta_{\alpha J})\sin\delta_{\alpha J}(\alpha J|f) \qquad (10')$$

if $E_g + h\omega$ exceeds the fragmentation threshold E_{ft}. This expression needs modifying if $E_g + \hbar\omega$ falls short of E_{ft}, as it does in the case of collisions, by a procedure outlined in the example of section 13.4.3.1.

the molecular axis by a quadrupole potential that decreases as r^{-3}. At larger distances the electron motion is instead controlled by its sharing of

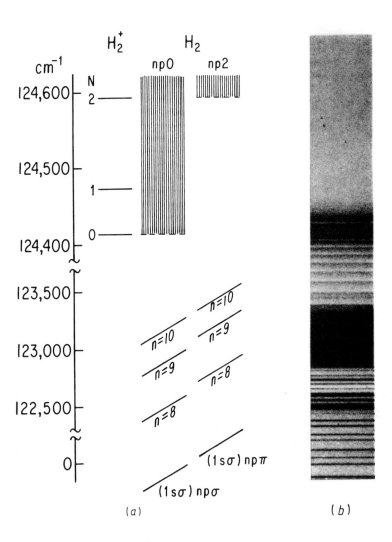

Figure 13.15. *Photoabsorption spectrum of molecular hydrogen in its state of lowest energy ($(1s\sigma)^2, v = J = 0$). (a) Schematic diagram of alternative pairs of excitation channels identifiable in the ranges of lower excitation and above ionization thresholds, respectively. The lower excitations are labelled by $|\alpha J\rangle$, where $\alpha \equiv (1s\sigma np\sigma$, or $1s\sigma np\pi)$ and $J = 1$. The ionization channels are instead labelled by $|fJ\rangle$ with $f \equiv (N = 0$ or $N = 2)$ and $J = 1$. The lower excitations are represented schematically by three pairs of levels with $n = 8, 9, 10$ and quantum defects $\mu_\sigma = 0.22$ and $\mu_\pi = -0.06$. The ionization levels are represented by continuous bands with different thresholds labelled by $N = 0$ or 2. (b) Observed spectrum extending over a range of about 100 cm^{-1} straddling the $N = 0$ threshold. Individual features are represented accurately by solutions $\{A_\sigma, A_\pi\}$ of the pair of equations (12) below the $N = 0$ threshold and of the single equation with $N = 2$ above that threshold.*

1071

Gerhard Herzberg

(German, b 1904)

Gerhard Herzberg was born and educated in Hamburg, and received his PhD in Darmstadt. He has since pioneered in molecular physics for nearly 70 years, producing an extraordinary volume of research, both experimental and theoretical, basic textbooks and the development of a first-class laboratory. Forced out of Germany, he worked and taught at the University of Saskatchewan, at the Yerkes Observatory, and finally at the NRC Institute in Ottawa which now bears his name. He received the Nobel Prize for Chemistry in 1970.

The whole body of molecular physics, primarily through the study of spectra, relies largely on his action, through his invention of experimental approaches, his detection of novel regularities, and his development of texts with extensive tables of data. The monumental series *Molecular Spectra and Molecular Structure* consists of three volumes on *Spectra of Diatomic Molecules, Infrared and Raman Spectra* and *Structure of Polyatomic Molecules*.

energy with rotation of the ionic core H_2^+ with angular momenta $N = 0$ or 2; the small rotational energy difference, $E_{t2} - E_{t0} = 0.022$ eV, looms large on the scale in figure 13.15(*a*) and of the electron motion close to its ionization threshold [35]. The combined action of these circumstances is treated in Box 13I.

The treatment of Herzberg's two-channel example extends readily to atoms and molecules with any number of channels, namely, to the majority of spectra, as noted in Box 13I. It is indeed normal that the state of an electron, or other fragment, emerging from a core still bears the characteristic of the combined system 'core + fragment'. This state evolves, however, as the fragment's radial distance increases, to assume the character of a loose association of fragment and core. It is thus generally appropriate to distinguish long-range channels (*f* or *i*) with loose association from short-range eigenchannels αJ.

The characteristic feature of the algebra in Box 13I, introduced by Seaton [27], lies in the occurrence of M-fold products of periodic functions of parameters ν_f (or equivalent) which depend in turn on the total energy E. The seeming capriciousness of spectral lines and resonances actually reflects the interplay of the periodic functions in equations (9′), (12) and (13). Knowledge of this feature helps to discriminate intrinsic from accidental features of observations.

A third excitation channel, namely, vibrational excitation, manifests itself in figure 13.15(*b*) mainly through the interloping dark bands, generated by combinations of lower, shorter-lived electronic excitations and of vibrational transitions. The latter are very intense because the removal of the excited electron from the molecule's ground state relaxes the bond between its atoms, thus inducing a major readjustment of the nuclear motion. The vibrational excitations have been fitted quantitatively within this framework, as detailed in the references quoted above. Additional applications of this approach to molecules, prior to 1985, have been reviewed in reference [37].

13.4.3.2. *Empirical parametrization of channel coupling.*

The use of experimental short-range parameters $\delta_{\alpha J} = \pi \mu_{\alpha J}$ and of the separation of ionization thresholds $E_{t2} - E_{t0}$ in section 13.4.3.1, together with graphical illustrations of the solution of equation (12), opened up an extensive series of spectral applications. The first of these consisted of a detailed analysis of the photoabsorption spectrum of xenon by K T Lu [38]. The Xe atom, and those of all noble gases except He, have a doublet ionization threshold $(t_{3/2}, t_{1/2})$ with easily measurable separation. Accordingly, the set of ν_f parameters of the relevant form of equation (13) reduces to two functionally related elements $\{\nu_{3/2}, \nu_{1/2}\}$ at all energies much lower than the thresholds for electron excitations of Xe^+. Fitting the wealth of lines observed in this spectrum afforded a quantitative determination of an extended set of parameters, analogues of those in equations (11–13).

It proved particularly effective here as it had in reference [35], to plot the energy of each spectral line, represented by its parameter pair $(\nu_{3/2}, \nu_{1/2})$ as a point on a graph with these coordinates. The dependence of equation (13) on trigonometric functions of ν_f made it possible to restrict the coordinate scales to their non-integral values, as illustrated by figure 13.16. The features of the curves drawn through the experimental points provided rich physical interpretations.

Spectra with multiple ionization potentials have been similarly analysed using alternative partial two-dimensional plots, combined with parameter fitting by computer operations. Such procedures have now become a major tool of spectral analysis. Indeed it has become practical to distribute spectroscopic information in terms of multi-channel theory.

BOX 13I: INTERPLAY OF SHORT-RANGE AND LONG-RANGE PARAMETERS

The channel coupling formulation of Box 13H, with coefficients $(\alpha J | f)$, turned out to be ideally suited to connect the short-$\{(\alpha J| \equiv (\sigma, 1), (\pi, 1)\}$ and long-range $\{|f) \equiv (N = 0, N = 2)\}$ parameters. The relevant $(\alpha J | f)$ coefficients, determined by simple angular momentum algebra, equal $\{\sqrt{(1/3)}, \pm \sqrt{(2/3)}\}$; the quantum defects $\{\mu_\sigma, \mu_\pi\}$, observed in low excitation spectra, yield the two eigenphases $\delta_{\alpha J} = \pi \mu_{\alpha J}$, whereas the two ν_f parameters $\{\nu_0, \nu_2\}$ are related to the threshold energies E_{t0}, E_{t2}. and to the electron's energy E by

$$\nu_f = \left(\frac{13.6 \text{ eV}}{E_{tf} - E} \right)^{1/2} \qquad f \equiv N = (0, 2). \qquad (11)$$

The occurrence of different pairs of channels, at short and long ranges from the H_2^+ ion core (see section 13.4.1), now plays a decisive role: whereas, in the single channel equation (9′), the parameters ν and μ_l belong to the same channel, here the short-range quantum defect μ_l is replaced by the pair of alternative short-range parameters (μ_σ, μ_π). By the same token, both short-range channels contribute to each long-range channel $|f)$, with respective amplitudes (A_σ, A_π). The condition $\sin \pi (\nu - l + \mu_l) = 0$ that removes the rising exponential in equation (9′) is now replaced by the *pair* of conditions

$$\Sigma_{\alpha = \sigma, \pi} A_\alpha (\alpha J | f) \sin \pi (\nu_f - l + \mu_\alpha) = 0 \qquad \text{for } f \equiv N = (0, 2) \quad (12)$$

where each sine function depends on parameters (ν_f, μ_α) of *different* channel sets.

This pair of linear equations, in the two amplitudes (A_σ, A_π), is solved by elementary algebra *if, and only if*, its coefficients satisfy the trigonometric condition.

$$\tfrac{1}{3} \sin \pi (\nu_0 + \mu_\sigma) \sin \pi (\nu_2 + \mu_\pi) + \tfrac{2}{3} \sin \pi (\nu_0 + \mu_\pi) \sin \pi (\nu_2 + \mu_\sigma) = 0.$$
$$(13)$$

CONTINUED

13.5. Collisions between atoms or ions [39]

The study of collisions, introduced in section 13.2.4, has expanded greatly since mid-century. The wide diversity of circumstances presented by

---

BOX 13I: *CONTINUED*

Solutions of equations (12) and (13) determine then the positions and intensities of the spectral lines as well as the intensity spectrum and angular distribution of photoelectrons ([26,35] and references therein).

The basic equation (13) remains relevant for electron energies E between the two ionization thresholds E_{t0} and E_{t2}, with a minor modification. Here the condition (12) for $f \equiv N = 0$ is no longer relevant, whereby the spectrum of energies E is continuous. The intensity of this spectrum is, however, modulated by equation (12) with $f = 2$, a modulation consisting of *seemingly capricious resonances*, corresponding loosely to the Rydberg series that would converge to the ionization threshold E_{t2} in the absence of channel coupling [36].

Thus equations (11) and (12) apply to larger channel sets $\{|f)\}$, to sums over many eigenchannels $(\alpha J|$ and to larger matrices of coefficients $(\alpha J|f)$.

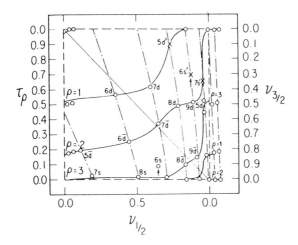

Figure 13.16. *Plot of observed level energies E in the Xe spectrum as open circles, each with a pair of coordinates $(\nu_{1/2}, \nu_{3/2})$ determined from the analogue of equation (11) with reference to the observed pair of series limits $(E_{t3/2}, E_{t1/2})$ as indicated in the text. The large number of points available proved sufficient to identify three lines ($\rho = 1,2,3$) on which all points lie, and thereby to evaluate all parameters of algebraic equations represented by those lines.*

this study nevertheless still leaves it in a rather fragmentary state, a view applying specifically to the subject of the present section. Electron collisions have been largely included in section 13.4, albeit mostly by implication. Collisions between molecular aggregates of atoms will be touched upon in section 13.6 on molecules.

Collisions between atoms or ions are governed by different circumstances depending on whether the atoms' speed is higher or lower than that of the relevant atomic electrons. (The *energy* of colliding atoms generally exceeds by far that of relevant electrons owing to the far higher mass of nuclei; recall in this context that a heavy particle can impart to a single electron at rest no more than twice its own speed.)

Collisions by very fast atoms are thus unlikely to transfer much energy to electrons. Collisions by very slow atoms or ions are also unlikely to transfer much of their own energy in a single process. Collisions with speed comparable to that of valence electrons have the highest probability of exciting their target; their correct treatment has become reasonably accurate only in the last decade. These three collision classes will be discussed in separate subsections. Collisions involving highly charged (stripped) ions will be dealt with separately in section 13.5.4.

13.5.1. *Projectiles faster than atomic electrons*

Long-range collisions of this class, mediated by the Coulomb field of charged projectiles, were introduced in section 13.2.4.1. Their average energy transfer to the target decreases approximately in inverse ratio to the squared velocity of the projectile owing to the shorter duration of the collision. Specific effects of interest may decrease much faster, of course, as the projectile's speed increases; an extreme example is seen in the electron capture by the projectile which would decrease as v^{-12}, according to elementary theory, but actually does so only as v^{-11} through the Thomas process described in section 13.2.4.3. Neutral projectiles interact only in grazing and close collisions. At high speeds, projectile and target constituents act basically as independent nuclei or electrons colliding with one another in binary processes as follows.

The impulsive character of fast collisions, stressed by Bohr (section 13.2.4.1), is shared by close- and long-range collisions; one or more constituent of the target, recoiling upon absorbing a quick shock, may react independently if the shock is sufficiently violent.

Attention has focused, in recent decades, on the following special processes.

(i) A variant of electron capture, in which a target electron fails to become attached to the projectile but travels with it at nearly equal speed and direction (Rudd–Macek effect).

(ii) Ejection of inner-shell electrons by the impact of fast ions.

(iii) Electron capture by highly charged ions; in this case the electron settles preferentially in excited states where its speed remains more nearly comparable to the ion's velocity.

(iv) Collisions leading to excitation of two (or more) electrons. This process, purely electronic for long-range collisions, may or may not involve repeated actions by the projectile.

(v) In contrast to long-range collisions, whose action is independent of the sign of the projectile's charge, close collisions differ, e.g. for protons or antiprotons which attract or repel the electrons, respectively.

13.5.2. *Projectiles slower than atomic electrons*

The effectiveness of slower collisions is restricted by the 'Massey criterion', introduced in section 13.2.4.1, which excludes energy transfers whenever the product of the collision's duration and of the transfer's oscillation frequency greatly exceeds unity. This limiting frequency drops, however, sharply for collisions at certain interatomic distances that bring the energies of two electron levels within the *colliding pair* into near coincidence (quasi-degeneracy). The collision is accordingly viewed in this case as a *molecular* process.

The occurrence of quasi-degeneracies, apt to generate collisional energy transfers, becomes obvious upon inspection of the 'Born–Oppenheimer potential curves'—introduced in section 13.2.5.1 and illustrated in figure 13.3—for the 'complex' formed by the colliding atoms. Degeneracies correspond to crossings and quasi-degeneracies to 'avoided crossings' of such curves. Sizeable studies of such excitations resulting from grazing collisions have been developed in recent decades, combining experiments and theory. One specific aim of these studies has been to observe and interpret the orientation and alignment of excited atoms induced by collision [40].

Close collisions tend to be more complex as atoms interpenetrate deeply, displacing whole shells of electrons. A dramatic demonstration of the role of level crossings emerged, however, in the 1960s, in experiments on $Ar^+ + Ar$ collisions at energies of the order of 50 keV, that is, at a speed far lower than the electrons' [41]. In these collisions the nuclei follow a rather definite trajectory, thus correlating the observed projectile deflection with the closest approach, r_0, of the two nuclei. The balance of momentum and energy transfers also determines the total energy Q transferred to electrons. Morgan's plot of the average $\bar{Q}$ as a function of r_0 showed a sharp jump from $\sim$100 eV to 600–700 eV as r_0 traverses the very narrow range from $\sim$0.025 to $\sim$0.023 nm (figure 13.17(a)). This sharp discontinuity was interpreted as the *onset of overlap* between the inner L-shells of the colliding atoms, in apparent contrast to the notion of each shell spreading over a broad range of radial distances.

Further details emerged soon from observations at St Petersburg detecting projectile and recoil ions in coincidence, and thereby resolving the jump into two steps of $\sim$250 eV each, corresponding to the sequential ejection of two L-shell electrons. Connection with theory was then

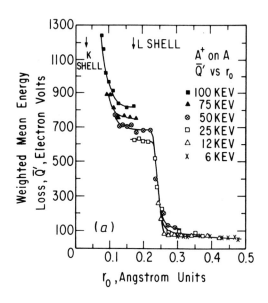

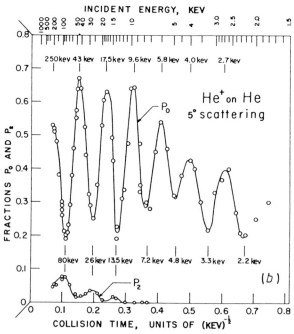

Figure 13.17. *Early demonstrations of novel phenomena in ion–atom collisions. (a) Energy transfer to electrons in $Ar^+ + Ar$ collisions as a function of internuclear distance, for various collision energies. (b) Incident ions He^+, colliding with neutral He gas atoms and emerging with 5° deflection, are selected according to their charge after collision: neutral, single or double. The fractions (P_0, P_1, P_2) thus selected are oscillating functions of the collision duration with nearly constant frequency.*

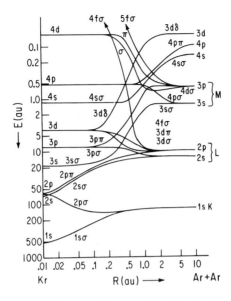

Figure 13.18. *Diabatic potential energy functions for the Ar$_2$ complex.*

established by W Lichten's construction of potential curves for the Ar$_2^+$ complex (figure 13.18), whose 'united atom' limit reduces the pair to a Kr$^+$ ion. Lichten identified a single level, occupied by just two electrons, which belongs to the L shell ($n = 2$) manifold in the 'separate atom' limit Ar+Ar$^+$, but gets 'promoted' [15], as r_0 decreases, to the $n = 4$ manifold of Kr$^+$. At the critical range of r_0, where $\bar{Q}$ jumps, this rising level *crosses* a bundle of descending levels that represent bonding valence states of the Ar$_2^+$ molecular ion; the degeneracy of the crossing thus shifts the electron pair into the valence shell and eventually to full ejection.

The striking phenomenon in figure 13.17(*a*) appears now as an early example of a multitude of level-crossing effects, ubiquitous in atomic interpenetrations, which promote electrons from inner to outer shells. The notion has been introduced recently that these effects may be usefully traced to a single (or multiple?) singularity of the mathematical representation of collisions as functions of the internuclear distance r_0 viewed as a complex variable [42].

13.5.3. Collision speed comparable to electron velocities

Early expectations that these collisions would favour electron transfers between projectiles and targets led theorists to anticipate that even repeated transfers could occur in a single grazing process. Actual observation of this phenomenon in He$^+$–He scattering with 5° deflection, at Connecticut in 1958, nevertheless attracted considerable attention [43] (figure 13.17(*b*)). Formation of the molecular complex He$_2^+$, as He$^+$ and

He merge at modest speed (but multi-keV energy), leads one electron to occupy with near equal probability either of two molecular energy levels, the lower one 'bonding' the two atoms, the higher one being instead repulsive (anti-bonding) as displayed in figure 13.3 for H_2^+. The energy gap between these levels rises to a few eV as projectile and target approach, decaying again as they separate. This circumstance leads the electron to oscillate between the two atoms with the beat frequency corresponding to the energy gap, as outlined at the end of section 13.2.1. The number of oscillations displayed in figure 13.17(*b*) represents the product of the oscillation frequency and of the collision's lifespan.

This notable observation coincided with the introduction, by D R Bates and collaborators, of systematic procedures for calculating the probability of electronic excitations resulting from similar processes, a development motivated at the time by interest in astrophysical applications. Projectiles are treated in these procedures as moving classically on straight, or nearly straight, trajectories. Atomic electrons are treated instead quantum mechanically, by perturbation methods under favourable conditions. More elaborate procedures are required to evaluate sizeable probabilities of single or sequential electron transitions. Specifically one calculates probability amplitudes $a_n(t)$ of observing, at various times t during the collision, the target and/or the projectile in their nth alternative 'closely coupled' state. The collision dynamics is then embodied in a system of differential equations governing the amplitudes $a_n(t)$ [44].

Applications of this approach have expanded progressively, utilizing the improvements in computer performance and responding to the rising requirements of plasma research, which involves highly ionized and colliding gases at thermal energies of the order of keV. Critical to the study of relevant collisions is the selection of distribution functions analogous to the $f_{kl}(r)$ (introduced in section 13.2.4.2 and again in section 13.4) pertaining to the relevant atomic or molecular state of projectiles and targets, and of their collision complexes. A recent review of these studies includes over 350 references [45].

These collision studies have dealt largely with grazing collisions and outer-shell excitations but have also reached deeper into atoms, treating for example the transfer of inner-shell electrons onto fast incident protons or onto highly charged stripped ions, viewed at times as analogues of the simple process of figure 13.17(*b*). We shall return to this subject in sections 13.5.4 and 13.7.

13.5.4. *Highly stripped ions*

Experimentation with multiply charged atomic ions started in about 1970, utilizing accelerators originally built (but grown obsolescent) for nuclear studies. Low-charge ions thus accelerated lost additional electrons upon traversing a foil; the higher charge thus attained could then amplify the energy gain by further acceleration followed by further stripping, at least in principle. Currently, alternative procedures can strip all electrons even from the atoms with highest nuclear charge.

The accelerated ions with intermediate charge ze (e.g. with $z \sim 10$) available in the 1970s and 1980s served initially as projectiles, more efficient than protons or α-particles, to study inner-electron ejection in ion–atom collisions. Accordingly, the circumstances of these collisions are generally analogous to those described in section 13.5.1. Electrons still clothing a stripped ion are also liable to excitation or ejection by interacting with a target. The study of ion–atom collisions was thereby enriched and pursued extensively. Collisions were classified according to the location and number of electrons actually involved. This multiplicity of alternative channels has hindered, of course, the development of any comprehensive theory.

The more recent availability of ions with high atomic number Z and clothed with only very few electrons has shifted recent attention progressively to a more straightforward task, namely, the collision and spectral study of few-electron systems in very strong Coulomb fields where relativistic conditions prevail. This study presents a specific challenge in that its relativistic multi-particle aspects are only amenable to perturbative, if very accurate, treatment. No unanticipated observation has emerged thus far in this field, but its study warrants attention [46].

Highly stripped ions are also normal constituents of plasmas at stellar temperatures, in the range 1–100 keV. Such plasmas are now generated in laboratories rather easily, if briefly, by the focused action of a pulsed laser on condensed matter surfaces; they are also maintained for longer intervals in devices developed for nuclear fusion. Their radiative emissions are of interest on one hand as useful indicators of the instantaneous state of a plasma and on the other hand as undesirable vehicles of energy dissipation. The spectroscopy of these emissions has expanded greatly in recent decades under the stimulus of plasma research. Collisions among stripped ions and with plasma electrons are, of course, important elements of plasma dynamics and evolution, as reported in Chapter 22.

13.6. Molecular physics [47]

The study of molecules has branched out since mid-century into a wealth of new directions, opened by technical developments. Novel observations of phenomena, specific advances of molecular spectroscopy, giant strides of quantum-chemistry calculations afforded by computer development, and progress on reactive collisions will be dealt with in separate subsections.

13.6.1. New avenues of experimentation
Much of the experimental progress resulted from improvements of seemingly straightforward procedures such as: (a) achieving, maintaining and measuring lower and lower pressures in experimental vessels by means of more powerful pumps, wall materials and seals, by about 1960; (b) producing, maintaining and monitoring more perfect and cleaner surfaces, whose study became possible in the 1970s; (c) improved preparation of material samples; (d) fast electronic controls of valves and of measuring devices (e.g. of mass spectrometers); (e) availability of intense, monochromatic and sharply tunable radiation sources (lasers) for spectroscopy; (f) development of shorter and shorter radiation pulses reaching now into the femtosecond (10^{-15} s) scale.

13.6.1.1. Cross-beam experiments. Chemical reactions had traditionally been produced and studied in bulk matter, generally gaseous or in solution. The production and observation of single collisions at the crossing point of unidirectional, monovelocity [48] beams of reactant molecules opened a new era of chemistry, beginning in the 1950s and fully established after 1960 [49]. Each collision product could then be collected and analysed in appropriate directions, thus determining dynamical parameters of each collision event.

13.6.1.2. Low-temperature beams from nozzle expansions. Sudden (adiabatic) expansion of a compressed gas into vacuum lets it squirt into a jet, converting most of its thermal energy into kinetic energy of unidirectional motion. Development of this technique in the late 1960s made available beams of desired composition at controlled kinetic temperatures of the order of 1 K [50]. At these temperatures, new molecular aggregates (van der Waals molecules and clusters of desired composition) form under the influence of their van der Waals attraction. It proved practical to expand buffer gases of low molecular weight, typically He or Ar, 'seeded' by traces of molecules of interest; fair-sized molecules could thus be observed with a few rare gas atoms stuck to them. These atoms could then be shaken off following spectral excitation of the carrier molecule [51]. A large variety of experiments has since been performed by this method.

13.6.1.3. Resonances. Resonant electron excitations, similar to those of atoms discussed in section 13.4, occur in molecules as well. A separate mechanism is afforded by suitable distributions of atoms and charges occurring in molecular structures, distributions that can hold an additional electron temporarily within the volume of a valence shell. A simple such mechanism rests on the non-negligible length/width ratio of diatomic molecules. This ratio allows a slow electron with orbital momentum $l = 2$ or higher to penetrate the centrifugal barrier

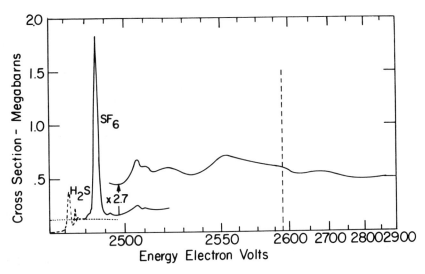

Figure 13.19. *Cross sections for photoabsorption by the K shell of the sulphur atom, near its threshold at 2.5 keV, in two alternative molecular combinations. The photoabsorption by SF_6 is boosted greatly by resonance resulting from the obstacle to S-electron escape through the surrounding, negatively charged, fluorine atoms. The integral of S-photoabsorption over the whole spectrum is equal for the two molecules, but its concentration near threshold in the case of SF_6 is outstanding. The vertical dotted line represents each of two different, but accidentally equal, contributions to photoabsorption: (a) the constant value of the K-shell photoabsorption in H_2S above its ionization threshold, (b) the contribution of photoabsorption by the six F atoms of SF_6 in this energy range.*

surrounding the molecule along its interatomic axis, thus reaching its internal Coulomb field more readily. The successive electron escape from this internal trap proceeds again more readily along the axis. The formation of such a short-lived molecular state is called a 'shape resonance'; the lowest one is known as the 'lowest unoccupied molecular orbital' (LUMO), a name contrasting with the 'highest occupied molecular orbital' (HOMO). Numerous such resonances have been observed at energies up to 30–40 eV above the valence levels [52,53].

This phenomenon is magnified in molecules with a central atom surrounded by 'electronegative' atoms that bear more than their share of electrons; the net charge on these atoms hinders the escape of any extra electron injected in their middle. Injection is provided here conveniently by photoexcitation of an inner shell, as illustrated in figure 13.19 by the contrasting K-shell absorptions of sulphur in the molecules SF_6 and SH_2 [54]. This striking observation was soon extended to other molecules, judiciously selected particularly by V I Fomichev and T M Zimkina in St Petersburg, and has recently become a frequent tool for studying the mechanics of molecular valence shells.

Extensive molecular resonances have also been observed in collisions of electrons with molecules, processes discussed in section 13.4 and illustrated in figure 13.12. E Lassettre in the 1950s and 1960s and P D Burrow more recently have been major contributors to these studies, which actually have a bearing on negative ion states of each target molecule.

13.6.1.4. Larger aggregates. The structure and the dynamics of the flimsy molecules held together by van der Waals forces, rather than by chemical bonds, have been studied by microwave spectroscopy, typically along the low-temperature beams within which they originate. The strength of the 'adsorption' bond by which a small molecule hangs onto a large one can thus be measured. New insight is also gained from the geometrical structures thus determined: For example, the dimer of the polar molecule HF might have been expected to form with each positively charged H atom next to the negatively charged F atom of the other molecule. One F atom appears instead to be bonded to both H atoms, tightly to one and loosely to the other, with the two bonds at an angle comparable to that of the water molecule H_2O, whereas the second F atom hangs loose [55].

Clusters, typically of monatomic substances, have been studied extensively since the 1970s, primarily as transition structures between molecules and bulk matter. Their cohesion energy per particle appears to oscillate with increasing size, being largest for atom numbers that afford a high-symmetry structure. A typical question concerns cluster reactions to external disturbances: is it elastic as for solids or yielding as for liquids? It is actually solid-like at lower temperatures and liquid-like at higher ones [56].

Polymers, i.e. aggregates of long and flexible molecular chains, were introduced in section 13.2.5, with reference to the rubber elasticity. They have been studied extensively since mid-century, with diverse compositions and properties, typically with regard to their relaxation after removal of an externally imposed strain.

Liquid crystals, i.e. aggregates of moderate-sized, often rod-like, molecules with important technical applications, have also been synthesized and studied extensively since the 1960–70s. P G de Gennes, a principal theoretician of both liquid crystals and polymers, reports on this subject in Chapter 21.

Biomolecules, primarily proteins and polynucleotides, also consist of chemically bonded chains of molecular groups more highly differentiated than for ordinary polymers. Knowledge of their structure was achieved in the 1950s through dramatic progress of x-ray diffraction of their crystalline form. Understanding of their delicate, critical and still rather mysterious properties—responses to external ionic charges, folding and unfolding in specific manners, efficient and controlled transport of electrons over nanometre ranges—remains fragmentary.

13.6.1.5. Observation of fast processes. Molecular processes develop on various, quite different, time-scales. Their observation and study have thus expanded in step with the progress of time-interval resolution of physical measurements. This resolution stood roughly at the nanosecond $(10^{-9}$ s) scale in about 1960, at picosecond $(10^{-12}$ s) in about 1970, and at femtosecond $(10^{-15}$ s) in the late 1980s. The nanosecond scale introduced 'flash-photolysis' and 'flash-spectroscopy' in physico-chemical research. The picosecond scale afforded observations of dissociation processes, of the approach to relaxation of the orientation and energy of molecules in solution; the femtosecond scale afforded fuller scope of relaxation phenomena.

13.6.2. *Molecular spectroscopy*
Molecular spectroscopy continued to develop after mid-century on its earlier path, yet expanded remarkably through the progress reported in sections 13.3, 13.4 and 13.6.1.

The treatment of multi-channel electronic excitations and ionizations, reported in section 13.4, more specifically through its illustration by the example of H_2 photoionization (section 13.4.3.1), proved very feasible for the molecular applications reviewed in reference [37]. Specifically relevant to molecules was the introduction of the quantum defect's dependence on internuclear distances R as the agent of specific transitions between vibrational states with distribution functions $|\chi_v(R)|^2$ and $|\chi_{v'}(R)|^2$

$$(v'|\mu_{\alpha j}|v) = \int_0^\infty dR \chi_{v'}(R) \tan[\pi \mu_{\alpha j}(R)]\chi_v(R). \qquad (14)$$

The availability of radiation sources with photons in the range $20 < \hbar\omega < 1000$ eV (section 13.3.1) has facilitated the detection of valence-shell responses to the ejection of inner-shell electrons. A surprising observation thus emerged by comparing the K-shell ionization thresholds of atomic and molecular oxygen, which differ by $\sim$10 eV ($\sim$540 eV for O, $\sim$530 eV for O_2), owing to relaxation of the valence electrons about the K-shell vacancy [57]. (This relaxation delays the vacancy's 'tunnelling' between the O_2 atoms beyond its lifetime, thus 'breaking the symmetry' of the homonuclear diatomic O_2.)

Molecular physics was greatly advanced by removal of the gap between the near infrared and microwave spectral ranges (section 13.3.2), affording full exploration of the vibrational and rotational spectra that serve to identify molecules, e.g. in outer space. Indeed the astrophysical role of molecules was hardly suspected until rotational frequencies appeared in radiofrequency spectra from astrophysical sources, in about 1960. Laboratory studies of molecular microwave spectra hinged initially mainly on measuring the power loss generated by introducing vapours in tunable cavities. The extension of vibrational spectra to approach the dissociation thresholds of H_2 and HD has been noted in section 13.4.2.2. Currently molecular spectra are often studied by passing a beam through waveguides that generate standing waves of accurately determined wavelength in the microwave or infrared range as indicated in section 13.3.2.

In contrast to the easy identification of optical emission or absorption in visible spectra, the detection of resonances in the lower frequency range hinges on preliminary, but quite accurate, knowledge of their probable location, owing to the limited practical range of tunability. A striking example of the difficulty in identifying specific spectra is afforded by the saga of the ion H_3^+, the smallest polyatomic structure, which was identified early in this century by mass spectroscopic analysis of discharges in hydrogen gas. Its spectral 'fingerprint' eluded detection until 1980, when successive estimates of its vibration frequencies and improved techniques led T Oka to its spectral detection at $\lambda \sim 4~\mu m$. Knowledge of its spectrum led then to abundant detection of H_3^+ in outer space, especially in the outer atmosphere of Jupiter [58].

13.6.2.1. Radiationless transitions. The motions of electrons and nuclei within a molecule have been viewed in section 13.2.5 as basically independent, owing mainly to the different scales of their respective energy spectra. (Electronic excitations are nevertheless generally accompanied by satellite vibrational excitations.) However, the higher-level density of polyatomic molecules, and of all molecules near their fragmentation thresholds, restricts the independence of electronic and nuclear motions, often drastically.

A prototype of radiationless transition between electronic levels, which transfers their energy difference to vibrational motions rather than to radiation, occurs commonly in organic molecules. The spins of ground-state electrons normally cancel each other forming a 'singlet' state, as indicated in section 13.2.5; their excitation by photoabsorption does not affect their spin state directly. On the other hand, the stability accruing to ground states by forming a spin singlet no longer holds for the excited electron. Indeed, a 'spin flip' by this electron—stimulated by the spin-orbit coupling (section 13.2.3.2)—reduces its Coulomb repulsion by the residual unpaired electron, owing to the centrifugal repulsion of 'triplet coupling' (Box 13A). (Excited electron states with very similar distributions in space generally occur with alternative mutual orientations of their spins, higher net spins then correspond to energy levels lower by amounts of the order of 1 eV, according to Hund's rule.) The energy thus released by electron spin flips is readily dissipated to vibrational motion in molecules of sufficient size or excitation.

This phenomenon is typical of organic *dye molecules* which absorb light of specific colours readily and fail to release it promptly by fluorescence coupled with a second spin flip. The energy is trapped instead in triplet or higher spin states, unable to dissipate quickly a larger amount of energy to vibration; it may be re-emitted eventually by delayed photoemission (phosphorescence) or dissipated slowly to other molecules in condensed media.

Analogous processes of radiationless transitions are increasingly common for higher and higher excitations of molecules. These phenomena escaped general attention prior to mid-century, when attention focused mainly on diatomics. Their emergence in the 1950s and 1960s represented a departure from the conventional wisdom viewing electronic and vibrational motions as distinct. General appreciation of the rates of electron energy conversion to vibration, rates that are modest yet frequently larger than for light emission, took hold eventually in the community, closing a chapter of doubts and debates.

13.6.3. Quantum chemistry. Theory and computations

The theory of molecular excitations and transformations (quantum chemistry), laid out in its essentials prior to mid-century (section 13.2.5), has been pursued vigorously. Its applications have expanded greatly, of course, stimulated by the pace of computer technology and by the amount of available effort. The basic ingredient of calculations is built from large sets of independent-electron distribution functions (orbitals) suitably combined with coefficients adjusted to optimize energy levels or analogous parameters. The optimization utilizes the Hartree–Fock method, configuration mixing and other variational procedures, as in the past. A bench-mark was achieved in the 1960s, by the Kolos–Wolniewicz calculation of the ionization threshold of H_2 molecules within 1 part in 10^5 of its experimental value. C Rothaan's introduction of analytic orbitals in the 1950s, affording easier interpretation and handling of electron distributions, leavened the field but wholly numerical descriptions of orbitals remain in use.

Still greater evolution occurred in the diversity of researchers' perceptions and attitudes regarding computations. Numerical results, as provided by current procedures, long reflected mainly the input data rather than the underlying physics. This subject was discussed in the closing lecture by C H Coulson to a 1959 conference, a lecture viewed now as historical [59]. Coulson noted the danger that the quantum-chemistry community would split between practitioners of massive computations and workers looking for greater transparency of results and approaches as well as for greater adherence to physical mechanisms.

This contrast is still apparent but has not led to dire consequences. It has been attenuated by the reduced cost of computing, both in terms of money and of personal effort. This reduction has encouraged heuristic computations as tools to explore, for example, the dependence of results on alternative inputs, an activity that has been expanding in recent years.

The construction of potential surfaces, requiring calculation of bond energies over multi-dimensional lattices of atomic configurations, has been particularly enhanced by computer progress. This activity was confined largely to tri-atomic systems for some years, but has since expanded rapidly to larger systems, for example Truhlar's group has dealt quantitatively with the seven-atom set in the reaction $CH_4 + OH \rightarrow CH_3 + H_2O$ [60]. Only three of these atoms (C, H, O) participate directly in the transfer of a single H atom, but the relaxation of the CH_3 residue is also relevant.

13.6.4. Reactive collisions

Reactive collisions have also been advanced greatly by computer developments, mainly through the construction of the potential surfaces outlined above. Basically, the calculation of a reactive collision consists of studying the evolution of a system of interest along a path on a potential surface, from a starting point where atoms are grouped, e.g. as AB+C (as introduced in section 13.2.5.3) to an end point with the structure A + BC.

The drastic schematization of such a process, described in section 13.2.3.5 as collinear, is now generally dispensed with, but section 13.6.3 noted that calculations extending beyond tri-atomic examples still remain infrequent. The central difficulty of this study lies in satisfying its implication that the potential surface should reflect realistically the deformation of atomic and molecular structure as reactants approach and intermingle into a complex.

A more subtle difficulty lies in verifying compliance with a main feature of the early concept of traversing a transition state in mid-collision, namely, that the traversal would proceed steadily forward without intermediate loops. Deviations from purely classical motion along a reaction path have been allowed for in past decades, for example by allowing for quantum tunnelling through potential barriers. Allowances for effects of quantum fluctuations of a system's motion astride and along a classical path appear to have been scarce.

Some progress towards treating the motion of several atoms quantum mechanically and more visually has been achieved by representing their positions on potential surfaces in polar coordinates [61]. (These coordinates are called 'hyperspherical' for 'surfaces' of more than three dimensions.) Thereby the coming together and eventual separation of reactants is represented by the variation of a single radial variable.

A valuable example treated successfully by this approach is provided by the reaction

$$H_2 + F = HF + H \tag{15}$$

studied by Y T Lee through detailed observation of its cross-beam collision products. The energy released by this exothermic process is found to remain largely concentrated in the vibrational excitation of HF to its $v = 3$ level, whereas H_2 was initially in its ground state. (This finding parallels the electronic excitation of an ion reported as item (iii) of section 13.5.1.) Quantum hyperspherical treatment of this reaction by J M Launay has reproduced the vibrational excitation successfully [62].

13.7. Inner-shell phenomena [63]

The production of vacancies in the inner shells of atoms or molecules, generally by photoabsorption or collision, serves two main purposes: (a) the generation of monochromatic x-rays by the oscillating current of an electron transition that fills the vacancy; (b) the study of secondary processes resulting from a similar, but *radiationless* (section 13.6.2.1) transition.

Vacancies, with the same consequences, can also arise as secondary effects of nuclear processes: (c) electron capture into a nucleus made unstable by excessive positive charge, accompanied by ejection of a neutrino; (d) internal conversion of nuclear energy through electromagnetic interaction analogous to 'photoemission by the nucleus + photoionization' of the inner shell. The occurrence of either process is manifested by the subsequent x-ray emission or radiationless transitions.

The rate of x-ray emission, proportional to the squared ratio of its electron displacement to the x-ray wavelength at lower photon energies, lags behind that of competing radiationless (Auger) transitions, until its proportionality to the cube of frequency prevails at $\hbar\omega > 10$ keV. Auger transitions occur instead generally within ~10 fs, regardless of their energy.

Specific aspects of the x-ray emission and Auger processes are outlined in sections 13.7.1 and 13.7.2, respectively. Section 13.7.3 deals instead with the role of inner electrons in properties of whole atoms and molecules encompassed by the generic names 'screening' and 'antiscreening'. The former refers to the reduction of nuclear attraction upon outer electrons by the negative charge of inner electrons. The latter refers to a less obvious influence of inner electrons that amplifies the action of outer electrons on nuclear orientations.

13.7.1. X-ray studies

The broad classification of atomic x-ray spectra, characteristic of each element, into K, L, M ..., series was established long before mid-century (section 13.2.3.5). Coster–Kronig transitions were also identified, connecting subshell levels with the same n and different l and/or j values (as defined in section 13.2.3.2); the energy separation of these levels reflects differences in their closest approach to the nucleus as represented by different values of their quantum defects μ_{lj}. The comparatively high intensity of Coster–Kronig emissions demonstrates the high intensity of the electron currents involved in transitions between levels of the same shell.

Attention has been directed more recently to the 'satellites' of most x-ray lines, that is to sets of lines with frequencies slightly lower than that of a regular line, whose emission is accompanied by one of many alternative secondary excitations. The measurement, classification and theory of these emissions have been advanced mainly by two schools guided by B Crasemann at Oregon and T Åberg in Finland.

The precision determination of characteristic x-ray wavelengths and photon energies has been an important element not only for providing specific standards but for interconnecting measurements performed on diverse scales. It involves, specifically, the x-ray diffraction by crystal lattices, the lattice spacing of the crystals, the Planck constant, etc. A recent paper in this field indicates a current accuracy of the order of 1 part in 10^6 [64]. The intensity of each x-ray emission reflects instead the distribution in space of the radiating current.

Precision experiments and theory on x-ray transitions between levels of highly stripped ions have attracted much attention recently, as reported in section 13.5.4.

'Resonant' production of an inner-shell vacancy, as a preliminary to x-ray emission, consists of exciting an inner-shell electron to a discrete level within, or slightly above, the valence shell of an atom or molecule. This process, facilitated by high-resolution synchrotron radiation, has attracted much attention recently as a tool to observe the influence of the excited electron on the valence shells, particularly in molecules, as noted in section 13.6. This influence is reflected, in part, by the subsequent x-ray emission.

13.7.2. *Auger emissions*

The Auger phenomenon consists of the escape from atoms of monoenergetic electrons, such as would result from photoionization by characteristic x-rays. Indeed their energy coincides with that of photoelectrons ejected by x-rays from the same atom, but is transmitted by direct interaction with an electron that fills an inner-shell vacancy. This transfer of excitation energy may be viewed as an example of the coupling between 'closed' and 'open' channels described in section 13.4, i.e. as an example of auto-ionization.

Beginning in the 1950s the study of Auger electrons was enhanced by K Siegbahn's design of ESCA electron spectrometers with high resolving power and luminosity and by its observation in gas rather than condensed phase, largely at W Mehlhorn's initiative. The analysis of fine structures and of satellite spectra, analogous to those of x-rays, thus became accessible. The Auger process is initiated, of course, by the production of an inner-shell vacancy, i.e. by photoionization or by collision as for the production of characteristic x-rays.

The combination of vacancy generation and of Auger emission in rapid succession (~10 fs) affords opportunities for further studies of atomic dynamics. Thus, vacancy production with low excess energy generates a rather slow electron that may be overtaken by the following, faster, Auger electron, thus affording exchanges of energy and/or angular momentum between these electrons (post-collision interactions). The production of a specific inner-shell vacancy by an incident beam generally impresses on the vacancy an alignment and/or orientation with reference to the beam direction; these geometric characteristics are then transmitted to the Auger electrons whose study proved more fertile than the corresponding study of characteristic x-rays. The greater flexibility of Auger-electron analysis extends to observing effects of resonant generations of vacancies mentioned at the end of section 13.7.1.

A comprehensive review of recent studies on Auger electrons from free atoms and molecules, including their detection in coincidence with recoil ions, is provided in reference [65]. Since the 1970s Auger emission has also emerged as an analytical tool of the presence and state of atoms or molecules adsorbed on surfaces.

13.7.3. *Screening and antiscreening*

The electric potential within an atom, at a radial distance r from its nucleus with charge Ze, can be represented as $(Z - s)e/r$ where the number s indicates the screening action of all the atomic electrons at radial distances $r' < r$. The value of s equals the sum over all bound electrons with indices (n, l) of the values of their distribution functions $|f_{nl}(r')|^2$ (introduced in section 13.2.3.1) integrated over the range $0 < r' < r$.

The determination of the functions $f_{nl}(r)$ for atomic ground states has been indicated in section 13.2.3. The resulting values of the screening function $s(Z; r)$ depend smoothly on the atomic number Z, in general. We note here, however, a distinct effect of the relativistic mass boost of K-shell electrons of high-Z elements, mentioned in section 13.2.1, that was identified by J P Desclaux and Y K Kim in the 1970s. This boost draws K electrons closer to the nucleus, thereby increasing their contribution to the screening of other electrons, especially of those in states with $l \neq 0$ that are kept away from the nucleus by the centrifugal force. The resulting modification of screening propagates domino-like from shell to shell, as l increases, as demonstrated by appreciable reductions of the quantum defects μ_l.

The 'antiscreening' effect stems from asymmetries, of electron distribution or of spin orientation, inherent to the structure of incomplete (open) valence shells (or subshells). The Coulomb interaction between each electron of such a shell and each electron of the underlying, spherically symmetric, 'closed' shells perturbs this symmetry. This perturbation, inversely proportional to the third, or higher, power of the distance of each electron pair, propagates—again domino-like—from shell to shell, eventually influencing the orientation of a nuclear spin. Its inverse dependence on electron distances magnifies the initially insignificant magnitude of the perturbation as it propagates to increasingly dense distribution of inner-shell electrons.

This phenomenon, and its name, was identified by R Sternheimer around 1950 in the context of the quadrupole coupling of valence electrons with nuclei, that is, between the elongations (or oblateness) of their respective electric charges, whose weak strength is nevertheless detected by nuclear magnetic resonance. The corresponding magnification of the interaction between the magnetic moments of the valence electrons and nuclei proved even more striking. Roughly speaking, electrons with parallel spins keep more nearly apart than those with antiparallel spins, as noted in section 13.2.1. Thereby electrons of each shell are shifted slightly towards smaller or larger radial distances depending on whether their spins are parallel or antiparallel to the spin of valence electrons. (A similar effect occurs for the magnetism of orbital currents.) In about 1960 it was noted that the magnetic field prevailing in the outer shell of an iron magnet ($\sim 30\,000$ G) is amplified by antishielding to tenfold strength in its action on Fe nuclear spins.

13.8. Spectral fingerprints of atoms and molecules [66]

The yellow colour of a flame has long been recognized as indicating the presence of a sodium salt. Nineteenth century spectroscopy traced this colour to emission of light with wavelength of ~ 580 nm, which has since been resolved into a doublet with the wavenumbers $16\,956.183$ and $16\,973.379$ cm^{-1}. Again in the early nineteenth century, Fraunhofer identified characteristic absorption lines in the solar spectrum; by the turn of the century helium was discovered in the solar spectrum, but the absorption by the H$^-$ ion and the ion's very existence were first noticed there in 1938. The unravelling of rare-earth spectra, to the extent of identifying the ground states of these chemically similar elements, was completed after mid-century.

The present section outlines the general character and the current availability of spectral information for diagnostic purposes.

A notable systematics governs the visible and near-visible spectra of atomic elements, which reflect the structure of their valence shells, particularly the number of their s and p ($l = 0, 1$) electrons. These spectra are accordingly similar for all elements of each column of Mendeleev's periodic system. A different systematics emerges instead for the far more complex spectra of the transition and rare-earth groups, elements that differ from one another by the number of their d or f ($l = 2$ or 3)

electrons. The distribution functions $f_{nl}(r)$ of d and f electrons peak at shorter radial distances; these electrons thus bear little influence on the chemical properties of each element but their interaction with s and p electrons enrich and complicate the spectra greatly. These features of atomic spectra bear also on the spectra of molecules containing each element, but systematic features are less apparent in molecules owing to their more complex structure.

Astrophysics has been, of course, both a principal consumer and a provider of spectral information, if limited in the past to the visible and near-visible range. This limitation afforded little evidence of the presence of molecules in space. The actual wealth of astrophysical molecular processes has emerged only since the 1960s through the evidence on rotational and vibrational spectra tapped by joining the infrared and radiofrequency ranges (section 13.3.2).

The progress of spectroscopy, mainly on either side of the optical range, rested until mid-century upon the dedicated effort of a modest number of individuals, who improved their measurements, extended their coverage and gladly relayed their results to users. A first comprehensive publication listing atomic energy levels was published by R F Bacher and S Goudsmit in 1932. The growing number of users led W F Meggers, in the late 1940s, to organize—with the support of E U Condon, Director of the US National Bureau of Standards—a standing programme to collect, analyse and produce extensive reports on that subject [67]. (Its last volume, on rare-earth spectra, appeared in 1978.) Dr Charlotte Moore Sitterly, an astrophysical spectroscopist, led that work to the end of her career.

Since that time, the growth of physics personnel, the extension of spectroscopy to far broader frequency ranges (section 13.3), the onset of plasma physics for nuclear fusion and other purposes (Chapter 22) and the spectacular expansion of astrophysics (Chapter 23) have occurred. These combined factors have transformed the task initiated by Meggers into a sizeable industry, growing on its own and in response to needs and opportunities. The NBS (now the National Institutes of Standards and Technology) still constitutes a pillar of this activity; neither it nor other institutions appear to take a comprehensive responsibility for the whole industry.

A useful survey of the present situation, with regard to astrophysics, has emerged from the International Astronomical Union's convening in 1991 a meeting on the subject, which led in turn to a published study of the *Needs, Analysis and Availability of Data for Space Astronomy* [68]. This study includes a report on the recent international 'Opacity Project', organized by M J Seaton to calculate atomic data bearing on the opacity of stellar envelopes

On the plasma side, support has been provided to extend atomic spectroscopy to the study of variously stripped ions. Here, however, the main diagnostic tool appears to consist of the easily identified and simpler spectra of ions isoelectronic to the (monovalent) alkali atoms introduced in section 13.2.3.1 (see also section 13.5.4).

13.9. The role of atoms in metrology and instrumentation [69]

The role of atoms in metrology and instrumentation has become prominent since mid-century. This role rests on two main considerations: (a) molecules of any chemical substance are *indistinguishable* from one another and so are all of their atomic or subatomic constituents; (b) each of their properties remains constant in time. Moreover, most species of atoms or molecules are readily available to experimenters, thus serving as convenient standards.

See also p 1233

These features emerged strikingly from the 1946 independent discoveries of nuclear magnetic resonance by F Bloch and by E Purcell. The spins of hydrogen nuclei in a water sample subjected to a magnetic field B precess about the field direction at a uniform rate proportional to B; this rate is readily detected and measured by 'tuning in' an induction circuit or a cavity. A water sample and common instruments can thus serve as standards to determine the absolute strength of unknown magnetic fields, less easily measurable otherwise.

A key circumstance complements here the role of atoms: frequency standards of high accuracy and precision are readily provided not only by electric technology, reaching now into the THz range, but also by atomic radiation sources. The international standard of time intervals has been provided since 1971 by a hyperfine frequency transition of ^{133}Cs nuclei in an atomic beam, namely, 9 192 631 770 Hz. Frequency measurements lend themselves to high precision owing to the linearity of typical electrical phenomena that afford superposing currents or voltages and beating (heterodyning) of their frequencies for easy observation and intercomparison. Access to high frequencies—up to 10^9 Hz at mid-century, far higher by the mid-1990s—provides long trains of oscillations with high-quality periodicity.

The accuracy of a periodic motion depends on the length of its oscillation train, which is extremely high for many spectral lines and is reflected in their observed line width as noted earlier (section 13.3.3). Spectral lines emitted by a gas are, however, broadened by the Doppler effect, i.e. by the apparent frequency shift of emissions by atoms moving at different speeds v in the observer's direction, a shift proportional to v/c. This effect is reduced for emissions from beams at very low temperature directed orthogonally to the observer.

Major efforts have been devoted for decades to the task of sharpening line profiles by controlling the temperature of radiation emitters. A further stage has been reached more recently by slowing down individual atoms through light pressures applied simultaneously from different directions, thus trapping, storing and displacing them as desired [70].

Such increasingly refined controls are currently leading to major strides in metrology.

Atomic standards of measurements have been replacing earlier ones gradually. The ^{133}Cs frequency–time standard was adopted officially when it proved more stable than the Earth's rotation rate [71]. Wavelength standards, anchored to interference fringes of atomic emissions from ^{86}Kr, replaced the original 'metre bar' as length units for a time. They have been replaced, in turn, by the distance travelled by light *in vacuo* in 1 s, thus being anchored to the frequency standard by a numerically defined value of the velocity of light c.

This evolving process reflects the continuing progress of technology in comparing measurements of physical quantities greatly differing in magnitude, often by 6–10 orders of magnitude, i.e. basically in the ratio of laboratory and atomic dimensions. These intercomparisons are done by stages of scaling, by factors of 100 to 1000 at a time. Each stage has two components: (a) the 'scoring' of discrete 'steps' such as the passage of wave crests, the turns of precessing motions, or discontinuities in an intensity–voltage plot; (b) the precise measurement of fractions of a single step. The scoring process reflects the nature of atomic elementary processes (section 13.2.2.1). Fractions of a turn are familiarly measured on graduated circles. Basically similar procedures are reported for precision measurements of widely different phenomena.

This universality of the atomistic underpinning of diverse phenomena has been documented further since the 1960s by the unanticipated contribution of the Josephson junction and of the quantum Hall effect to basic metrology. Both of these phenomena involve frictionless conduction in bulk materials. The Josephson junction connects materials in their superconducting state, which implicitly extends properties of single atoms and molecules to macroscopic systems, as pointed out by F London in 1935; the quantum Hall effect deals with electrons confined in one direction but free to drift across it. These surprising contributions to metrology are outlined below.

A Josephson junction consists of a nanometre thick layer of insulating material separating two superconductors held at a small potential difference of V volts. Complementing this constant potential by an AC component of frequency $2eV/\hbar$ causes electron *pairs*, in a singlet spin state, to tunnel through the junction as predicted by B D Josephson in 1962. However, surprise greeted the later publication of a thorough study by B N Taylor *et al* [72], which showed that the resonance frequency $2\,eV/\hbar$ is *fully independent* of the nature and detailed properties of the two superconducting materials. The study thus documented that the pair of materials behaves just like a *single* atom whose electrons are transferred from one to another stationary state by resonant radiation, fully verifying the atomistic stationary character of superconducting states. (Notice, incidentally, that injection of an electron pair into a

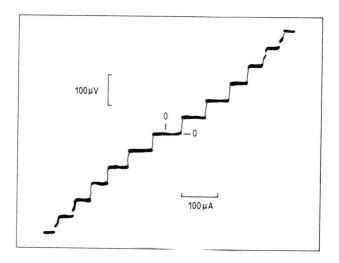

Figure 13.20. *An oscilloscope display of the voltage–current relation for a Josephson junction exposed to 36 GHz microwaves. The sequence of identical supercurrent steps affords precision measurements of the ratio 2e/h.*

system with macroscopic capacitance is hardly perceptible, in contrast to its addition to an atom or to the recently developed 'mesoscopic' systems.) Frequency measurements of the Josephson effect provide the present standard of electric potential.

The Josephson effect can be displayed in alternative fashions, of which the above outline appears conceptually simplest while others appear more striking. If the junction is exposed to microwaves of constant frequency ω and to an increasing potential V that drives through it a supercurrent of intensity i, the $i(V)$ plot rises *stepwise* whenever 2 eV traverses a *multiple* of the photon energy $\hbar\omega$, as shown in figure 13.20. This plot illustrates how the number of successive steps can be scored so as to interconnect quite different potential scales precisely, e.g. the μV scale of this figure and the voltage scale of electrical engineering. Anchoring engineering voltages to the voltage scale of the Josephson phenomenon in this figure, a task performed by Taylor at NBS shortly thereafter, rested on this process.

The normal Hall effect is displayed by an electric current flowing along a conductor strip subjected to a magnetic field orthogonal to the strip. This field pushes the current towards one side of the strip, generating a Hall potential U_H, i.e. a voltage difference across the strip corresponding to a transverse electric field E_H whose strength balances that of the magnetic action. The magnitude and sign of U_H are reciprocal to the concentration and sign of the charge carriers in the conductor; indeed measurements of U_H serve generally to assess that concentration and sign. The quantum Hall effect, which occurs under very special circumstances, measures similarly a fundamental atomic parameter, namely, the fine structure constant α introduced in section 13.2.1.

The carriers in the strip labelled 'surface channel' in the inset of figure 13.21 are electrons squeezed into a narrow layer at the interface of two semiconductors at kelvin temperatures. (The figure is drawn from the original quantum Hall report [73].) The squeezing stems from a 'gate voltage' V_g that controls the electron distribution among energy levels whose spectrum consists of a few energy bands, each band corresponding to one 'Landau level' of motion in a strong magnetic field. The electron density, per unit area, in each filled band is fixed by the magnetic field's strength. Figure 13.21 shows how the Hall voltage U_H remains constant as V_g increases through each gap between a filled and an empty level; the potential U_{pp} *along* the strip remains at or near zero in the gap. Additional electrons move into the layer as V_g enters a new band, causing U_H to decline; U_{pp} rises instead sharply as a new level gets filled. Each *step* value of U_H equals a fixed factor divided by the parameter α and by the number of filled levels, a number that determines the electron density throughout the layer. Here again, as in the Josephson junction studies, we observe the occurrence of level steps of macroscopic quantities, whose value reflects the value of atomic parameters.

Both of these phenomena afford tools for the precise measure of macroscopic variables. The same function is performed by the observation of fixed or moveable sequences of interference or diffraction patterns, whether of microwave, optical or x-ray radiation, whose shift reflects the corresponding fine displacements of mirrors, gratings or crystals. Such displacements are controlled at this time to picometre accuracy. An additional important tool for such fine adjustments has been provided by piezoelectric crystals whose length varies at rates of the order of 0.1 nm V^{-1}.

13.10. Optical handling of atomic systems and transformation of light by atoms

Technologies have arisen since mid-century that channel atomic motions into desired paths through optical actions. Conversely atoms serve currently to transform radiations. These activities have grown into a discipline and an industry loosely identified as 'quantum optics'.

This development was sparked in 1950 by A Kastler's conception of 'optical pumping', an action that transfers light polarization (circular or linear) to atoms in gaseous or condensed phases [74]. More recently one has learned how to steer, cool and trap atomic particles, as anticipated in section 13.9, and to achieve and study multi-photon processes. These several aspects are outlined in the following subsections.

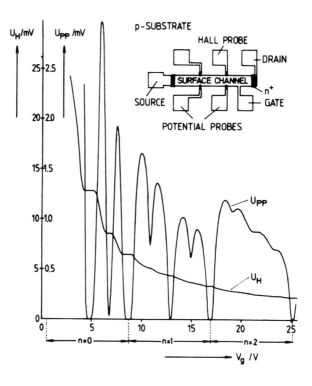

Figure 13.21. *Recordings of the Hall voltage U_H and of the voltage drop U_{pp} between the potential probes as functions of the gate voltage V_g at the temperature 1.5 K, constant magnetic field 18 T and source drain current 1 μA. The inset shows a top view of the device, with length 400 μm, width 50 μm and 130 μm distance between the potential probes. The flat portions of the U_H plot play a metrology role analogous to that of the steps in figure 13.20.*

13.10.1. *Optical pumping*

As a prototype process, consider monovalent sodium atoms in their ground state, subjected to a magnetic field pointing in the $\hat{z}$ direction. Atoms with downward spin direction ($j_z = -\hbar/2$) have lower energy than those with opposite spin; thermal collisions nevertheless keep both spin orientations statistically in near balance. Consider now the action of yellow light—in resonance with atomic excitation—propagating along $\hat{z}$ with positive circular polarization, thus imparting to atoms the angular momentum $j_z = \hbar$ per absorbed photon. Each photoexcitation event is followed by re-emission of a photon (fluorescence) with unspecified polarization. The angular momentum imparted by the light thereby shifts the balance of spin orientation towards the $j_z = \hbar/2$ ground-state component, thus 'pumping up' the spin population.

An analogous effect occurs in the absence of the external field, owing to the alternative mutual orientation of the electron's and nuclear spins. If the electron spin was initially pumped in the presence of a magnetic field, removal of this field would allow the hyperfine interaction (Box 13C) to share part of the electron's acquired orientation with the nucleus. Spin polarization of electrons is also transferred to nuclei of other elements, e.g. Xe, in the course of collisions [75].

The ensuing rich phenomenology has been studied extensively [76], achieving widespread application to orient the spins of nuclei and other particles in accelerator beams and in their targets.

13.10.2. Cooling, steering and trapping of atoms

The generic process through which electric fields steer atoms or molecules consists of polarizing them—i.e. of inducing a dipole moment—and thereby attracting them towards positions of lower potential energy [77]. A more specific process imparts to each atom or molecule the momentum $\hbar\omega/c$ of each absorbed photon $\hbar\omega$ in the direction of a light beam; the momenta lost in the successive fluorescence point in average directions thus leaving the absorbers with a net momentum in the beam's direction. 'Cooling' is achieved through this process by tuning the light frequency a little below the resonance of absorbers at rest; the Doppler effect of an absorber's motion causes the light to be absorbed only by atoms moving with more than average thermal velocity opposite to the beam's direction.

A convenient arrangement for steering, and cooling, atoms consists of setting up a standing wave of light resonant or quasi-resonant with an atomic sample. Atoms traversing this wave are strongly polarized and/or excited and thereby driven towards the wave troughs where the light intensity and the atoms' energy are lowest, an arrangement largely developed and utilized in D E Pritchard's laboratory.

'Trapping' is achieved generally by steering atoms into a small volume by converging light beams or static fields. Particularly effective for metrology have proven 'Penning traps' that hold single charged particles (electrons or ions) for long times during which their motion is controlled precisely by oscillating electromagnetic fields. Charge/mass ratios and magnetic moments of such particles have thus been measured with extreme precision, especially by H G Dehmelt and his school.

13.10.3. Multi-photon processes

The phenomena of photo-absorption and -emission by atoms have been introduced in section 13.2.2 as weak processes, whose treatment is elementary, on the grounds that each relevant mode of a radiation field spreads over a volume of space far larger than atoms. The electric field strength corresponding to the 'photon' unit of excitation of each mode is thus extremely small compared with intra-atomic field strengths.

This feature has been altered by the introduction of lasers, whose very principle lies in multiple excitation of specific radiation modes. The action of early lasers, with modest intensity, could still be treated as weak, in the sense that it did allow for the resonant absorption of two or more photons of the same mode in a single elementary process, if only with a small probability.

The introduction of this class of processes nevertheless opened a new field of physics, dubbed *non-linear optics* (Chapter 18), under primary leadership by N Bloembergen in the early 1960s. Simultaneous absorption of two or more photons affords bypassing the selection rules of previous spectroscopy and attaining high excitations by lower-frequency sources. The small probability of multi-photon processes is, however, enhanced whenever the energy of a fraction of the absorbed photons approaches that of an intermediate atomic level, affording thus a 'quasi-resonant' stepping stone.

The 'perturbative' treatment of multi-photon processes grew progressively inadequate as laser intensities increased, particularly with the introduction of short-pulse lasers that release energy stored in a condenser within a short time, currently even in the femtosecond range. An early indication of high-probability multi-photon processes emerged from the discovery of photoionized electrons ejected with alternative kinetic energies differing by one or more photons (above threshold ionization).

The ratio of radiation field strength to the strength of atomic fields (the latter of the order of 100 V nm^{-1} in atomic valence shells) was stated to be exceedingly small in ordinary processes. It has now been boosted even past unity by the combination of several elements: strong laser sources, pulse compression into short intervals, focusing the source image into a spot a few μm across. This ratio attains unity for power fluxes of the order of 10^{17} W cm^{-2}. Still higher fluxes have been seen to pulverize atomic structures.

A distinct process competes with photoabsorption proper as the ratio of field strengths approaches unity, namely, the 'quiver motion' characteristic of the free electrons' response to an AC field of strength E and frequency ω. Its amplitude, $eE/m\omega^2$, is particularly large at the low frequencies of powerful infrared lasers and may even exceed the size of excited atoms. The free-electron aspect of this phenomenon contrasts sharply with the photoabsorption's dependence on intra-atomic forces, which is required to preserve the balance of both energy and momentum (section 13.3.3).

These contrasting features of diverse yet comparably strong actions of intense radiation pulses complicate their joint treatment. Successful, but fragmentary, quantitative evaluations of specific processes have been reported, but no coherent theory has been developed thus far. Whether the more comprehensive and flexible approach of section 13.11.2 will prove adequate remains to be seen.

13.10.4. *Transformations of light by atoms*

The achievement of multi-photon absorption processes has opened the door to multiplying, and subdividing, radiation frequencies. A transparent example is afforded by the process called 'four-wave mixing'. Basically, an atomic or molecular level excited by absorption of several photons may return to the ground state by emitting a single photon with energy and frequency equal to the sums of those absorbed. This process is, however, restricted by the requirement of parity conservation (Box 13A). Each photoabsorption step generally involves transitions between atomic states of opposite parity that display an oscillating, odd parity, dipole moment. A closed cycle of photo-absorption and photo-emission must thus involve an even number of steps. The earliest opportunity for frequency multiplication is thus afforded by triple absorption followed by a single emission.

Parity conservation may be bypassed by introducing an asymmetry. Frequency doubling is thus typically achieved by light propagation through 'optically active' media, with non-centrosymmetric structure, or—selectively so—at interfaces. An atom, excited to a level $\hbar\omega$ and then *traversing* a cavity (Chapter 18) tuned to support oscillations with frequency $\omega/2$, can yield its excitation to the second excited level of the cavity in a process dubbed 'parametric amplification'.

Combinations of manifold analogous processes have afforded a wealth of applications, many of them bearing on precision metrology or affording novel measurement procedures. These endeavours now occupy a major portion of atomic–molecular science.

13.11. A current overview

Studies of the diverse topics outlined in the previous sections are currently developing over expanding areas and scales which in turn are increasingly being subdivided into smaller branches. This progress, fuelled by the rising input of personnel and of technological innovation, remains constrained conceptually and computationally by reliance on the original independent-particle model of atomic processes. This model implies that interparticle correlations are to be either evaluated and described laboriously for each particle pair, triplet, etc, or else taken into account 'behind the scenes' by computer evaluation of interactions among large basis sets of single-particle 'orbitals'. Such procedures, well suited to their earlier goal of treating a few particles at a time, have grown increasingly laborious and opaque as applied to larger and integrated systems.

The desirability of more flexible models, utilizing appropriate 'collective coordinates', has been apparent for some time. Progress towards this goal has been slow, because innovation implies identifying and overcoming novel hurdles, with benefits that remain elusive and distant for years. Section 13.11.1 will outline phenomenological elements whose theoretical formulation would encompass a large fraction of all atomic–molecular processes. Section 13.11.2 will introduce the hyperspherical approach to multi-particle phenomena, which displays the required flexibility.

13.11.1. *A comprehensive phenomenology*

An illusory contrast between collision and spectral phenomena has been indicated in section 13.1. To remove its source, consider that the formation and the fragmentation of a complex, in the course of a 'close' collision (section 13.2.4), are actually *reciprocal* processes governed by the *same dynamics*, regardless of differences between the initial and final reactants [78]. The common element of formation and fragmentation lies in the evolution of the complex between any of its compact states and one or another of its alternative fragmented states. This evolution and its inverse are naturally represented by reciprocal parameters.

The excitation of an atom, molecule, or even of a cluster, into a spectrum of alternative levels, following the absorption of a photon, may be viewed as the evolution of a complex towards fragmentation, an evolution powered, for example, by the injection of radiation energy into an initial ground state. Whether this evolution can proceed to complete fragmentation in one channel, *A*, or be aborted by meeting a potential barrier, depends on whether the available energy does or does not exceed the threshold for full fragmentation in that channel. If it does not, the evolution towards fragmentation in *A* is inverted, but may again proceed to completion in another channel *B* with lower threshold.

The formation of standing waves by reflection on a barrier in channel *A* then generates resonances in the spectrum of fragments in channel *B*, as described in section 13.4; this phenomenon is called 'auto-ionization' or 'autodissociation' of discrete levels of the 'closed' channel *A*. The continuity of parameters as functions of energy through thresholds has been stressed in sections 13.3.4 and 13.4.

Photoprocesses may also be viewed as forming a complex consisting of atomic target + radiation energy, much as a negative ion is formed by the combination of an incident electron with a neutral atom. On the other hand, no complex is formed in long-range collisions (section 13.2.4) with large impact parameters, in which the Coulomb field of a charged projectile transfers energy to a distant target much as radiation does (section 13.2.4.1). We thus see that complex formation may or may not involve an extensive structural rearrangement.

The chief common element in the evolution of complexes between their compact and fragmented states lies in the contrasting symmetry features of these states. A complex in its compact state displays central symmetry by rotating freely about its centre of mass. A fragmented state displays instead symmetry about the axis through the centres of mass of separate fragments. This feature was stressed by A R P Rau in the

1970s, jointly with its analogue in the desorption of a particle from a surface. The contrasting symmetry will also lead us to identify the main dynamical element of the evolution of a complex.

13.11.2. Dynamics of the evolution of complexes in the hyperspherical approach

The elements of this dynamics are viewed here as analogues of those introduced in section 13.2 for the dynamics of a single electron. The confinement of an electron within a volume of radius a was stated to boost its kinetic energy by an amount proportional to $1/a^2$. Similarly the confinement of a whole complex into a compact state of radius R about its centre of mass boosts the total kinetic energy of its constituents by a corresponding factor, to be defined in the following. A complex formed by collision or photoabsorption has generally a radius smaller than its equilibrium value. The corresponding excess kinetic energy will drive its subsequent evolution towards fragmentation.

The motion of a single excited electron was resolved, in section 13.2.3, into radial and rotational components. The radial component, represented by a density factor $|f_{nl}(r)|^2$, is governed by the nuclear attraction and by the radial component of the kinetic energy, i.e. by a factor proportional to $1/a^2$ or its equivalent. The rotational component is represented by a directional distribution $|Y_{lm}(\theta, \varphi)|^2$ (Box 13B), whose index l indicates its number of lobes, i.e. in essence the degree of its directional confinement; the kinetic energy associated to this confinement also tends to push the electron outwards, being accordingly often represented as a centrifugal potential proportional to $l(l+1)/r^2$.

The rotational confinement of a multi-particle complex exhibits analogous features, even though more complex than for a single electron. One element of this complication was faced before mid-century in evaluating the rotational energy of multi-electron levels (section 13.2.3.2), a goal attained by the algebraic procedures mentioned in Box 13D. The introduction of hyperspherical coordinates has extended those procedures since the 1960s to evaluate the energy of both rotationally and radially correlated motions of multi-particle states.

13.11.2.1. Hyperspherical coordinates.

Hyperspherical coordinates were introduced in the 1930s (and described in the physico-mathematical text by Morse and Feshbach) mainly to provide a more explicit description of the correlated motion of two electrons in the helium atom. V Fock utilized them in the 1950s to discover a previously overlooked aspect of He dynamics. Figure 13.22 shows how they serve to represent the joint position of the He's electron pair by a hyper-radius $R = (r_1^2 + r_2^2)^{1/2}$ and five angles; four angles identify the directions of the two electrons, while the fifth one represents the ratio $r_2/r_1 = \tan \alpha$ of their respective distances from the nucleus. Wannier's analysis of the double-electron

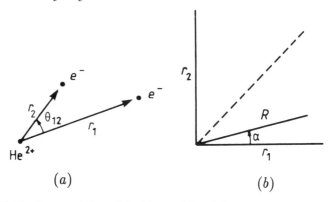

Figure 13.22. *Representation of the joint position of the electron pair in a He atom by five hyperspherical coordinates. The plane containing the nucleus and the electrons is identified by three standard Euler angles (not shown). (a) The two-electron configuration with respect to the nucleus is identified by their radial distances (r_1, r_2) and the angle θ_{12}. (b) The pair of radial distances is identified in polar coordinates (R, α).*

fragmentation of He (section 13.4.2.1) centred on the value $\alpha = \pi/4$ in figure 13.22 as identifying the critical path of the pair towards double ionization at near-threshold energies.

These coordinates allow resolving of the evolution of an He state into a hyper-radial motion, represented by a distribution $|f_{n\lambda}(R)|^2$, and a hyper-rotational motion, represented by harmonic distribution functions in the five angles $|Y_{\lambda\mu'\mu''\mu'''\mu''''}(\theta_1, \varphi_1, \theta_2, \varphi_2, \alpha)|^2$. (The set of indices $\{\mu', \ldots, \mu''''\}$ will be indicated by μ in the following, the set $\{\theta_1, \ldots \alpha\}$ by ω.) Thereby one views radial correlations, represented by the dependence of $|Y_{\lambda\mu}(\omega)|^2$ on α, as fully analogous to their dependence on the angular correlations represented through the angle θ_{12} between the directions (θ_1, φ_1) and (θ_2, φ_2). The harmonics $Y_{\lambda\mu}(\omega)$ are constructed explicitly by standard analytical procedures.

The radial kinetic energy of the electron pair depends on R^2, as it did on the parameter a^2 of equation (1); the rotational kinetic energy is proportional, as for a single electron, to a quadratic polynomial in λ divided by R^2. The Coulomb potential energy between the three helium constituents, averaged over their hyperangular distribution, is represented by an algebraic function of the indices $\{\lambda, \mu\}$ divided by R.

At variance with the case of a single electron, the energies of the hyper-radial and hyper-rotational motions and of the Coulomb interactions do not remain independent as functions of R. Their interdependence has been studied extensively over the last 25 years with notable, if partial, success.

The key effect of the hyperspherical formulation of He mechanics has been to represent implicitly all correlation dynamics among the three constituents of He—one nucleus and two electrons—by algebraic functions of the parameters $\{\lambda, \mu\}$ and of the single radius R.

This representation extends straightforwardly to all atomic–molecular systems.

It suffices to view the positions of N particles as represented by a single point in a $3(N-1)$-dimensional space with the origin of its coordinates at the centre of mass. (He yields $N=3$ and $3(N-1)=6$.) The hyper-radius R needs redefining for particles of different mass; since it represents the radius of inertia for He (regarding the nucleus as infinitely heavy) it may be replaced by the square root $\sqrt{I}$ of the moment of inertia about the systems' centre of mass or, with reference to the total mass M, by $R=(I/M)^{1/2}$. Angular and radial correlations are then represented through harmonics $Y_{\lambda\mu}(\omega)$ of a set $\{\omega\}$ with $3N-4$ elements.

The analytical geometry of these multi-particle spaces was developed largely by Russian nuclear physicists in the 1960s and 1970s, including the all-important transformations among alternative sets $\{\omega\}$, properly viewed as Jacobi coordinates of analytical mechanics [79]. An historical incident appears relevant in this context: molecular theorists, meeting shortcomings of the Born–Oppenheimer approximation (section 13.2.5.1) near a dissociation threshold, bypassed them through an analytical transformation, later identified as leading to hyperspherical coordinates. Similarly, extensive efforts to develop a 'correct' form of the translation factors mentioned in section 13.5.3, are fully bypassed in the hyperspherical frame.

13.11.2.2. The dynamics of the fragmentation process. The dynamics of the fragmentation process hinges primarily on the evolving balance between the complex' kinetic and potential energies, as it does for all mechanical systems. For quantum aggregates of nuclei and electrons, the kinetic energy is proportional to $1/R^2$ and the potential energy to $1/R$, with the 'radius' R defined above as $(I/M)^{1/2}$, because all potential energies are Coulombic. The coefficients of $1/R^2$ and $1/R$ are algebraic functions of the indices λ and $\{\mu',\mu'',\ldots\}$ introduced above. Fragmentation will introduce separate parameter sets $\{R_1,\lambda_1,\mu_1\}$ and $\{R_2,\lambda_2,\mu_2\}$ for the fragments. The total angular momentum J is conserved in the process.

The earmarks of compact states are small values of R and the lowest values of the λ index compatible with a given J value and the maximal correlations—angular and radial—which are characteristic of compactness. (Since the functions $Y_{\lambda\mu}(\omega)$ are, in essence, polynomials of degree λ in the variables ω, their dependence on ω sharpens as λ increases.) Small values of R boost, of course, the kinetic energy factor $1/R^2$ in contrast to the $1/R$ dependence of the potential energy.

The earmark of fragmentation channels—i.e. of each among the alternative channels—consists of a set of $\{\lambda,\mu\}$ indices that makes the potential energy stable—i.e. maximum, minimum or saddle—under small variations of the indices. (This requirement of stability is illustrated by obvious physical examples; for example, dissociation of a molecule leads generally to stable or metastable fragments.) The process of fragmentation is thus described by the evolution of the set of $\{\lambda, \mu\}$ values from the lowest values of λ, earmarks of compactness, to the values of $\{\lambda, \mu\}$ that characterize any specific fragmentation channel. This evolution is paced by the increasing access to larger values of λ afforded, at constant total energy, by the decline of the rotational energy $[\lambda + (3N-5)/2]^2/R^2$ with increasing R.

These features of the evolution, common to most atomic and molecular processes, were pointed out in 1981 [80], in the context of a theoretical procedure that had proved valuable in calculating the lower doubly excited levels of He. However, that procedure failed to yield the observed approach to double ionization along the fragmentation axis $\alpha = \pi/4$ (figure 13.22 and section 13.4.2.1), by disregarding the limited access to high λ values mentioned above [81]. The key role of this limitation for stabilizing states at high potential energies was indeed overlooked until the late 1980s. The progressive access to distribution functions $Y_{\lambda\mu}(\omega)$ with higher λ indices, as R increases, also illustrates how the state of a complex grows progressively less compact, acquiring sharper features in its correlated variables ω. Sharpness of these features is, of course, essential to the contrasting localization of alternative channels on the space of ω variables.

Appreciation of these aspects of the evolution of complexes has encouraged the development of procedures, leading to its description through analytical functions of the $\{\lambda, \mu\}$ indices at increasing values of R. This development, carried out thus far for simple systems, has verified the expectation that the distribution of the indices $\{\lambda, \mu\}$ concentrates along fragmentation channels with characteristic features [82].

References

[1] The evidence for the spin itself, i.e. for an intrinsic angular momentum of each electron, is reported in section 13.2.3.

[2] The occurrence of these units emerges from considerations analogous to those that determine the size of atoms. For monochromatic radiation of frequency ω, the energy within a small volume oscillates periodically between its electric and magnetic forms; it thus mimics the energy of a mechanical oscillator, with mass m, force constant $k = m\omega^2$ and amplitude A, which oscillates between its potential and kinetic peak values $kA^2/2$ and $m\omega^2A^2/2$. Quantum mechanics sets a lower limit to the energy of *any* variable thus confined, as indicated at the outset of section 13.2.1.

[3] Fermi E 1932 *Rev. Mod. Phys.* **4** 87

[4] Heitler W 1936 *Quantum Theory of Radiation* 5th edn (Oxford: Clarendon)

[5] The splitting of Ag beams into two components was discovered in the early Stern–Gerlach experiments prior to quantum mechanics, its implications remaining unclear.

[6]   Wigner E P 1931 *Gruppen Theorie und ihre Anwendungen* (Braunschweig: Vieweg) (Engl. Transl. 1959 *Group Theory* (New York: Academic))

[7]   Van der Waerden B L 1932 *Die Gruppentheoretische Methode in der Quantenmechanik* (Berlin: Springer)

[8]   Racah G 1949 *Phys. Rev.* **76** 1352; 1965 *Group Theory and Spectroscopy (Springer Tracts in Modern Physics 37)* (Berlin: Springer) and references therein

[9]   Wigner E P 1959 *Group Theory* (New York: Academic)

[10]   Condon E U and Shortley G H 1935 *The Theory of Atomic Spectra* (Cambridge: Cambridge University Press)

[11]   Mott N F and Massey H S W 1933 *The Theory of Atomic Collisions* (Oxford: Oxford University Press) (5th edn 1965)

[12]   The probability of impulsive collisions decreases as the projectile's velocity increases because the projectile acts for a shorter time. On the other hand the collision becomes elastic at low velocities. The probability of inelastic collisions thus appears to peak for $b\omega/v \sim 1$. Analogous considerations have often been presented under the name of '*Massey criteria*'. Their simplistic application may, however, be vitiated when the amount of energy $\hbar\omega$ required by a transition within the target depends on the impact parameter b. An instance of such strong dependence is described in section 13.5.2.

[13]   Bethe presents his treatment as a weak (Born) approximation that yields a probability amplitude of momentum transfer $\hbar k$ *linear* in the Fourier coefficient $1/k^2$ of equation (3). In this respect his calculation of the projectile's Coulomb field action on each electron is actually exact, in both quantum and classical mechanics. Indeed Rutherford's classical formula yields a probability of momentum transfer between colliding charges proportional to $1/k^4$, with the same coefficient as Bethe's.

[14]   Bohr N 1948 The penetration of atomic particles through matter *Kgl. Via. Selsk. Mater.-Fys. Medd.* **18** 8

[15]   Mulliken R 1930 *Rev. Mod. Phys.* **2** 60 and 506; 1931 *Rev. Mod. Phys.* **3** 89; 1932 *Rev. Mod. Phys.* **4** 1

[16]   Herzberg G 1950 *Spectra of Diatomic Molecules* (Princeton, NJ: Van Nostrand); 1954 *Infrared and Raman Spectra* (Princeton, NJ: Van Nostrand); *Electronic Structure and Spectra of Polyatomic Molecules* (Princeton, NJ: Van Nostrand)

[17]   Dibeler V H and Reese R M 1964 *J. Res. NBS* A **68** 409

[18]   Lukirskii A P, Rumsk M A and Smirnov L A 1960 *Opt. Spectrosc.* **9** 262

[19]   Weissler G L 1956 *Encyclopedia of Physics* vol 21, ed S Fluegge (Berlin: Springer) p 306

[20]   Fano U and Cooper J W 1968 *Rev. Mod. Phys.* **40** 441

[21]   Comer P 1970 *Ann. Rev. Astron. Astrophys.* **8** 269

[22]   Polar molecules in interstellar space receive instead energy from collisions and dissipate it by radiation.

[23]   Available from the US National Institute of Standards and Technology, Report NBSIR 87-3597 by M J Berger and J Hubbell.

[24]   Schulz G J 1963 *Phys. Rev. Lett.* **10** 104

[25]   Buckman S J and Clark C W 1994 *Rev. Mod. Phys.* **6** 539
        Compton R N 1993 *Negative Ions* ed V A Esaulov (Cambridge: Cambridge University Press)

[26]   Fano U and Rau A R P 1986 *Atomic Collisions and Spectra* (Orlando, FL: Academic) ch 7 and 8

[27]   Seaton M J 1983 *Rep. Prog. Phys.* **46** 167 and references therein

[28]   Wannier G 1953 *Phys. Rev.* **90** 817

[29] Fano U 1983 *Rep. Prog. Phys.* **46** 258

[30] Buckman S J, Hammond P, King G C and Read F H 1983 *J. Phys. B: At. Mol. Phys.* **16** 4219

[31] The term '*R* matrix' was introduced by Wigner, in nuclear physics, in about 1948 to denote 'short-range' adaptations of 'reaction matrices', called *K* in scattering theory, which contain the same information as the transition matrices $(i|T|f)$ in symmetrized form. The *K* and *T* matrices are related by $K = T/(1 + iT)$ and $T = K/(1 - iK)$; the transition element $T_l = \exp(i\delta_l) \sin \delta_l$ of Box 13F corresponds in this notation to $K = \tan \delta_l$, as easily verified. The matrix $(i|K|f)$ is similarly obtained from $(i|T|f)$ by matrix algebra. Short-range *K* matrices may be denoted by $K^{(s)}$ instead of *R*.

[32] Greene C H, Fano U and Strinati G 1983 *Phys. Rev.* A **28** 2209; 1988 *Phys. Rev.* **38** 5953; *Rev. Mod. Phys.*

[33] Lutz H O, Briggs J S and Kleinpoppen H (ed) 1985 *Proc. NATO Advanced Study Institute S Flavia, Italy* (New York: Plenum) p 51

[34] A parallel independent experiment was done by S Takezawa.

[35] Herzberg G and Jungen Ch 1972 *J. Mol. Spectrosc.* **41** 425

[36] The behaviour is represented analytically by replacing the product πv_0 in equation (13) by an index $-\Delta$, where Δ represents the phase shift of the photoelectron exiting from the atom in the open channel $f \equiv N=0$. The equation thus modified is readily solved to yield the value of Δ as a function of the energy *E* or of its equivalent value of v_2, from equation (11). The modulation of the photoelectron intensity is represented by the derivative $d\Delta/d\pi v_2$. The modulations of the density of spectral lines and of their intensities, displayed in figure 13.15 below the lowest threshold E_{t0}, is conveniently viewed as an appendage of the pattern prevailing above the threshold and reflecting the presence of the closed channel $f \equiv N = 2$.

[37] Greene C H and Jungen Ch 1985 *Adv. At. Mol. Phys.* **21** 51

[38] Lu K T 1971 *Phys. Rev.* A **4** 579

[39] The content of this section 13.5 derives largely from conversations with C D Lin (Kansas State University) and from references provided by him.

[40] Andersen N O, Gallagher J W and Hertel I V 1988 *Phys. Rep.* **165** 1

[41] Morgan G H and Everhart E 1962 *Phys. Rev.* **128** 662

Afrosimov V V, Gordeev Yu S, Panov M N and Fedorenko N V 1965 *Zh. Tekhn. Fiz.* **34** 1613ff

Kessel Q C, Russek A and Everhart E 1965 *Phys. Rev. Lett.* **14** 484

[42] Ovchinnikov S U and Solov'ev E A 1986 *Sov. Phys.–JETP* **63** 538

[43] Ziemba F P and Everhart E 1959 *Phys. Rev. Lett.* **2** 299

[44] In the process of electron capture from the target into a projectile's field, the analytical description of the electron's motion must display explicitly its absorption of the necessary momentum. This requirement was met initially by Bates' group by inserting an *ad hoc* translation factor. The correct form of this factor has been discussed for decades in dozens of contributions. Only in the 1980s was the *ad hoc* approach itself found to be inadequate, an observation not yet widely known to which we shall return in section 13.10.

[45] Fritsch W and Lin C D 1991 *Phys. Rep.* **202** 1

[46] Richard P, Stockli M, Cocke C L and Lin C D (ed) *VIth Int. Conf. on Physics of Highly Charged Ions (AIP Conf. Proc. 274)* (New York: AIP)

[47] Discussions, data and critical readings of this section have been provided by R S Berry, D H Levy and other colleagues.

[48] Velocity selection by filtration through rotating wheel slots is costly in terms of beam intensities. Experimental art lies in finding suitable compromises.

[49] Herschbach D R 1966 *Adv. Chem. Phys.* **10** 319

[50] Anderson J B, Andres R P and Fenn J B 1966 *Adv. Chem. Phys.* **10** 275

[51] Smalley R E, Levy D H and Wharton L 1976 *J. Chem. Phys.* **81** 5417

[52] Dehmer J L, Parr A C and Southworth S H 1976 *J. Chem. Phys.* **64** 3266
See also, Nenner I *Handbook on Synchrotron Radiation* ed G V Marr (Amsterdam: North-Holland) p 355

[53] The retention of electrons by centrifugal potentials is also apparent in atomic photoionization of high-l inner shells, typically in $4d \rightarrow 4f$ transitions of rare-earth atoms (section 13.7).

[54] La Villa R and Deslattes R D 1966 *J. Chem. Phys.* **44** 4399

[55] Howard B J, Dyke T R and Klemperon W 1984 *J. Chem. Phys.* **81** 5417

[56] Haberland H (ed) 1994 *Clusters of Atoms and Molecules* (Berlin: Springer)

[57] Bagus P S and Schaeffer H F III *J. Chem. Phys.* **56** 224

[58] Oka T 1992 *Rev. Mod. Phys.* **64** 1141

[59] Coulson C A 1960 *Rev. Mod. Phys.* **32** 170

[60] Truong T N and Truhlar D G 1990 *J. Chem. Phys.* **93** 1761

[61] Manz J 1986 *Commun. At. Mol. Phys.* **17** 91

[62] Launay J M and Lepetit B 1988 *Chem. Phys. Lett.* **144** 346

[63] This section owes much to advice and material provided by Professor W Mehlhorn (Freiburg) and by R D Deslattes (NIST) on x-ray measurements.

[64] Mooney T, Lindroth E, Indelicato P, Kessler E G and Deslattes R D 1992 *Phys. Rev.* A **45** 1531

[65] Mehlhorn W 1990 *X-Ray and Inner Shell Processes (AIP Conf. Proc. 215)* ed T A Carlson *et al* (New York: AIP) p 465

[66] Major contributions to the material of this section and of the preceding section 13.2.3 have been provided by Dr William C Martin (NIST).

[67] Moore C E 1971 *Atomic Energy Levels (NSRDS-NBS 35, vol I-III)* (Gaithersburg, MD: NBS)
Martin W C, Zalubas R and Hagan L *Atomic Energy Levels (NSRDS-NBS 60, vol IV)* (Gaithersburg, MD: NBS)

[68] Smith P L and Wiese W L (ed) 1992 *Atomic And Molecular Data for Space Astronomy (Lecture Notes in Physics 407)* (Berlin: Springer)

[69] The content of this section derives largely from extended conversations with Dr R D Deslattes (NIST) and from references and illustrations provided by him.

[70] Cohen Tannoudji C N and Phillips W D 1990 *Phys. Today* **43** 33

[71] Petley B W 1985 *The Fundamental Physical Constants* (Bristol: Hilger) p 15ff

[72] Taylor B N, Parker W H and Langenberg D N 1969 *Rev. Mod. Phys.* **41** 375

[73] von Klitzing K, Dorda G and Pepper M 1980 *Phys. Rev. Lett.* **45** 494

[74] Kastler A 1950 *J. Phys. Radium* **11** 255

[75] See, for example, Schaefer S R, Cates D G and Happer W 1990 *Phys. Rev.* A **41** 6063

[76] Fano U and Macek J H 1973 *Rev. Mod. Phys.* **45** 553

[77] This process parallels the magnetic steering displayed in figure 13.3, except that electrons have an intrinsic spin dipole magnetic moment, whereas atoms are normally unpolarized. Many molecules, e.g. H_2O, have an intrinsic electrical moment, that is usually averaged out by thermal rotation.

[78] The probabilities of a collision starting from complex formation in channel *A* followed by fragmentation in channel *B* and of its reciprocal collision,

       $B \rightarrow A$, may nevertheless differ owing to differing densities of states
       for alternative reactants.

[79] Smirnov Yu F and Shitikova K V 1977 *Fiz. Elem. Chast. At. Yadra* **8** 847
     (Engl. Transl. 1977 *Sov. J. Part. Nucl.* **8** 344)

[80] Fano U 1981 *Phys. Rev.* A **24** 2402

[81] This failure has since been remedied by S Watanabe through an extensive
     calculation of the necessary corrections.

[82] Fano U and Sidky E 1992 *Phys. Rev.* A **45** 4776; 1993 *Phys. Rev.* **47** 2812
     Bohn J 1994 *Phys. Rev.* A **49** 3761; *Phys. Rev.* A **50** 2893; *Phys. Rev.* A **51**
     1110

Chapter 14

MAGNETISM

K W H Stevens

14.1. Introduction

14.1.1. *Magnetism prior to the twentieth century*

At the beginning of the nineteenth century the phenomenon of magnetism, which had been known for many centuries, was assumed to be due to the existence of magnetic dipoles, pairs of closely spaced equal and opposite magnetic poles, a concept first introduced by Gilbert in the sixteenth century. By the end of the century this picture had almost entirely disappeared and magnetism was then regarded as due to tiny circulating electric currents, each of which would set up a magnetic field which would be indistinguishable from the field of a magnetic dipole. Thus magnetism and electricity had become different aspects of a unified description, which made free use of the field concept, that at each point in space there would be electric and magnetic fields which would have magnitudes and directions. In fact for both there were two fields, denoted in the case of magnetism by B, the *magnetic induction*, and H, the *magnetic intensity*, two vector quantities related by $B = H + 4\pi M$, where M denoted the *intensity of magnetization*, which also had magnitude and direction. (This is the relation in the cgs system of units then in use.) At any point in a vacuum M was taken to be zero; B and H then became identical, which raised the question of which to use and what to call it. The convention chosen was to use H and call it the magnetic field.

The introduction of B, H and M was the outcome of attempts to formulate the theory without assumptions about microscopic mechanisms, and yet incorporate the idea that the field inside a magnetized specimen is not necessarily the same as the external field creating the magnetization. It has led to a good deal of confusion for teachers and students, not least in reconciling this continuum theory with atomistic models. One of the problems is Lorentz's demonstration

1111

that when magnetic effects are ascribed entirely to currents the average magnetic field, h, (taken over all space, inside and outside atoms) is the same as B. Since these various problems are man-made, and evaporate in a consistent application of Lorentz's methods and their successors in atomic physics, I shall avoid using B and H as far as possible, except in free space, where they are identical, and in experimental curves where the originators have used them. The magnetization, M, however, is a fundamental concept and its microscopic interpretation plays a central role in this story.

For most of the nineteenth century magnetic phenomena were regarded as static, and the advent of Maxwell's electromagnetic equations, late in the period, made little difference. Even now most investigations of magnetic phenomena which involve electromagnetic waves use frequencies for which the wavelengths are long compared with some relevant dimension of the sample of interest.

It was mainly due to Faraday that in the second half of the nineteenth century it had at last been realized that there were several different types of magnetic behaviour. Lodestone had long been known for its magnetism and had been turned to practical use in compasses. Iron too had also been found to have magnetic properties, for it was attracted to lodestone, but after these two discoveries centuries were to go by before any further additions were made. The next significant step occurred in the eighteenth century when cobalt and nickel were also found to have magnetic properties similar to those of iron, though it was not until the nineteenth century that there was reasonable certainty that the properties of cobalt could be attributed to the pure element. So at the beginning of the twentieth century there were just three known elements, iron, cobalt and nickel, some of their alloys and lodestone which had strong magnetic properties.

Iron and its alloys had, of course, become extremely important in the development of structural engineering; the fortunate combination of its strength and magnetic properties, together with Faraday's discovery of electromagnetic induction, had resulted in the establishment of the electrical power industry. Thus by 1900 some quite advanced electrical machinery existed along with a range of sensitive instruments for measuring electrical currents and voltages. The combination of high-field electromagnets and improved instrumentation meant that it had become possible to make measurements of magnetic properties which would have been quite impossible in the early part of the previous century.

Faraday, about the middle of the 1840s, also initiated a general study of the magnetic properties of gases, liquids and solids. This work produced no further examples of strong magnetic materials but it did show that weak magnetism appeared to be a universal property. As a result he could begin dividing the weakly magnetic materials into two classes, according to whether they were repelled from a region of high magnetic field or attracted into it. The former were said to be *diamagnets* and the latter *paramagnets*; a diamagnetic substance, when placed in a magnetic field, develops a magnetic moment in the opposite direction to the field whereas a paramagnetic substance develops a moment in the direction of the field. To these classes we must add the *ferromagnets* (Fe, Co, Ni), but not lodestone, which is nowadays classed as a *ferrimagnetic*, and the *antiferromagnets*, both of which are much later discoveries.

It is also possible to subdivide the various magnetic solids according to whether or not they are electrical conductors. This separates lodestone from the ferromagnetic metals and alloys as well as separating the diamagnets from most of the paramagnets. For the normal metals, those which do not show ferromagnetism, the paramagnetism shows little or no temperature dependence, whereas for most of the insulators it is diamagnetism which shows little temperature dependence. Thus the electrical behaviour separates the paramagnets from the diamagnets except for a small class, mainly the insulating salts of the 3d transition elements and the rare-earths, which show pronounced temperature-dependent paramagnetism.

In describing the historical development a classification based on electrical properties has seemed to be slightly more convenient than that introduced by Faraday. Table 14.1 gives some idea of the range of topics to be covered, and the main theoretical concepts. The century itself will be treated in three periods—before the introduction of quantum mechanics (1900–25), up to the end of World War II and the subsequent years of reconstruction (1925–50), and the rest of the century (1950 onwards). In the third period, as well as the extension of the concept of antiferromagnetism to ferrimagnetism, two major new fields were opened, the study of the rare-earths and the actinides, and nuclear magnetic resonance. The first can be incorporated in the general scheme, but the second is so different that it has seemed best to regard it as a separate topic and describe it separately. Then towards the end of this period developments in technology opened further fields of study, two based on random systems, spin glasses and amorphous metals, and another on the properties of very thin films. These also are best treated as separate topics.

14.2. The period 1900–25

14.2.1. Diamagnetism

The investigations of magnetic properties, begun by Faraday, were gradually extended, particularly by Curie [1], with the result that by the end of the first quarter of this century it had been found that the majority of gases (except O_2, NO, NO_2 and ClO_2), liquids and solids were diamagnetic, the exceptions being the ferromagnets, most compounds containing ions of the 3d elements (Ti, V, Cr, Mn, Fe, Co, Ni and Cu) and

Table 14.1. *The top part of the table lists the topics of special interest in two periods, 1900–50 and 1950 onwards. The bottom part, under 'Theoretical concepts', numbers the main theoretical ideas. Their numbers are attached to the topics in the top half, indicating where they have been used.*

	1900–50	1950 onwards
Insulators	Diamagnetism 7 Curie–Weiss paramagnetism 1, 2	Antiferromagnetism 2, 4 Ferrimagnetism 2, 4
Conductors	Temperature-independent paramagnetism 3 Ferromagnetism 2, 3	Rare-earth metals 2, 3, 4
		Nuclear magnetism 5 Spin glasses 4 Amorphous alloys 6 Thin films 2, 3, 6
Theoretical concepts	1. Crystal fields 2. Exchange interactions 3. Band theory 4. Spin-Hamiltonians 5. Bloch equations 6. Correlated electrons 7. Closed shell ions	

the normal metals, the non-ferromagnetic conductors. It had also been established that their susceptibilities, usually denoted by χ, and defined as the ratio of the magnetization to the applied magnetic field in the weak-field limit, were independent of temperature and of the strengths of the fields then available. Weber, in 1854, had already suggested that molecules could be divided into two classes according to whether or not they had resultant magnetic moments due to circulating currents. On the supposition that an applied field would induce circulating currents in molecules which otherwise had no such currents, the characteristics of diamagnetism could be explained. The molecular model was by no means generally accepted, and Weber's views were fulfilled only with Lorentz's electron theory of matter, initiated in 1892 and given a firm basis by Thomson's discovery, in 1897, of what we now know as the electron.

Among the many phenomena Weber analysed was the Zeeman effect in spectroscopy. Voigt [2] and Thomson [3] had already tried to develop an electronic theory of magnetism, with limited success, and it was Langevin [4] who pointed out that the Zeeman effect and diamagnetism were different aspects of the same phenomenon. The application of a magnetic field would modify all orbital motions by inducing precessions about the field direction, so creating changes in each molecular magnetic moment. Furthermore, the theory could be used to estimate the average sizes of the electron orbits from the experimental results, and Langevin noted that the orbits were small enough to be contained within molecules.

Lorentz went further, suggesting that the induced precession was due to the force on each electron from the electric field set up as the magnetic field was increased from zero, and that there would need to be a massive and immovable positively charged sphere present to maintain electrical neutrality. (It is remarkable that although planetary motion was well understood there seems to have been a marked reluctance to envisage molecules as mostly empty structures.) He also expressed concern that the precession, once it had been set up, seemed to persist indefinitely, which appeared contrary to physical experience: explaining why was to dog magnetic theory until the advent of quantum concepts.

Lorentz's misgivings were only too well founded since, as van Leeuwen [5] showed in 1919, a system of charged particles governed by classical mechanics can have no magnetic properties in a state of thermal equilibrium. This was remarked on by Bohr [6] in his doctoral thesis but became widely known only through van Leeuwen's work.

14.2.2. Quantum concepts

The justification of Langevin's model came with the Bohr [7] theory which supplemented the Rutherford model of the atom, a central, massive and positively charged nucleus about which the electrons were circulating as in a tiny planetary system, with a quantum postulate which restricted the possible atomic energies to a number of discrete values. In particular an atom or ion would have a lowest energy level from which there could be no decay. Thus the problem of motion with no dissipation in energy disappeared.

The initial impact of the Bohr theory was, however, much greater in spectroscopy than in magnetism, where an extensive theoretical structure was constructed and used to interpret a wide range of experimental observations. Three quantum numbers were introduced, one associated with each degree of spatial freedom, with relations between them. (The position of any particle can be described in a variety of ways using three 'coordinates'. Those chosen for an orbiting electron were its distance from the nucleus, and two angles, one of which was in the plane of its motion and the other gave the angle of the normal to the plane relative to some fixed direction.) In due course these were found to be insufficient and a fourth quantum number was added, associated with the assumption that the electron was spinning about its axis.

Magnetic moments had already been associated with two of the original quantum numbers, those related to angular variables, and it was natural to associate magnetic moment with the spin of the electron: the only strange feature, apart from the spin taking a half integer quantum number, was that from the study of the Zeeman effect it appeared that the magnetic moment associated with the spin angular momentum was twice that associated with orbital momentum. To complete the picture it was also necessary to assume [8] that no two electrons could have the same four quantum numbers (Pauli's exclusion principle).

The restriction imposed by this last assumption had direct consequences for magnetism because it led to the picture of the shell structure of atoms and ions and the recognition that in the diamagnetic insulating compounds the electronic structures were those of ions with all the occupied shells completely filled. Thus all the electrons could be regarded as paired off; with neither a resultant orbital nor a resultant spin angular momentum there would be no net magnetic moment. Furthermore, it was realized that the paramagnetic insulators all contained ions with electrons in incompletely filled shells and that non-zero magnetic moments could well be associated with these.

The diamagnetism of closed shells was accounted for by the precession of the orbits in the presence of a magnetic field, an extension of Lorentz's proposal [9]. Then the temperature independence followed because the electronic arrangement was so stable. To unpair any of the electrons by, say, moving one of them to an unoccupied orbital would need far more energy than was available from thermal fluctuations. So such transitions would not occur, a conclusion which also explained why these diamagnets were insulators.

With just the Bohr theory no further progress was possible.

14.2.3. *Paramagnetic insulators*

In 1895 Curie reported that a number of substances showed susceptibilities which as well as differing from diamagnetism in the direction of the induced moment also differed in having a strong temperature dependence; the results could be fitted to the relation $\chi = C/T$, where T is the absolute temperature and C is a constant which depended on the nature of the sample. The relation is now known as Curie's law. As the temperature range was extended it was found that a better fit was usually obtained with the Curie–Weiss law, $\chi = C/(T - \Delta)$, where Δ was positive for some substances and negative for others. There were no measurements down to temperatures of the order of the magnitudes of the Δs, so the corrections to Curie's law were small. In 1905 Langevin put forward an explanation of the Curie form using the assumption that not only could an external magnetic field induce precessions but it would also tend to line up any permanent magnetic moments. The alignment would not be complete because of the randomizing effects of temperature, which could be taken into account using Boltzmann's statistics. At low field strengths the magnetization M is proportional to the applied field, but in very strong fields it saturates at M_0, when all individual moments are aligned with H. The detailed expression for the variation of M with H is still known as Langevin's function

$$M/M_0 = \coth a - 1/a$$

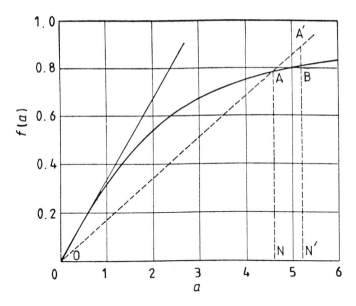

Figure 14.1. *Langevin's graph of the function* $\coth a - 1/a$ *(the curved solid line).*

where $a = \mu H/kT$, μ being the magnetic moment of a molecule (figure 14.1). As with Langevin's theory of diamagnetism this analysis also falls foul of van Leeuwen's theorem, but is restored by Bohr's assumption of stable orbits giving rise to permanent molecular moments.

Debye [10], who had based his theory of polar molecules on Langevin's work, modified the details of the argument to take account of the concept of spatial quantization that had developed from Bohr's atomic theory. The permanent magnetic moment could not be assumed capable of pointing at any angle to H, but only at a few discrete orientations. The theory was expected to be only strictly applicable when the magnetic interactions between different molecules could be neglected, as in gases, but from the observation that the law held more widely it could be inferred that the restriction was probably unnecessary.

This conclusion received strong support from Lorentz's treatment of the interaction of electric dipoles. He showed by direct calculation that if the molecules in a spherical sample are arranged in a cubic array they produce no internal field; each dipole is acted upon by the external field as if all the others were absent. The same was expected to apply to the magnetic case, for small fields and high temperatures, particular as the strength of the magnetization is then usually so small that there is a negligible difference between the spherical sample and a sample of any other shape (the restriction to a cubic array was not expected to be of any significance).

In spite of this result Weiss [11] took the remarkably bold step

1117

of extending the Langevin theory by assuming that the field acting on a molecular magnet would be the externally applied field plus a substantial extra field proportional to the magnetization. This led him to substitute $H + \lambda M$ in place of H in the Langevin expression for the paramagnetic moment, taking λ to be positive. The expression for the magnetization for a sample obeying Curie's law, $M = CH/T$, was changed to $M = C(H + \lambda M)/kT$, which could be re-arranged to give $M = CH/(T - \lambda C)$, the same form for the susceptibility, M/H, as that of the Curie–Weiss law, when Δ was identified as λC. Weiss must have realized that if this was to be the way of accounting for the Curie–Weiss law it was quite certain, from the known values of the Δs, that they could not be due to the known interactions between the magnetic dipoles. He was therefore led to postulate that some other interaction, of unknown origin, must be present, which he presciently suggested might arise from the electrical forces between the molecules.

See also
p 545

Even more remarkably, when he examined the modified formula of Langevin, without making approximations about the temperature and the strength of the magnetic field, he found that quite a different behaviour emerged at low enough temperatures and with H put to zero. There were two solutions for M. It would either be zero or it would have a finite value which depended on T, and the solution with the finite M would have the lower energy. This suggested that a system obeying the Curie–Weiss law with a positive Δ could be expected to develop a spontaneous moment at low enough temperatures. The theory also showed that this moment would grow from zero as the temperature was further lowered from a critical value (now usually referred to as the Curie temperature).

That a spontaneous magnetization might arise can be inferred by supposing that a uniform magnetization has developed in the absence of an applied field. Weiss's assumption would then imply that there would be an internal field in the direction of this magnetization which would create a magnetization in the direction of the internal field. A self-consistency argument would then suggest that the initial spontaneous magnetization should be identified with the magnetization it actually produced.

It may seem surprising that the basis of Weiss's result can be understood by such a simple argument, but like most of the advances in physics something has to initiate events. Similar reasoning is now used in a variety of contexts, often under a description which includes the term 'self-consistency'. It seems that this was the first example of its use in a solid-state context, where it provided an explanation of a solid-state phase transition from paramagnetism to ferromagnetism.

The implications for experimental physics were considerable. It had already been suggested, by Curie, that ferromagnets would become paramagnetic at high enough temperatures. Now it seemed

Pierre Ernst Weiss

(French, 1865–1940)

Physics has always exploited its discoveries and has also had room for the experimentalist whose interest is in facts rather than the current notions of theoreticians. Pierre Ernst Weiss, who was born in Mullhouse, France, spent almost the whole of his professional life working in magnetism. Faraday had suggested that all materials would have magnetic properties and it was Weiss, using the most sensitive equipment then

available, who confirmed this suggestion, exploring, with his students, a large number of substances over extended ranges of temperature. Beginning in Zurich he eventually moved to Strasbourg, in 1919, where he created a centre for magnetic studies which still exists. His discoveries showed the limitations of current theory, which he proceeded to extend, by assumptions which, although they must have been anathema to the theoreticians of the day, explained his observations and produced the first theory to explain how phase transitions could occur in solids. In his ideas he was far ahead of his time. His ideas can be found, along with many experimental results, in *Le Magnetisme* (1931) [12], a book with G Foëx as co-author, which, incidentally, makes no use of quantum mechanics!

that examples were to hand of paramagnets which would become ferromagnetic at low enough temperatures. However, the necessary low-temperature facilities were not available to test this, so advantage was taken of another prediction, that the paramagnetic magnetization would not remain proportional to the field strength as the field increased and that the departures from linearity would be easier to observe as the temperature approached that of the phase transition, which was predicted to occur when the temperature equalled Δ.

All this was a considerable stimulus to the development of low-temperature facilities, and indeed the interest in producing ever lower temperatures and research in magnetism have all along been closely linked. Nevertheless the experimental demonstration of a paramagnetic salt becoming ferromagnetic at low temperatures failed, in spite of there being a number of positive Δs. Weiss and Foëx [12] in 1931 gave a table of values about half of which were positive. In fact it is now known that ferromagnetism is uncommon in the salts. It seems probable that the observations were made on samples containing unidentified impurities. A tiny amount of unrecognized ferromagnetic iron could have invalidated the observations.

Having accounted for the $1/(T - \Delta)$ part of the Curie–Weiss law there remained the question of C, which reflected the magnitudes of the ionic magnetic moments. They should have been obtainable from the Zeeman splittings of the ground states of the ions. However, it was found that in most cases the values obtained for C did not agree with the experimental values. (Nor did the Bohr theory explain the Zeeman observations.)

This was the position prior to the introduction of electron spin and its anomalous magnetic moment. Once it was included the problems over the Zeeman splittings largely disappeared, but this did not clear up the problem of the C values. Then it was realized that much better agreement could be obtained if the magnetic moments in the ionic compounds were assumed to be entirely due to electron spin, so creating the problem of accounting for the disappearance of orbital contributions in the crystals. The Bohr theory gave no answer, though the position was somewhat improved when Hund [13] showed that the Zeeman observations on a range of rare-earth ions could be used to explain most of their Cs. Van Vleck and Frank [14] subsequently showed that his failure on two of the ions had a simple explanation.

A failure to obtain a fit to experimental results because some feature has been accidentally omitted is unfortunate, but when the error is subsequently corrected there is an almost overwhelming impression that one's understanding of what is happening must be correct. Since the rare-earth ions were known to have magnetic properties due to electrons in 4f shells, which were well inside the ions whereas the magnetic properties of the 3d transition ions were known to be due to electrons on the outsides of the ions, it was an obvious step to assume that in some way the difference between the two families arose because the 3d ions were more sensitive to their crystallographic environment. The Bohr theory, however, gave no help.

14.2.4. *Ferromagnetism*

Although the Weiss theory explained how ferromagnetism might occur when a paramagnet is cooled sufficiently there was a problem with applying this argument, because freshly prepared iron at room temperature has no magnetic moment. To cope with this situation Weiss first assumed that a typical iron sample would consist of a random assembly of single crystals. He then assumed that in each single crystal the direction of the spontaneous moment would be in a direction determined by its crystal structure. A random distribution of crystallites would therefore have no net moment. In a uniform external magnetic field only a few of the crystallites would have their moments lying parallel to this field, so he next considered those crystallites with the opposite alignment assuming that on increasing the field from zero nothing much would happen until the external field reached a critical value, when the magnetization would suddenly completely reverse. In fact, of course, the bulk of the crystallites are not aligned in either of these directions and for these he assumed that those lying near the field directions would not be affected but those lying in the 'wrong' direction would eventually reverse their moments, when the component of the applied field in the direction of their axes reached the critical value. Thus the overall picture, on gradually increasing the external field, would show evidence of moment reversals taking place successively as the field reached the critical value for the different crystallites. Eventually all the wrongly aligned moments would be reversed, after which there would be no further increase in overall moment. On reducing the field nothing much would happen and the system would retain its magnetization until the field reached zero. On increasing it again, in the opposite direction, moment flips would begin as the axial component of the field reached the critical value. In this way Weiss was able to explain how an initially unmagnetized sample would acquire a moment in an applied field and how the moment would change under field cycling [15].

Field cycling behaviour (hysteresis, the term introduced by Ewing in 1881) had long been observed experimentally and although this model, when worked out in detail, did not fully reproduce the observations there was enough similarity between its predictions and what was being observed to give confidence in the basic idea (figure 14.2). For many years, however, there was no direct way of viewing the distribution of the magnetization, though there was evidence of sudden changes in it as an applied field was altered, through the Barkhausen [16] effect—a sudden change in the magnetic flux threading a coil in series with an earphone produced a click.

A complete theory of hysteresis was of much less importance than the need to have information about the hysteresis properties of the ferromagnetic elements and alloys. This was complicated because evidence was accumulating that the magnetic behaviour was probably being affected by the presence of unidentified impurities in both the elements and the alloys, and by inhomogeneities in their structures. (The mechanical properties of iron, for example, are strongly dependent on the amount of carbon present and the heat treatments that have been carried out.) The reason for the interest in hysteresis was a technical one, the need for magnets with optimum properties. There were basically two different requirements, permanent magnets and core materials for transformers. For permanent magnets the need was to have as large a magnetization as possible in high field and a large critical field. With

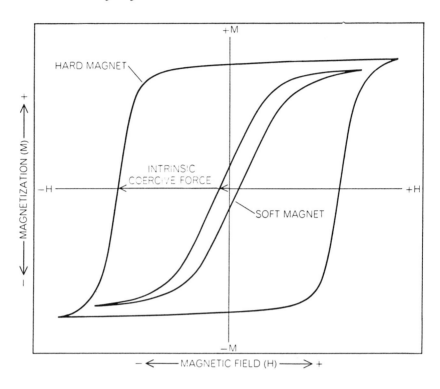

Figure 14.2. *A schematic illustration of the hysteresis behaviour of a hard magnetic material, the large loop, and a soft magnetic material, the small loop. The initial magnetization curve contains a part which joins the origin to the loop.*

the former a large magnetization would then be retained on reducing the high field to zero and the latter would ensure its stability against the random fields to which the magnet would inevitably be exposed. Stability against temperature fluctuations was also a requirement. In the transformer applications the magnetization would be repeatedly taken round a hysteresis cycle and a simple calculation showed that in each cycle there would be an energy loss determined by the area of the hysteresis loop. Apart from heating the core, which meant that some form of cooling might be necessary, it was undesirable from the point of view of efficiency, so the need was to make the area of the hysteresis loop as small, and the magnetization as large, as possible. As a result ferromagnets were broadly classified by their hysteresis loops; 'hard' ferromagnets having large-area hysteresis loops were potentially useful for permanent magnets, and 'soft' ferromagnets having small areas were potentially useful in transformers. The only way to find suitable materials for the different uses was by trial and error, with the result that while many alloys were classified as above, relatively few were used. In the transformer applications the fact that the ferromagnets were also conductors was a disadvantage, because the magnetic flux changes induced current flows in them, resulting in ohmic energy losses. As the number of hysteresis cycles per second and the magnitudes of the induced voltages increase with frequency, ferromagnetically cored transformers have been confined to low-frequency applications.

Although there was considerable interest in the technical applications there were also academic aspects of ferromagnetism, particularly in the relationship between magnetic moment and angular momentum. Indeed it is interesting to read, in an account of this investigation by Galison [17], of Einstein's interest from 1905 onwards and of the controversy which arose amongst the experimentalists, Richardson, Barnett, Einstein and de Haas, Beck and co-workers. It arose over whether or not the *gyromagnetic ratio*, the ratio of the magnetic moment to the angular momentum, in appropriate units, was 1, as expected from Ampère's model of current loops, or some other value. Galison actually gives a diagram in which the values from the different investigations are plotted against their dates, showing the wild fluctuations which were obtained from about 1915 until the middle 1920s, when the ratio more or less settled down near 2. No explanation could be offered until 1925 when Goudsmit and Uhlenbeck [18] suggested that the electron had a spin and that to account for spectroscopic observations the associated magnetic moment had to be twice that expected for ordinary angular momentum. The ratio, in appropriate units, was denoted by g and the value, which is very close to 2, has since been known as the anomalous g-value of the electron spin.

These experiments showed that the magnetic moments of ferromagnets arise almost entirely from the spins of unpaired electrons. Furthermore, from the saturation moments, measured in high magnetic fields when it could be assumed that all the local moments were aligned, the total number of aligned electrons could be determined. Surprisingly the numbers were not simply related to the number of atoms, being for example 0.6 for the unpaired electrons per atom in nickel; and this in a lattice in which all the nickel atoms seem to be identically situated. It was a long time before an explanation was forthcoming.

It was during this period that single crystals of ferromagnets were first made, with Beck [19] the pioneer in the study of their magnetic properties. Little of value was discovered until later, and the important work will be described in due course.

See also p 1134

14.3. The period 1925–50

14.3.1. *Quantum theory of ions*

By the early 1920s doubts were beginning to creep in about the merits of the Bohr theory, particularly in the context of spectroscopy, where it was being severely tested. A comprehensive 300-page review by Van Vleck [20] in 1926 was probably the last to appear before the theory was entirely replaced by quantum mechanics. This development had enormous consequences for most of physics, including, of course, magnetism. There is no need for a comprehensive account of it, but as some of the results of its application to the spectroscopy of ions had immediate consequences for magnetism an account of these seems desirable.

The description of an ion which it gave was, in some ways, similar to that in the Bohr theory. In a first approximation each electron was described by a wavefunction which involved four quantum numbers, identical to those in the Bohr theory, with the same restriction that no two electrons could have the same set. The energies of the individual electrons were also similar, so the same shell structures emerged; but in the concept of angular momentum differences particularly relevant to magnetism became apparent.

Angular momentum is a difficult concept in quantum mechanics because of the role played by the uncertainty principle. In classical mechanics it is represented by a vector quantity which can be regarded as having components along three orthogonal axes, $0x$, $0y$ and $0z$. In quantum theory the measurement of any one component upsets the others, so an angular momentum cannot be specified as in classical physics. The position is not quite as bad, however, as the above statement might suggest, for although the three components cannot be known simultaneously it is possible to know one component and the total magnitude. The component, in magnetism, is usually taken as that in a direction determined by an external field. There is then no possibility that the moment is completely aligned in this direction, for this would imply that the components at right angles are also known to be zero. So there is always an angle between the direction of the total moment and the direction of the field. In describing a moment as aligned the meaning is that the projection of the total moment onto the field direction has its maximum allowed value. Quantum mechanics confirms the conclusion of the Bohr theory, that an ion which has the right number of electrons to fill the shells up to definite values of n and l, leaving the shells outside this empty, possesses neither orbital nor spin angular momentum and therefore no magnetic moment. Quantum theory was thus able to account, in much the same way as the Bohr theory, for the temperature-independent diamagnetism of a wide range of insulators.

The next application was to the salts showing the Curie–Weiss type of paramagnetic susceptibility, ionic insulators which contain ions of either the iron or the rare-earth groups, with electrons in partially filled shells. Hund's [21] analysis of spectra, even before the advent of quantum mechanics, showed that in the lowest energy arrangements the electrons in the partially filled shells have, as far as possible, aligned spins. (When a shell is more than half-filled the exclusion principle does not allow all the electrons to have parallel spins.) Quantum theory provided an explanation in terms of the Coulomb interactions between the electrons and the effects of the exclusion principle. It also showed how the spectral properties could be used to estimate the energy needed to reverse one spin against the others. This result was far larger than any previous estimates and provided evidence for an interaction of ferromagnetic character between spins, of electrical origin as had been postulated much earlier by Weiss.

About the same time Heisenberg [22] and Dirac [23] independently studied a problem in which N electrons were placed in N different orbital wavefunctions, leaving the spin of each electron free to take either of two spin orientations ($m_s = \pm\frac{1}{2}$, in units of $\hbar$). They were able to show that the energies of such a model would be exactly the same as would be found for a different model, one in which there were N spins of $\frac{1}{2}$ coupled by what is now known as a ferromagnetic Heisenberg exchange interaction, and that the energy would be lowest when all the spins were aligned in the same direction. An external magnetic field would cause all the magnetic moments to align themselves in its direction. Here was another example of a ferromagnetic interaction and again it had arisen from the Coulomb interaction between electrons and the need to satisfy the exclusion principle.

Before ending this account of some of the results of quantum theory mention must be made of Kramers' [24] theorem, because it is frequently used in magnetism. He proved that, if the number of electrons is odd and no external field is present, every energy level will contain an even number of states of the same energy. In many cases this is the only degeneracy which remains and its importance is that if a magnetic field is applied it can be expected that the two levels will diverge linearly. When the splitting is measured it gives the magnetic moment to be associated with each state. (One state has its moment parallel to the field and the other antiparallel, so the level separation gives the energy needed to reverse the moment against the field.)

14.4. Paramagnetism

14.4.1. *Adiabatic demagnetization*
In 1926 Debye [25] and Giauque [26] independently pointed out that the rise in temperature found on applying a magnetic field to paramagnetic salts indicated that the reverse process could be used to obtain cooling in the helium temperature range beyond the limit which had then been reached. The proposal was rapidly followed up, though a good deal of preparatory work was first required. The argument involved no quantum theory, but only thermodynamic reasoning and experimental results. It was more a matter of choosing an optimum material. Kurti

and Simon [27] showed that recent results on the specific heat of the rare-earth salt gadolinium sulphate indicated that it would be an even more promising material than they had first expected. This was confirmed by Giauque and MacDougall [28] who used it to reach a temperature of 0.53 K, beginning at 3.4 K with a field of 8000 gauss (the Earth's field is about 1 gauss). Almost immediately afterwards de Haas, Wiersma and Kramers [29] reached 0.05 K using potassium chromic alum.

In zero field and at low temperatures each ion in an assembly of identical Kramers ions will be in one or other of two states, with equal probabilities. On applying a field the states will be symmetrically split; having equal populations is then not the thermal equilibrium arrangement so some ions will move from the upper to the lower energy level and the energy released will appear as heat. It is removed by conduction and the overall energy is reduced. The assembly is now thermally isolated and the field is reduced slowly enough to maintain thermal equilibrium at all times. As the splitting of the states decreases, ions return from the lower to the upper state, for which they need energy. This comes from the thermal energy of the lattice, and so the crystal is cooled.

The experimental demonstration of the cooling was not easily accomplished. A major problem arises from heat leaks, particularly if the time to equalize the populations, at the final stage, is long. A full account has been given by Hudson [30].

14.4.2. Crystal field theory

Adiabatic demagnetization focused attention on the need to understand more about the low-lying energy levels of magnetic ions in crystals and on the way in which thermal equilibrium is established following a disturbance of the level populations. This was met by crystal field theory.

The basic idea [31] was that the electrons of importance in a magnetic ion, those in a partially filled shell, are exposed to an electric field from adjacent negatively charged diamagnetic ions, normally identical, six in number, and arranged octahedrally (figure 14.3). The first theoretical analyses, based on a crystal field which was not that of a regular octahedron, led to problems [32] which were only resolved when Penney decided to use what is now known as a cubic potential, that due to a regular octahedron. This had the required property that at a distance r from the magnetic ion the potential is largest along lines to nearest neighbours and weakest in intermediate directions. Such a potential hinders the free orbital motion of the electrons of an isolated ion and results in their energy levels being split and their states being modified.

The theory was extended to take account of small departures from octahedral symmetry, the presence of spin–orbit coupling and the inclusion of external magnetic fields. The first produced a further lowering of the symmetry, which generally meant that all orbital

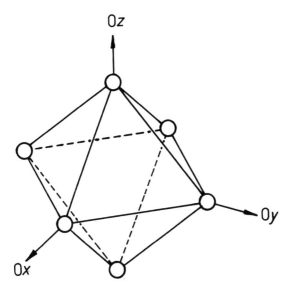

Figure 14.3. *The six circles represent the positions of the nuclei of neighbours in an octahedral arrangement. The axes, which are mutually at right-angles, join the centre of the octahedron to the vertices and pass through the magnetic ion.*

degeneracies of the free ion were resolved and all orbital motion was *quenched*, leaving no orbital magnetic moment. This explained why spin-only moments were observed in the paramagnetism (figure 14.4).

The wavefunctions which go with these energy levels are not easy to represent diagrammatically, for generally they describe many-electron functions. However, when there is only a single 3d electron in the ion they are easier to picture, and an example of how quenching arises is shown in figure 14.5. For a more complicated ion it is simplest to accept that the charge distribution has lobes which in the lowest state are set so as to minimize the energy. Detailed theory is needed to calculate the orientation of least energy.

When the spin–orbit interaction was included it was found that the orbital magnetic moment was then not quite quenched and that the effect was as if the *g*-value of the electron had been changed by about 10% from the normal value (approximately 2). The theory was able to predict the sign of the departure from 2, which varied with the ion, and also that it would be directionally dependent. That is, the observed Zeeman splittings would vary with the direction of the magnetic field relative to the crystallographic directions.

The same basic theory was also applied to the rare-earth ions, with the assumption that the crystal fields would be much weaker and that the effect of the spin–orbit coupling would be stronger. The result was that crystal field effects were expected to be negligible except at very low temperatures, a conclusion which fitted in well with the knowledge

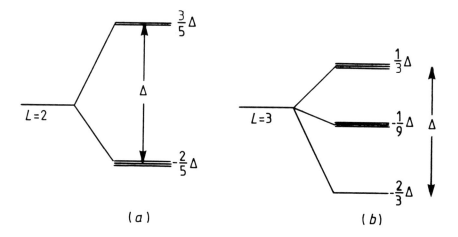

Figure 14.4. *In an octahedral environment the five-fold degeneracy of a free ion with L = 2 is split into two coincident states at $(3/5)\Delta$ and three coincident states at $(-2/5)\Delta$, where Δ is the overall splitting. Its sign depends on the number of d electrons and is positive for $3d^1$ and $3d^6$ and negative for $3d^4$ and $3d^9$. (For clarity the levels which are actually coincident are shown with small splittings.) The pattern for L = 3 has three coincident states at $(1/3)\Delta$, three at $(-1/9)\Delta$ and a single level at $(-2/3)\Delta$. Δ is positive for $3d^3$ and $3d^8$ and negative for $3d^2$ and $3d^7$. $3d^5$ has L = 0 and so shows no splitting. Its pattern has therefore been omitted. The Δ used in this figure should not be confused with the Δ in the Curie–Weiss law.*

that their susceptibilities could be obtained from observations of the free ion moments. When, eventually, low-temperature susceptibility results became available, on the sulphates of praseodymium and neodymium, it was found that χ tended to a constant for the former and to infinity for the latter as the temperature tended to zero. These conclusions were in agreement with the predictions of crystal field theory. So it seemed that it was a viable theory for both the iron and the rare-earth salts. (Later work has shown that six-fold coordination of neighbours is uncommon for rare-earth salts and that an arrangement with nine neighbours is much more common. Thus the satisfactory explanation of the low-temperature susceptibilities of these two salts probably owes less to the choice of crystal field than to the fact that praseodymium is a non-Kramers ion, and so has a singlet lowest state, whereas neodymium is a Kramers ion and so has a doublet lowest.)

The only remaining need, as far as susceptibilities were concerned, was to account for the Curie–Weiss Δs, for the Cs were now explained. Weiss had already attributed the Δs to interactions between ions, and the Heisenberg–Dirac (H–D) theory had produced exchange couplings which could be re-interpreted as arising from the spins of magnetic ions in quenched orbital states. Van Vleck [33] had therefore no hesitation

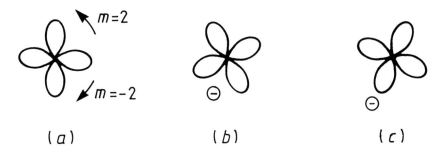

Figure 14.5. *(a) The instantaneous projection of a typical d-wave function onto a horizontal plane. The state with 2 units of angular momentum has this shape rotating anticlockwise about the normal to the plane. The state with −2 units is its clockwise rotating companion. (b) and (c) show what happens if a single negative charge is placed somewhere in the horizontal plane. The rotational motion is arrested and in the lowest energy state the electron is in a wavefunction which places it as far away as possible from the hindering charge. That is, in the state shown in (b). As there were initially two states and the number is unaltered by the presence of the charge there is a second, higher energy state, in which the electron is much closer to the negative charge, (c). The states with 2 and −2 units of angular momentum can be regarded as wave motions about the vertical axis. (b) and (c) can be regarded as standing wave patterns constructed from them. Having no angular motion they have no magnetic moments.*

in attributing the Δs to exchange interactions between the spins of the magnetic ions. In 1937 he gave a theory for a set of such spins interacting through nearest-neighbour exchange interactions and confirmed that the susceptibility would then take the Curie–Weiss form and that Δ would be proportional to the exchange constant. It would therefore be positive for the ferromagnetic exchange of the H–D theory. So the origins of the Δs were established, except that their signs, on the whole, were turning out to be opposite to those of the H–D theory.

The expression *exchange interaction* seems, by this time, to have become fairly firmly embedded in the literature of magnetism, both in the context of the paramagnetic salts and in the ferromagnetism of iron, cobalt and nickel. This was undoubtedly due to the initial concept of Weiss, though it is curious that where the theory had been most developed the experimental evidence was that the sign was not that of ferromagnetism, and where it was least developed it was.

14.4.3. Antiferromagnetism

The study of salts with negative Δ was dominated, from 1932 onwards, by the work of Néel, who pioneered the concept of antiferromagnetism. The term is now applied to such salts when they are below a phase transition which occurs at the *Néel temperature*, T_c ($\sim$ −Δ). Bitter [34] described the transition by an extension of the Weiss theory of ferromagnetism, using the negative Δs to justify the choice of a model in which the sign of the exchange tended to make neighbouring spins

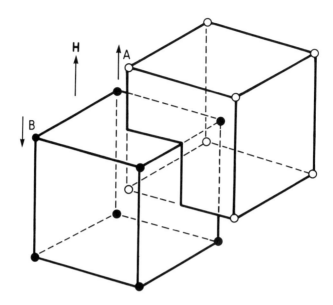

Figure 14.6. *The body-centred cubic lattice can be thought of as two interpenetrating simple cubic lattices, one unit cell of each being shown here. If there is a spin pointing up at each open circle, e.g. A, and a spin pointing down at each solid circle, e.g. B, the result is an antiferromagnetic system. Each spin is surrounded by a shell of 8 oppositely pointing spins.*

align antiparallel rather than parallel to one another. To cope with this difference he assumed that the magnetic lattice could be regarded as composed of two interpenetrating sublattices, A and B, which were magnetized in opposite directions (figure 14.6). In this example each spin is surrounded by 8 others of opposite orientation—A by 8 Bs and B by 8 As.

If the aligning forces are short-ranged the internal field at each spin is due entirely to its neighbours and in the same direction relative to its own moment as in the Weiss theory. In an external field H, weaker than any internal exchange fields, and in the direction of the moment at an A site, the internal field at A is in this same direction and is of the form $H + \gamma M_B$ (M_B and γ are both positive) while at a B site the field in the direction of the B spin is $\gamma M_A - H$, with M_A and γ again positive. The total moment in the direction of H is $M_A - M_B$.

The various moments and fields are then substituted in the Langevin expressions. At high temperatures and small H the theory predicts a Curie–Weiss type of susceptibility with a negative Δ and, at low enough temperatures, a phase change such that M_A and M_B would take finite values even when H is put to zero. Unlike the ferromagnetic case there is no net magnetic moment, and in fact a susceptibility can be defined even below the phase transition. Like the ferromagnetic case the theory does

John Hasbrouck Van Vleck

(American, 1899–1980)

The first book to apply quantum mechanics to solid-state physics was published in 1932 by J H Van Vleck, a Professor of Theoretical Physics at Wisconsin and a tenth-generation American. It was not his first major publication, for in 1926 he had published *Quantum Principles and Line Spectra*, which discussed the application and limitations of Bohr's correspondence principle.  Nevertheless it is quite remarkable how quickly he realized the significance of the new theory and set about using it. Indeed, so thorough was his understanding that over 60 years later, his *Theory of Electric and Magnetic Susceptibilities* is still regarded as the 'bible' of magnetic theory.

In 1934 he moved to Harvard, where he was to remain for the rest of his life, and from 1951 to his retirement in 1969 he held the Hollis Chair of Mathematics and Natural Philosophy, the oldest endowed scientific chair in North America. Throughout this period, and afterwards in retirement, the steady flow of important scientific papers never ceased, and it gave great pleasure to his many friends and admirers when his career was crowned in 1977 by the award of the Nobel Prize for Physics.

not predict the direction of alignment in zero external field. If, however, there is some interaction, neglected in this analysis, which determines the alignment, the susceptibility in an external field will differ from that predicted, and will depend on the angle between the external field and the direction of alignment. For parallel alignment of moment and field, the susceptibility was predicted to fall to zero linearly with temperature from its value at the phase transition, and for a perpendicular alignment to remain constant. A consequence is that, as an external field is increased from zero, the magnetic energy depends both on the strength of the field and its orientation relative to the magnetization. The lowest energy, in a high enough field, is obtained with the alignment perpendicular to the field direction, as had previously been found by Hulthen [35], who had used a different theory to investigate the way in which the low-temperature alignments would alter with temperature. He had suggested that on gradually increasing a field aligned in the direction of the magnetization a value would be reached at which the moment would change its direction to a perpendicular alignment. Such an effect was not observed until 1961, in MnF_2 [36].

At the time of this analysis no antiferromagnet had been positively identified, though in 1936 Néel [37] had suggested, because of a specific heat anomaly near 350 °C, that manganese, a metal, might be antiferromagnetic. The first positive identification [38], MnO in 1938, was closely followed [39] by Cr_2O_3. The basic experimental difficulty was that nothing very drastic happens at the Néel temperature. No large magnetization develops and there is only a slight change in the susceptibility. There had been indications of phase changes in thermal measurements, but these could hardly be regarded as indicating magnetic phase changes. Agreement with theoretical predictions about susceptibilities were in the end much more convincing.

14.4.4. *Relaxation*

As a result of the above developments it must have seemed that the magnetism of the transition group salts was well on the way to being understood. Attention was also being given to the question of what determined the rates at which level populations, and therefore magnetic moments, could be changed. The process was well understood in the spectroscopy of ions, where transitions between energy levels are induced by radiation of specific frequencies. But while there was no direct experimental evidence about the very closely spaced energy levels of magnetic ions it could be estimated, by extending the theory to very low frequencies, that thermodynamic equilibrium could not be restored in a reasonable time after demagnetization had the process needed to rely on transitions associated with electromagnetic radiation.

Attention therefore turned to a new mechanism, energy exchanges between ions and lattice vibrations. The result was a sequence of suggestions about detailed mechanisms and estimates of the times which they would give. Each was weeded out in turn, for not giving fast enough relaxation, only to be replaced by an even more complicated theory. The climax was reached in 1940, when Van Vleck [40] extended crystal field theory, with all its ramifications of spin–orbit interactions and external magnetic fields, by including the modulation of the crystal field by vibrational motion of the surrounding ions. This predicted that there should be a wide variation in relaxation properties depending on how the energy level patterns were split by the static crystal field and whether the ions had an odd or an even number of electrons. From his estimates of the rates it seemed that the theory would account for the relaxation process, though its complication and the time of its publication did not encourage detailed tests. It was, in fact, many years before the theory was generally accepted as correct. Apart from the initial delay due to World War II it took some time to realize that the measured relaxation times were often not those of the ion being examined. A given crystal can often contain undetected impurities which relax rapidly, and mechanisms exist whereby the energy of a slow relaxing ion can be rapidly transferred to a fast relaxing ion and thence to the lattice, so short-circuiting its own relaxation processes.

The problem of understanding the phenomenon of relaxation did not inhibit its experimental investigation which was started in 1936 by Gorter and Kronig [41]. The basic idea was to superimpose a time-dependent magnetic field on a steady field applied to a paramagnet. The simplest conceptual arrangement is one in which the additional field is switched periodically from h to $-h$. Then, if the switching rate is slow, the magnetization in the field $H + h$ will settle at $\chi(H + h)$ before there is a change to $H - h$, when it will settle to another steady value, $\chi(H - h)$. As the rate of switching is increased a point will be reached at which the magnetization will have so little time to adjust to its equilibrium value before the field is again switched that it will tend to stay at the average value, χH. Thus by gradually increasing the rate of switching it should be possible to observe the change in behaviour and so obtain an estimate of the time to establish equilibrium as a function of the steady field and of the temperature. In fact at the time there was relatively little experience of step-wise field switching, and sinusoidally oscillating fields were used as a matter of course. At low frequencies the magnetization should then vary in phase with the oscillating field, but as the frequency increases a phase difference should develop along with a decrease in magnitude of the response. From such observations it should be possible to estimate the relaxation time. A limit was set, however, on the range of relaxation times which could be measured by the range of available frequencies. In the first experiments, Gorter used frequencies in the range 10–30 MHz, later extended to 78 MHz, with small amplitude fields (8 gauss). Full accounts of the method were published in 1947 by Gorter [42] and more recently by Standley and Vaughan [43].

14.5. Conductors

14.5.1. *Normal metals*

Little need be said here about the paramagnetism of normal metals for in 1927 Pauli had used the standard quantum theory treatment of a particle in a box to account for it. At absolute zero all the electrons are spin-paired and at a finite temperature the number of spins which can be freely oriented is proportional to T. So all that is necessary, at least in an elementary version of the theory, is to replace the C in the Curie law by a multiple of T and so obtain a temperature-independent paramagnetism. In 1930 Landau [44] showed that there was a diamagnetic contribution

See also
p 1348

because an applied field would also modify the orbits, and there the position rested until interest was revived, after World War II. As this is described in detail in Chapter 17 nothing more need be said here.

14.5.2. *Ferromagnets*

The advent of quantum mechanics had little initial impact on the study of ferromagnetism, which was a field of considerable technical importance. Many elegant experimental studies were carried out, with a view to throwing light on such phenomena as hysteresis and what were known as easy and hard directions of magnetization (i.e. why the moments lined up in specific crystallographic directions and how far they could be moved into other directions with externally applied magnetic fields). Single crystals were much in demand for such studies. Thus Honda and Kaya [45] reported measurements, in 1928, on carefully prepared oblate spheroids made from single crystals of iron, and similar measurements on disc-shaped crystals of nickel were reported by Sucksmith *et al* [46], and on cobalt by Kaya [47]. While it is difficult to assess the intrinsic value of the results obtained from the careful shaping, which seems to have been dictated by the classical concepts of demagnetizing factors, such experiments gave important technical information about magnetic anisotropy, which had not been included in the original theory. Thus iron was found to have the [100] directions as easy directions of magnetization, nickel the [111] and cobalt its hexagonal axis. It seems that a single crystal was automatically assumed to be a single domain, an assumption that was later questioned by Landau and Liftshitz [48] who pointed out that, in the absence of an external field, a rod-shaped specimen with its easy direction of magnetization along the axis could lower its energy by breaking into two oppositely directed domains. Thus in crossing the rod there would first be a region where the magnetization would be along the axis in one direction, followed by another region in which the magnetization was in the opposite direction. It was not known how abruptly the change occurred, nor what would happen if a field were to be applied along the axis, though it could be expected that the favourably oriented domain would grow at the expense of the unfavourably oriented one. The details of this were clearly important in the context of hysteresis, but how the Bloch wall, the region between the two oppositely oriented domains, would change when the external field altered was difficult to determine.

A new method of investigation emerged from the work of Bitter [49] who had shown that if finely divided magnetic powder was distributed over the surface of a ferromagnet it would tend to be attracted towards any domain boundaries which were present. However, the use of this method called for polished surfaces and suitably chosen suspensions of powders, and before it could come into general use a good deal of effort had to be devoted to optimizing these. Even then the techniques

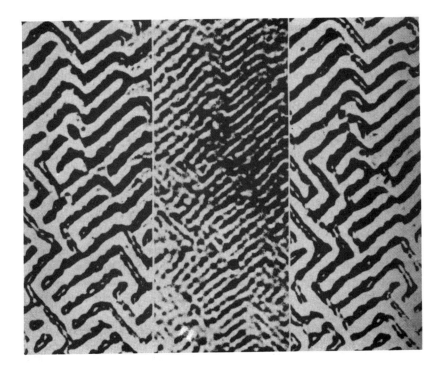

Figure 14.7. *Many different domain patterns have been observed, of which these are examples.*

offered little scope for observing movements of domain boundaries. Nevertheless the technique eventually became widely used [50], many beautiful domain patterns being seen and analysed (figure 14.7).

Another area of study was the behaviour of ferromagnets based on iron, cobalt and nickel above their Curie temperatures, to determine whether they then showed a Curie–Weiss type of susceptibility and whether the C corresponded to a whole number of spins at each site. As the Curie temperatures were all well above room temperature, so limiting the accessible ranges for paramagnetism, it was difficult to reach firm conclusions. The position was different with gadolinium, which had been shown to be a ferromagnet [51] in 1935. Its Curie temperature is at room temperature and the measurements indicated that it had seven aligned spins at each site in both phases. Its properties could be understood in terms of the Weiss model, a spin of $\frac{7}{2}$ at each site and an H–D form of exchange between the spins at adjacent sites. For iron, cobalt and nickel the Weiss theory continued to find favour, with the ferromagnetic alignment attributed to 'exchange interactions'—a useful phrase to describe the origin of Weiss's macroscopic internal field.

Various attempts were made to provide a microscopic model which would also account for the 0.6 unpaired spins per site in nickel. An early example from 1933 is the work of Stoner [52], who used the concept of energy bands which was proving so useful for normal conductors. The basic idea was that since an external field produced a temperature-independent paramagnetism in a normal metal a strong exchange interaction would, as in the Weiss theory, produce ferromagnetism. Shortly after, Mott [53] suggested that there might be two overlapping bands, one based on the atomic 4s states and the other on the 3d states, and that the electrons were distributed between the bands. The conduction could then be associated with the 4s electrons, which are on the outsides of the atoms, and the magnetism with the electrons in the 3d band, which would be similar to the electrons responsible for the moments in the 3d ions. This idea allowed the possibility of a fractional number of magnetic ions per site, but it left unexplained why they were aligned. Similar ideas were applied to the magnetic alloys as well as the pure metals in attempts to reveal systematic behaviour, with limited success. The stumbling block was that although it seemed fairly certain that the ferromagnetism was due to the Coulomb interactions between the electrons combined with the exclusion principle there was no satisfactory way of showing this. (Even if it had been possible it would not have explained why there are easy directions of magnetization, a comment which also applies to the Weiss theory and its extension to antiferromagnetism.)

Such a deficiency had been rectified in crystal field theory, where the introduction of spin–orbit couplings had produced anisotropic g-values, so showing that magnetic properties can be anisotropic. In an attempt to explain ferromagnetism by an extension of crystal field theory, Van Vleck [54] suggested that the spin–orbit interactions would also modify the form of the H–D exchange interaction and as a result a ferromagnet would have easy and hard directions of magnetization. While this was almost certainly the germ of the explanation of an important property of ferromagnets there was already enough difficulty in dealing with the Coulomb interactions without having to take on the spin–orbit interactions as well. So while both the origin of ferromagnetism and the reason for its anisotropy were thought, in principle, to be understood no fully convincing theoretical explanations were forthcoming.

14.6. 1950 onwards

World War II interrupted most magnetic research, and recovery took much longer in some countries than others, but once recovery had begun there were few further breaks, except those due to the discovery of new techniques and new lines of development in existing topics. In describing the second half of the century it has seemed best to follow particularly interesting lines as long as they were progressing rapidly, rather than describe the general field of magnetism in any particular period, except in the last decade.

A significant change in the attitude to research in magnetism came about after the war. Having known some of the pre-war experimentalists and read about the work of many of the others, I have concluded that on the whole they were much more comfortable with the ideas of classical magnetism than with those of quantum mechanics. Enough satisfaction could apparently be derived from developing experimental techniques, observing properties and applying the result of the classical theory of magnetism. Such interpretations as needed quantum mechanics could be safely left to a small number of theoreticians.

With the ending of the war the number of students in physics departments increased greatly, and this in due course led to considerably increased research in magnetism. The newcomers had usually taken comprehensive courses in quantum mechanics and seemed generally to be unattracted by simply making experimental observations, so much so that it became almost *de rigueur* to attempt a quantum theoretical explanation of all observations. Indeed some of the papers on what might be supposed to have been about experimental work contained so much theory that one is led to wonder whether the distinction between theoretician and experimentalist had almost disappeared. Later the pendulum was to swing even further, for now it is almost impossible to have a paper solely on experimental work accepted, while papers which are entirely theoretical have little difficulty.

The beginning of this period saw the development of two new techniques, magnetic resonance and neutron scattering. Both were destined to have a major impact on research in magnetism. The former could be used to study both electronic and nuclear magnetism and shared a similar theoretical basis, the Bloch [55] equations, which had originally been proposed for use in nuclear resonance. The experimental techniques were, however, somewhat different, with the result that not only did the two fields develop differently but the field of electronic resonance also divided. For convenience, therefore, several fields of study will be described as if they were in separate compartments, as indeed they were from the experimental point of view, though theoretically they had much in common. Since neutron scattering developed into a major tool in wide areas of science it is treated in some detail elsewhere. Here it will be assumed that the reader is familiar with its general features and its impact on magnetism will be mentioned as appropriate.

14.7. Electron paramagnetic resonance

14.7.1. Iron group ions

In 1945, in Kazan, Zavoisky [56] observed what was known initially as *paramagnetic resonance*, though later the description *electron spin resonance*

(ESR) came into use as an alternative. He applied a uniform magnetic field, the main field, to salts containing divalent copper or manganese ions, both of which have an odd number of electrons, so that there were Kramers degeneracies to be lifted. He then applied a small oscillating magnetic field at right angles to the main field, knowing from the quantum-mechanical selection rules, that if the quantum, $h\nu$, of the oscillating field equalled the separation between the split Kramers levels, energy should be absorbed from the oscillator and the loss might be observable. He chose a frequency of 120 MHz for the oscillating field and looked for an absorption as the main magnetic field was varied. Rather surprisingly, for the experimental arrangement was by no means an optimum, he found absorption curves, almost as broad as the range of available main fields. These were the first resonance results to be reported for magnetic ions in crystals and the only experiments which Zavoisky is known to have carried out.

In 1946 Cummerow and Halliday [57] published the results of a similar experiment, also on a manganese salt, using a much higher frequency from a source which had been developed during the war. At much the same time similar work began in Oxford [58], at room temperatures and an even higher frequency ($\sim 10^{10}$ Hz). A number of absorption lines were obtained by slowly changing the main field, which had a magnitude of a few thousand gauss and was produced by an electromagnet. The widths of the lines were of the order of several hundreds of gauss, much narrower than those observed by Zavoisky.

It was also significant that in quite a number of crystals which contained Kramers ions no resonances were found. An explanation came from Van Vleck's theory of spin-lattice relaxation which predicted that if an ion had a set of closely spaced orbital levels the spin-lattice relaxation time could be exceptionally short. It could also be expected, on general grounds, that the linewidth would depend inversely on the relaxation time, so the obvious next step was to lengthen the relaxation time by going to lower temperatures.

This became the province of Bleaney and Penrose, and in a very short time all the missing resonances, except from titanium, had been found. Advantage was taken of the variable valencies shown by these atoms. Thus if resonance could not be observed with one valency it was usually possible to find another compound in which the ion had a different valency. So one way or another each ion could be found with an odd number of electrons and then the Kramers degeneracy came in useful. (Resonances from Ti^{3+}, which has an exceptionally fast relaxation time, were later found.) A survey of the results published [58] in 1948 demonstrated that *electron paramagnetic resonance* (EPR) was an important new technique. For the next few years there was intense activity in its study, particularly at Oxford, where the investigations had got off to a flying start, due to the availability of the necessary microwave equipment. Gradually more groups joined in and now EPR has become a standard technique in chemistry and biology as well as in solid-state laboratories.

From crystal field theory it was found, as expected, that most of the resonances occur at frequencies close to the free electron spin resonance in the same main magnetic field, and that they depended on the orientation of the field relative to the crystallographic axes. (For experimental reasons the frequency was held constant and the main field was varied in magnitude and orientation.) What was surprising was the strange behaviour of the linewidths as the direction of the main field was altered. Then, in 1948, Penrose went to Leiden, where Gorter was anxious to set up resonance work because of the information it could provide about the potential of magnetic crystals for use in adiabatic demagnetization. On returning briefly to England for Christmas, Penrose reported that he had carried out resonance experiments on a non-magnetic magnesium Tutton salt in which about 5% of the zinc ions had been replaced by copper ions. It had been expected that such a dilution would sharpen and weaken any copper resonance lines; sharpen because the interactions between the ions were reduced and weaken because of the smaller number of magnetic ions. Unexpectedly, instead of there being one sharper line there was a pattern of four lines (later recognized to be eight). Penrose had revealed for the first time in a solid a process known to occur in the spectroscopy of free ions, a splitting of energy levels due to an interaction between the magnetic electrons of a Cu^{2+} ion and its nuclear spin. Further investigations showed that the hyperfine splittings, as they are known, varied considerably with the field direction and that it was the underlying unresolved hyperfine structure in the fully concentrated salts which led to the strange behaviour of the linewidths. Unfortunately Penrose died, at the early age of 28, before he could pursue his results and it fell to Gorter to write a brief account of them [59]. The discovery of nuclear hyperfine structures resulted in a major re-direction of the resonance work at Oxford; henceforth almost all the resonance studies were on magnetically dilute crystals.

The first attempt, by Abragam and Pryce [60], to produce a theory of these nuclear hyperfine structures was unsuccessful, for they had not appreciated that the magnetically active spins in the 3d shell induce magnetic moments in the s electrons and so produce extra indirect interactions with the nuclear spin. Once this mechanism was introduced a satisfactory explanation emerged [61].

14.7.2. *3d ions: spin-Hamiltonians and paramagnetic resonance*
There are classes of compounds which have similar structure and, particularly with transition-metal ions, such a class is produced when one transition ion is replaced by another, leaving the rest of the constituents unaltered. The class is often named after the first person to describe the

structure (e.g. Tutton) or a particular gem stone that is a member of the same class (e.g. garnet). Resonance work frequently relies on finding a class in which magnetic ions replace non-magnetic ions, for this is a convenient way of obtaining magnetically dilute crystals. Such crystals are usually then described by the magnetic impurity rather than the non-magnetic ion it is replacing. Thus Penrose's sample would be described as a diluted copper rather than a magnesium salt.

The class called Tutton salts have the property that the surrounding six non-magnetic ions do not form a regular octahedron because an opposite pair are further away from the central ion than its other four neighbours. They therefore provided an opportunity to study a range of different magnetic ions in a tetragonal environment and to compare these with another family, such as the fluosilicates, in which two opposite faces of the octahedron are separated to produce an environment having three-fold symmetry. The initial interpretations, though complicated and lengthy, used a direct application of Van Vleck's crystal field theory. However, Van Vleck [62] had shown, along the lines of the H–D exchange calculation, that for one specific magnetic ion it was possible to derive an operator which involved only the angular momentum operators of the total spin and a few parameters to describe the lowest-lying crystal field levels. Since these were just the levels being examined in the resonance experiments it was quickly realized [63] that the same concept could be applied to most of the ions and that the forms of these so-called spin-Hamiltonians could be tentatively written down with just a knowledge of the total spin of the ion and the symmetry of its surroundings. To interpret his measurements the experimentalist needed to know the properties of spin angular momentum operators, but beyond that it was only a matter of choosing a few parameters. This procedure was remarkably successful. For a given ion, resonances could be observed for magnetic fields at virtually all angles to the crystallographic axes and a wealth of results could be obtained and fitted to a consistent model. Thus spin-Hamiltonians, though initially introduced via crystal field theory, came to be established in their own right as an economical, and physically meaningful, description of experimental results.

The spin-Hamiltonians for ions which have $S = \frac{1}{2}$ are generally very simple, for unless there is a nucleus with a spin, only three parameters, usually denoted g_x, g_y and g_z, are needed to account for the resolution of the two-fold Kramers degeneracy by a magnetic field. (With three-fold and four-fold symmetries the form is even simpler, for two of the three are equal.) With nuclear spins present the forms become slightly more complicated, though they take a common form for all nuclear spins. For ions which have electron spins greater than $\frac{1}{2}$ and no nuclear spins, the spin-Hamiltonians usually need extra terms to describe what are known as zero-field splittings, that is small splittings of energy levels which are present even before a magnetic field is applied.

A main role for the theoretician was to explain the origins of these parameters, using crystal field theory, which involved a fair amount of quite complicated mathematics. Thus for a given class it seemed of interest to show that keeping the crystal field parameters much the same would nevertheless give the observed spin-Hamiltonian parameters, which, on the whole, showed considerable variations from ion to ion.

In 1952 I showed [64] that the theory could be greatly simplified by introducing the concept of equivalent operators, a step which eliminated much of the complication of crystal field theory. Indeed it then became very similar to spin-Hamiltonian theory, for crystal field theory can be regarded as a sequence of steps showing how the degenerate states of a low-lying energy level of an isolated ion are first partially resolved to give a low-lying orbital level, and then how the spin degeneracy of this is lifted. Spin-Hamiltonian theory cuts out the first step and assumes that the orbital resolution has already occurred. After that an operator in spin variables suffices to describe the resolution of the remaining degeneracies. The equivalent operator technique deals with the first step by replacing the crystal fields by expressions in orbital momentum operators and so makes it similar to the second step, with orbital replacing spin for the angular momentum variables. This formalism is now widely adopted, so much so that crystal fields as such are seldom given explicitly. Instead an expression in equivalent operators, each multiplied by a parameter, is written down and the parameters are chosen to fit the observations. These parameters are often called crystal field parameters, though it is now established that there is actually no need to have a crystal field model to justify their introduction. This has led to some confusion in the literature over how to describe them, for *crystal field* has been attached both to the parameters in the spin-Hamiltonian and to those in the equivalent operators.

14.7.3.   4f ions: the rare-earths
With the growth in understanding of the 3d ions attention turned to the rare-earths which had become more readily available, as a by-product of intense metallurgical work for the atom-bomb project. The main problem with applying the resonance technique was that these ions generally had extremely short relaxation times so that it was necessary to work at liquid helium temperatures. The programme began with a study of the rare-earth ethyl sulphates, compounds which had already been studied for possible use in adiabatic demagnetization experiments. In a very short time a host of new results were obtained and reviewed [65].

The obvious assumption was that crystal field theory would also account for these, though initially there was little supporting evidence, for the only published work on rare-earths had assumed that, like the iron group ions, the dominant crystal field was due to an octahedron of neighbours, but much weaker in strength. This had seemed satisfactory at the time, but a later attempt to explain the susceptibility (actually the Faraday rotation, to which it is proportional) of some rare-earth ethyl sulphates had created misgivings and indeed Van Vleck told me in a letter dated 7 February 1951, that he doubted whether crystal field theory could be used for the rare-earths. However, new resonance results on a number of Kramers ions were making it clear that the crystal fields were quite different from those of iron group ions and that crystal field theory could be used. In particular the g-values were much more anisotropic and there was no way in which this could occur with an approximately octahedral crystal field. Nor was it consistent with the crystal structure, which showed that the magnetic ions had twelve nearest neighbours. Treating the crystal field of such an arrangement was a major problem, and indeed it was to circumvent this that the equivalent operator version was introduced. After that the algebraic difficulties disappeared. However, there was one particular feature, a sixth-order crystal field parameter B_6^6, which was new and for which there was no direct evidence from most of the resonance results. In due course Gd^{3+} was examined. This is an ion with a high spin, $\frac{7}{2}$, which presents no problem for crystal field theory, for there is no orbital degeneracy in the free state. All that should be found is a straightforward Zeeman splitting of the spin degeneracy into eight equally spaced levels. In fact what was actually found was quite different. In the absence of a magnetic field the eight spin states were split into four Kramers doublets and the resonance experiment showed that there was a six-fold symmetry axis. It did not, at least in any simple way, justify the introduction of a B_6^6 crystal field parameter into the spin-Hamiltonian, for all such parameters should been zero. It did, however, show that indeed the six-fold symmetry of the neighbours existed and it also focused attention, once again, on the problem of understanding the zero field splittings of S-state ions, a problem which had already arisen with the iron group examples and is still not resolved.

Even though the resonance results could be explained using equivalent operators there still remained too much ambiguity in the values of the parameters. This arose because for most of the ions the theory predicted that there would be an extensive pattern of low-lying levels, doublets for the Kramers ions. To determine the parameters precisely it was necessary to known the properties of all the levels, but at liquid helium temperatures it was expected that only the lowest of these would be populated. The best that could then be done was to supplement the resonance measurements with susceptibility results. Later, inelastic neutron scattering proved to be a valuable technique for determining these quite closely spaced energy levels and now the uncertainty in the parameters is much reduced.

The investigation of the rare-earths by resonance methods provided a wealth of information about a group of elements which were previously quite unfamiliar to most physicists, and it is appropriate to acknowledge the contribution made to this field and to magnetism in general by F H Spedding, of Iowa State University. Thanks to him these elements became available in substantial quantities and in separated forms, leading to a growth in interest in all their properties, going far beyond what has so far been described. They are now used extensively in technological applications.

EPR was also used to study ions of the higher transition groups, though on the whole there was not the same diversity of results because many of the ions are found in covalent complexes and show no resonances. An interesting exception was the $IrCl_6$ complex [66] in which the magnetic electrons were found to be interacting with the spins of all seven nuclei, showing that the electronic orbits spread over the whole complex. Each magnetic electron of the nominally iridium ion was estimated to spend approximately 30% of its time near the chlorine sites. Crystal field theory provided no framework for explaining this and it was necessary to turn to a molecular orbital type of theory, such as was first described in 1935 by Van Vleck [67]. This was an indication that all might not be well with crystal field theory. Nevertheless in its spin-Hamiltonian form it continued to meet all requirements, as is readily apparent from Abragam and Bleaney's [68] very full account of the resonance work and its associated theory up to 1970.

14.7.4. *Exchange interactions*

We now return to salts of the iron group. It had been realized that in a magnetically dilute crystal a statistical accident was likely to produce a few pairs of magnetic ions on adjacent lattice sites. However, before this could be explored by spin resonance Guha [69] suggested that an anomalous susceptibility in copper acetate might be due to the presence of pairs of Cu^{2+} ions. This was confirmed by resonance methods [70]. Resonance showed three closely spaced levels, an excited set with an effective spin of 1, above a non-magnetic singlet ground state. This is just what was to be expected from isolated pairs of Cu^{2+} ions, each with spin of $\frac{1}{2}$, coupled by an antiferromagnetic exchange interactions. Further analysis showed the mechanism to be an exchange interaction, dominantly that of the H–D model though with the opposite sign, with a smaller part due to coupling of the spin directions to the lattice structure.

The understanding of exchange interactions was in its infancy, though ever since the H–D analysis it had been assumed that there would be such interactions between nearest-neighbour magnetic ions, and various observations had been interpreted on this basis. Copper acetate gave direct evidence for it and confirmed that it could be of antiferromagnetic character. Also, at about the same time neutron diffraction showed [71] that exchange interactions could be strongest between magnetic ions which were not nearest neighbours, particularly if they are separated by non-magnetic ions.

Explanations were put forward under a number of descriptions (e.g. direct exchange, superexchange) but a satisfactory and systematic theory had to wait until about 1959 when Anderson wrote several papers, including a review of previous work [72].

14.7.5. Problems with crystal field theory

Although crystal field theory was being assumed necessary for the derivation of spin-Hamiltonians there were one or two clouds on the horizon. The iron group cyanides, covalently bonded complexes, had been investigated, and for these it seemed necessary to have a theory like that used for the $IrCl_6$ complexes. Then Feher [73], using a complicated technique known as *Endor*, produced evidence that the electrons of the so-called ionic magnetic ions were straying far from the nucleus, which again could not be explained by crystal field theory. Another cloud was that Kleiner [74] had calculated the coefficient of the cubic part of the crystal field using quantum mechanics rather than the classical theory used by Van Vleck, and had come out with the opposite sign which, had it been correct, would have ruined all agreement between theory and experiment. Perhaps not unnaturally this result was largely ignored. (A later 'improved version' [75] restored the original sign but with a value which seemed far too small.)

A potentially much more serious criticism was that in crystal field theory electrons are being distinguished, for those on the neighbouring ions are regarded as simply producing contributions to the crystal field whereas those on the magnetic ion are given a full quantum-mechanical treatment. A fundamental property of electrons was therefore being violated. The criticism came to the fore because of the good progress that was being made in understanding metals. While crystal field theory took account only of local variations of potential around the ion to which an electron was attached, in a metal the potential is periodic in space, and Bloch's theorem shows that electrons are no longer confined to the vicinity of any particular nucleus. There is no problem in incorporating their indistinguishability and the exclusion principle into the basic theory. The dilemma was that the fully concentrated magnetic crystals are also periodic, so it seemed only reasonable that their electrons should also move through the whole lattice, a feature that is not contemplated in crystal field theory. Slater [76] expressed the view that it was only a matter of time before band theory replaced crystal field theory. It had already been used with modest success by Stoner and Mott, in an attempt to explain the ferromagnetism of iron, cobalt and nickel, and it seemed reasonable that the paramagnetism of the transition group salts could be understood in similar terms. There would be no need to distinguish between electrons nor to use concepts like the crystal field. Slater had undoubtedly put his finger on a weak spot, but the implication that all would come right through band theory was far too optimistic; it has still not happened [77]. It has, however, proved possible, comparatively recently, to reformulate the theory of magnetic insulators, retaining the periodicity of the lattice and keeping the electrons indistinguishable, and to show that this leads to each ion being described by a spin-Hamiltonian of the usual form together with terms that describe exchange-like interactions between the effective spins [78]. This work did not use the concept of electrons in bands. If electrons are regarded as moving independently in a common periodic potential then bands are inevitably introduced. However, the concept of a periodic potential is an approximation, and another one, presumably better for the insulating magnets, gives results much more in line with observations.

The result of not realizing this was that for some 25 years from the early 1950s magnetism was divided into two camps, one wedded to crystal fields and the other to band theory, with the result that the literature of magnetism is sprinkled with comments about the need to decide, in accounting for some observation, whether to use a localized description (a crystal field model) or a delocalized description (a band theory model). The crunch came when the rare-earth metals were studied, for it was then found that the best way to explain their conductivity was to invoke bands for the outer electrons and, for the magnetism of the 4f electrons, to use crystal field concepts. While there is still no satisfactory theory which allows spin-Hamiltonians for localized moments and band states for conduction, at least each camp is now using the other's concepts.

14.7.6. *Relaxation processes*

The study of magnetic ions as low-concentration impurities in non-magnetic hosts had really only been possible by working at low temperatures, because the intensity of the absorption increased as the temperature was lowered, so helping to offset the loss due to dilution, but it also raised other difficulties, one being that of saturation. The basic idea of the resonance experiment was to induce absorption of electromagnetic energy by raising an ion in a low energy state to one of higher energy. But if this was all that happened it would not have been long before all the ions would have been lifted to the higher state, in which case no further absorption would occur. That something of this kind could happen had been observed and it had become part of the resonance technique to vary the incident power to check that *saturation*, as it was known, was not occurring. As the resonance technique became established interest grew in the relaxation processes which tended to prevent saturation occurring.

See also
p 177

In the spectroscopy of isolated ions there are two processes which can prevent saturation, both associated with electromagnetic radiation and famously described by the Einstein A and B coefficients—A for spontaneous emission and B for the stimulated exchange of energy between the ion and incident radiation of the correct frequency. In optical spectroscopy spontaneous emission usually dominates. However, in EPR the observations gave relaxation times which were far too short to be due to spontaneous emission, nor could stimulated emission be invoked because this would simply return any absorbed energy to the exciting wave and so reduce the overall absorption. Almost from the beginning, therefore, it was assumed that the answer was to be found in Van Vleck's [40] theory of relaxation, which replaced the modes of the electromagnetic spectrum by the spectrum of the lattice vibrations. However, the theory had not been tested in detail, though the prediction that the relaxation times would be temperature dependent was obviously correct.

14.7.7. *Acoustic paramagnetic resonance*

If magnetic ions do relax by stimulated emission of lattice vibrations then it should also be possible to excite transitions by means of monochromatic vibrational waves. There were, however, several problems to be overcome. At a frequency of 10^{10} Hz lattice waves have a wavelength approximating to that of visible light and could be expected to be highly attenuated. Furthermore, no way was known of generating such monochromatic waves, until in 1959 Jacobsen *et al* [79] showed that it was possible to use the piezoelectric properties of quartz to generate monochromatic lattice waves of the same frequency from electromagnetic waves; with these he and his colleagues demonstrated what is now known as *acoustic paramagnetic resonance* (APR). Even in the first experiments resonances were seen which had not been observed by EPR in the same sample. The inference was that some unidentified magnetic centres were present which EPR had missed. The experimental technique was quite different from that used in EPR because of the very short wavelength, but nevertheless it was successfully developed and many resonances that EPR had missed were later detected. (The detection of a resonance, whether by EPR or by APR, does not in itself identify its origin; hyperfine structure, if present, is a great help for it will usually identify the nucleus.)

Looking back at the EPR results it could be seen that most of the information came from Kramers ions, which were predicted by the Van Vleck theory to be much more weakly coupled to the lattice than the non-Kramers ions. So there was an obvious reason why APR should differ, for it would be most effective for non-Kramers ions. The specific detection of such ions confirmed an impression that had already come from work on relaxation processes in EPR, that Kramers ions were relaxing by a process which was not included in the Van Vleck theory—cross-relaxation—which was a well-established process in the field of nuclear resonance [80]. Basically it involves an ion in an excited state dropping to a lower one while a similar ion in the lower state moves to a higher one, a process in which both energy and total spin are conserved. A fast relaxing and different ion can be substituted for one of the ions and, provided their resonance lines overlap, energy can be taken from an excited A ion, a slow relaxing Kramers ion, and used to excite a fast relaxing B ion which then rapidly de-excites by transferring the energy to the lattice. A small number of undetected B ions can completely alter the relaxation characteristics of the A ions. It was the demonstration that such fast relaxing ions were indeed present which explained why the relaxation observations on what should have been slow relaxing ions had played havoc with the testing of the Van Vleck theory. (It is unusual for a theory to exist for something like a quarter of a century before it is proved to be correct.)

14.7.8. Co-operative phenomena in magnetic insulators—antiferromagnetism

Until the early 1950s there were two experimental methods available to investigate antiferromagnetism. The most direct was the measurement of magnetic susceptibility on, if possible, single crystals, as a function of temperature and field direction relative to the crystalline axes. The second used the Faraday effect, the rotation of the plane of polarization of light when it propagates in the direction of the magnetic field, the magnitude of effect having been shown to be proportional to the susceptibility [81]. When the new technique of EPR was developed it was naturally of interest to examine what would happen, in a fully concentrated crystal, to a resonance line as the temperature was lowered through an antiferromagnetic phase transition.

The first experiments [82] were on powdered Cr_2O_3, which has a Néel temperatures near 40 °C. While a resonance was observed above the transition, below the transition it was difficult to decide whether it had disappeared completely or moved outside the range of available fields. The same phenomenon was soon observed in other salts, and explained independently by Kittel [83] and Nagamiya [84], who showed that the resonance had probably moved to a much higher frequency.

Their theory was based on the Bitter model of an antiferromagnet, with the added concept that the sublattice magnetizations were not simply lined up in antiparallel directions but were actually precessing, as for electron spins. Thus the moment of one sublattice would be precessing in the effective internal magnetic field due to its exchange interactions with the other sublattice and vice versa. A similar model had already been introduced for ferromagnets, also by Kittel [85], the difference in the antiferromagnets being that the two-sublattices precess in opposite directions. As a consequence two resonances could be expected, both with the same frequency in zero applied field. This was verified in 1959 by Johnson and Nethercot [86] who took a good deal of care over their choice, MnF_2, as a suitable antiferromagnetic. They found resonant absorption, rather as in EPR, except that no external magnetic field was needed. The resonant frequency was temperature dependent because of the temperature dependences of the two sublattice magnetizations. This may be regarded as the first definitive observation of what is now known as antiferromagnetic resonance.

See also p 460

A large change in the experimental techniques came in 1951, when Shull *et al* [71] used the elastic scattering of neutrons in studies on a number of Mn^{2+} salts, obtaining their magnetic structures, the magnitudes of the magnetic moments and their orientations relative to the crystal axes. In the simplest examples their results confirmed the Néel picture of two oppositely directed interpenetrating ferromagnetically ordered sublattices, as modelled in Bitter's theory. The experiment relied on the neutrons, which have magnetic moments due to their spins, being scattered from the static magnetic array in much the same way as x-rays are Bragg scattered from the electronic charges in a crystal. The neutrons could therefore be used to determine the magnetic structure and, in particular, show whether this was the same or different from the charge structure. It came as no surprise to find the two were not the same in an antiferromagnet, though it was a considerable surprise to find, in MnO for example, that the strongest antiferromagnetic couplings were not between nearest-neighbour magnetic ions but between next nearest neighbours, magnetic ions separated by non-magnetic O^{2-} ions (figure 14.8). The nearest Mn^{2+} ion which has a spin which is antiparallel to that of a chosen Mn^2 ion is reached through an intervening O^{2-} ion.

Since then many antiferromagnets have been discovered, with a variety of sublattice structures, and neutron scattering has become the favoured technique for its examination.

14.7.9. The Heisenberg–Dirac model and spin waves

See also p 1010

Ever since the introduction of the H–D model of exchange coupled spins there has been a continuing interest in using it to show theoretically that such a system would show a phase transition at low enough temperatures. It is deceptively simple-looking and so has been particularly interesting to theoreticians. The discovery of antiferromagnetism, which in the theory simply meant that the ferromagnetic pattern of energy levels was inverted, provided a stimulus to further studies. However, it would be out of place in this account to attempt a review of such progress as there has been, which is quite limited, so I shall content myself with a reference [87] and some observations. There is extremely little evidence that the model applies to any known system. Ferromagnetic insulating salts are quite uncommon and in the antiferromagnets there are known to be alternating magnetic

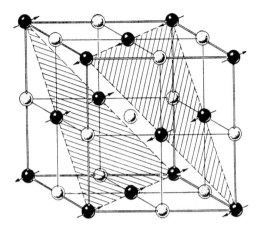

Figure 14.8. *In MnO the moments which are antiparallel have an intervening oxygen ion. The structure shows that the strongest exchange interactions must therefore be between next nearest neighbours and that they are of antiferromagnetic character.*

moments at the sites of the order of magnitude of those of unpaired electrons. The H–D model predicts that at absolute zero there would be no moment at any site, which suggests that a realistic low-temperature model needs to include further (small) interactions. On the other hand the model does seem to have the potential for predicting the existence of a cooperative phase transition and the temperature at which it occurs.

Alongside the rigorous work on the H–D model there has been a large effort devoted to approximate models, which also can be traced back to the H–D model of a ferromagnet, the ground state of which has all the spins aligned parallel to one another (identical m_s values) to create a macroscopic moment. If each s is taken to be $\frac{1}{2}$ and its magnetic moment is taken to be the Bohr magneton μ_B, then for N spins the total moment will have a projected magnitude of $N\mu_B$ which, in accordance with the uncertainty principle, means that the total moment has a magnitude of $[N(N + 1)]^{1/2}\mu_B$. For large N, therefore, the total magnitude and its projection are virtually aligned and so are like classical moments. In the absence of any external field they can point in any direction. However, in the presence of a field, H, they can align themselves at any one of a large number of angles to the field direction to give rise to a set of equally spaced energy levels, the separation being $\mu_B H$—a macroscopic Zeeman effect. The total moment precesses like a gyroscope around the field direction (figure 14.9). There is also another set of energy levels with the same spacings but with a slightly smaller total moment. This arises from a spin arrangement in which one of the spins has been reversed, which makes the largest projected value $(N - 1)\mu_B$, a value which also occurs for one of the θ values in the case when all the spins are all parallel.

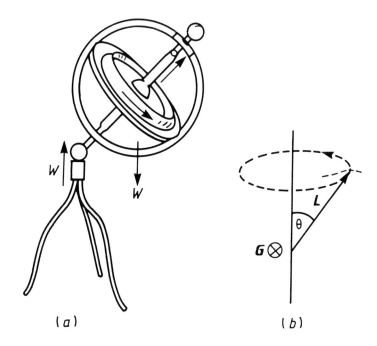

(a)   (b)

Figure 14.9. *In the classical gyroscope (a) the couple, G, is provided by the weight W and an equal and opposite force acting on its point of support. G creates an angular acceleration in its own direction, which is at right angles but in the same plane as the angular momentum L of the fly wheel. The result is that there is a precession about the vertical at an angle θ. Due to frictional forces the angular motion is damped and θ increases until the whole precessional motion collapses. For a magnetic moment (b) the couple G is created by an external magnetic field and the angular momentum L is usually due to aligned electron spins. There is no damping and the precessional angle θ remains constant. However, θ is restricted to a finite number of values, two in the case of a single electron, and the classical damping is replaced by discontinuous changes in θ being allowed. The energy changes associated with these are quantized, for the energy levels are equally spaced.*

When the arrangements are examined in detail it is found, in typical quantum-mechanical fashion, that although the moment in the direction of the field is less than the fully aligned value, the spin reversal is not at a specific site. Rather it must be regarded as propagating, with a range of possible velocities, so that it is equally likely to be at any site. A phrase has been invented to describe these motions—*spin deviation waves* or simply *spin waves*—and various attempts have been made to visualize them (figure 14.10).

The mode which is in phase everywhere corresponds to the case when no spin has been reversed, so nothing is being propagated. In an extended system it is, of course, possible to envisage that two or more spin wave-packets have been excited and that they are in different parts of the sample. They should therefore be able to travel a long way before

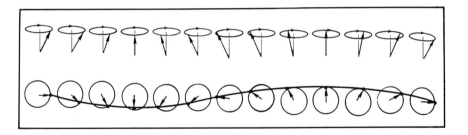

Figure 14.10. *A ferromagnetic spin wave, a normal mode of the spin system, is described semiclassically as a precession of each spin about the magnetization direction with the same frequency. The top diagram shows the spins viewed from the side and the bottom diagram shows the view from above. Due to the rotation the wave appears to move with a wavelength determined by the difference in angle between adjacent spins. Energy is needed to change this angle so the energy associated with a spin wave depends on its wavelength. Most wave motion has a finite number of wavelengths and forms a 'wave-packet'.*

encountering one another. It seems reasonable to suppose that quite a lot of spin waves can exist at the same time without much mutual interference.

With such diagrams it is easy (and sensible) to forget the reasoning which has led to them, because they are physically convincing and therefore likely to be independent of any mathematical, or semi-mathematical derivation. Such waves will probably exist for any ferromagnet and possibly for any cooperative magnetic system, including antiferromagnets. The really relevant thing is to demonstrate their existence.

The Kittel theory of antiferromagnetic resonance can be regarded as an illustration of a theory of two motions which have the same phase throughout each sublattice, so they correspond to the non-propagating modes of the ferromagnet. The first propagating mode was observed [88] by inelastic neutron scattering in 1961. A little later Collins [89] observed 42 spin-wave modes in MnO and used his results and spin-wave theory to estimate the nearest and next nearest exchange interactions, commenting that this was the only technique for observing spin-wave modes. The theory has been fully reviewed by Keffer [90]. A later investigation [91] on CoF_2, which from EPR was expected to have a complicated spin-Hamiltonian, produced evidence for a temperature dependence of the spin-wave modes as the Néel temperature (38 K) was approached, as well as evidence for their existence above the Néel temperature. The latter was unexpected, though the former could have been anticipated from the Kittel theory (but not from the version of the H–D model which takes the fully aligned sublattices as a starting point).

By the end of the 1960s the interest in crystalline antiferromagnetism seems to have been waning, probably because a wide range of other magnetic phenomena was being generated, particularly those connected with the technical uses of new insulating materials which, unlike the antiferromagnets, possessed permanent magnetic moments. So this section will be closed with a brief mention of some special cases of antiferromagnetism.

The rare-earth chromites show two Néel temperatures [92], one corresponding to an antiparallel alignment of chromium moments and a second due to an alignment of rare-earth moments. Then a number of crystals which were expected to be antiferromagnetic, e.g. Fe_2O_3, Ni_2O_3 were found to be weakly ferromagnetic. Two independent mechanisms were introduced [93] to account for this. The two-sublattice concept was retained with the assumption that they did not have antiparallel alignments, either because there were zero-field splittings due to crystal fields with different distortion axes for the different sublattices, or because of the presence of antisymmetric exchange interactions, which might arise from the combination of spin–orbit interactions and lattice distortions. The former were familiar from EPR studies, whereas the significance of the latter was not appreciated until it was invoked by Moriya [94] to account for slight departures from antiparallel alignments.

14.8. Ferromagnetism and ferrimagnetism
It is convenient to begin this section by grouping ferro- and ferrimagnetism together, for the experimentalists had long found difficulty in distinguishing between them. Theoretically there was no difficulty, nor would there have been experimentally if the significance of electrical properties had been appreciated. Rather it seems that the experimentalists, particularly those brought up using pre-war techniques, regarded the advent of ferrimagnetism as an addition to the range of ferromagnets, one which provided a new set of materials to which to apply their existing experimental methods.

14.8.1. *Ferromagnetic resonance*
This position changed when Griffiths returned to Oxford at the end of World War II and began using the microwave equipment there for an investigation of the high-frequency permeabilities of the ferromagnetic metals. Using a resonant cavity, one end wall of which was made of the ferromagnetic metal, he attempted to measure one of the sources of energy loss in a ferromagnet at a much higher frequency than had previously been available. In trying to isolate the loss he decided to apply a steady magnetic field in the plane of the wall and as a result discovered [95] a strong resonance as the magnitude of the field was varied (figure 14.11). This occurred near the value expected from the relation $h\nu = 2\mu_B H$, appropriate for an electron in the internal field of classical magnetic theory. The experiment was basically the same as that

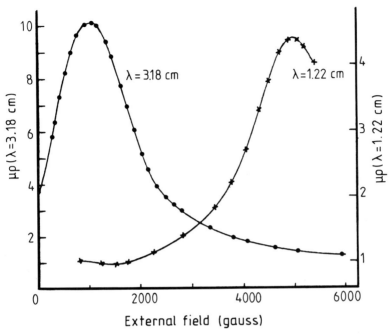

Figure 14.11. *The ferromagnetic resonances first observed by Griffiths.*

being used in the early days of EPR. It was, though, quite clear that the resonance was not due to a set of independent electrons, both from its intensity and because they would not be showing ferromagnetism. Instead, all the spins were imagined to be precessing at the same rate about the direction of the internal field, as in figure 14.10, but remaining parallel to one another as they did so.

Apart from the indirect observation in the gyromagnetic effect, the first direct observation of precessional motion seems to have been in atomic and molecular beams by Rabi [96] in 1937, then (1946) in solids in the macroscopic resonance of Griffiths' ferromagnetic end walls and about the same time in water, in the nuclear moments of protons, by Bloch *et al* [97] and independently, by Purcell *et al* [98]. Resonance could thus be induced in atomic beams, in ferromagnets, in paramagnets and in nuclei, a remarkably wide range.

The original theoretical explanation of Griffiths was corrected by Kittel [85], and then re-derived using microscopic equations by Van Vleck [99]. As a result it was realized that ferromagnetic resonance provided a way of using single crystals to determine the magneto-crystalline anisotropy energies, macroscopic expressions which were introduced to account for the preferred directions of magnetization found in ferromagnetic crystals.

14.8.2. Magnetic domains

The concept of a macroscopic anisotropy energy can probably best be understood by considering another technique for exploring ferromagnets, which had, by the late 1940s, been brought to a mature stage, the study of domain structures by means of Bitter patterns. In a very short time a large number of domain patterns were photographed and showed that a uniform magnetization is rarely to be found in a ferromagnet with no external field applied. Rather, the magnetization consists of domains of uniform magnetization. So two questions came to the fore. What determined the direction of the moment within a given domain and what determined the domain patterns?

See also p 1134

For the most part the theory needed to answer these questions was couched in macroscopic terms, for the quantum theory of ferromagnetism in conductors was still in an insufficiently developed state. This was not regarded as a handicap, but just the opposite, for the ideas of classical magnetism, with a few *ad hoc* assumptions, had been quite successful so far and there was no reason to suppose it would now fail. So the technique was to write down, for a uniformly magnetized region, an energy expression which depended on assumed directions and magnitudes for the magnetization, the magnetic anisotropy energy. The values actually taken up would then be those which minimized the total energy, subject to any forces across the boundaries between the uniformly magnetized regions. There also needed to be terms associated with exchange and since H–D theory gave an energy which depended solely on the angle between two spins it was given a classical interpretation, as an energy which remained constant as the direction of the magnetization rotated and which was proportional to the square of its magnitude. The anisotropic energy already had a classical explanation in terms of demagnetizing factors, the only problem being that the associated energies seemed to be far too small, though the angular dependence seemed to be correct. So the simplest thing was to accept the measured values for their coefficients and attribute them to microscopic anisotropic exchange, while describing them macroscopically. In more sophisticated treatments further energies were included, such as those which couple the dimensions of the domain to its magnetization, for it had been found that magnetization was accompanied by volume changes. Armed with these energy expressions it was possible to use various measurements to sort out the magnitudes of the unknown parameters.

The Bitter patterns showed that in many cases the sample was not a single domain. The model therefore had to be enlarged to include energy changes as one domain grew at the expense of another. Also the domain walls made their contribution to the energy; within these the magnetization changes direction so the isotropic part of the exchange plays a significant role, for adjacent spins are no longer parallel. So complicated did the domain patterns and the theory become that their study, for many of the experimentalists, became more like a study of art forms, while for the theoretician they became more like a nightmare.

The further question of how to understand their time dependence, the changes in shapes and sizes of domains and the motions of domain walls during a hysteresis cycle was even more of a problem. The Bitter technique, being static, was little help. A new avenue was opened by Lee *et al* [100], who introduced the use of the transverse Kerr effect, which relied on the observation that when a plane-polarized beam of light is reflected from the surface of a magnetized sample the plane of polarization is rotated. The experimental method used changes in the angle of rotation of the polarization of a narrow beam of light to monitor changes in the magnetization at the spot from which the beam was being reflected. It was not entirely what was wanted, because it did not follow the motion, for example, of a domain wall. This was to come later, particularly for ferrimagnets, as will be discussed in due course. Domain techniques have been reviewed by Craik and Tebble [101] and Dillon [102].

14.8.3. *Ferrimagnetism and ferromagnetism*

It is now convenient to turn to ferrimagnetism, which for many years had hardly been distinguished experimentally from ferromagnetism, a situation which was to continue for long after it had been shown that ferrimagnetism was fundamentally closer to antiferromagnetism. The reason was, of course, that the two shared the important property of having macroscopic magnetic moments, and this was of more importance than the way in which it came about. Thus any property of one which depended on the macroscopic moment could be expected to occur with the other.

In particular ferrimagnets showed [103] the equivalent of ferromagnetic resonance as well as the more classical phenomena of hysteresis and domain structures.

With such a similarity to ferromagnets it might have been expected that some of the ferrimagnets would simply have replaced some of the existing ferromagnetic metals and alloys in specific applications. However, this is not what happened, because the ferrimagnets had one property which made them significantly different from the ferromagnets: they were electrically insulating. Thus instead of rivalling the ferromagnets they complemented them, and so greatly extended the frequency range in which magnetized materials could be used.

The first inkling that ferrimagnets could have valuable uses came in 1946 when Snoek [104], working at the Philips Laboratories in Holland, announced that *ferrites* had been produced with strong magnetic properties, high electrical resistivity and low hysteresis loss. (Ferrite is used to describe a whole class of oxides.) Hilpert had made some of them in 1909 but had concluded that his were of no commercial value, so until

See also
p 1330

Snoek's announcement there had been little interest in their magnetic properties, though de Boer and Verwey [105] had been interested in their electrical properties (see Chapter 17).

14.8.4. Crystal structures

The magnetic properties of the ferrimagnets are strongly linked with their crystal structures. For example, Fe_3O_4 is basically a face-centred cubic lattice of O^{2-} ions with Fe ions in interstitial positions. There are 96 of these in the unit cell, 64 having four and 32 having six adjacent O^{2-} ions, with only 24 Fe ions to be accommodated. Eight of these are present as Fe^{2+}, which has four unpaired electrons, and 16 are present as Fe^{3+} , which has five unpaired electrons. So not only does iron occur with two different magnetic moments, it also has a choice of sites with quite different crystal fields. The actual distribution is determined in the formation of the crystal, and this and the states of ionization of the ions at the various sites can only be determined afterwards, which is not easily done. There is also some evidence that the choice is not the same for all unit cells.

Standley [106] lists the crystallographic data for a range of similar compounds, the spinels (with the general formula $M^{2+}Fe_2^{3+}O_4^{2-}$ or $M^{2+}O^{2-}$, $Fe_2^{3+}O_3^{2-}$, where M is a divalent metal ion) the garnets and the perovskites. Among these there are many examples, including lodestone, that are both ferrimagnetic and insulating.

By this stage the reader will probably have concluded that the oxides form a large family of complicated substances, a view which is not uncommonly held, particularly when it is realized that in addition to their number they are all highly refractory and so difficult to prepare with a given composition. It is therefore reasonable to ask why they are given any attention? The best answer seems to be that in the late 1940s they were indeed of very limited interest. Now they are all around us, though we may not realize it. To give a few examples, we rely on particular members of the ferrites and some closely related families for the internal aerials of our radio sets, the recording media for our tape and video recorders, and the permanent information stores in our computers and various 'cards'. And these are just a few of the uses of specially selected ferrimagnetic insulators.

A perhaps more pertinent question should therefore be, how is a ferrimagnet chosen for a particular application? In the early days such a question could hardly have been asked, let alone answered, for it was only after the properties of many of them had been explored that the possible applications were identified. Now it can be answered: from a pool of painstakingly acquired information. Thus the story of the development of ferrites is a good counter-argument to a claim that research should be consumer directed. Would the need for a radio small enough to be portable ever have produced even one of the necessary components, a ferrite aerial?

For all that the macroscopic magnetic properties of ferrimagnets are similar to those of ferromagnets, microscopically they are closer to antiferromagnets. An interesting consequence is that there is no reason to suppose, even for one which can be regarded as composed of two antiparallel magnetic sublattices, that the moments on the two sublattices should be similar either in magnitude or temperature dependence. Thus the thermal behaviour of the net macroscopic moment of a ferrimagnet may be quite different from that of a ferromagnet. This was described by Néel [107] in an account of his original work, which includes a number of illustrations of what might occur. Probably the most interesting prediction was that as the temperature is raised the net magnetization might decrease to zero and then rise in the opposite direction. While this does not happen in the majority of ferrimagnets it was demonstrated by Gorter and Schulkes [108] in some mixed ferrites. The temperature at which the net moment falls to zero is now known as the compensation point.

For the most part attention will, from now on, be focused on the uses which arose for ferrites, garnets etc, which are electrical insulators.

An immediate consequence of the high resistivity was apparent in the resonance experiments. In ferromagnetic resonance in a conductor the microwave radiation can only penetrate a very small distance (*skin effect*) which is why Griffiths used thin films, supported by the non-magnetic wall of a standard microwave resonator. The number of unpaired spins which could partake in the resonance was therefore limited. With an insulator the position was quite different, for it could be expected that it would be virtually transparent to microwave radiation unless there was unexpectedly large absorption due to its magnetic properties, which might include absorption from domain wall movements and rotations of magnetization, as well as the resonance. The non-resonant effects were generally found to be small which meant that large samples could be studied and, if necessary, placed anywhere in the resonator.

The transparency was exploited in an investigation of the propagation of electromagnetic waves along the direction of an external magnetic field. The theory was developed by Polder [109] who confirmed that the Faraday rotation of the plane of polarization would have the interesting property that if for one direction of propagation through a finite sample it was rotated through a particular angle then on the propagation in the opposite direction the rotation would be through the same angle. That is, if it corresponded to a right-handed screw rotation for one direction of propagation it corresponded to a left-handed screw for the other. This property is exploited in a number of devices [110], such as the *isolator* (figure 14.12). It was also found [111] that when certain ferrites were exposed to high microwave powers they would radiate microwaves at a

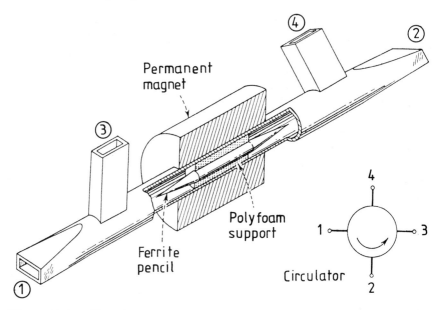

Figure 14.12. *The polarization of a wave incident at 1 is rotated through 45° and emerges at 2. A similar wave incident at 2 emerges at 3. So waves travelling in opposite directions are separated. The device can be used in radar systems, to separate outgoing waves from their reflections. (A rotation of order 90° can be achieved in distances of order 1 cm.)*

frequency double that of the incident radiation, a result which led to a good deal of interest in non-linear effects. For example Pippin [112] soon showed that two incident waves of differing frequencies could produce both sum and difference frequencies, and Suhl [113] showed that it should be possible to use the non-linear properties to produce a parametric amplifier for microwaves, the first demonstration being given by Weiss [114]. Since any amplifier can be made to oscillate by feeding back part of the output to the input this was later turned into an oscillator.

14.8.5. *Magnetic recording*

As long ago as 1898 Poulsen filed a patent which described an instrument which would record and reproduce sound through the orientation of magnetic domains, and a picture of his device is to be found on p 13 of *Magnetic Recording* by Lowman [115]. When it was demonstrated at the Paris Exposition of 1900 the interest was so great that he was awarded the grand prize. However, the practical obstacles to be faced before magnetic recording reached the present stage were immense, though the basic ideas have hardly changed. Before the ferrites came along thin strips of metallic ferromagnets were investigated; it was hoped that a strip, on emerging from a magnetizing coil, would retain a record of the time-dependent current in the exciting coil in the form of a variation of the local permanent magnetization along its length. Unfortunately, because of the conductivity of the strip, on entering the coil the induced magnetization took time to build up and as it emerged took time to settle to its remanent value, so it did not accurately represent the instantaneous current. Another problem arose because the easiest experimental arrangement was to magnetize the strip perpendicularly to its plane whereas the marked preference of a strip is to have the magnetization in its plane.

The advent of ferrites (which in this context includes all ferri- and ferromagnetic insulators) changed the position, since among them were materials which would respond much more rapidly than metallic films to changes in magnetizing fields. There were plenty of other difficulties, but at least the goal seemed sufficiently near in the late 1940s to justify a good deal of industrial research activity.

A modern tape consists of particles which have anisotropic magnetic properties such that the magnetization is uniaxial. The particles are then dispersed in some suitable non-magnetic medium and supported on a non-magnetic film. In some manufacturing processes, the tapes are exposed to strong magnetic films in the plane of the tape to orient the particles, after which they are cemented in position and demagnetized, for example, by heating to above the Néel temperature. On its passage through the recording head the tape passes through a field which first rises from, and then returns to, zero. It is left with a residual magnetization determined by the field cycle. For good reproduction this needs to be proportional to the current exciting the coil, which is unlikely to occur of its own accord, since the magnetization cycle is highly non-linear. It has been found that the imposition of a much stronger field, which oscillates at a frequency higher than that which is being recorded, gives better reproduction.

There is a good deal of technical knowledge and experience about the best magnetic materials to use, the production of suitable films and the best recording techniques. The range of materials in use is reviewed by Bate [116], and the techniques of recording by Lowman [115].

14.8.6. Bubble domains

In 1957 Dillon [117] reported that it was possible to transmit visible light through thin sections of the ferrimagnet, yttrium–iron–garnet. This opened the possibility of using Faraday rotation to see domain patterns with a microscope. If for one direction of magnetization the rotation of plane-polarized light is right-handed then for the other it is left-handed, and the two should be distinguishable using an analyser. The first photographs showing domain structures obtained by this new technique, in a film which was not in a magnetic field, soon followed [118]. Each domain appeared as a snake-like figure, winding its way across the sample, adjacent 'snakes' being oppositely magnetized in the direction normal to the film. The patterns could readily be changed by the application of a magnetic field and it is noticeable that each domain preferred to end at a boundary of the film rather than to close on itself. From then on the technique received considerable development and there is an extensive literature on it (figure 14.13).

It was suggested that instead of storing information in a static form on a tape and reading it by moving the tape, one might store it on a stationary tape and move the information. Bobeck [119] has given a preliminary account of progress towards this end. A large number of possible ferrimagnets were studied, among which were some where the serpentine domains of Dillon closed on themselves. Then, when a uniform magnetic field was applied, in a direction perpendicular to the tape, the regions with a magnetization in the direction of the field would grow at the expense of those magnetized in the opposite direction. For certain ranges of fields there would appear to be small circular domains with one sign of magnetization immersed in a sea with the opposite magnetization. From their appearance when viewed through the optical system they came to be known as magnetic bubbles. In their physical properties they were similar to small cylindrical magnets, which meant that they could be moved fairly easily by non-uniformities in the applied magnetic field. The study of bubble domains and their dynamics became a field of considerable interest [120]. By applying the correct fields near the edge of the tape, and by other techniques, it was possible to nucleate closed domains and so create bubbles, which could afterwards be moved and arranged as needed. So the problem of moving information along a stationary tape was well on the way to being solved, if the local concentration of bubbles could be varied as required and the whole pattern moved as one.

The direction of the development then changed when it was found that bubbles and their absence offered a way of providing high-density permanent stores for 'noughts' and 'ones', and that it was also possible to add and subtract them and so use them as computing elements [15]. Not much, though, seems to have come of this because of the competition from semiconductors.

14.8.7. *Rare-earth metals and their alloys*

There had been an interest in the magnetic properties of the rare-earth metals even before World War II, but as they became more readily available, and in much better separated forms, it became clear that the doubtful purity of the materials made a good deal of the pre-war work unreliable. Indeed many of the early EPR studies in insulators showed resonances from rare-earth ions which were not supposed to be there, indicating in particular that gadolinium impurities would have significantly distorted earlier macroscopic magnetic measurements.

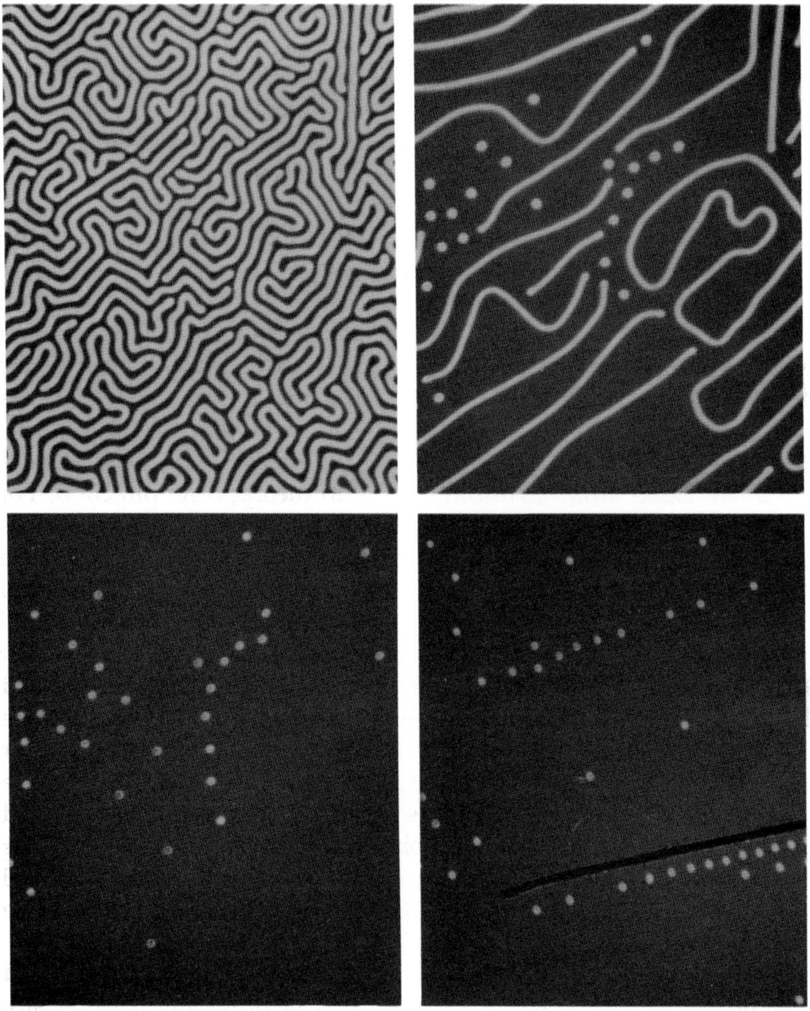

Figure 14.13. *Emergence of magnetic bubbles in a thin wafer of magnetic garnet is demonstrated in this sequence of photomicrographs taken at the Bell Telephone Laboratories. The magnetic domains in the specimen rotate polarized light in different directions depending on whether the internal magnets in the crystal point up or down. By adjustment of a polarizing filter, domains with the same orientation can be made to look either bright or dark. When no external magnetic field is present (top left), the domains form serpentine patterns, with domains of opposite magnetization occupying equal areas. (The apparent departure from equality here is an artifact of the exposure.) When an external magnetic field is applied perpendicularly to the specimen (top right), the domains that are magnetized in the opposite direction shrink and in a few cases contract into bubbles. Although these circular domains are called bubbles, they are actually stubby cylinders viewed from the end. A further increase in the external bias field (bottom left) converts all the remaining 'island' domains into bubbles. With a 'soft' magnetic wire whose magnetization is polarized by the external field one can move the bubbles around freely (bottom right). Because the bubbles repel one another they tend to maintain a certain minimum separation. Nevertheless, they can be packed with a density of more than a million per square inch.*

Nevertheless the pre-war work was indicative of interesting magnetic properties, for gadolinium had been found to be ferromagnetic and it seemed probable that all the members of the family would be of interest. So when they became more readily available, in increasingly pure forms, their study was taken up in earnest.

Unlike most metals, the paramagnetic susceptibility of neodymium varied strongly over a range of temperatures in the vicinity of room temperature, and the variation was very much like that found in salts containing trivalent neodymium ions [121]. Since the explanation of the susceptibilities of these salts was falling nicely into place as a result of the free-ion Zeeman studies and the interpretations of EPR, the results on the metal were similarly interpreted on the assumption that all the magnetism was due to localized inner-shell 4f electrons.

There was no inkling of the complexity yet to come. The first indication came from neutron diffraction studies of polycrystalline samples of erbium and holmium [122]. As the study developed a panoply of exotic magnetic configurations was revealed in the elements and, in due course, in their alloys. Legvold [123] has given a detailed account of this and various other physical properties together with diagrams illustrating the various types of magnetic ordering. Below some critical temperature, which is usually below room temperature, some of the elements (e.g. gadolinium) show moments aligned as in typical ferromagnets, in others (e.g. terbium) there are aligned moments in parallel planes but with the directions rotating through a fixed angle from plane to plane, and in still others (e.g. holmium) there are combinations of the above so that some components of magnetization rotate from plane to plane while other components point perpendicularly to the plane (figure 14.14). As if this was not enough the arrangement in any one element could also change with temperature.

See also
p 991

A good deal of information was also found by inelastic neutron scattering, as described in Chapter 12. An incident neutron would lose some of its energy and change its direction of motion, showing that an excitation of known energy and momentum had been created in the magnetic system. While this was, experimentally, similar to the excitation of a spin wave in a more conventional ferromagnet there was some hesitation over using such a description in such complicated magnetic structures. The excitations could hardly be regarded as due to simple reversals of electron spins. They came to be referred to as *magnetic excitons* or *magnons*.

It had long been accepted that the rare-earth ions are unique in having partially filled 4f shells which are well inside the electronic charge distribution. This concept seems to have been readily accepted for the elementary metals. So they too had immobile electrons in partially filled 4f shells, which meant that the conductivity had to be associated with the outer electrons, which were, presumably, in conduction bands. Little

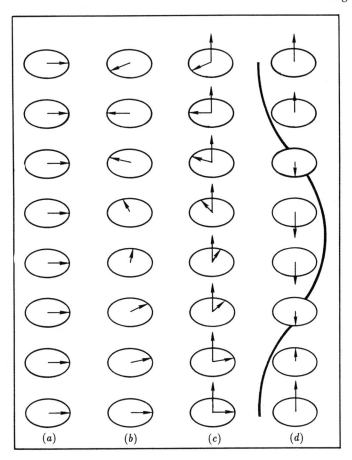

Figure 14.14. *Magnetic structures of the heavy rare-earths. The moments in a particular hexagonal layer are all parallel. In the basal-plane ferromagnet, (a), all moments are aligned along the direction of magnetization. The moments in the helix, (b), rotate by a specific angle between neighbouring planes. The cone, (c), is a combination of a helix and an axial ferromagnetic component, so that the total moment rotates on the surface of a cone. In the longitudinal wave structure, (d), the c-axis component varies sinusoidally, while that in the basal plane is disordered. These structures can all be understood as a combination of the exchange interaction, which produces long-range periodicity along the c axis, and the crystal fields, which tend to orient the moments.*

notice sees to have been taken of Slater's view, at least as far as the 4f electrons were concerned, that in a periodic structure all electrons should be regarded as in band states. Rather the accepted view was that the 4f electrons could be distinguished and treated by crystal field theory, with the 'ions' bathed in a sea of conduction electrons which itself provided a major contribution to the crystal fields.

The description of the metals then became quite similar to those of the salts, for their Curie–Weiss-like susceptibilities at high temperatures could readily be understood without a need for a detailed description of the exchange interactions. However, explaining the temperatures at which the cooperative phases set in was more of a problem because they are much higher than those of the salts. Some years before there had been an interest in exchange interactions between manganese ions in a non-magnetic host and out of this had come the idea of an indirect exchange interaction due to conduction electrons. This idea was adopted for the rare-earth metals. When a conduction electron is in the vicinity of a localized moment it is assumed that there will be an exchange interaction between its spin and the local moment. Then, as the conduction electron moves away, some memory of the correlation will be retained, and this will be passed on to a second localized moment, so producing an indirect correlation between the two localized moments. The resultant coupling is now usually described as the RKKY interaction (using the initials of four of the main contributors, Ruderman and Kittel [124], Kasuya [125] and Yosida [126]). Unlike exchange in insulators it was predicted that its magnitude would fall off more slowly with the local moment separations and show oscillations in sign. The magnons could then be pictured as excitations of the coupled system of localized moments and induced moments in the cloud of conduction electrons [127].

To find out more about the crystal fields the rare-earth elements were dissolved in non-magnetic host metals of the same structure. This reduced the exchange interactions and allowed the low-lying patterns of crystal field splittings to be observed directly by the inelastic scattering of neutrons [128]. Not only did this work yield the crystal field parameters but it showed there was little variation from one rare-earth to the next.

This work showed clear evidence of unquenched orbital moment and created problems in describing the exchange interactions, which could hardly be expected to have the same form as when the orbital moment is quenched. There was also the complication of taking account of the crystal field splittings. The only hope seemed to be to use the neutron dispersion results to provide detailed descriptions of the exchange interactions and the success achieved has been described by Mackintosh [129].

It seems probable that had the work on the rare-earth elements not been so successful the division of opinion about which description to use, localized or non-localized, would have continued, but with these examples, where there were undoubtedly periodic lattices, it became increasingly difficult, though not impossible (according to Freeman [130] on whom Slater's mantle appears to have fallen) to maintain that they could only be understood on the basis of band theory. The corresponding difficulty with the insulators was resolved without the introduction of bands, and it seems not improbable that the same will occur with the conductors. Undoubtedly the simple concept of crystal fields will disappear, though this has not stopped various workers from trying to estimate them using various electrostatic-like models of host metals, such as gold and aluminium. In most of this work the 4f electrons are distinguished from the rest, though not by Schmitt [131] who found that there was an exchange-like contribution to the crystal field.

I took up the problem of how to obtain equivalent operator forms in place of the crystal fields, while retaining indistinguishability, without success [132]; and though Dixon and Wardlaw [133] later managed to clarify some of the issues, they did not entirely resolve them.

While theoretical progress has been limited the same has not been true experimentally, for it has been found that alloys containing rare-earths have a diversity of uses. On this topic a few remarks and references must suffice. The remarks are based on three articles in *Magnetism in the Nineties*, which look back as well as forward in time. The first, by Strnat and Strnat [134], gives an account of the work on rare-earth–cobalt alloys and lists their uses. Some are of considerable industrial importance, particularly where very powerful permanent magnets are needed and expense is not a restriction. The magnets in use are not simple things—the authors remark that 'Real magnets are more complex than the intermetallics identified as potential hard-magnetic materials. They are not stoichiometric compounds; there are always several phases present; the metallurgical microstructure is typically complex and not in thermodynamic equilibrium. Some magnets are even composites containing nonmagnetic binders.'. The second article, by Herbst and Croat [135], describes another important family, the neodymium–iron borates, and the third, by Buschow [136], describes the permanent magnet materials of type $RT_{12-x}M_x$, where T can be any of Co, Ni and Mn and M any of Al, Si, Ti, V, Cr, Mo, W and Re. All three papers give a large number of references and a further source is a volume in the series known as Landolt-Börnstein [137].

14.9. The changing pattern
By about the middle of the 1970s a good deal of the steam resulting from the immediate post-war developments in electron resonance, ferrites and rare-earths seems to have been dissipated, so although developments in all of them continued, interest was beginning to turn to new directions. For twenty or more years Slater's view, that with periodic systems their understanding would necessarily come from band theory, had been ignored and as a result that part of magnetism which used localized concepts with great success had become divorced from the magnetism which used band theory, which had hardly progressed at all. This was in remarkable contrast with practically all the rest of solid-state physics where band theory had come to be the dominant approach. So the traditional ferromagnets iron, cobalt and nickel had come to be 'the piggy in the middle', explicable neither by band theory nor by localized moments. Furthermore, with the subsequent demonstration that the

localized models needed neither to distinguish electrons nor to give up periodicity the common sense approach of staying with concepts which worked seemed to have paid off.

The above remarks have slightly exaggerated the position for although band theory was having enormous success in wide areas of solid-state physics there was a growing sense that something more was sometimes needed. It would be out of place to go into a lot of detail here so a few remarks must suffice. There were two basic problems. The first was that given a band structure, say that of sodium, it was possible to work out the probabilities that, at a given lattice site, the number of 3s electrons would be 0, 1 or 2. The isolated sodium atom has only one such electron and the expectation was that this would be by far the most probable occupation number in the solid, and yet the values derived from band theory were sufficiently different to give cause for concern. The idea was creeping in that it might be necessary to add something to band theory to produce restrictions in the possible local expectation numbers. One technique [138] was to introduce localized functions derived from Bloch functions and characterized by a parameter commonly called the Hubbard U. (A tribute to John Hubbard who wrote a number of important papers on this topic in the 1960s and died at an early age.)

The second problem was that in band theory almost all the electrons are spin-paired, so there was limited scope for magnetism to arise from unpaired electron spins, and yet the number of examples where local moments were being found was increasing. Neutron scattering had indicated that even in chromium and manganese there were small localized spin magnetic moments, with similar results on some of the actinide elements. (The actinides were, at one time, expected to be magnetically similar to the rare-earths, having unpaired electrons in 5f orbitals, but since these electrons are on the outsides of the ions there was the alternative possibility that they would be much more sensitive to their environment. Sorting out what actually occurs has been far from simple, particularly when there is also the possibility that some or all of the electrons might be in the nearby 6d orbitals. The magnetism of the actinides has received a lot of attention, but this will not be reviewed here, except to say that there seems to be little that is systematic about the magnetism of these elements and their compounds.) The de Haas–van Alphen effect, which has proved so powerful in the analysis of electron behaviour in non-magnetic metals, has also given strong support to Stoner's conception of itinerant magnetism. As explained in Chapter 17, the effect provides a detailed picture of the band structure of conduction electrons. Gold was the first to show that the conduction electrons in iron and nickel form two different assemblies, each containing spins of one orientation. The difference in the number of electrons in the two assemblies accounts fairly well for the saturation magnetization. This

See also p 1348

is a complex subject with many unresolved problems concerning the distinction between localized and itinerant electrons, such as to keep alive the issues which troubled Slater. A review has been given by Lonzarich [139].

The problem of describing localized charge distributions, where the changes in magnitude on moving from one site to another are smaller than the electronic charge, led to the introduction of the concept of charge fluctuations, as well as spin fluctuations for describing local magnetic moments when the moments are less than that of a single spin. These are useful expressions but they can also be misleading to the casual reader, for they suggest that charge or moment at a site is changing with time, whereas often the change is with position and is static in time, but as technical terms they are here to stay.

An upshot of the various extensions to band theory has been a growth in interest in the magnetism of conductors, so much so that the study of itinerant magnetism is becoming one of the most active topics in the field; perhaps the most remarkable point is that quite a lot of this is actually about the disappearance of magnetism as the temperature is lowered.

14.9.1. The Kondo effect

To give an example, many years ago de Haas *et al* [140] observed a minimum in the electrical resistance of gold at about 5 K, a similar effect being subsequently found in magnesium [141] and other elements. There was no explanation of this quite unexpected effect, though it was suggested that it might be due to impurities. In due course this was confirmed, as it became increasingly clear from work on non-magnetic metals containing deliberately incorporated 3d elements, such as manganese, that localized magnetic moments were involved. (By this time temperature-independent magnetism from band electrons and/or filled inner shells of electrons was regarded as universal in normal metals. So, any conductor which had magnetic properties solely of this kind was regarded as non-magnetic.) The paramagnetism with the impurities present was Curie-like and so was obviously due to localized moments. Then it was found that at about the temperature of the resistance minimum the paramagnetism disappeared.

See also p 1342

This phenomenon led to a lot of theoretical work, under the description *Kondo effect*, which acknowledged the contribution to the explanation which Kondo made in a series of papers in the 1960s [142]. The physical explanation was that as the temperature decreased the arrangement of the conduction electrons would change and that a degree of charge and spin localization would develop near a given impurity, such that the localized magnetic moment in the conduction cloud would actually cancel the localized moment on the impurity. It was as if the spin angular momentum of the impurity had become coupled to an equal spin angular momentum induced in the cloud of conduction electrons to give a total spin of zero, so forming a non-magnetic cluster. Thus an interest which was originally in electrical conduction came to be associated with the disappearance of magnetism and the extension of itinerant magnetism to take account of what became known as *strongly correlated electron systems*.

14.9.2. *Intermediate valence and heavy fermions*

In the early 1970s a rather similar magnetic phenomenon was found, the disappearance of magnetic moments in certain fully concentrated conducting compounds of rare-earths at low temperatures. This too might have been regarded as indicating that the conduction electrons had developed magnetic moments, which precisely cancelled the localized moments from the 4f electrons, had it not been that it was difficult to accept that it could happen at all the rare-earth sites in a crystal. (There is a related question as to what happens to the Kondo effect in a normal metal when the number of magnetic impurities is increased.)

To illustrate the direction taken by the explanation it is useful to go back to the rare-earth metals and ask why there are itinerant conduction electrons as well as electrons localized in partially filled 4f states. The simplest explanation is that, for the postulated electronic arrangement to be in equilibrium, it must require a positive amount of energy to take an electron out of the 4f shell at a given site and place it at the top of the Fermi distribution of the conduction electrons. Similarly it must also take a positive amount of energy to move an electron from the top of the Fermi distribution to a 4f orbital, so increasing the number of 4f electrons at a typical site. In other words it must be energetically unfavourable to alter the number of 4f electrons at any site.

Suppose now that in some rare-earth conductor a small amount of energy could actually be gained by moving a 4f electron from a particular site and placing it at the Fermi level. Then this should happen, for it will lower the energy. But having granted this, why not allow a second electron to move from another site? This too could happen. The transferred electrons are delocalized and as the process is continued and electrons from successive sites are moved to the conduction band the later ones will have to go into increasingly higher-lying band states, for the lower ones, which were previously empty, will now be occupied by electrons which have already been transferred. So less energy will be gained and the process will eventually stop, when the energy needed for the transfer becomes positive. But if the number of electrons so transferred is less than the total number of 4f sites it would seem that there is now a mixture of sites, some of which have lost an electron from the 4f shell and others which have not. There is then the problem of reconciling this picture with the quantum requirement that crystallographically identical sites must have identical charges.

Experiments on some rare-earth sulphides seemed to indicate that such transfers had occurred and that all the rare-earth sites had, somehow, remained crystallographically identical. To add to the difficulty the conduction electrons were then found, at low temperatures, to develop localized magnetic moments which cancelled those in the 4f shells. So there seemed to be something like a Kondo effect in a fully concentrated system. There would have been localized moments had no transfers occurred and probably there still are, at high enough temperatures, even though transfers to conduction states had occurred.

The phenomenon as a whole was described as *intermediate valency* and, whether or not the above physical description is valid, it was quickly recognized that its theoretical treatment presented a major problem in quantum mechanics. The basic phenomenon, the disappearance at low temperatures of a magnetic moment, can hardly be regarded as an exciting event in magnetism, any more than other recent topics involving strongly correlated electrons—Kondo lattices [143] and heavy fermions [144], a rather similar phenomenon which is particularly associated with actinide conductors. Yet magnetism is the soil from which these new developments have sprung, and it is experts in magnetism who are most familiar with the available theoretical techniques.

A problem with most of these new developments is a shortage of experimental techniques which can pin-point what is happening, the realization of which serves to emphasize how fortunate it was, particularly for the insulators, that resonance techniques were available. The position is much less favourable for conductors, for with bulk specimens something which will penetrate a metal is needed. Neutrons can be used and so can elastic waves, though only over a limited frequency range. Electromagnetic radiation is restricted to very low and very high frequencies. X-rays can excite electrons from bound states to high-lying conduction states, so high in fact that the electrons emerge from the conductor with energies that can be measured. This technique, the photoemission of electrons, has been used, but so far these and similar methods have not been as easy to interpret as resonance experiments.

Probably the best way of regarding these efforts to extend itinerant magnetism is to expect that they will result in much of magnetism disappearing as a separate topic in solid-state physics. Indeed it is probably only because magnetism was a well-developed subject before the relevance of band theory was fully appreciated and that, when it was, the main developments were dominated by crystal field theory, that it remained separate. Whether the whole of magnetism will disappear into solid-state physics is another matter, for just as the traditional topics, which focused on crystalline materials, seem to be close to being unified, new fields, involving non-periodic lattices, are being opened. There is much interest in amorphous magnetic materials and very thin films which, though periodic in two dimensions, are far from periodic in the third.

14.9.3. *Amorphous magnetism and thin films*

There is a long history, in magnetism, of the experimental study and use of magnetic alloys, which can be classed under the general heading of random systems. Similarly in EPR the magnetic ions are often widely dispersed in magnetic hosts, and so provide another example of randomness, though one in which the consequences of randomness are not obvious. The position changes as the concentration of magnetic ions is increased, particularly if the exchange interactions are antiferromagnetic, as they often are. For low concentrations there will probably be no phase transition corresponding to the onset of antiferromagnetism, but as the concentration is increased it becomes increasingly probable that some kind of phase transition will occur. The study of some of these systems comes under the heading of spin glasses, which are usually taken, at least in theoretical studies, to imply either a random spatial arrangement of spins interacting with short-range exchange interactions or a random occupation of lattice sites with longer-range exchange interactions which vary in sign, as in the RKKY theory of impurity spins in a conductor. Both models present great theoretical difficulties and have excited a good deal of interest in the last decade, particularly over the question of how to recognize a phase transition if it should occur; or indeed whether there will be a transition at all. A simple example of just three spins mutually coupled antiferromagnetically serves to illustrate the sort of problem that can arise. It is already difficult to decide how these would align at low temperatures because they cannot all be antiparallel, a situation described as *frustration* [145].

The experimentalist can fairly readily be provided with amorphous magnetic samples, for there is a variety of techniques for making them, and they can be either insulating or conducting. So there is no shortage of examples to study. On the other hand the theoretician, while he can invent a variety of models, is not able to get very far with most of them. So both experimentalist and theoretician can be kept occupied, but whether, for any specific case, there is a convincing match between experiment and theory is a matter I do not feel competent to decide. It is probably not important at the present stage, because magnetism has usually progressed quite well simply on the basis of experimental investigations, with theory sometimes providing a stimulus and sometimes an *a posteriori* explanation. Here there is an extra stimulus, a technological interest in producing isotropic magnetic materials.

The other big development is the study of thin films, which may be of the order of 100 atomic layers in thickness, crystalline or amorphous, magnetic or non-magnetic. Such films need to be supported, and a variety of materials can be used as substrates, either for single films or for sandwiches of films having different compositions. The possibilities are almost endless, and enough has already been done to show that a rich variety of new properties will be found. Thus it has been possible to grow a layer of an insulating ferrimagnet on top of a normal non-magnetic conducting layer, and demonstrate that changing the magnetization in the ferrimagnet changes the resistivity of the conductor. Such an arrangement may well find use in the reading heads of tape recorders.

I am aware that no references have been given in this section and for this there are two reasons. One is that so many workers have contributed that it is difficult to decide on priorities and the other is that I find it difficult to make an assessment of which of the many possibilities have the greatest potential for development. I hope, therefore, that my deficiency can be compensated by a reference to the articles in *Magnetism in the Nineties* [144], which is primarily aimed at predicting what is going to happen in magnetism in the rest of this century. For the immediate past there are the papers given at the International Conference on Magnetism, I.C.M. 91 [146], one of a regular sequence held at intervals of approximately three years.

14.10. Nuclear magnetism

14.10.1. Nuclear magnetic resonance (NMR)

At about the time that ferromagnetic and electron paramagnetic resonance were being introduced a related interest was being developed in resonances associated with nuclei. In 1946 two papers appeared [97, 98] which were to set the scene for a whole new development in magnetism. It was known that a substantial number of nuclei possess nuclear spins and associated nuclear magnetic moments which, as with the electrons, are tied to the spin directions. So in an external magnetic field it could be expected that such a nucleus would precess about the field direction with a frequency determined by the strength of the field and the nuclear g-value, the ratio of the magnetic moment to the angular momentum in appropriate units. Also that an oscillating transverse field, of the right frequency, would induce transitions between the nuclear spin orientations (the moment projections in the direction of the field). Indeed the basic concepts of the various magnetic resonance phenomena are so similar that they can reasonably be regarded as variants of the same concept. That more unification has not occurred can be attributed to differences in experimental technique. Nevertheless some contact has been maintained, particularly on the theoretical side, so in describing NMR the differences rather than the similarities will be stressed.

The magnitudes of nuclear spins are of the same order as those of the electron, but because the nuclear masses are something like 1000 times greater the nuclear magnetic moments are less than the electronic moment by a similar factor. Thus a magnetic field which gives an EPR resonance in the microwave region, at say 10^{10} Hz, is likely to give a nuclear resonance at a frequency near 10^7 Hz, where the experimental techniques are substantially different. In both, the strength of the absorption depends on the magnitude of the magnetic moment, the total number of spins in the sample and the difference between the number of these which are oriented parallel and antiparallel to the magnetic field, which introduces a factor of $\mu H/kT$, where μ is the magnetic moment and T is the temperature. In introducing any new technique it always seems difficult to observe any responses to begin with, and it is therefore remarkable that nuclear resonances were first observed at about the same time as electron resonances, for all the above numerical factors, except the total number of spins, are far less favourable. The apparent imbalance was largely offset by the higher detection sensitivity at the lower frequency, though the choice of sample was also important.

The first nuclear resonance experiment in a solid, by Gorter and Broer [147] in 1942, was unsuccessful. However, by 1946 resonances due to protons, hydrogen nuclei, had been reported in paraffin and water and once these had been found the way was open to making improvements in the experimental techniques and moving to other nuclei. From the beginning there was an interest in relaxation phenomena, energy exchanges between the spins and energy losses to the environment, which cannot readily be described in terms of the concept of Zeeman split levels. Instead they were described phenomenologically, in the Bloch [55] equation, which is an extension of the classical and quantum-mechanical equations of gyroscopic motion. The experiment is regarded as inducing a change in the precessing magnetization, which can be divided into two parts, a constant part in the direction of the field and another part which is rotating in the perpendicular plane (see figure 14.10). The latter was regarded as the resultant of all the spin precessions, and so it would have a persistent non-zero value as long as the exciting field continued to be present. However, if the excitation is removed both components of magnetization can be expected to change. The steady part should alter because the resonance phenomenon will have changed the initial thermal distribution of up and down moments (the constant part in the field direction). So there will be a change back to the original thermal equilibrium distribution, characterized by a time, T_1, known as the spin-lattice relaxation time. The transverse moment, which would have been zero before the experiment began, will decay back to zero by spin–spin relaxation processes, mutual spin-flips of oppositely aligned spins which destroy the coherences in the precessions of the spins. This decay was characterized by another time, T_2. For a liquid there is, of course, no lattice and the T_1 and the T_2 process are related. Two magnetic dipoles close together interact, and if they are at fixed positions, as in a solid, this interaction accounts for T_2. However, in a liquid the nuclei are moving and the motion of one dipole creates a time-dependent field acting on the others, which is much like that of the oscillating field in the experiment. It can therefore induce spin reversals as well as mutual spin-flips, and so contribute to T_1 as well. Its similarity in this respect to spin-lattice relaxation in EPR may be taken to indicate the great potential of NMR for studying motion in liquids.

The lower frequencies used in NMR give it an advantage over EPR for the study of metals, since the fields penetrate further; NMR has proved valuable here and has found a wide use mainly because the steady field which the nucleus experiences can be different from that which is applied, due to moments induced in the electron clouds. These investigations have helped in the study of local moments in conductors. A similar effect in molecules has been widely taken up in chemistry, for it has provided a means of identifying different chemical complexes. Thus the chemical shift, as it is known, can be used to distinguish between the protons in the CH_3, CH_2 and OH groups, which all occur in ethyl alcohol. The electronic arrangements in the groups are different, so the fields at the protons in the different groups are also different, even though they are all in the same external field. The resonances are very close, so the individual lines must be so narrow that they do not overlap. This presents more of a problem in solids than in liquids, where the motional narrowing, as it is called, helps to keep the lines narrow, but even so the resolution may be lost if the external field is not sufficiently uniform over the sample. Resonances from the same group in different parts of the sample will then occur at different field strengths and artificially broaden the lines. This high-resolution NMR, of which an example is shown in figure 14.15, therefore requires a highly uniform external field. The intensity of each line is proportional to the number of protons in each group, which can also be useful.

The reasoning can be inverted, for if the applied field is non-uniform in a known way, and all the nuclei of interest have identical surroundings, the intensity of the resonance line at a particular field value gives a measure of the number of nuclei in that particular field. If it could be arranged that every tiny cube in the sample experienced a different applied field, the line shape would give the spatial density of the nuclei. Furthermore, if the same nucleus occurs in two different environments and the non-uniform field can be given a time variation there is the possibility of determining the spatial distribution of each. Demonstrating this was a major challenge because the more the nuclei are divided up into packets which resonate at different field values the less they are resonating at any given field strength, and the weaker the resonance lines. Nevertheless the challenge has now been met, but only after a sequence of quite remarkable concepts was proposed and the relevant experiments performed.

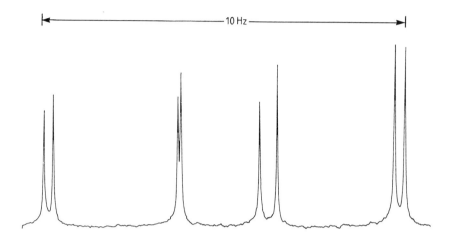

Figure 14.15. *Example of extremely sharp proton resonances at 200 MHz in a magnetic field of 4.7 T. The whole spectrum shown occupies a band of 10 Hz and the lines are typically 0.04 Hz wide at half-peak. The spectrum was obtained with a WP 200 spectrometer (Bruker, Spectrospin Ltd), the sample being a dilute solution of ortho-dichlorobenzene, $C_6 H_4 Cl_2$ in deuterated acetone.*

The success is largely due to Hahn's [148] idea of substituting pulses for the steady intensity of radiation used up till then for observing the resonant absorption. While this may not seem to be much of a change, or even perhaps a step backwards, because it reduces the energy to which the spins are exposed, the ramifications have been enormous. In the first place the Bloch equation could be solved, to sufficient accuracy, by what amounted to a geometrical type of reasoning, by considering how a rotating observer would describe what was occurring. It then became possible, particularly for Hahn and his associates, to see how to design sequences of pulses to increase the experimental sensitivity. One development was the production of 'echoes'. A spin system, suitably excited would produce an 'echo' at a later time, a response in a receiver coil, which, depending on the sequence, would give values either for T_1 or T_2. Other sequences would transfer spin polarizations between low- and high-abundant spin systems and so allow the resonances of low-abundance spins to be detected indirectly. These techniques, when combined with time variations of the so-called steady main field, produced magnetic resonance imaging, which is now finding wide use in the medical field by revealing the density (and/or the relaxation time) distribution of protons and other nuclei in human organs, by what is thought to be a non-invasive technique. It is confidently expected that, in due course, it will be possible to follow some of the important chemical reactions which take place in the body, by combining magnetic resonance imaging (MRI) and chemical shifts. (A reaction results in changes in the concentrations of different chemical groups.)

The literature on nuclear magnetic resonance is very extensive, and only a brief survey has been possible here. For further reading there are books by Andrew [149], Abragam [150] and Mansfield and Morris [151], which between them cover most of its aspects.

14.10.2. *Nuclear demagnetization*

After a considerable amount of effort the demagnetization of suitable magnetic crystals produced cooling to temperatures well below those which can be obtained with liquid helium, so extending the available range down to about 0.01 K, with the hope of an extension to 0.001 K [30]. To someone unfamiliar with low-temperature work such a small temperature decrease probably seems a small reward for so much effort, at least until it is realized that in low-temperature physics the important parameter is the reciprocal of the temperature. With the boiling point of liquid helium being near 4 K, a drop to 0.01 K changes $1/T$ from about 0.25 to about 100, which puts the extension in quite a different light. Reaching 0.001 K would give 1000 for $1/T$ and so would represent another major extension. A two-stage demagnetization process was therefore considered by Kurti, with the first using electronic moments and the second nuclear moments [152]. Eventually, however, the first stage became unnecessary when the development of dilution refrigerators, based on the properties of mixtures of helium isotopes allowed starting temperatures of the order of 0.01 K to be maintained indefinitely.

From about 0.01 K the direct demagnetization of nuclei looked promising, for the factors which had favoured electronic demagnetization from an initial temperature of 2 K could then be seen to be similar to those for nuclei starting from 0.01 K, but this was by no means enough, for in working at such low temperatures there were many obstacles to be overcome. It would be out of place to go into all the details, which can be found in two papers, describing work at Lancaster [153] and Bayreuth [154]. Both groups cooled copper down to about 10^{-5} K, and more recently the nuclei in Cu have been cooled to about 10^{-7} K, at which extremely low temperature there was a transition of the nuclear spin array to an antiferromagnetic structure.

14.11. Concluding remarks

A century of magnetism is a long period to describe in an article of limited length, and from direct familiarity with only part of the period. I do not expect that all readers will agree with my selection or with my presentation. I hope though that my deficiencies will not have upset any of the many friends I have made through some forty years of interest in magnetism, and that I have managed to show less expert readers why we have found magnetism so interesting. Those of us who have been privileged to work in the field have found it fascinating, as

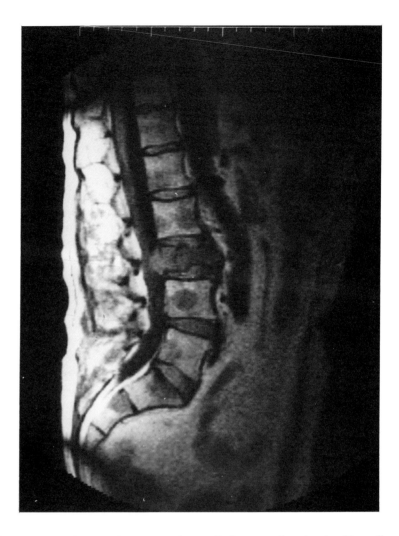

Figure 14.16. *A magnetic resonance image of a human spine, showing (i) a collapsed vertebrae in the centre of the spinal column, and (ii) tumours within vertebrae, where the tumours replacing normal marrow are shown by the dark areas. The spinal cord and nerve roots are also clearly visible to the left of the spinal column.*

well as being a specialism which we can confidently feel has contributed enormously to the well-being of society. From compasses and magnets to pick up pins, it has progressed through the development of the electrical power industry to information storage and retrieval in many forms. Now magnetic resonance imaging seems set to change whole fields of medical practice, without, as far as is known, introducing any undesirable side-effects (figure 14.16).

See also
p 1923

Not a bad record, surely, for something which throughout has been driven by curiosity!

Acknowledgments

I would like to acknowledge the help and interest shown to me by Brebis Bleaney in the early stages of the project, and my colleagues in Nottingham throughout. In particular I would like to thank John Fletcher, who made his substantial collection of early books on magnetism available to me and allowed me to draw on his wide knowledge of physics, John Owers-Bradley who tutored me on adiabatic demagnetization and several other topics, Peter Morris for his guidance about nuclear magnetic resonance and Brian Worthington for figure 14.16.

References

[1] Curie P 1895 *Ann. Chim. Phys.* **5** 289
[2] Voigt W 1902 *Ann. Phys., Lpz* **9** 115
[3] Thomson J J 1903 *Phil. Mag.* **6** 673
[4] Langevin P 1905 *Ann. Chim. Phys.* **5** 70
[5] van Leeuwen J H 1919 *Dissertation* Leiden
[6] Bohr N 1972 *Collected Works* (with English translations) vol 1, ed J R Nielson (North-Holland: Amsterdam)
[7] Bohr N 1913 *Phil. Mag.* **26** 1
[8] Pauli W 1925 *Z. Phys.* **31** 765
[9] Pauli W 1920 *Z. Phys.* **23** 201
[10] Debye P W 1926 *Phys. Z.* **27** 67
[11] Weiss P 1907 *J. Physique* **5** 70
[12] Weiss P and Foëx G 1931 *Le Magnétisme* (Paris: Libraire Armand Colin)
[13] Hund F 1925 *Z. Phys.* **33** 85
[14] Van Vleck J H and Frank A 1929 *Phys. Rev.* **34** 1494
[15] Bobeck A H and Scovil H E D 1971 *Sci. Am.* June p 78
[16] Barkhausen H 1919 *Phys. Z.* **20** 401
[17] Galison P 1987 *How Experiments End* (Chicago: University of Chicago Press) ch 2
[18] Goudsmit S and Uhlenbeck G E 1925 *Naturwissenschaften* **13** 953
[19] Beck K 1918 *Vjschr. Naturf. Ges. Zurich* **63** 116
[20] Van Vleck J H 1926 *Bull. Natl Res. Counc.* **10** 9
[21] Hund F 1927 *Z. Phys.* **43** 788
[22] Heisenberg W 1926 *Z. Phys.* **38** 411
[23] Dirac P A M 1926 *Proc. R. Soc.* A **112** 661
[24] Kramers H A 1930 *Proc. Amsterdam Acad.* **33** 959
[25] Debye P 1926 *Ann. Phys., Lpz* **81** 1154
[26] Giauque W F 1927 *J. Am. Chem. Soc.* **49** 1864
[27] Kurti N and Simon F 1933 *Naturwissenschaften* **21** 178
[28] Giauque W F and MacDougall D P 1933 *Phys. Rev.* **43**
[29] de Haas W J, Wiersma E C and Kramers H A 1934 *Physica* **1** 1

[30] Hudson R P 1972 *Principles and Application of Magnetic Cooling* (Amsterdam: North-Holland)
[31] Van Vleck J H 1932 *Electric and Magnetic Susceptibilities* (Oxford: Oxford University Press)
[32] Van Vleck J H 1950 *Am. J. Phys.* **18** 495
[33] Van Vleck J H 1937 *J. Chem. Phys.* **5** 320
[34] Bitter F 1938 *Phys. Rev.* **54** 79
[35] Hulthen 1936 *Proc. Am. Acad. Sci.* **39** 190
[36] Jacobs I S 1961 *J. Appl. Phys.* **32** 61S
[37] Néel L 1936 *Ann. Phys.* **5** 232
[38] Bizette H, Squire C F and Tsai B 1938 *C. R. Acad. Sci., Paris* **207** 449
[39] Foëx G and Graff M 1939 *C. R. Acad. Sci., Paris* **209** 160
[40] Van Vleck J H 1940 *Phys. Rev.* **57** 426
[41] Gorter C J and Kronig R 1936 *Physica* **3** 1009
[42] Gorter C J 1947 *Paramagnetic Relaxation* (Amsterdam: Elsevier)
[43] Standley K J and Vaughan R A 1969 *Electron Spin Relaxation Phenomena in Solids* (London: Hilger)
[44] Landau 1930 *Z. Phys.* **64** 629
[45] Honda K and Kaya S 1928 *Sci. Rep. Tohoku Univ.* **17** 1157
[46] Sucksmith W, Potter H H and Broadway L 1928 *Proc. R. Soc.* A **117** 471
[47] Kaya S 1928 *Sci. Rep. Tohoku Univ.* **17** 1157
[48] Landau L and Liftshitz E 1935 *Phys. Z.* **8** 153
[49] Bitter F 1932 *Phys. Rev.* **41** 507
[50] Bozorth R M 1951 *Ferromagnetism* (New York: Van Nostrand)
[51] Urbain G *et al* 1935 *C. R. Acad. Sci., Paris* **200** 2132
[52] Stoner E C 1933 *Phil. Mag.* **15** 1018
[53] Mott N F 1935 *Proc. Phys. Soc.* **47** 571
[54] Van Vleck J H 1937 *Phys. Rev.* **52** 1178
[55] Bloch F 1946 *Phys. Rev.* **70** 460
[56] Zavoisky E 1945 *Fiz. Zh.* **9** 211 245
[57] Cummerow R L and Halliday D 1946 *Phys. Rev.* **70** 433
[58] Bagguley D M S, Bleaney B, Griffiths J H E, Penrose R P and Plumpton B I 1948 *Proc. Phys. Soc.* **61** 542, 551
[59] Penrose R P 1949 *Nature* **163** 992
[60] Abragam A and Pryce M H L 1949 *Nature* **163** 992
[61] Abragam A and Pryce M H L 1950 *Proc. Phys. Soc.* A **63** 409; 1951 *Proc. R. Soc.* A **205** 135
[62] Van Vleck J H 1939 *J. Chem. Phys.* **7** 61
[63] Abragam A and Pryce M H L 1951 *Proc. R. Soc.* A **205** 135; 1952 *Proc. R. Soc.* A **206** 164, 173
     Stevens K W H 1963 *Magnetism* vol 1 (New York: Academic) p 1
[64] Stevens K W H 1952 *Proc. Phys. Soc.* **65** 209
[65] Bleaney B and Stevens K W H 1953 *Rep. Prog. Phys.* **16** 108
[66] Owen J and Stevens K W H 1953 *Nature* **171** 836
[67] Van Vleck J H 1935 *J. Chem. Phys.* **3** 807
[68] Abragam A and Bleaney B 1970 *Electron Paramagnetic Resonance of Transition Ions* (Oxford: Clarendon)
[69] Guha B C 1951 *Proc. R. Soc.* A **206** 353
[70] Bleaney B and Bowers K D 1952 *Phil. Mag.* **43** 372; 1952 *Proc. R. Soc.* A **214** 451
[71] Shull C G, Strausen W A and Wollan E O 1951 *Phys. Rev.* **83** 333
[72] Anderson P W 1963 *Magnetism* vol 1 (New York: Academic) ch 2
[73] Feher E R 1956 *Phys. Rev.* **103** 500
[74] Kleiner W H 1952 *J. Chem. Phys.* **20** 1784

[75] Freeman A J and Watson R E 1960 *Phys. Rev.* **120** 1254
[76] Slater J C 1953 *Rev. Mod. Phys.* **25** 199
[77] Cyrot M 1982 *Magnetism of Metals and Alloys* (Amsterdam: North-Holland)
    de Haas W J *et al* 1933 *Physica* **1** 1115
[78] Gondairi K-I and Tanabe Y 1966 *J. Phys. Soc. Japan* **21** 1527
    Stevens K W H 1976 *Phys. Rep. C* **24** 1
    Brandow B 1977 *Adv. Phys.* **26** 651
[79] Jacobsen E H, Shiren N S and Tucker E B 1959 *Phys. Rev. Lett.* **3** 81
[80] Bloembergen N, Shapiro S, Persham P S and Attman J O 1959 *Phys. Rev.*
    114 445
[81] Van Vleck J H and Penney W G 1934 *Phil. Mag.* **17** 9
[82] Trounson E P *et al* 1950 *Phys. Rev.* **79** 542
[83] Kittel C 1951 *Phys. Rev.* **82** 565
[84] Nagamiya T 1951 *Prog. Theor. Phys. Japan* **6** 342
[85] Kittel C 1948 *Phys. Rev.* **73** 155
[86] Johnson F M and Nethercot A H Jr 1959 *Phys. Rev.* **114** 705
[87] Caspers W J 1989 *Spin Systems* (Singapore: World Scientific)
[88] Riste T and Wanic A 1961 *J. Phys. Chem. Solids* **17** 318
[89] Collins M F 1964 *Proc. Int. Conf. on Magnetism (Nottingham, 1964)* (London:
    Institute of Physics/Physical Society) p 319
[90] Keffer F 1966 *Handbuch der Physik Band XVIII/2* p 1
[91] Martel P, Cowley R A and Stevenson R W H 1967 *J. Appl. Phys.* **39** 1116
[92] Aleonard R, Panthenet R, Rebouillat J P and Vevrey C 1968 *J. Appl. Phys.*
    39 379
[93] Moriya T 1963 *Magnetism* (New York: Academic) ch 3
[94] Moriya T 1960 *Phys. Rev.* **117** 91
[95] Griffiths J H E 1946 *Nature* **158** 670
[96] Rabi I I 1937 *Phys. Rev.* **51** 652
[97] Bloch E, Hansen W W and Packard M 1946 *Phys. Rev.* **70** 474
[98] Purcell E M, Torrey H C and Pound R V 1946 *Phys. Rev.* **69** 37
[99] Van Vleck J H 1950 *Phys. Rev.* **78** 266
[100] Lee E W, Callaby D R and Lynch A C 1958 *Proc. Phys. Soc.* **72** 233
[101] Craik D J and Tebble R S 1961 *Rep. Prog. Phys.* **14** 116
[102] Dillon J F 1963 *Magnetism* vol III (New York: Academic)
[103] Bagguley D M S and Owen J 1957 *Rep. Prog. Phys.* **20** 304
[104] Snoek J L 1947 *New Developments in Ferromagnetic Materials* (Amsterdam:
    Elsevier)
[105] de Boer J H and Verwey E J W 1937 *Proc. Phys. Soc. A* **49** 59
[106] Standley K J 1962 *Oxide Magnetic Materials* (Oxford: Clarendon) ch 3
[107] Néel L 1948 *Ann. Phys., Paris* **206** 49
[108] Gorter E W and Schulkes J A 1953 *Phys. Rev.* **90** 487
[109] Polder D 1949 *Phil. Mag.* **40** 99
[110] Nicolas J 1980 *Ferromagnetic Materials* vol 2 (Amsterdam: North-Holland)
    243
[111] Melchor J L *et al* 1957 *Proc. Inst. Radio Eng.* **45** 643
[112] Pippin J E 1956 *Proc. Inst. Radio Eng.* **44** 1054
[113] Suhl H 1957 *Phys. Rev.* **106** 384
[114] Weiss M T 1957 *Phys. Rev.* **107** 317
[115] Lowman C E 1972 *Magnetic Recording* (McGraw-Hill: New York)
[116] Bate G 1980 *Ferromagnetic Materials* vol 2 (Amsterdam: North-Holland)
    p 381
[117] Dillon J F 1957 *Phys. Rev.* **105** 759
[118] Dillon J F 1958 *J. Appl. Phys.* **29** 1286
[119] Bobeck A H 1967 *Bell Syst. Tech. J.* **46** 1901

[120] Bobeck A H and Della Torre E 1975 *Magnetic Bubbles* (Amsterdam: North-Holland)
       Malozemoff A P and Slonczewski J C 1979 *Applied Solid State Science: Advances in Materials and Device Research (Supplement 1: Magnetic Domain Walks in Bubble Materials)* ed R Wolfe (New York: Academic)
[121] Elliot J F, Legrold S and Spedding F H 1954 *Phys. Rev.* **94** 50
       Bates L F, Leach S J, Loasby R G and Stevens K W H 1955 *Proc. R. Soc.* B **68** 181
[122] Koehler W C and Wollan E O 1955 *Phys. Rev.* **97** 1177
[123] Legvold S 1980 *Ferromagnetic Materials* vol 1, ed E P Wohlfarth (Amsterdam: North-Holland) ch 3
[124] Ruderman M A and Kittel C 1954 *Phys. Rev.* **96** 99
[125] Kasuya T 1956 *Prog. Theor. Phys.* **16** 45
[126] Yosida K 1957 *Phys. Rev.* **106** 893
[127] Kasuya T 1966 *Magnetism* vol IIB, ed G T Rado and H Suhl (New York: Academic) p 215
[128] Rathmann O and Toubourg P 1977 *Phys. Rev.* B **16** 1212
       Toubourg P 1977 *Phys. Rev.* B **16** 1201
[129] Mackintosh A R 1977 *Phys. Today* June p 23
[130] Freeman A J 1972 *Magnetic Properties of Rare Earth Metals* ed R J Elliott (London: Plenum)
[131] Schmitt D 1979 *J. Phys. F: Met. Phys.* **9** 1759
[132] Stevens K W H 1976 *Magnetism in Metals and Metallic Compounds* ed J T Lopuszanski, A Pekalski and J Przystawa (New York: Plenum) p 1
       Stevens K W H 1977 *Crystal Field Effects in Metals and Alloys* ed Furrer (New York: Plenum)
[133] Dixon J M and Wardlaw R S 1986 *Physica* **135** 105
[134] Strnat K J and Strnat R M W 1991 *Magnetism in the Nineties* ed A J Freeman and K A Gescheidner Jr (Amsterdam: Elsevier) p 38
[135] Herbst J F and Croat J J 1991 *Magnetism in the Nineties* ed A J Freeman and K A Gescheidner Jr (Amsterdam: Elsevier) p 57
[136] Buschow K H J 1991 *Magnetism in the Nineties* ed A J Freeman and K A Gescheidner Jr (Amsterdam: Elsevier) p 79
[137] Landolt-Börnstein 1962 *Eigenschaften der Materie in Ihren Aggregatzuständen* (Berlin: Springer) p 9
[138] Hubbard J 1966 *Proc. R. Soc.* A **296** 82
[139] Lonzarich G G 1980 *Electrons at the Fermi Surface* ed M Springford (Cambridge: Cambridge University Press) p 225
[140] de Haas W J, Wiersma E C and Kramers H A 1934 *Physica* **1** 1
[141] MacDonald D K C and Mendelssohn K 1950 *Proc. R. Soc.* A **202** 523
[142] Kondo J 1969 *Solid State Phys.* **23** 183
[143] Lacroix C 1991 *Magnetism in the Nineties* ed A J Freeman and K A Gescheidner Jr (Amsterdam: North-Holland) p 90
[144] Adroja D T and Malik S K 1991 *Magnetism in the Nineties* ed A J Freeman and K A Gescheidner Jr (Amsterdam: North-Holland) p 126
       Steglich F 1991 *Magnetism in the Nineties* ed A J Freeman and K A Gescheidner (Amsterdam: North-Holland) p 186
[145] Binder K and Young A P 1986 *Rev. Mod. Phys.* **58** 801
[146] ICM '91 *J. Magn. Magn. Mater.* **104–107, 108**
[147] Gorter C J and Broer L J F 1942 *Physica* **9** 591
[148] Hahn E L 1950 *Phys. Rev.* **80** 580
[149] Andrew E R 1955 *Nuclear Magnetic Resonance* (Cambridge: Cambridge University Press); 1970 *Magnetic Resonance* ed C K Coogan, N S Ham, S N Stuart, J R Pilbrow and G V H Wilson (New York: Plenum) p 163

[150] Abragam A 1961 *The Principles of Nuclear Magnetism* (Oxford: Oxford University Press)

[151] Mansfield P and Morris P 1982 *Advances in Magnetic Resonance* Supplement 2, ed Waugh (New York: Academic)

[152] Kurti N *et al* 1956 *Nature* **178** 450

[153] Bradley D I, Guenault A M, Keith V, Kennedy C J, Miller I E, Mussett S G, Pickett G R and Pratt W P 1984 *J. Low Temp. Phys.* **57** 359

[154] Gloos K, Smeibide P, Kennedy C, Singaas A, Sekowski P, Mueller R M and Pobell F 1988 *J. Low Temp. Phys.* **73** 101

Chapter 15

NUCLEAR DYNAMICS

David M Brink

15.1. Background

From the time of the discovery of the neutron, physicists became increasingly interested in studying the properties of nuclei. This chapter describes the development of nuclear structure physics from about 1936 until 1975. The main focus is on the decade following the end of World War II. Many people contributed to experimental and theoretical advances, but here we concentrate on the most significant developments which opened up new fields of research.

The year 1932 was a turning point in the development of nuclear physics. On 28 April 1932, the Royal Society held an important meeting on the structure of atomic nuclei. Lord Rutherford delivered the opening address [1] reviewing the progress made in the previous three years, and spoke about the discovery of the neutron by James Chadwick [2] and the deuteron by Urey, Brickwedde and Murphy. He also reported the results of the first nuclear disintegration experiments with artificially accelerated ions carried out by J Cockcroft and E Walton [3]. These were the first of a series of dramatic developments which transformed our understanding of atomic nuclei. Some of these advances are described in detail in Chapter 2 by Abraham Pais and Chapter 5 by Laurie M Brown. We mention them again here to provide a background for the subsequent developments in nuclear dynamics. The book *Inward Bound* by Abraham Pais [4] is a fascinating source of information about the history of events of this period.

In 1932, most nuclear physicists supposed that the nucleus of a heavy element consisted mainly of alpha particles with an admixture of a few free protons and electrons. Lord Rutherford spent a part of his lecture to the Royal Society discussing that model and the difficulties associated with it. He suggested that 'an electron cannot exist in a free state in a stable nucleus but must always be associated with a proton or other possible massive units. The indication of the existence of the neutron in certain nuclei is significant in this connection.'.

In June 1932, just four months after Chadwick had published his discovery of the neutron, Werner Heisenberg [5] submitted the first of a series of papers in which he developed the basic ideas of the neutron–proton model of nuclear structure. He assumed that neutrons obey the laws of Fermi statistics and have spin $\frac{1}{2}$, arguing that these assumptions were necessary in order to explain the Bose statistics of the nitrogen nucleus and empirical results on nuclear moments. These developments are described by Abraham Pais in section 2.12 of Chapter 2.

In his address to the Royal Society in 1932, Lord Rutherford described the development of positive ion accelerators by research workers at the Cavendish Laboratory in Cambridge, UK, the Department of Terrestrial Magnetism in Washington DC and the University of California in Berkeley, and announced the results of the first experiments using accelerators for nuclear studies carried out at Cambridge. The other two groups performed similar experiments during the next few months. Until that time most information on the structure of nuclei had come from experiments with alpha particles from natural radioactive sources. Accelerators formed an additional line of attack which had many advantages. The intensities were much greater and the energies of the particles could be varied at will. A contemporary picture of the impact of accelerators on the development of nuclear physics can be obtained from the review article of M Livingston and H Bethe [6] published in 1937. Detailed information about the events described in the next few paragraphs can be found in their review. Pais gives interesting historical insights in section 17b of his book [4].

It had been expected that energies higher than the potential barrier associated with the Coulomb repulsion between the target and projectile would be required to produce a nuclear reaction. In 1928 G Gamow [7] and E Condon and R Gurney [8] introduced the quantum-mechanical theory of alpha decay which predicted that there was a certain probability of penetration of a potential barrier by particles with considerably less energy than required to go over the top of the barrier (cf Chapter 2, section 2.11). Cockcroft and Walton [9] realized that this might make a nuclear reaction possible with relatively low-energy protons.

J Cockcroft and E Walton had begun their work in 1929 at the Cavendish Laboratory. In 1930, using a half-wave rectifier to obtain a constant potential and a single-section accelerator tube, they produced protons with an energy of 300 keV. Then they developed a more powerful machine with a voltage multiplier and a two-section tube and obtained protons with energies up to 700 keV. The first disintegration experiments were performed with this machine in 1932 on the reaction

$$^7\mathrm{Li} + \mathrm{p} = {}^4\mathrm{He} + {}^4\mathrm{He}.$$

Ernest O Lawrence

(American, 1901–58)

Ernest O Lawrence was born in Canton, South Dakota, on 8 August 1901 of parents who were Norwegian immigrants. In 1919 he joined the university of South Dakota and graduated in 1922 with a BA degree in Chemistry. Then he spent periods at the Universities of Minnesota, Chicago and Yale. He was awarded his PhD at Yale in 1925 and obtained a National Research Fellowship which allowed him to stay on there for two years. He married Mary Kimerly Blumer in 1923.

In 1927 Lawrence accepted a faculty position as Associate Professor at Berkeley. He was strongly attracted to nuclear physics and was stimulated by the experiments of Rutherford. He saw that new advances could be made if ways could be found to accelerate charged particles to high velocities. As a result of his efforts and those of his colleagues the first cyclotron was operating at Berkeley in 1932. He became the Director of the Radiation Laboratory in 1935 and was awarded the Nobel Prize in 1939.

Lawrence had an intuitive feeling for physics and was a natural inventor. In the end, his unusual powers of leadership, his enthusiasm and personality were more important than his physics. He died on 27 August 1958.

E O Lawrence of the University of California began to work on the design of the cyclotron in 1929. By 1930 he had a small model working and obtained protons with an energy of 80 keV. In 1932 Lawrence and S Livingston had constructed a larger machine which produced protons with energies up to 1 MeV.

Several years earlier, in 1926, G Breit had investigated the Tesla coil as a means of producing high voltages. He was joined by M Tuve at the Department of Terrestrial Magnetism in Washington DC, and in the years up to 1930 made great progress in the design of accelerator tubes. However, the Tesla coil was not satisfactory as a high-voltage source because of the pulsed nature of the potential and the fluctuating value of the peak voltage. At Princeton in 1929, R van de Graaff built the first electrostatic generator of the type named after him. Tuve and his collaborators realized the virtues of the van de Graaff generator as a high-voltage source, and in 1932 they built a small model with a sectional metal and glass accelerating tube which had been used previously in the Tesla coil work. They published the first results of nuclear disintegration experiments with ions accelerated up to 600 keV by this machine in 1932. The results of experiments performed with these accelerators over the next few years are collected together in the 1937 review article of Livingston and Bethe [6].

An event of considerable importance was the discovery of artificial radioactivity by Irène and Frédéric Joliot-Curie [10] in 1934. Not only was this the first example of the creation of a new isotope, it was also a discovery of considerable practical importance. It provided a convenient label for identifying unstable isotopes, and was a precursor of Enrico Fermi's important experiments on neutron capture reactions. It also opened up the possibility for the use of radioisotopes for chemical and medical research (cf Chapter 2, section 2.12).

15.1.1. Nuclear masses and the liquid-drop model

The liquid-drop model was invented by George Gamow to describe a number of basic properties of nuclei. We begin the story with the first chapter of his book *The Constitution of Atomic Nuclei and Radioactivity* [11], which was published in May 1931 just before the discovery of the neutron. The book is a slim volume of 114 pages and the author aimed to give 'as complete an account as possible of our present experimental and theoretical knowledge of the nature of atomic nuclei'. Gamow was well placed to do this job. He had developed the theory of alpha decay in 1928 and had shown that the phenomenon could be understood as a quantum-mechanical tunnelling process. Niels Bohr was impressed with Gamow's work and invited him to spend the academic year 1928–9 in Copenhagen (figure 15.1). Bohr also arranged for him to visit Cambridge in January and February 1929, and he presented his theory at a meeting of the Royal Society in London in February [12]. He spent the academic year 1929–30 in Cambridge and then went back to Copenhagen. He interacted with Bohr, H Casimir, F Houtermans, L Landau and N Mott while he was in Copenhagen.

When Gamow wrote his book, detailed information about nuclear masses was beginning to emerge. In 1919 F Aston published an account of his mass spectrograph. His first results confirmed F Soddy's prediction of the widespread occurrence of isotopes of the elements. His instrument could measure masses with an accuracy of 1 part in 1000. The measurements showed that the atomic mass of every isotope was approximately a whole number (the mass number A) on a scale where the mass of the common isotope of oxygen was 16. Other workers

Enrico Fermi

(Italian, 1901–54)

Enrico Fermi was born in Rome on 29 September 1901. His parents came from the Po valley and his father had a Civil Service position with the Italian Railways. Fermi became committed to physics at the age of 15, and his unusual grasp of the subject amazed his examiners when he was admitted to the Scuola Normale in Pisa. He graduated in 1922 and spent a short time in Göttingen and Leiden before being appointed to the new chair of Theoretical Physics at the University of Rome in 1926. From 1922 onwards he had strong support from Senator Corbino, politician, scholar and scientist who was the Director of the Physics Institute at the University of Rome. He published his work on Fermi statistics in 1926 and his theory of beta decay in 1934.

Fermi started to form a research group in 1927 to study nuclear physics and in 1934 started experiments on neutron-induced radioactivity. In July 1928 he married Laura Capon. He was awarded the Nobel Prize in 1938. He left Italy for Columbia University immediately afterwards and arrived in New York on 2 January 1939. At Columbia he started to study chain reactions. He went to Chicago in 1942 to direct the experiments which lead to the first chain-reacting pile. After the war he was a professor at the Institute of Nuclear Studies at the University of Chicago where he remained until he died on 29 November 1954. Fermi had unusual gifts of concentration and a passion for clarity. In all his work he emphasized the physical content of theories rather than their formal aspects.

entered the field and there were advances in techniques for producing ion sources for heavy metals. Aston had developed his second mass spectrograph by 1927. With the new instrument an accuracy of 1 part in 10 000 could be achieved for light isotopes, and 1 part in 5000 for heavy isotopes. Interesting historical information is contained in an article on mass spectrometry prepared by K Bainbridge in 1951 [13].

Figure 15.1. *Niels Bohr, pictured beside a Cockcroft–Walton accelerator.*

By 1928 the atomic weights of about 250 isotopes were known. They were precise enough to measure mass defects (the differences between the exact atomic weights and the mass numbers) to an accuracy of about 20%. The mass defects were interpreted by a model which assumed that a nucleus was composed of protons and electrons. The idea, as stated by Rutherford [14] at a Royal Society meeting in 1929, was as follows 'The difference in mass between the free and nuclear proton is ascribed to a packing effect, i.e. to the interaction of the electromagnetic fields of the protons and electrons in the highly condensed nucleus. On modern views we know that there is a close relation between mass and energy (Einstein's relation $E = mc^2$). The free proton has a mass 1.0073 while the proton in the nucleus has a mass very nearly 1. This apparently small loss of mass means that a large amount of energy has been emitted in the entrance of the free proton into the structure of the nucleus, an amount equal to about 7 MeV.'. Aston's measurements showed that the binding energy per proton was almost constant throughout the entire periodic table. The significance of this important fact was first recognized by Gamow who used it in his formulation of the liquid-drop model.

We know nowadays that the density distribution of matter in a nucleus is approximately constant in its interior and falls rapidly to zero at the nuclear surface. The radii of all nuclei may be represented approximately by the formula

$$R = r_0 A^{-1/3} \tag{1}$$

where A is the mass number and r_0 is approximately 1.2 fm (1 fm = 10^{-15} m). The nuclear volume is approximately proportional to A or, equivalently, the density of 'nuclear matter' in the interior of a nucleus is nearly constant throughout the whole periodic table.

The position was not nearly so clear in 1931. Something was known about the size of light nuclei from the scattering of alpha particles and heavy nuclei from alpha-decay lifetimes. As early as November 1928 Gamow had pointed out that the radius of the aluminium nucleus obtained from alpha scattering and the radii of heavy nuclei deduced from alpha-decay lifetimes were consistent with the formula in equation (1). There was hardly sufficient data to prove Gamow's hypothesis, but the idea was consistent with the model of the nucleus which he presented at the Royal Society discussion in February 1929.

Gamow proposed that alpha particles played an important role in nuclear structure. He argued that nuclei contained a few 'loose protons' and a number of 'loose electrons' but that most of the protons and electrons in a nucleus were 'packed up' into alpha particles. At the time there was no way to understand the bewildering behaviour of electrons in nuclei. He writes in his book [11] 'For some unknown reason, although electrons in the nucleus behave in a peculiar and obscure way, this does not affect very much the laws governing the motion of the nuclear alpha particles and protons.'. There is more about the 'electron problem', especially in the context of beta decay in the chapters by Abraham Pais and Laurie M Brown of the present book (Chapters 2 and 5). Gamow was able to make progress by ignoring the nuclear electrons. Of course the 'electron problem' disappeared after the discovery of the neutron by Chadwick in 1932.

Gamow assumed that the alpha particles in a nucleus behaved like hard spheres interacting by a strong attractive force with a strength decreasing rapidly with distance. Such a collection of alpha particles would have properties similar to a small drop of liquid. On the basis of this model the binding energy and volume of a nucleus should be proportional to the number of alpha particles, a result which was consistent with Aston's mass measurements.

The alpha-particle model was soon to be superseded. In February 1932, Chadwick's letter giving evidence for the existence of the neutron appeared in *Nature*. Just four months later in June 1932, Heisenberg published the first of a series of three papers in which he put forward the model which formed the basis of later investigations of nuclear structure. He suggested that nuclei were composed of protons and neutrons and that nuclear structure could be described using the laws of quantum mechanics in terms of the interaction between the nuclear particles. But Gamow's ideas about nuclear matter carried over into the new theory. His assumption that the nuclear volume was proportional to the mass number was used by W Heisenberg [5] in 1932 for his discussion of nuclear stability. E Majorana [15] wrote in his 1933 paper on the neutron–proton model of the nucleus 'Thus one finds in the centre of an atom a sort of matter which has the same property of uniform density as ordinary matter. Light and heavy nuclei are built up from this matter and the difference between them depends mainly on their different content of nuclear matter.'.

See also p 372

In his third paper, Heisenberg assumed forms for the neutron–neutron, neutron–proton forces and applied the Thomas–Fermi method to study the properties of nuclear matter. He imagined the nucleus to be a gas of freely moving particles obeying Fermi statistics and held together by the nuclear forces, and tried to find a self-consistent effective potential. Heisenberg's theory was the first application of the independent particle model to nuclear structure and was a precursor of the shell model. Heisenberg used the theory to obtain an approximate expression for the binding energy of nuclear matter as a function of the mass number A. There was a difficulty with the theory: Heisenberg found that the nuclear binding energy as a function of the atomic weight A increased more rapidly than A^2 if the interaction was attractive for all values of the inter-nucleon separation. This result was in contradiction to the mass measurements of Aston. The constant density of nuclear matter and the proportionality of the binding energy to A are important facts which impose strong conditions on nuclear forces. Forces which reproduce these properties satisfy the 'saturation condition'. Heisenberg's force did not produce saturation.

In order to satisfy the saturation condition, Heisenberg had to assume that the nuclear force became repulsive at small distances. Another way out of the difficulty was proposed by Majorana in March 1933. Heisenberg was guided by an analogy when he was trying to find a suitable interaction between nuclear particles. He treated the neutron as a composite particle and assumed an exchange interaction similar to the one responsible for the binding of the hydrogen molecular ion. Majorana doubted the validity of this analogy and preferred an alternative approach. He wanted to find the simplest law of interaction which would produce saturation. He came up with the exchange interaction named after him and showed that it satisfied the above criterion. During the next few years, a number of physicists tried to calculate the properties of nuclear matter using forces derived from the theory of light nuclei. All of these attempts were unsuccessful and in their 1936 review article Bethe and Bacher concluded that 'the statistical [i.e. Thomas–Fermi] model was quite inadequate for the treatment of nuclear binding energies'. In fact it is very difficult to calculate the properties of nuclear matter from a realistic nucleon–nucleon interaction and Bethe and others spent the next half-century grappling with this problem.

In 1935 C von Weizsäcker [16] proposed a semi-empirical method for calculating nuclear energies. He assumed a form for the nuclear energy which was indicated by the statistical model, and which could be derived from simple qualitative considerations based on a liquid-drop description. However, in the formula certain constants were left arbitrary and were determined from experimental data. von Weizsäcker's approach was very successful and gave strong support to the liquid-drop model of nuclear structure.

15.2. Nuclear dynamics as a many-body problem

Heisenberg [5] imagined the nucleus to be a gas of protons and neutrons obeying Fermi statistics, moving independently but held together by strong nuclear forces. This Fermi-gas model was the standard picture of nuclear structure for two or three years, but it went out of favour in the mid-1930s. The change of viewpoint was a consequence of Fermi's neutron capture experiments and their interpretation by Bohr.

Soon after Chadwick had found the neutron and the Joliot-Curies had discovered artificial radioactivity, Fermi decided that neutron-induced reactions would be an ideal tool for studying nuclei. He had collected together a powerful group of young physicists in Rome and they decided to investigate artificial radioactivity produced by neutron bombardment [17]. The story is told in Laura Fermi's book [18] and in an article by E Segrè [19] written after Fermi's death in 1955. Experimental nuclear physics was a new field to the Rome group and several of its members had visited laboratories around the world [19] to learn techniques. The laboratories visited included those of R Millikan in Pasadena, P Debye in Leipzig, O Stern in Hamburg and Lise Meitner in Berlin. The first positive result of their neutron experiments was obtained with a fluorine target in March 1934. By July the Rome group had investigated about 60 elements. Of these, more than 40 could be activated by neutrons. In October the group found that the rate of neutron capture was greatly increased by the presence of materials containing hydrogen. They explained this in terms of the slowing of the neutrons to thermal energies by collisions with protons in the hydrogen and by the rapid increase in reaction cross sections at low velocities.

The first results on reaction rates seemed very erratic until it was realized that there were many sharp resonances in the cross sections. These results were a mystery until Bohr [20] formulated his compound nucleus theory of nuclear reactions. The advent of the compound nucleus model is described by Pais in Volume 1 of this history and by Stuewer in his article on Niels Bohr and nuclear physics [12]. Stuewer describes the movement away from the independent particle model of nuclear structure. In September 1934, Bethe gave a seminar in Copenhagen in which he presented a single-particle theory of nuclear reactions. From the time when the results of the first artificial disintegration experiments came out Bohr had wondered about the reaction mechanism. He was not happy with Bethe's approach. Six months later C Møller, who had just returned to Copenhagen from Rome, reported on the latest neutron capture experiments. Apparently, after about half an hour, Bohr suddenly took to the floor and offered a new interpretation of Fermi's results. By the end of 1935 Bohr had developed his compound nucleus theory of neutron-induced reactions [20].

See also
p 123

The change of viewpoint is described in the introduction to Bethe's 1937 review paper [21] on nuclear reactions. Bethe writes 'Bohr was the first to point out that every nuclear process must be treated as a many-body problem. It is not at all permissible to use a one-body approximation, particularly in the case of heavy nuclei.'. He goes on to say that when a neutron or proton falls on a target nucleus, it interacts strongly with the individual nuclear particles in the target. As a result the energy, which is initially concentrated in the incident particle, will soon be distributed among all the particles of the compound nucleus which consists of the particles in the original nucleus plus the incident particle. Bethe expressed the idea as follows 'Each of the particles of the "compound nucleus" will have some energy, but none will have sufficient energy to escape from the rest. Only after a comparatively long time the energy may "by accident" be again concentrated on one particle so that this particle can escape.'.

Although nuclear processes are governed by quantum laws, Bohr recognized that a classical picture could describe the essential physics of compound nucleus formation. He had small models made in the workshop of the Copenhagen Institute to illustrate his theory. They consisted of circular wooden dishes (representing the nucleus) filled with small ball-bearings (representing the nucleons). In 1937 Bohr lectured on the compound nucleus model in a number of universities in the USA, always with the help of his wooden model (figure 15.2). He described how it worked in an article in *Science* [22] in 1937 'If the bowl were empty, then a ball which was sent in would go down one slope and pass out on the opposite side with its original energy. When however there are other balls in the bowl, then the incident one will not be able to pass through freely but will divide its energy first with one of the balls, these two

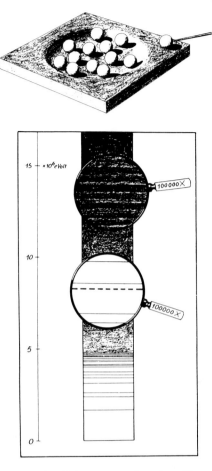

Figure 15.2. *Bohr compound nucleus diagrams.*

will share their energy with others, and so on until the kinetic energy is divided among all the balls.'. According to quantum mechanics a bound nucleus has stationary energy states. Bohr recognized that a nucleus with an excitation energy of several MeV should have a series of closely spaced energy levels. These would have essentially the same character below and above the dissociation energy for break-up into a neutron (or proton) and a residual nucleus.

In 1935 Bethe and Fermi had developed theories of neutron capture based on an independent particle picture. They wanted to understand why slow neutrons could be easily captured by many nuclei. They represented the interaction between a neutron and a nucleus by a potential well and calculated the capture cross section by quantum mechanics. In the limit of low neutron velocities, they found that the capture cross section was inversely proportional to the neutron velocity. This result was in agreement with some aspects of the experimental data but not with others.

Breit and Wigner, working at Princeton in 1936, proposed a mechanism [23] in which the neutron was captured into a quasi-stationary (virtual) energy level, which could then decay either by re-emitting a neutron or by gamma emission. Their result was the famous Breit–Wigner formula for the reaction cross section of a resonance reaction. Bethe described the idea as follows 'If the energy of the incident particle is equal, or nearly equal, to one of the energy levels of the compound nucleus, the probability of the formation of the compound nucleus will obviously be much greater than if the energy of the particle falls in the region between two resonance levels. Therefore we shall find characteristic fluctuations of the yield of every nuclear process with energy, from high values at the resonance energies to low values between resonance levels.'. In spite of the complexity of the compound nucleus, the simple Breit–Wigner formula for the reaction cross section gave a quantitative account of its variation with energy. The resonance energies, total widths and partial widths were parameters which could be fitted to experimental data.

Physicists soon understood that the Breit–Wigner theory brought Bohr's semi-classical picture of the compound nucleus into agreement with the postulates of quantum mechanics. Jensen was in Copenhagen at the time and described the impact of the Breit–Wigner formula in his Nobel Prize Lecture in 1963 'It originated outside Copenhagen, but it soon could be seen on every blackboard of Niels Bohr's institute.'.

The compound nucleus picture fitted in well with the liquid-drop model. It implied a strong coupling between the nucleons in a nucleus and seemed to rule out any treatment in terms of independently moving particles. In his 1936 paper Bohr wrote 'In contrast to the usual view, where the excitation is attributed to an elevated quantum state of an individual particle in the nucleus, we must in fact assume that the excitation will correspond to some quantized collective motion of all the nuclear particles.'. This idea was elaborated by Bohr and Kalkar in 1937. They proposed [24] that there could be excited states of a nucleus corresponding to the surface oscillations of the nuclear 'liquid drop'.

Bohr's compound nucleus model was invented to describe the experimental results of Fermi and his collaborators on neutron capture reactions. The Rome group had set out to study neutron capture in all the elements they could lay their hands on. In particular, by bombarding uranium with neutrons they hoped to produce elements with atomic numbers larger than 92. Irène Joliot–Curie in Paris and Lise Meitner and Otto Hahn in Berlin also took up this line of research. From the beginning the results for uranium were confusing. Many radioactive periods were found and radiochemical methods were used in an attempt to assign atomic numbers to the reaction products without any clear conclusion.

Hahn and Meitner had been collaborators off and on for almost 30 years and had done important work on the beta and gamma spectra of radioactive nuclei. Fritz Strassmann, a young analytical chemist, joined their uranium research project and contributed to the radiochemical analysis. The chemistry was difficult because of the minute size of the samples and also because elements like radium and barium with quite different mass numbers have very similar chemical properties. In June 1938, about four years after the project had started, Meitner, who was Jewish, had to leave Berlin. Until then she had been protected from Nazi persecution because of her Austrian nationality, but when the German army marched into Austria on 12 March 1938 her position changed overnight. She moved to Copenhagen (where her nephew Otto Frisch was working in the Niels Bohr Institute) and then, in July, to Stockholm to join K Siegbahn's laboratory. Hahn and Strassmann continued their studies in Berlin and soon found that a group of at least three radioactive products, formed from uranium under neutron bombardment, were chemically very similar to barium. At first they thought that they had produced isotopes of radium, but further experiments forced them to conclude that isotopes of barium ($Z = 56$) were formed as a consequence of bombardment of uranium ($Z = 92$) with neutrons.

Hahn and Strassmann wrote up their results for publication [25] and sent the news to Meitner on 19 December 1938. She and Frisch spent Christmas together on the west coast of Sweden. They discussed the outcome of the Berlin experiments and arrived at an interpretation which was published [26] in a note to *Nature* on 11 January. They wrote 'On account of their close packing and strong energy exchange, the particles in a heavy nucleus would be expected to move in a collective way which has some resemblance to the movement of a liquid drop. If the movement is made sufficiently violent by adding energy, such a drop may divide itself into two smaller drops.'. After division the drops would repel each other and gain a kinetic energy of about 200 MeV. The whole 'fission' process could be described in an essentially classical way.

See also p 129

On returning to Copenhagen, Frisch conceived an experiment to confirm the 'fission' interpretation. He measured the ionization produced by the highly charged fission fragments. The experiment, which he finished within a week, was consistent with the fission hypothesis, and the results were published in *Nature* [27] on 18 February 1939. Frisch discussed the fission hypothesis with Bohr early in January. Bohr accepted it immediately. A few days later he set off on a visit to the United States taking the news with him. It made an instantaneous impact and on 29 January there was even a note in the New York Times about the discovery. While in America, Bohr collaborated with John Wheeler to produce a detailed theory of the fission process. Their paper [28] was submitted to *Physical Review* at the end of 1939.

15.3. The effects of World War II

World War II had a profound impact on the post-war development of nuclear physics. This section focuses on two points. The first is to outline how the wartime activities led to the creation of national laboratories in the post-war period. The second is to mention some specific areas where there were major advances in experimental techniques and in the understanding of nuclear dynamics resulting from the wartime programme.

Before the war there was a massive migration of scientists. Many physicists, chemists and biologists left Germany, and other continental European countries for the United States, for Britain and for different countries throughout the world. Many of these scientists were later involved in defence work. During the war others moved from the UK, Denmark and France to the USA or Canada to participate in defence programmes. These scientists were based in various laboratories in the USA and Canada and established contacts which persisted and developed into collaborations when the scientists returned to their home institutions at the end of the war. This had a profound influence on the way in which nuclear physics progressed in the post-war period.

Scientists in the United States began to think about a nuclear defence programme early in 1939 [29, 30]. Fermi had moved to the United States at the end of 1938 and joined the Physics Department of Columbia University. After Niels Bohr had brought the news of the discovery of fission to New York in January 1939, Fermi and Herbert Anderson, with the collaboration of the head of the physics department, Professor Pegram and the director of the cyclotron laboratory, J R Dunning, began to make fission experiments with the Columbia University cyclotron [18]. They were joined by the Hungarian physicist Leo Szilard and the Canadian Walter Zinn. Quite soon they became concerned that fission could be the basis of a nuclear weapon and in March 1939 made the first attempts to establish contacts with the government. In October President Roosevelt appointed an Advisory Committee on Uranium. In December 1941, just before Pearl Harbour, there was a decision by the United States to make an all-out effort in atomic energy research. This was soon to become the huge Manhattan Project. This was after the United States Government had been shown, in July 1941, the British Maud Report which demonstrated how and why an atomic bomb was possible [31, 32].

Almost immediately a number of nuclear physics defence laboratories were set up. Already in June 1940 President Roosevelt had established the National Defense Research Council, with Vannevar Bush as its head. In October 1941 MIT was chosen as the site for the new radiation laboratory for radar research. Nuclear programmes started at the Metallurgical Laboratory of the University of Chicago in the first half of 1942, at Los Alamos, New Mexico, early in 1943, at the Clinton

Figure 15.3. *The pile in Chicago, in which Fermi achieved, for the first time on 2 December 1942, a man-made chain reaction. (To the left, Dr H W Newson, a member of the original Fermi team.)*

Laboratories in Oak Ridge in late 1943 and at the Hanford Engineering works in Washington State in 1944. From April 1942 until summer 1944 Fermi was based in Chicago.

In 1941 plutonium was discovered at Berkeley. The story of the discovery is described in section 15.8 of this chapter. It was recognized that neutrons produced in the fission of ^{235}U might allow the conversion of natural uranium into plutonium. Work on plutonium was continued at the Berkeley cyclotron until the early part of 1942 and then transferred to the Metallurgical Laboratory at Chicago. At the same time research on reactor physics began. The first atomic pile (nuclear reactor) began operating in a squash court in Chicago in December 1942 (figure 15.3). The trial atomic bomb was exploded at Alamogordo, in New Mexico, on 16 July 1945 and the two nuclear bombs were dropped on Japan in August 1945, just before the end of the Pacific war.

After the end of the war, the United States Government decided to continue nuclear research both for military and civil purposes. Some of the wartime laboratories were converted to National Research Laboratories for the development of nuclear power and other nuclear applications. The Metallurgical Laboratory in Chicago was the precursor of the Argonne National Laboratory. Los Alamos and Oak Ridge also developed into National Laboratories. They had strong physics divisions where important research in nuclear physics was carried out.

Research laboratories were also founded in other countries. The origin of the Chalk River Laboratory in Canada and Harwell in England is particularly interesting. Some British scientists had ideas about nuclear weapons soon after the discovery of fission. Margaret Gowing tells the story in her book *Britain and Atomic Energy 1939–45* [31]. The Maud Committee was set up in 1940 to report on the feasibility of the uranium bomb. There was a period of intense and productive work by British scientists over 15 months which resulted in the presentation of the Maud Report in July 1941. In 1939–40 a French group under J F Joliot were working on ideas for a nuclear reactor project in France and had succeeded in obtaining a supply of heavy water from Norway for use as a moderator. Just before the fall of France two French nuclear physicists, H von Halban and L Kowarksi, succeeded in escaping to England with 185 kg of heavy water.

Contacts between British and American groups began with a visit of Sir Henry Tizard and John Cockcroft to Washington DC in the autumn of 1940. Canada also became involved with the British project as a supplier of uranium and also because Canadian physicists were interested. Canada was considered as a possible site for a British atomic energy project. After the Maud report had been submitted there was a question of cooperation between the British and American projects. There were many difficulties and misunderstandings and the cooperation problem took two years to resolve. The final result was the Quebec Agreement of August 1943. The immediate result was that groups of British scientists joined various American projects at Los Alamos, Oak Ridge and Berkeley. Even before the Quebec Agreement the British and Canadians and the French group started a laboratory in Montreal to continue work on a heavy-water moderated reactor. In the end, after complex negotiations between the United States, Canada and Britain, there was an agreement in April 1944 that a project to build a heavy-water reactor in Canada should go ahead. This led to the establishment of the Chalk River Laboratory which had a staff of Canadian, French and British physicists. It started late, the small pilot reactor ZEEP had not even gone critical by the end of the Pacific war, but the project became the cornerstone of the British and Canadian post-war atomic energy programmes. British scientists gained experience at building and operating reactors. Cockcroft went straight from being director at Chalk River to being director at Harwell. British and Canadian physicists worked closely together over a number of years even after the establishment of the Harwell Laboratory. Harwell was the leading nuclear physics laboratory in Britain for a number of years in the post-war period.

During the war many technical advances were made and new

knowledge was acquired. The advances were particularly striking in several fields, notably in neutron physics, and in isotope production and separation. The defence work focused on military applications but after the war the methods evolved could be applied in other areas. Neutron sources developed out of all recognition during the war period. In the early years of the war Fermi and other physicists at the Columbia Cyclotron developed neutron sources 100 000 times more intense than the ones they had used in Rome, by bombarding beryllium with deuterons. Later, in 1943–4, at the Metallurgical Laboratory in Chicago they had neutron sources from the nuclear reactor ('pile') which were 100 000 times stronger than the ones in Columbia. With these sources they were able to produce macroscopic amounts of radioactive isotopes. The emphasis was on the production of plutonium, but the knowledge gained there could be applied later to the production of other radioisotopes for research purposes and for applications in medicine and industry.

High-intensity cyclotrons could be used directly for the production of significant amounts of radioisotopes. This was done first be Seaborg and his collaborators in Berkeley where they produced about 0.5 micrograms of plutonium by bombarding a uranium target with 3500 microampere hours of deuteron current. Later on it became a standard method for producing other radioisotopes.

In the course of the defence programme neutron cross sections were measured for many elements with high-intensity neutron sources. A collection of cross sections was compiled in 1945 for the Manhattan Laboratories by Goldsmith and Ibser [33] at the Metallurgical Laboratory of the University of Chicago. An updated version of this compilation was eventually published in the *Review of Modern Physics* in 1947 [33]. In the fission of uranium many unstable isotopes are produced. About 60 fission chains were identified and studied during the war. Many of the isotopes produced were isolated and their masses and lifetimes were measured. The results were eventually published in the Plutonium Project Report in 1946 [34].

15.4. Technical advances

15.4.1. *Techniques available in 1930*

Theoretical advances in physics have often been associated with the development of new experimental techniques. Since nuclear physics began there has been a continuous effort to find new sources of projectiles for nuclear reactions and new methods for detecting reaction products. This section describes the history of some of the breakthroughs made in the period 1930–65. The book *Radiations from Radioactive Substances* by Rutherford, Chadwick and Ellis [35] gives a detailed account of the techniques available in 1930 just before the discovery of the neutron and the invention of accelerators in 1932.

Until 1932 projectiles used in nuclear reaction studies were from natural radioactive sources. One popular material was polonium (RaF) which could be electrochemically separated from parent materials and supplied a pure source of monoenergetic alphas with an energy of 5.30 MeV. Higher energy alphas (7.68 MeV) were obtained from RaC'. One of the most useful gamma-ray sources was ThC" which has a single strong line of 2.62 MeV.

In 1930 four kinds of devices were in common use for the detection of charged particles. They were photographic plates, zinc sulphide scintillators, Geiger counters and the Wilson cloud chamber. Becquerel made his original discovery of radioactivity by observing its action on photographic plates. In the period up to 1930, and even beyond, photographic plates were used as detectors in various kinds of particle spectrometers. They could also be used in another way. Alpha particles and other charged particles leave tracks in photographic plates, and studies initiated by Kinoshita [36] in 1910 showed that charged particle tracks were made clearly visible by the succession of grains along the track and that reactions occurring in the plate could be studied by measuring the tracks.

The scintillation method for counting alpha particles with zinc sulphide as a scintillator was devised by Regener [37] in 1908. It was used in the original experiments on alpha scattering by Geiger and Marsden [38] in 1908–9 and zinc sulphide was still an important detector in 1930. The scintillations were counted visually using a microscope with a large numerical aperture to increase their apparent brightness. The book of Rutherford, Chadwick and Ellis has a section on methods for determining the efficiency of an observer in counting scintillations.

The first steps towards electrical counting methods were made by Rutherford and Geiger in 1908 [39]. They made use of a property discovered by Townsend, that an ion moving in a strong electric field in a gas at low pressure, produces an avalanche of fresh ions by collision with gas molecules. By suitable adjustment of the conditions the ionization due to a single alpha particle yielded a measurable electric current. Geiger subsequently developed variants of the method which gave much greater sensitivity and the Geiger counter became one of the most important detection devices for charged particles up until about 1950. The importance of the Geiger counter increased when electronic counting methods were introduced at the end of the 1920s. At the same time other kinds of ionization chambers were developed detecting individual charged particles and measuring their energies (Ward, Wynn-Williams and Cave 1929 [40]). The Geiger counter could also detect gamma rays.

The expansion method resulted from the discovery by Wilson [41] that ions produced in gases by the passage of charged particles act as nuclei for the condensation of water vapour when a certain supersaturation condition is reached. Wilson found by an application of this method that tracks of individual alpha and beta particles could be clearly seen and photographed. The Wilson cloud chamber was especially important for the early studies of nuclear reactions by Blackett in 1925 [42] and others. Particle energies could be measured in a cloud chamber by using range–energy relations.

The energies of charged particles were measured by deflection in magnetic fields. The first measurements on alpha particles were made by Rutherford in 1903 and there was a steady improvement in techniques over the next few decades. The discovery of monoenergetic groups of electrons emitted by beta radioactive substances was due to the work of von Baeyer, Hahn and Meitner [43] in 1911. They studied the deflection of beta rays in a weak magnetic field. In fact these homogeneous groups of electrons were not the disintegration electrons at all, but were internal conversion lines associated with gamma transitions. These groups proved to be of the greatest importance for measuring gamma-ray transition energies. The first focusing beta-ray spectrometer was invented by Danysz in 1913 and developed by Rutherford and Robinson [44, 45]. By 1950 many sophisticated beta spectrometers were available. Gamma-ray wavelengths were first measured directly using crystal diffraction by Rutherford and Andrade in 1914 [46].

Initially, alpha-particle energies were more difficult to measure because stronger magnetic fields were needed to produce a significant deflection. Techniques improved and in 1914 Rutherford and Robinson [45] were able to measure the energies of alpha particles from RaC with an accuracy of better than 1%.

15.4.2. *Accelerators*

Over the years the cyclotron and the synchrocyclotron have been amongst the most important accelerators for nuclear physics research. Their history was reviewed by Livingston in 1952 [47] and more information about the early period can be found in his article. The book by Livingston and Blewett [48] is an interesting source of information about accelerators in general up to 1962.

The cyclotron was developed in Berkeley, California, by Ernest O Lawrence. The first machine built by Lawrence and Livingston in 1931 had pole faces 11″ in diameter and produced a 1.2 MeV proton beam. Soon it was superseded by one with poles of 37″ diameter which could produce 16 MeV alpha particles. Lawrence and his co-workers continued with the design of larger instruments and in 1939 completed the 60″ machine which could produce 8 MeV protons, 16 MeV deuterons and 38 MeV alpha particles (figure 15.4). The Berkeley 60″ machine was the first of the big cyclotrons and set the pattern for others of this size. In 1952 there were 14 cyclotrons in England and Europe and 21 in the United States. Most of them were in scientific laboratories and supported

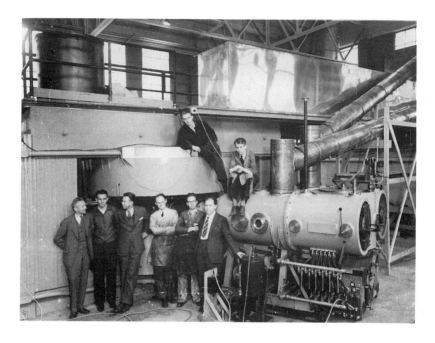

Figure 15.4. *The 60" cyclotron as installed at the Radiation Laboratory. Left to right top; Alvarez and MacMillan: left to right bottom; Cooksey, Lawrence, Thornton, Backus, Salisbury.*

research in nuclear physics or produced radioactive tracer isotopes for research in chemistry, biology or medicine.

There is a relativistic limitation on the energy of a fixed frequency cyclotron which was reached in the 60" machines. In 1945 McMillan [49] suggested the use of frequency modulation to overcome the relativistic limitation. Veksler in the USSR discovered important stability properties of the orbits in such a machine. It could be shown that when the applied frequency is decreased slowly the particles follow this change adiabatically so as to gain energy slowly and remain in resonance. The first of the frequency modulated machines was the 184" synchrocyclotron which began operating in Berkeley at the end of 1946. It produced 350 MeV protons, 200 MeV deuterons and 400 MeV alphas. By 1952 there were seven synchrocyclotrons operating throughout the world. At the end of World War II there were four cyclotrons in Japan, but they were destroyed by the United States occupation forces.

There was a further advance in 1959 with the invention of the spiral ridge cyclotron. This is a fixed-frequency machine and relativistic variation of mass with energy is compensated by a magnetic field which varies with angle and radius in a particular way. One of the first of its kind was an 88" machine built in Berkeley in 1962. The average beam intensity in a spiral ridge machine is 100 times larger than in a synchrocyclotron.

Another type of accelerator which has been widely used for nuclear physics research is the van de Graaff electrostatic generator. There is an interesting review by van de Graaff, Trump and Buechner [50] which gives a picture of the state of the art in 1948 just fifteen years after the first experiments were made with a van de Graaff machine. The first large machines were completed at the Department of Terrestrial Magnetism of the Carnegie Institution by Tuve *et al* in 1935 and at MIT by van de Graaff and his colleagues in 1936 (figure 15.5). They had terminal voltages of about 3 MeV. The potential developed in an electrostatic machine is limited by corona breakdown and it was found that the maximum potential developed could be increased by operating the machine under pressure in a gaseous insulating medium. Herb and his associates at Wisconsin obtained remarkable results with a generator operating in an atmosphere containing freon. By 1948 sulphur hexafluoride was the favoured insulating gas.

P H Rose wrote an interesting article on the origin of the tandem accelerator in a book dedicated to L W Alvarez [51]. The information in this paragraph comes from there. Rose writes that the private company High Voltage Engineering (HVE) was founded in 1946 to build and market electrostatic generators. Denis Robinson was the president and Robert van de Graaff and John Trump were associated with it. HVE built a very effective 6.5 MeV machine for the Oak Ridge National Laboratory in 1951. With this success HVE stole the initiative from university and government laboratories and for 20 years was the world leader in direct current accelerators. In the mid-1950s they developed the tandem accelerator which became one of the most widely used machines for low-energy nuclear physics research. In a tandem, negative ions from an ion source are accelerated towards the positive high-voltage terminal. Inside the terminal the ions pass through a region of higher gas pressure where collisions with the gas molecules strip off electrons converting the negative ions into positive ions. The acceleration continues by repulsion of these positive ions away from the high-voltage terminal. The idea of the tandem had been patented by Bennett in 1950. The concept was reinvented by Alvarez who promoted it enthusiastically and encouraged Robinson to develop it as a new HVE product. The first machine was sold to the Chalk River Laboratory and delivered in 1958. It had a terminal voltage of 5 MeV and could accelerate protons to 10 MeV and heavy ions to a higher energy. The machine was simple to install and use. It proved to be very popular and a total of 62 were built and sold to laboratories throughout the world.

Two other kinds of accelerators have had a significant impact on nuclear structure physics. They are the betatron and the linear accelerator. A third, the synchrotron, has been very important for particle

Figure 15.5. *The 2 m van de Graaff generator at the Carnegie Institution. Merle Tuve is in the suit.*

physics, but not as much for nuclear physics.

The betatron was invented by D Kerst (cf reference [48]) at the University of Illinois and in 1940 he built a 2.3 MeV machine in his laboratory there. It was a circular machine with a special magnetic field configuration and particles were accelerated by magnetic induction. The design was based on detailed theoretical calculations by Kerst and Serber. Once the basic principle had been understood these machines were very easy to construct and could accelerate electrons up to much higher energies. Kerst transferred his base of operations to the General Electric Company (GEC) Research Laboratories where a 20 MeV machine was completed in 1942. Westendorp and Charlton of GEC continued the development with a 100 MeV machine in 1945. Commercial production was started at GEC and in other companies. The commercial machines were mainly in the 20 MeV range and were built to meet the rising demand from research laboratories, hospitals and industrial plants. Kerst returned to Illinois to build a 300 MeV model, the largest and possibly the ultimate betatron. Nowadays betatrons are hardly used for nuclear physics research but in the 1940s they provided the first source of high-energy gamma rays. These were x-rays (bremsstrahlung) produced when the high-energy electrons impinged on a tungsten target. The energy could be varied by changing the betatron energy. In 1947 experiments with high-energy gamma rays from the 100 MeV betatron at GEC lead to the discovery of the collective giant dipole resonance in nuclei.

The first proton linear accelerator was built by L Alvarez and his group in Berkeley in 1947. There is a description of the construction of this accelerator by W Panofsky in reference [52]. A low-energy linear accelerator had been built by Sloan and Lawrence in 1931. Interest was renewed towards the end of World War II due to the many advances which had been made in the development of very high frequency power sources. In fact both Lawrence and Alvarez had worked on power sources and transmitters for microwave radar at the MIT Radiation Laboratory from 1940 to 1943 [53]. One of the things which motivated Alvarez to push ahead with linear accelerators were cost estimates for high-energy accelerators. He argued that the cost of a cyclotron would increase roughly with the third power of the energy, while for a linear accelerator the energy–cost relation would be approximately linear. Thus eventually, as one went to higher energies, the linear accelerator would replace the cyclotron on economic grounds. This argument was correct, but irrelevant after the invention of the strong focusing synchrotron. Alvarez's second motivation was the availability of surplus radar equipment after the war. The machine was built and operated successfully to produce a 32 MeV proton beam. It was used for fundamental proton–proton scattering experiments and for studying various nuclear reactions. Several other linear proton machines were built, the most notable being the 'Meson Factory', LAMPF, at Los Alamos which was the leading tool in medium-energy nuclear physics for many years.

Linear accelerators proved to be much more important for electrons than for protons. The linear accelerator is particularly suited to producing high-energy electrons. In a circular machine the continual acceleration of the electrons towards a centre by the magnetic field produces enormous radiation losses. These are avoided in the linear machine. Another advantage of the linear accelerator is the excellent collimation of the emergent beam. Stanford has always been a centre for high-energy electron linear accelerators. Initially the driving force was W W Hansen who in the mid-1930s was thinking about this kind of machine. It became a practical possibility with the advent of magnetrons developed for radar applications. Work on these accelerators also began at the end of World War II and a 2 m machine was constructed in 1948 which produced a 4 MeV electron beam. Progress in those early days was described in a review article by Fry and Walkinshaw in 1948 [54]. The Stanford Mark I machine [48] was built in 1947. It was 12 ft long, was powered by magnetrons and produced a 6 MeV beam of electrons. Then the Stanford team were able to develop high-powered klystrons which had much better phase stability properties. The Stanford Mark II was powered by these klystrons and yielded 35 MeV electrons. This was the prototype for the Mark III, designed for electron energies up to 1 GeV, and consisted essentially of 30 Mark II accelerators arranged in a line. It was finished in 1960 but was built in stages and was used at lower energy for important nuclear structure studies by Hofstadter and collaborators from 1953 onwards.

15.4.3. *Neutron sources*

Between 1934 and 1942 the standard neutron source was a sealed tube containing Be powder and Ra gas. The radon is alpha active and the alphas produce neutrons from the reaction

$$^4\text{He} + {}^9\text{Be} \rightarrow {}^{12}\text{C} + \text{n}.$$

Most of the experiments of Fermi and collaborators, Hahn, Meitner and Strassmann and others were all made with this kind of source. Radon and its daughters have a high gamma emission which contaminates the neutron source. This contamination can be removed by replacing the radon by polonium as the source of alpha particles.

The development of the fission reactor in 1942 provided a new, high-intensity neutron source. Cutting a small hole in the shielding of a reactor allows a beam of neutrons to be extracted for experimental purposes. The high neutron flux from a reactor is particularly useful for the production of radioisotopes by neutron capture.

15.4.4. *Measuring devices and detectors*

Magnetic spectrometers measure the momenta of charged particles by deflection in a suitable arrangement of magnetic fields. Already by the beginning of World War II the designs had been developed to quite a high level of sophistication and used in mass spectrometers and in beta-ray spectrometers. After the war spectrometers began to be used in association with accelerators for the analysis of the products of nuclear reactions produced by the accelerator beams; also solutions to accelerator problems contributed to the design of spectrometers and vice versa. In particular the Kerst–Serber [48] theory of orbit stability in a cyclotron or betatron was used in the design of spectrometers and the principle of the Siegbahn double-focusing beta spectrometer [55] was adapted to the design of spectrometers for accelerators.

Another technique which is now of historical interest is crystal diffraction spectroscopy for measuring gamma-ray wavelengths. The method was first used by Rutherford and Andrade [46] in 1914, and was later developed by other workers. The most successful device was the curved crystal focusing spectrometer proposed by DuMond and Kirkpatrick in 1930 [56]. A spectrometer based on this principle was designed in 1947 and built by DuMond and his collaborators. It could measure wavelengths of gamma rays in the 0–500 keV range with an accuracy of better than 0.01%. Now semiconductor counters can measure gamma-ray energies with comparable accuracy.

Several major advances in detector technology were made after World War II. One of these was the invention of proportional counters. These arose from a refinement of the Geiger counter. If an avalanche counter was carefully designed and operated under controlled conditions, then it could act as a detector for beta or gamma rays and, at the same time measure the energy of the radiation. The status of proportional counters in 1950 is reported in the book by D H Wilkinson [57]. By 1960 this kind of counter had been largely replaced by semiconductor counters.

A method of counting ionizing radiations by photoelectric measurement of the fluorescent light pulses emitted by certain organic substances was first pointed out by Kallman in 1947. It was a development of the old scintillation method, but the scintillations were counted electronically by a photomultiplier rather than the eye. Researchers immediately began to search for scintillating materials which would work well in conjunction with photomultipliers. In 1948 Hofstadter [58] found that sodium iodide crystals activated by thallium could be used as an efficient detector of gamma rays. Hofstadter was not searching at random. He knew the properties which a suitable phosphor should possess. The importance of impurities was known to Rutherford and his colleagues and in section 126 of reference [35] there is a statement 'It is well known that pure zinc sulphide does not scintillate when bombarded with alpha particles nor does it fluoresce under the action of light. In order to sensitize the zinc sulphide, the insertion of a minute quantity of impurity, such as copper, is necessary.'. The phosphorescent properties of alkali halides had already been studied in the 1930s. Sodium iodide which had a good stopping power for gamma rays and could be grown into large crystals seemed to be a promising choice. Until 1949 the Geiger counter had been the most commonly used detector for gamma rays. Within the space of a couple of years it had been replaced by the sodium iodide detectors, and counting efficiencies had been increased by a factor of 1000. Many important experiments made in the early 1950s would have been impossible without the new detectors. Sodium iodide detectors are still in common use as gamma detectors, often in combination with semiconductor detectors.

The scintillation detector combined high detection efficiency with moderate energy resolution. If high energy resolution was required then a beta-spectrometer or a crystal diffraction technique still had to be used. This situation changed with the introduction of semiconductor counters based on silicon or germanium. References [59] and [60] review the development of these detectors. Already in 1949 McKay [61] had made the first steps towards designing a semiconductor counter based on germanium and in 1953 he explored the use of silicon. A group at Purdue [62] was also actively interested. The turning point was a successful application of germanium junctions in a low-temperature nuclear physics experiment by Walter, Dabbs and Roberts in 1958 [63]. Then a number of laboratories in Europe and North America began intensive development of semiconductor particle detectors. These were in common use by 1962. Next came lithium drifted germanium, Ge(Li), gamma-ray detectors which were used widely by 1967. For the past 25 years most high-resolution detectors for charged particles and gamma rays have been based on semiconductor technology. The Ge(Li) detectors for gamma rays are less suitable than sodium iodide scintillators when the highest efficiency is required but the energy resolution is better by more than an order of magnitude and line widths of 1 keV can be obtained. The energy resolution of silicon particle detectors is not as good as for magnetic spectrometers, but for most purposes such high resolution is not required.

15.5. Shell structure in nuclei

The 1936 review article of Bethe and Bacher [64] gives an account of nuclear structure theory just four years after the discovery of the neutron. It was recognized that nuclear forces saturated so that nuclear binding energy and the nuclear volume were approximately proportional to the mass number A. The variation of nuclear binding energies with neutron and proton numbers N and Z was understood at the level of the Weizsäcker mass formula. Physicists knew that at least two different kinds of force could produce saturation. One possibility was a van der Waals type force which was attractive at large distances and repulsive at small distances. Another was an exchange interaction. There was also a pairing effect which made nuclei with N and Z both even somewhat more stable than odd nuclei. The 1936 review discussed the theory of the deuteron and included calculations of neutron–proton scattering, the capture of neutrons by protons and the photoelectric dissociation of the deuteron. Experimental data on these reactions was becoming available from the new accelerators built around 1932. The review also contained a section on the theory of the triton, He^3 and the alpha particle. The authors concluded that the binding energies of these nuclei were consistent with a short-range force between nucleons. The neutron–neutron force had to be approximately equal to the proton–proton force (charge symmetry). Experimental binding energies were available from mass measurements and nuclear reaction data from the new accelerators.

After an analysis of the Fermi-gas model for nuclear matter, Bethe and Bacher discussed a Hartree or 'shell' model for the structure of heavier nuclei. The model assumed independent motion of individual protons and neutrons. The authors wrote 'This assumption can certainly not claim more than moderate success as regards the calculation of nuclear binding energies. However, it is the basis for a prediction of certain periodicities in nuclear structure for which there is considerable experimental evidence.'. This evidence related to the strong binding of O^{16} and to breaks in the abundance of elements when plotted against $I = N - Z$. There was some evidence for anomalies at $N = 82$ and $N = 126$. The regularities were noticed by Bartlett [65], Elsasser [66] and Guggenheimer [67]. In 1933–4 Elsasser considered the energy levels of neutrons and protons in a potential well. The levels were grouped together in 'shells'. Neutrons and protons filled up the levels in order of increasing energy, taking into account the restrictions of the Pauli principle. Whenever a shell was completed a particularly stable nucleus could be expected. When a new shell started the binding energy of the newly added particles should be less than for the particles completing the previous shell.

The Elsasser [66] shell model predicted closed shells for N or $Z = 2$, 8 and 20. These fitted in with the binding energy and abundance data. However, the model failed to give closed shells at $N = 82$ and $N = 126$. Not long after this Niels Bohr formulated his compound nucleus theory. This was a strong interaction model and seemed to be incompatible with independent particle motion. As a result the shell model was abandoned for more than a decade.

The jj-coupling shell model was discovered independently by Maria Goeppert Mayer [68] and Haxel, Jensen, and Suess [69] in 1949. There are interesting historical perspectives in the Nobel Prize Lectures of Mayer [70] and Jensen [71] and in lectures delivered by D Kurath [72] and H A Weidenmüller [73] at a symposium held at the Argonne National Laboratory in 1989 to celebrate the 40th anniversary of the shell model. Maria Mayer studied with Max Born in Göttingen. After taking her doctorate she married an American, Joseph Mayer, and moved to the United States in 1930. There she could not get a job because of antinepotism rules in universities, but managed to continue studying physics with the help of various associates. In 1946 Joseph Mayer moved to the University of Chicago and Maria was able to get a half-time position at the newly established Argonne Laboratory. Hans Jensen studied in Hamburg and during the 1930s spent some time at the Niels Bohr Institute in Copenhagen. After the war he had a position in Hannover and began collaborating with Haxel and Suess.

Around 1947 Maria Mayer worked with Edward Teller on the

origin of the elements and, as a first step, began analysing data on isotopic abundances. Haxel, Jensen and Suess also became intrigued by regularities in nuclear binding energies as evident, for instance, from the abundance distribution of the elements. In 1948 there was much more information available than there had been in 1934 and the picture was clearer. In 1948 Maria Mayer [74] published evidence for the particular stability for the numbers 20, 50, 82 and 126. Part of the evidence was the small neutron absorption cross sections for targets with 82 or 126 neutrons as measured by D J Hughes at Argonne. The ideas of Elsasser and others in the 1930s were revived and the search was on for a suitable shell model. Several ideas were tried out and then Mayer and Jensen came up, independently, with the jj-coupling shell model which fitted the observed 'magic numbers' (figure 15.6). Maria Mayer wrote in her Nobel Prize Lecture [70] 'At that time Enrico Fermi had become interested in the magic numbers ... One day as Fermi was leaving my office he asked: "Is there any evidence of spin–orbit coupling?" Only if one had lived with the data as long as I could one immediately answer "Yes, of course and that will explain everything.". Fermi was sceptical, and left me with my numerology.'. Later in the same lecture she said 'After a week when I had written up the other consequences carefully Fermi was no longer sceptical. He even taught it in his class in nuclear physics.'.

Maria Mayer's 1949 paper already noted some of the predictions of the shell model. One related to the spins and parities of odd-A nuclei. Already in 1949 some were known. In the next few years many spins and parities of both ground and excited states were measured. Gamma-ray angular correlation was one technique for measuring spins of nuclear states. This method was being developed around 1948, but after the advent of sodium iodide scintillation counters the experiments became much more practicable. Measured spins and parities could be compared with shell-model predictions.

One class of excited states was isomeric levels. They had very long lifetimes: hours, days or even years. The explanation was that the spins of the isomeric states and the ground state were very different and the gamma transition rates were hindered by angular momentum selection rules. The pattern of levels in the jj shell model was such that nuclei in particular regions of the periodic table were expected to have isomeric states. These regions were called islands of isomerism. The multipolarities and hence the lifetimes of the de-excitation gamma transitions could be predicted. These predictions of the jj shell model were in striking accord with experimental data.

A necessary condition for the validity of the shell model was that the mean free path of a nucleon inside the nucleus should be large compared with the nuclear radius so that the single-particle orbits could be well defined. This appeared to be in contradiction to Bohr's theory of nuclear

Figure 15.6. *Maria Goeppert-Mayer and Hans Jensen.*

reactions. In fact the success of Bohr's theory had encouraged theorists to abandon the shell-model description of nuclear structure. In his 1936 paper [20] Bohr explicitly rejected theories of nuclear structure based on independent particle motion. Referring to the shell model he wrote '... it is, at any rate, clear that the nuclear models hitherto treated in detail are unsuited to account for the typical properties of nuclei for which, as we have seen, energy exchanges between individual nuclear particles is a decisive factor'. He continued later in the same paragraph 'In the atom and in the nucleus we have indeed two extreme cases of mechanical many-body problems for which a procedure of approximation resting on a combination of one-body problems, so effective in the former case, loses any validity in the latter where we, from the very beginning, have to do with essentially collective aspects of the interplay between the constituent particles.'.

In 1949 the shell model was accepted immediately in spite of the apparent conflict with Bohr's views stated in the last paragraph. There were probably two reasons for this: (i) the model could explain the magic numbers and (ii) it gave wavefunctions for nuclear states which could be used to make predictions which could be tested by observations using new experimental techniques. Bohr was also convinced by the new results and in an unpublished note [75] which is reproduced in volume 9 of his collected works he argued that a deeper understanding of the quantum-mechanical many-body problem was needed to reconcile the two points of view.

In the meantime another piece of evidence for independent particle motion emerged. This was the optical model of low-energy neutron or proton scattering proposed by Feshbach, Porter and Weisskopf [76] in 1954. They argued that a neutron incident on a target nucleus might be absorbed to form a compound nucleus, or it might pass through with its motion modified due to its average interaction with the nucleons in the target. The optical model was sometimes called the cloudy crystal ball model. 'Cloudy' referred to absorption, while 'crystal ball' alluded to the refraction of the path of the neutron by the average field. The average field in the optical model had the same origin as the average potential in the shell model and the real part of the optical potential should be similar to the shell-model potential. The absorption was represented by an imaginary part in the optical potential. Bohr's compound nucleus model corresponds to a limiting case where the absorption is strong and the nucleus is 'black'. New experimental results on elastic scattering confirmed the predictions of the optical model and showed that the absorption corresponded to a relatively long mean free path for a nucleon inside a nucleus. It was long enough to permit the definition of shell-model orbits.

Another mystery in the shell model was the origin of the spin–orbit interaction. It had to be there to get the right magic numbers but for several years there was no clear idea about where it came from. The question was partially resolved by advances in the experimental and theoretical understanding of the nucleon–nucleon force. According to Yukawa the force between nucleons was due to meson exchange. The only strongly interacting meson known in 1949 was the pion and the nucleon–nucleon force produced by pion exchange had a tensor component but no spin–orbit part. At the time there were attempts to derive a spin–orbit force as a second-order effect of the tensor force. The resulting spin–orbit effect was too weak. The situation changed in the early 1960s when the rho- and omega-mesons were discovered. These were vector mesons with spin $S = 1$. A generalization of the Yukawa theory to vector mesons produced a nuclear force with a spin–orbit component. Throughout the 1950s there was a systematic effort to measure the nucleon–nucleon force by doing proton–proton and neutron–proton scattering experiments. An analysis by Stapp *et al* [77] in 1957 found clear evidence for the existence of a spin–orbit component in the nuclear force. This would explain at least a part of the spin–orbit force in the shell model. Even now there is no clear consensus as to whether it is the final answer.

Since 1949 the shell model has been generalized to enable it to describe properties of complex nuclear states and has given rise to tremendous progress in the understanding of nuclear structure and reactions. Some of this has been reviewed in a lecture by Talmi [78] at the 40th anniversary symposium for the shell model in 1989. He wrote 'It is no longer necessary to justify the shell model on "theoretical grounds". The proof of the shell model is in the agreement of its predictions with experiment and that is established without reservation.'. Perhaps not all nuclear physicists would agree with this statement, but it does represent a widely held point of view.

15.6. Collective motion in nuclei

Niels Bohr wrote about nuclear collective motions in several paragraphs of his 1936 paper on the compound nucleus theory. He saw it as a natural feature of a situation where nucleons in a nucleus interacted strongly with each other. He imagined the nucleus as a small drop of liquid, or even as a solid. In either case it could have various vibrational modes. In his 1937 review paper on nuclear dynamics Bethe reported on a calculation by Bohr and Kalckar on the spacing of states in a highly excited nucleus. The excitations were built up from surface and volume vibrational modes. Bohr and Kalckar also calculated the thermal properties of the liquid drop with this model. The liquid-drop model of nuclear fission due to Frisch and Meitner and later to Bohr and Wheeler involved large amplitude collective motion.

15.6.1. Giant resonances

A new kind of collective motion was discovered in 1948. The experiments were made by Baldwin and Klaiber [79] and the interpretation was due to Goldhaber and Teller, and Migdal [80]. They used the term 'dipole vibration' for the collective mode. Later it was called a 'giant dipole resonance' or GDR. The experiments were made at the General Electric Company laboratory at Schenectady, New York with gamma rays from one of the new betatrons built by Kerst and collaborators. Baldwin and Klaiber measured the photo-fission cross sections for uranium and thorium, and the photo-neutron cross sections for carbon and copper. The cross sections had the appearance of a high-frequency resonance. The resonance energies were 30 MeV in carbon, 22 MeV in copper, and 16 MeV in uranium and thorium. The results were remarkable in that nuclei from completely different regions of the periodic table had similar resonances. The resonance energy decreased regularly with increasing A; also the resonances were very similar for reactions with quite different final states.

Goldhaber and Teller [80] suggested that the gamma rays excited a collective motion in the nucleus in which the protons moved in one direction while the neutrons moved in the opposite direction. This was the origin of the name 'dipole vibration'. They also argued that the width of the resonance was due to the transfer of energy from the orderly dipole vibration into other modes. That is the width was due to a process analogous to damping by friction. The reaction had two stages. The first was absorption of a photon to form a compound nucleus, while the second was the decay of the compound nucleus into the final channel. The preferred decay of the compound nucleus was fission for heavy nuclei and neutron emission for lighter ones. Goldhaber and Teller were able to estimate the energy integrated photo-absorption cross section, with results consistent with the experimental data. Experiments on gamma-ray induced reactions made over the next few years confirmed the universal character of the giant dipole resonance.

A few years later Wilkinson [81] proposed a shell-model theory of the giant dipole resonance. He argued that the nuclear states which contribute to the dipole resonance are those formed by promotion of a single nucleon into the next higher shell. In all cases the angular momenta of the promoted particle and that of the 'hole' left behind were coupled to a total $J = 1$. The overall parity was negative. The energies of all these states clustered together around the observed resonance energy. The photon absorption excited a particular linear combination of these shell-model particle–hole states. The coherent superposition of the shell-model states can be shown [82] to correspond closely to the collective Goldhaber–Teller mode. This is an example of a case where the collective and shell-model descriptions are equivalent. In 1961 Thouless [83] showed that the random-phase approximation could give a systematic theory of vibrational states like the giant dipole resonance.

The giant dipole resonance is an isovector mode because the neutrons and protons move out of phase. In 1972 the isoscalar giant quadrupole mode was discovered in proton inelastic scattering experiments [84]. Later, in 1977, the isoscalar giant monopole, or breathing mode, was found using inelastic alpha-particle scattering. The neutrons and protons move in phase in these isoscalar modes. The quadrupole and monopole modes have the same universal properties as the giant dipole resonance. They exist in nuclei throughout the periodic table and their properties vary in a systematic way with N and Z. Spin giant resonances have also been found in many nuclei, but their properties are more dependent on shell structure. They can be excited by proton or electron inelastic scattering. Bertrand and Bertsch reviewed the state of the field in 1981 [85].

15.6.2. *Low-energy collective modes*

The shell model of Mayer and Jensen was based on the idea that protons and neutrons in a nucleus moved in a spherical shell-model potential representing their average interaction with the other nucleons. In 1949 Townes, Foley and Low (cf reference [87]) collected all the existing information about the electric quadrupole moments of odd nuclei. The quadrupole moments had been measured in the hyperfine structure of atomic spectra. Townes and his co-workers found evidence for shell structure in a plot of Q/R^2 against the atomic number Z (Q is the quadrupole moment and R is the nuclear radius). The plot showed a regular shape related to the filling of the individual particle levels and passed through zero at the closed-shell values. The quadrupole moment was negative just below the closed shell and positive above as predicted by the shell model. The plot had a broad peak between the closed shells $Z = 50$ and $Z = 82$. There was an unexpected result. At its maximum Townes *et al* estimated that the experimental Q/R^2 was 35 times larger than the value expected for an odd proton.

In a paper published in 1950 James Rainwater [86] argued that the quadrupole moment data gave a strong indication that the basic nuclear shape for these nuclei was not spherical, but corresponded to a distortion of the whole nucleus into a spheroidal shape. He also showed that, while the individual particle motion of the nucleons favours a spherical shape at closed shells, away from closed shells the centrifugal pressure of the odd nucleons tends to produce a deformed equilibrium shape. Rainwater was a professor in the Physics Department of Columbia University and Aage Bohr (Niels Bohr's son) held a research fellowship in the same department during the 1949–50 academic year (figure 15.7). They shared an office during that year [87]. Bohr was also interested in the large quadrupole moments of nuclei between closed shells and studied the possibility that deformed nuclei might have rotational spectra. He worked out some of the consequences of a model for odd nuclei in which the odd nucleon was coupled to the rotations of a deformed core containing all the other nucleons. In particular he studied the influence of deformations on nuclear magnetic moments [88].

Aage Bohr returned to Copenhagen in the autumn of 1950 and continued to work on collective motion. He was joined by Ben Mottelson who came to the Bohr Institute with a fellowship from Harvard. As far as one can judge from the published record, it was in the second half of 1952 that things began to fall into place. In June of that year Mottelson gave a lecture in which he outlined the consequences of a model with surface oscillations coupled to easily excitable single-particle modes. There is a reprint of that lecture in the book of Alder and Winther on Coulomb excitation [89]. Progress at this point was influenced by the appearance in July 1952 of an extensive review on 'Nuclear isomerism and the shell model' by Goldhaber and Hill [90] in which they collected together a mass of information on measured level schemes and electromagnetic transition rates in many nuclei. In November 1952 Bohr and Mottelson submitted a letter to *Physical Review* [91] on electric quadrupole isomeric transitions in a number of even nuclei. They remarked that the measured electromagnetic transition rates were more than a factor of 100 faster than shell-model predictions, but could be explained if the nuclei were deformed. They gave an interpretation in terms of rotational states of

(a)

(b)

(c)

Figure 15.7. *(a) Aage Bohr, (b) Ben Mottelson and (c) James Rainwater.*

deformed nuclei.

Bohr and Mottelson submitted another letter to *Physical Review* in March 1953 [92] which was focused specifically on rotational motion. A deformed rotating nucleus with N and Z even was expected to have a band of states with excitation energies $E = AI(I + 1)$ where the spin $I = 0, 2, 4, 6, \ldots$ and A is a constant proportional to the inverse of the moment of inertia. The spectrum of an odd nucleus could be obtained by coupling single-particle states to the rotating core. In the

March letter they showed the first examples of rotational spectra in two isotopes of hafnium with $A = 178$ and $A = 180$. These and other results were presented in a more extensive paper in 1953 [93]. Within the next year many examples of rotational spectra were found. For example in July 1953 Asaro and Perlman [94] submitted a letter to *Physical Review* reporting patterns in the alpha-decay spectra of even nuclei which could be interpreted by the rotational model.

At the same time the Copenhagen group was thinking of other ways in which the collective model could be studied experimentally. In May 1953 Mottelson [89] suggested that Coulomb excitation would be an ideal experimental tool for studying low-energy collective modes in nuclei. He proposed to 'bombard atomic nuclei with charged particles whose energy is well below the barrier. The probability of penetrating the barrier would not be appreciable, and the nucleus would experience only the Coulomb field. Thus the excitation resulting from the pulse of electromagnetic energy would reflect the properties of the target nucleus free from complication by the nuclear force between target and projectile.'. Coulomb excitation had been considered before, first by Weisskopf in 1938 and later by Ter-Martirosyan (cf reference [89]) in 1951. There were also two experiments by Barnes and Ardine, and by Lark-Horowitz and collaborators in 1939 in which Coulomb excitation had been observed. The different situation in 1953 was that theoreticians and experimentalists were looking for a powerful method for investigating collective motion. Coulomb excitation offered a promising approach.

The first of the new generation of Coulomb excitation experiments were made in 1953 by Huus and Zupancic [89, 95] in Copenhagen and by McClelland and Goodman [96] at MIT. Both used protons as projectiles and NaI counters as gamma-ray detectors. Both groups observed strong Coulomb excitation. After this there was an explosion of activity. Temmer and Heydenburg [89, 97] at the Department of Terrestrial Magnetism, Carnegie Institution of Washington DC used alphas as projectiles. The Coulomb fields were stronger, the barriers higher and the cross sections larger. McGowan and Stelson at Oak Ridge [89, 98] combined Coulomb excitation with gamma-ray angular distributions to measure spins of excited nuclear states. The theory and the first experimental results were set out in a big review article by Alder, Bohr, Huus, Mottelson and Winther in 1956 [89, 99]. By then Coulomb excitation experiments had been made on almost 150 different target isotopes. The Nobel Prize Lectures of Bohr [100] and Mottelson [101] contain interesting information about all these developments.

In 1955 [102] Nilsson made an important step towards unifying the rotational model and the shell model. There was already a precursor of the idea in Rainwater's original paper where he calculated the individual nucleon states in a spheroidal potential well. Following Rainwater, Nilsson contended that the shell-model potential representing the average interaction of a nucleon with a nucleus, should be non-spherical in a deformed nucleus. The single-particle wavefunctions calculated in an appropriate deformed potential could be combined with the rotational model to make predictions about the properties of deformed nuclei with odd N or Z. This model, and more refined versions which followed it, were very successful. The Nilsson model gave a static picture of deformed nuclei. Dynamics was introduced by Inglis in 1954 with his remarkable 'Cranking Formula' [103].

The study of low-energy collective motion has been one of the most fruitful fields in nuclear structure physics. It has been characterized by a productive interaction between theory and experiment and many new and interesting phenomena have been found and continue to be found. The theory has been particularly effective in accounting for properties of nuclei with large deformations. A recent development has been the discovery of superdeformed nuclei at the Daresbury Laboratory by Twin and his collaborators in 1985. These are nuclear states where the major-to-minor axis ratio is 2:1. They have some unusual properties which are thought to reflect closed-shell effects which can occur in strongly deformed Fermi systems. Such effects had been predicted by Strutinsky [104] in 1966 and were observed in fission isomers. The first fission isomer was found by Polikanov and his colleagues in Dubna. Strutinsky's predictions about the properties of fission isomers were confirmed by Migneco and Theobald in 1968 and by other groups. The fission isomer story was reviewed by Bjornholm and Lynn [105] in 1980 and references can be found there.

15.6.3. *The interacting boson model*

Bohr and Mottelson's theory of nuclear structure has sometimes been called the geometrical collective model. It had its origins in Niels Bohr's image of nuclei as small liquid drops which had excited states related to the motion of the nuclear surface. In the case of deformed nuclei the motions were rotations and associated vibrations about the non-spherical equilibrium shape. Some nuclei have a spherical equilibrium shape. In the geometrical collective model the excited states would have a vibrational character. According to the ideas of the geometrical model the excited states of even nuclei with a few nucleons outside closed shells have vibrational spectra. Those with many nucleons outside closed shells are rotational. In between there are transitional nuclei with more complicated spectra related to anharmonic vibrations.

In 1977 Arima and Iachello [106] invented a new theoretical approach to the study of collective motion in nuclei. It is called the algebraic collective model as opposed to the geometrical collective model of Bohr and Mottelson. The algebraic model has its origin in the shell model. Nucleons near the Fermi surface in a nucleus interact and form pairs. The pairs act like bosons. Collective states are built out of these bosons.

Group theory is the mathematical apparatus of the interacting boson model, and the theory has several limiting cases where its predictions can be written in a closed form by using algebraic methods. The theory has been very successful in describing the spectra of even nuclei, especially those of the transitional nuclei mentioned in the last paragraph.

15.7. Nuclear scattering and reactions

15.7.1. *The compound nucleus*

A central idea of Bohr's compound nucleus theory was to divide a nuclear reaction into two stages. The first step was the formation of the compound nucleus and the second was its disintegration into the products of the reaction. At moderate excitation energies, reaction cross sections showed a resonance structure. Individual resonances corresponded to quantum states of the compound nucleus. The theory evolved in two ways. One was a statistical approach which averaged over the resonant structure, the other aimed at a complete characterization of the detailed energy dependence.

Weisskopf and Ewing [107] were able to develop a quantitative theory of energy-averaged cross sections based on the compound nucleus idea by assuming that the two stages of the reaction could be treated as independent processes. The mode of disintegration was assumed to depend on the energy, angular momentum and parity of the compound system but not on the specific way in which the compound nucleus was formed. The theory of Weisskopf and Ewing made some simplifying assumptions about energy and angular momentum dependence but was very effective and is still used today. A more complete theory was produced by Hauser and Feshbach in 1952 [108] and formed the basis of sophisticated statistical model theories which have been used to analyse experimental data.

The resonant structure of reaction cross sections was first worked out by Breit and Wigner [23] in 1936. A more detailed approach which formed the basis of a rigorous theory of nuclear reactions was given by Kapur and Peierls in 1938 [109]. This theory separated an internal region where the constituents of the compound nucleus were interacting strongly, and an external region where the reaction products were separating freely. The two regions were connected by boundary conditions. The structure of the compound nucleus could be analysed by focusing on the internal region. An extensive development of the formalism was made by Wigner and his collaborators in 1946 [110]. Wigner's presentation differed from that of Kapur and Peierls in the form of the boundary conditions. There is an extensive review of the subsequent developments by Lane and Thomas [111].

15.7.2. *Direct reaction theories*

There were new ideas in the air about both nuclear structure and nuclear reactions in the immediate post-war period. The shell model of Mayer and Jensen appeared in 1949, and the first direct reaction theories almost immediately afterwards. A direct reaction is a fast process and does not involve the formation of a compound nucleus intermediate state. The reaction products are produced as soon as the projectile interacts with the target.

A direct reaction mechanism was first suspected in the case of (d,n) reactions by Lawrence, Livingston and Lewis in 1933 [112] and the first direct reaction theory was proposed by Oppenheimer and Phillips in 1935 [113]. There were many other measurements of (d,p) and (d,n) reactions in the pre-war period [21] and they were known to have large cross sections. Experiments made after the war with 200 MeV deuterons from the Berkeley synchrocyclotron gave yields which were much too large to be accounted for by Bohr's compound nucleus model. Other experimenters reported forward peaking in the angular distribution. This did not fit with Bohr's compound nucleus theory which predicted that angular distributions should be symmetric about 90°. Bohr's compound nucleus model was clearly not working and theorists began to think of new approaches to reaction theory. One of the first was suggested by Serber in 1947 [114].

In 1950 and 1951 angular distributions for several (d,p) and (d,n) reactions with resolved final states were measured. Experiments were performed by Burrows, Gibson and Rotblat and by Holt and Young in 1950. They made measurements with ^{16}O and ^{27}Al as targets and found that the angular distributions were characterized by a large forward peak and one or more lesser peaks, very reminiscent of a diffraction pattern. In 1950 Butler [115], working with Peierls in Birmingham, proposed a theory for (d,p) reactions and used it to analyse the data of Burrows and his collaborators. By fitting the angular distribution to the experiment he was able to determine the angular momentum of the transferred neutron in the residual nucleus. In Butler's theory the target nucleus 'stripped' the neutron from the deuteron and the proton continued on its way. For this reason (d,p) and (d,n) reactions came to be called stripping reactions. Butler's theory went through several refinements. In the end the most successful approach was the distorted wave Born approximation (DWBA). Its systematical mathematical formulation was given by Horowitz and Messiah in 1953.

Stripping reactions interpreted by Butler's theory or its successor, the DWBA, were immediately recognized as an excellent spectroscopic tool. The sensitivity to the angular momentum transfer was used to give information about the spins of nuclear states reached by stripping reactions. Moreover the probability for stripping was proportional to the probability of the target nucleus accepting a single neutron or proton, so the magnitude of the cross section gave information about the single-particle structure of nuclear states. Stripping reactions with deuterons and other projectiles have continued to be important for spectroscopic studies until the present day.

Inelastic scattering of nucleons, alpha particles or other projectiles is another class of direct interactions. The Butler theory could be modified to calculate this kind of reaction but was not so successful. A version of the distorted wave Born approximation could be formulated for inelastic scattering and was applied to a case of proton inelastic scattering by Levinson and Banerjee [116] in 1957. It was the first application of the full DWBA theory to a nuclear reaction and was quite successful, though the reason for the success was not fully understood. Physical insight resulted from a series of scattering experiments with 40 MeV alpha particles in 1959. They were made by Blair and his colleagues at the University of Washington cyclotron [117]. The elastic angular distributions showed a series of regularly spaced diffraction maxima as would be expected from a black nucleus model. In 1959 Blair [118] developed a diffraction theory for inelastic scattering with extremely good results. He realized that inelastic scattering angular distributions were sensitive to the transparency of the target nucleus to the projectile. The black nucleus model was a good approximation for alpha particles but not for nucleons. After this insight DWBA could be used with confidence for various kinds of projectiles, provided care was taken in choosing the optical potential which generated the distorted waves.

One of the big problems in compound nucleus theory was to separate the internal compound nucleus region, which was interesting from the nuclear structure point of view, from the external region which was dominated by simple kinematics. Another problem was to provide a unified description of compound nucleus and direct reactions. In 1958 Feshbach [119] invented a new approach based on a projection-operator technique which was convenient from the mathematical point of view and allowed a clear physical separation of the prompt (direct reaction) and time-delayed (compound nucleus) component of a reaction amplitude.

15.7.3. *Electron scattering*

The first high-energy elastic electron scattering experiments were made with the Stanford linear accelerator by Hofstadter in 1953 [120]. They were continued over the following few years and the results were reviewed by Hofstadter in 1956. The electron scattering is sensitive to the distribution of electric charge in a nucleus. In order to investigate the details of the charge distribution the electrons must have a de Broglie wavelength which is small compared with nuclear dimensions. This means they must have energies in the range 200–500 MeV.

The first Stanford experiments in 1953 were made with 116 MeV electrons. This was soon increased to 180 MeV, and 550 MeV electrons were available by 1956. By that time the techniques were good enough to give accurate values of the radius of the charge distribution and of the surface diffuseness [121].

15.8. New isotopes and new elements

Most nuclei of atoms existing on the Earth are stable. An element may have several stable isotopes and about 270 stable isotopes are known. They are clustered along the valley of stability on a plot of Z against N. A number of radioactive isotopes also exist in nature. Most of them either have a lifetime longer than the age of the earth, or occur in a radioactive decay chain deriving from long-lived isotopes. A few radioactive isotopes are produced in small quantities by cosmic rays. An example is ^{14}C which is used for dating organic materials. Nuclei of isotopes beyond mass number $Z = 82$ are all radioactive and are unstable against alpha decay or spontaneous fission. This instability is due to the electrostatic repulsion between protons, and nuclei become more unstable as Z increases. Lighter nuclei off the stability line may be beta active. On the proton-rich side they decay by positron emission or electron capture. On the neutron-rich side they decay by electron emission. Beta-active isotopes usually have half-lifetimes in the range microseconds to years. They are long enough for their properties to be studied. When nuclei become very proton rich or very neutron rich they become unstable against proton or neutron emission, respectively. Such nuclei are said to lie beyond the proton or neutron drip lines and have very short lifetimes. All the isotopes considered in this section are radioactive, but lie within the proton and neutron drip lines.

The first isotopes, ^{13}C and ^{30}Si, were the artificially radioactive nuclei produced by the Joliot-Curies in 1933. Fermi and the Rome group found many more in their experiments on neutron reactions. By 1937 about 200 artificially radioactive isotopes had been produced using various nuclear reactions. In 1960 this number had increased to about 800. Many of the new isotopes were fission fragments. Their masses and lifetimes had been measured and in many cases there had been studies of the spectrum of excited states.

15.8.1. Transuranic elements

One of the most exciting developments in the history of nuclear physics has been the artificial preparation of transuranium elements, that is elements with atomic numbers larger than 92. They are all unstable and decay by alpha or beta emission or by fission. The lifetimes of some transuranium elements are very long, but not long enough for them to have survived in nature even if high concentrations had existed at the time of the formation of the Earth's crust. The one exception is the isotope ^{239}Pu of plutonium which has been found in minute quantities in pitchblende from Canada and the Congo. It is thought to have been produced by neutron capture in uranium with neutrons produced by spontaneous fission of uranium. There are many historical references in the book of Hyde, Perlman and Seaborg [122]. References to earlier reviews are quoted there. Other interesting and more recent references are Hyde *et al* [123] and Seaborg and Loveland [124].

The story began with the neutron capture experiments of Fermi and his co-workers in Rome in 1934. They had isolated a 13 min activity from a uranium sample after neutron irradiation and separated it chemically from all elements of atomic number 82 to 92. They concluded that the 13 min activity was an isotope of element 93. Further experiments by the Rome group and by groups in other countries showed that the situation was very complex and the matter was not resolved until after the discovery of fission by Hahn and Strassmann at the end of 1938. Their work cleared the way to further progress on the transuranic elements.

McMillan at the University of California focused on an activity with a 23 min half-life which had been identified with an isotope of uranium by Meitner, Hahn and Strassmann in 1937. Eventually in 1940 McMillan and Abelson [125] were able to announce the discovery of the first transuranium element (figure 15.8). It had atomic number $Z = 93$ and was named neptunium from the planet Neptune which is the first planet beyond Uranus. Klaproth who discovered uranium in 1789 had given the element this name in honour of the discovery of the planet Uranus in 1781 by Herschel. The identification of neptunium was delayed because its chemical properties were different from those which had been expected. The 23 min activity was due to the decay of ^{239}U into ^{239}Np. The ^{239}Np was itself unstable and decayed by beta-decay into the isotope ^{239}Pu of plutonium with $Z = 94$. The positive identification of this isotope of the new element was hampered by its long half-life, but was eventually accomplished by Kennedy, Seaborg, Segrè and Wahl [126] in 1941 in a major experiment with a 16 MeV deuteron beam from the Berkeley 60" cyclotron. The experimenters succeeded in isolating a 0.5 microgram sample of ^{239}Pu and found that it fissioned when bombarded with slow neutrons.

It was recognized that the slow neutron fission of ^{235}U might allow the conversion of appreciable quantities of natural uranium into plutonium. It was then that the secret plutonium project was set up by the Uranium Committee. The history was recorded in the book *Atomic Energy for Military Purposes* by Smythe in 1945. After 1940 all work on transuranium elements was classified and results were not published until after 1945.

Transuranium elements have many isotopes. Some were found in accelerator experiments and others with reactor experiments. Once one isotope had been produced its chemistry could be studied and the information so obtained helped with the identification of other

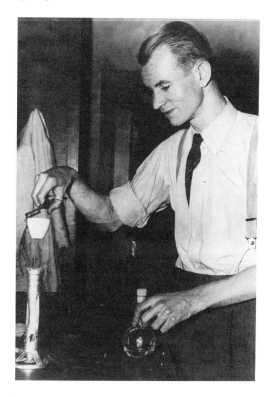

Figure 15.8. *Edwin McMillan in his laboratory at the time of the discovery of neptunium.*

isotopes. The next elements were found in 1944–5 after the advent of nuclear reactors. By that time ^{239}Pu existed in large enough quantities to be used as a target in accelerator or reactor experiments. The first isotope of curium (^{242}Cu, $Z = 96$), named after Marie and Pierre Curie, was prepared by Seaborg, James and Ghiorso in mid-1944 by bombarding a plutonium target with alpha particles at the Berkeley cyclotron. An isotope of americium (^{241}Am, $Z = 95$) was produced in the Chicago Metallurgical Laboratory as a product of neutron irradiation of plutonium.

The general strategy for producing new elements was to select the heaviest target nuclei possible, then add a few nucleons to them. Thus the first isotopes of the next two elements, berkelium ($Z = 97$) and californium ($Z = 98$), were made in 1949–50 by using americium and curium as targets and alpha particles as projectiles. Microgram quantities of the target material were sufficient for the experiments. One of the interesting consequences of the November 1952 'Mike' thermonuclear explosion was the discovery of elements 99 and 100, einsteinium and fermium, in the resulting debris. In addition to those new elements and many new isotopes of plutonium, americium, curium, berkelium and californium [127] were identified.

By 1992 the formation of all elements up to $Z = 109$ had been established and there were some indications that elements with Z up to 112 might be produced [128]. Most of these transuranium elements were first produced at Berkeley, some at Dubna and some at Darmstadt. As Z increases the elements become more and more unstable against fission and alpha emission. For example, the most stable californium isotope ($Z = 98$) has a half-life of 900 yr, while the most unstable isotope of hahnium ($Z = 104$) is about 40 s.

The availability of heavy-ion accelerators has changed the tools and strategies available for the formation of new elements. Until 1955 new elements were made by selecting the heaviest available target and adding a few nucleons. With heavy-ion projectiles, lighter more convenient targets can be used [124]. Element 107 was produced at the GSI Laboratory in Darmstadt by the reaction

$$^{209}\text{Bi} + {}^{54}\text{Cr} \rightarrow {}^{262}107 + \text{n}.$$

In 1966 Myers and Swiatecki used an elaborate liquid-drop mass formula with empirical shell corrections to study the fission barriers of transuranium elements. Their calculations suggested that nuclei near the magic numbers $Z = 114$ and $N = 184$ should be stabilized by closed-shell effects and have higher fission barriers than lighter nuclei. The general trend for nuclei with larger A to have shorter lifetimes for fission and alpha decay might be reversed and some of these 'superheavy elements' could even be stable. This suggestion started a programme to try to produce superheavy elements in heavy-ion reactions and to search for them in nature. So far none have been found, but the heaviest transuranium elements that have so far been produced are not so very far away from the predicted superheavy region.

At the time of going to press, the production of the element with $Z = 110$ at GSI was reported by Hofmann *et al* [129].

15.9. Creation of the elements

Progress in understanding the origin of the elements is reviewed in the Nobel Prize Lectures of Bethe [130] in 1967 and Fowler [131] in 1983. There are also some recent textbooks [132, 133]. The material for this section comes mainly from references [130, 131]. The question of the origin of the elements is closely connected with the production of energy in stars. In 1929 Rutherford wrote [134] 'It is natural to suppose that the uranium on our Earth has its origin in the Sun.'. In 1929 Atkinson and Houtermans concluded that in the interior of stars the temperature was high enough to allow nuclear reactions to occur. Gamow's book *Atomic Nuclei and Nuclear Transformations* [135] written in 1936 had a chapter on the relative abundance and the origin of the elements. He wrote about thermonuclear reactions in stars and noted that neutron capture would be an effective way of building up heavy elements.

It has been known for a long time that most stars are composed of hydrogen and helium with less than 1% of heavier elements. Thus if nuclear reactions are responsible for the production of energy, then hydrogen and helium must be involved. A scheme (the pp process) for producing helium from protons was suggested by von Weizsäcker in 1937 and calculated by Bethe and Critchfield in 1938. In the first step two protons interacted to produce a deuteron, a positron and a neutrino. This was a weak interaction process with a very small cross section; but enough to be effective on an astronomical timescale. The deuterons quickly interacted further and in the end produced ^{4}He. Stimulated by a conference organized in Washington DC by Gamow in 1938, Bethe produced an alternative scheme, the carbon–nitrogen cycle, for building up helium from hydrogen. In this process the carbon acted as a catalyst and was regenerated at the end of the cycle. It turned out that both processes were important. The pp process is now known to be the mechanism which operates in the sun while the carbon–nitrogen cycle dominates the energy production in a massive star during the hydrogen-burning stage.

In an attempt to find a way of synthesizing heavier elements Gamow and his collaborators [136] suggested that the early universe could have played the role of a gigantic fusion reactor. All atomic nuclei could be produced from an initial mix of protons and neutrons by neutron capture, beta decay and other nuclear reactions. The idea ran into a difficulty because there were no stable nuclei with mass 5 or 8. In 1939 Staub and Stephens [137] working in the Kellogg Radiation Laboratory at Caltech found that ^{5}He was not stable. The same was later shown to be true of ^{5}Li. In 1949 Hemmendinger [138] found that ^{8}Be was also unstable. The mass gaps at $A = 5$ and $A = 8$ were a problem for element synthesis not only in the big bang, but also in the interior of stars.

Helium could be synthesized from hydrogen by the pp process or the carbon–nitrogen cycle. The next step was to find a mechanism for building heavier elements out of helium. There was a problem in bridging the gap between ^{4}He and ^{12}C because of the unstable nuclei at mass 5 and 8. In the end the problem was solved by Fred Hoyle [139]. He was convinced that ^{12}C nuclei could be formed from the fusion of three alpha particles in red giant stars. This reaction required the simultaneous collision of three alpha particles, a very unlikely event except for the fact that it is favoured by a double resonance. The effect of the resonance in alpha–alpha scattering was calculated by Salpeter [140] during a visit to the Kellogg Radiation Laboratory in 1951, but his estimate for the fusion rate for three alpha particles into ^{12}C was still too small. Hoyle realized that the process would be speeded up if ^{12}C had an excited state just above the threshold for break up into three alpha particles. Hoyle visited the Kellogg Laboratory early in 1953 and questioned the staff about the possible existence of his state. At the time it was not known but Dunbar *et al* [141] looked for the state in a nuclear reaction and found it almost exactly where Hoyle had predicted it to be.

Once the mass gaps at $A = 5$ and 8 had been passed, heavier elements could be built up by a variety of nuclear reactions. In Hoyle's classic papers stellar nucleosynthesis of elements up to the iron group was achieved by charged particle reactions; heavier elements could be produced by neutron capture as suggested by Gamow [135, 136]. There are two versions to the neutron capture process. The s-process was suggested by Gamow in 1935. It is slow. The time interval between successive captures is generally longer than the beta-decay lifetimes and the pathway for building up heavier nuclei is along the valley of stability. The r-process is a rapid capture process. Two or more neutrons may be captured before beta decay and it tends to produce neutron-rich nuclei. It was used as a mechanism for heavy-element production in supernovae in the paper of Burbidge, Burbidge, Fowler and Hoyle [142] in 1957.

The solar neutrino problem has been a puzzle for more than 20 years. There are various side branches to the pp chain for hydrogen burning. In 1958 Fowler and Cameron [143] showed that ^7B and ^8B should be produced in small amounts as byproducts of the pp chain. The beta decay of these nuclei produces neutrinos which should be energetic enough to be detected through their interaction with ^{37}Cl to form radioactive ^{37}Ar. The idea was suggested by B Pontecorvo and L Alvarez in the 1940s. In 1970 an experimental set-up was constructed by R Davis and collaborators. It is located deep underground in the Homestake Gold Mine in Lead, South Dakota and started taking data in 1970. Only about one quarter of the number of neutrinos expected have been found. This is a problem for the standard stellar models.

The idea of using nuclear methods to measure astronomical timescales was proposed by Rutherford [134] in 1929 in a note to *Nature* on the origin of actinium and the age of the Earth. He assumed that actinium was a member of a decay chain originating from actino-uranium, a then unknown isotope of uranium with mass number 235. By making use of the known abundances of isotopes in the different radioactive decay chains he was able to estimate that the half-life of actino-uranium (^{235}U) should be about 0.42 billion years and argued that the Earth could not be older than 3.4 billion years. This was the first attempt to use nuclear processes to measure cosmological timescales.

We conclude this chapter with another example relating nuclear physics and the cosmos. The isotope of californium with $A = 254$ was first identified in the debris of the 1952 thermonuclear explosion. The half-life was measured to be 55 days. Baade and Minkowski studied the decay in the light intensity of supernovae and found that two kinds of behaviour could be distinguished. All type I supernovae had remarkably similar light curves and about 100 days after the birth of the supernova the brightness decreased linearly on a logarithmic scale with a half-life of about 50 days. In 1956 Baade, Burbidge, Hoyle, Burbidge, Christy and Fowler [144] suggested that the energy source for the light could be the decay of ^{254}Cf which is produced by the r-process in the explosion of the supernova. The isotope with $A = 254$ decays by fission and releases a lot of energy. Baade [145] has collected information about observations of historical supernovae and suggested that the supernovae observed by Tycho Brahe in 1572 and Kepler in 1604 had the same decay period. It was soon recognized that the story was more complicated. The isotope ^{254}Cf is indeed produced in supernova explosions, but not in large enough quantities to explain the light curves. Isotopes of some medium-mass elements with half-lives in the region of 50 days give the major contribution to the effect noticed by Baade and Minkowski. The nucleus ^{59}Fe is particularly important. It has a half-life of 45 days and decays by beta emission. Each decay releases an energy of 1.57 MeV. This is small compared with the energy released in the fission of ^{254}Cf, but the ^{59}Fe is produced in much greater quantities [146].

References

[1] Rutherford E 1932 *Proc. R. Soc.* A **136** 735

[2] Chadwick J 1932 *Nature* **129** 312

[3] Cockcroft J D and Walton E T S 1932 *Proc. R. Soc.* **137** 229

[4] Pais A 1986 Possible existence of a neutron *Inward Bound* (Oxford: Oxford University Press)

[5] Heisenberg W 1932 Über den Bau der Atomkerne *Z. Phys.* **77** 1; 1932 *Z. Phys.* **78** 156; 1933 *Z. Phys.* **80** 587

[6] Livingston M S and Bethe H A 1937 Nuclear physics C. Nuclear dynamics, experimental *Rev. Mod. Phys.* **3** 245

[7] Gamow G 1928 *Z. Phys.* **51** 204; 1929 *Z. Phys.* **52** 510

[8] Condon E U and Gurney R W 1928 *Nature* **122** 439; 1929 *Phys. Rev.* **33** 127

[9] Cockcroft J D and Walton E T S 1930 *Proc. R. Soc.* A **129** 477

[10] Curie I and Joliot F 1934 *C. R. Acad. Sci., Paris* **198** 254, 561

[11] Gamow G 1931 Un nouveau type de radioactivité *The Constitution of Atomic Nuclei and Radioactivity* (Oxford: Oxford University Press)

[12] Stewer R 1985 Niels Bohr and nuclear physics *Niels Bohr a Centennial Volume* ed A P French and P J Kennedy (Cambridge, MA: Harvard University Press)

[13] Bainbridge K T 1960 Charged particle dynamics and optics, relative isotopic abundances of the elements, atomic masses *Experimental Nuclear Physics* vol 1, ed E Segrè (New York: Interscience)

[14] Rutherford E 1929 *Proc. R. Soc.* A **123** 373

[15] Majorana E 1933 Über die Kerntheorie *Z. Phys.* **82** 137

[16] von Weizsäcker C F 1935 *Z. Phys.* **96** 431

[17] Fermi E, Amaldi E, D'Agostino O, Rasetti F and Segrè E 1934 *Proc. R. Soc.* **146** 483

[18] Fermi L 1955 *Atoms in the Family* (London: Allen and Unwin)

[19] Segrè E 1955 Fermi and neutron physics *Rev. Mod. Phys.* **27** 257

[20] Bohr N 1936 Neutron capture and nuclear constitution *Nature* **137** 344

[21] Bethe H A 1937 Nuclear physics B. Nuclear dynamics, theoretical *Rev. Mod. Phys.* **9** 69

[22] Bohr N 1937 Transmutations of atomic nuclei *Science* **86** 161

[23] Breit G and Wigner E 1936 Capture of slow neutrons *Phys. Rev.* **49** 519

[24] Bohr N and Kalkar F 1937 On the transmutations of atomic nuclei by impact of material particles, I: general theoretical remarks *Mater.-Fys. Medd. Dan. Vidensk. Selsk.* **14** No. 10

[25] Hahn O and Strassmann F 1939 *Naturwissenshaften* **27** 11

[26] Meitner L and Frisch O R 1939 Disintegration of uranium by neutrons: a new type of nuclear reaction *Nature* **143** 239

[27] Frisch O R 1939 Physical evidence for the division of heavy nuclei under neutron bombardment *Nature* **143** 276
Bohr N 1939 Disintegration of heavy nuclei *Nature* **143** 330

[28] Bohr N and Wheeler J A 1939 The mechanism of nuclear fission *Phys. Rev.* **56** 426

[29] Smyth H D 1946 *Atomic Energy for Military Purposes* (Washington, DC: US Government Printing Office)

[30] Hewlett R G and Anderson O F Jr 1962 *The New World: History of the United States Atomic Energy Commission* (University Park, PA: Penn State University Press)

[31] Gowing M 1964 *Britain and Atomic Energy 1939–45* (New York: Macmillan)

[32] Gowing M 1985 *Niels Bohr, a Centenary Volume* ed A P French and P J Kennedy (Cambridge, MA: Harvard University Press) p 266

[33] Goldsmith H H, Ibser H W and Feld B T 1947 Neutron cross sections of the elements *Rev. Mod. Phys.* **19** 259

[34] Siegel J M (ed) 1946 Plutonium project report *Rev. Mod. Phys.* **18** 513; 1946 *J. Am. Chem. Soc.* **68** 2411

[35] Rutherford E, Chadwick J and Ellis C D 1930 *Radiations from Radioactive Substances* (Cambridge: Cambridge University Press)

[36] Kinoshita S 1910 *Proc. R. Soc.* A **83** 432

[37] Regener E 1908 *Verh. Deutsch. Phys. Ges.* **19** 78, 351

[38] Geiger H 1908 *Proc. R. Soc.* A **81** 174
Geiger H and Marsden E 1909 *Proc. R. Soc.* A **82** 495

[39] Rutherford E and Geiger H 1908 *Proc. R. Soc.* A **81** 141–61

[40] Ward F A B, Wynn-Williams C E and Cave H M 1929 *Proc. R. Soc.* **125** 715

[41] Wilson C T R 1913 *Phil. Trans. R. Soc.* **193** 289

[42] Blackett P M S 1925 *Proc. R. Soc.* A **107** 349

[43] von Baeyer O and Hahn O 1910 *Phys. Z.* **11** 488
von Baeyer O, Hahn O and Meitner L 1911 *Phys. Z.* **12** 273, 378

[44] Rutherford E and Robinson H 1913 *Phil. Mag.* **26** 717

[45] Rutherford E and Robinson H 1914 *Phil. Mag.* **28** 557

[46] Rutherford E and Andrade E N 1914 *Phil. Mag.* **27** 854; 1914 *Phil. Mag.* **28** 263

[47] Livingston M S 1952 *Ann. Rev. Nucl. Sci.* **1** 157, 169

[48] Livingston M S and Blewett J P 1962 *Particle Accelerators* (New York: McGraw-Hill)

[49] Livingston M S 1959 *Phys. Today* **12** 18
McMillan E M 1959 *Phys. Today* **12** 24

[50] van de Graaff R J, Trump J G and Buechner W W 1946 *Rep. Prog. Phys.* **11** 1

[51] Rose P H 1987 The tandem accelerator: workhorse of nuclear physics

Discovering Alvarez; Selected Works of L W Alvarez ed W P Trower (Chicago, IL: University of Chicago Press)

[52] Panofsky W K H 1987 Building the proton linear accelerator *Discovering Alvarez; Selected Works of L W Alvarez* ed W P Trower (Chicago, IL: University of Chicago Press)

[53] Johnston L 1987 The war years *Discovering Alvarez; Selected Works of L W Alvarez* ed W P Trower (Chicago, IL: University of Chicago Press)

[54] Fry D W and Walkinshaw W 1948 Linear accelerators *Rep. Prog. Phys.* **xii** 102

[55] Siegbahn K (ed) 1955 *Beta- and Gamma-ray Spectroscopy* (Amsterdam: North-Holland)

[56] Du Mond J W M and Kirkpatrick H A 1930 *Rev. Sci. Instrum.* **1** 88

[57] Wilkinson D H 1950 *Ionisation Chambers and Counters* (Cambridge: Cambridge University Press)

[58] Hofstadter R 1948 *Phys. Rev.* **74** 100; 1949 *Phys. Rev.* **75** 796

[59] Miller G L, Gibson W M and Donovan P F 1962 *Ann. Rev. Nucl. Sci.* **12** 189

[60] Tavendale A J 1967 *Ann. Rev. Nucl. Sci.* **17** 73

[61] McKay K G 1949 *Phys. Rev.* **76** 1537

[62] Orman C, Fan H Y, Goldsmith G J and Lark-Horowitz K 1950 *Phys. Rev.* **78** 646

[63] Walter F J, Dabbs J W T and Roberts L D 1958 *Bull. Am. Phys. Soc.* **3** 181

[64] Bethe H A and Bacher R F 1936 Nuclear physics A. Stationary states of nuclei *Rev. Mod. Phys.* **8** 82

[65] Bartlett J A 1932 *Phys. Rev.* **41** 370; 1932 *Phys. Rev.* **42** 145

[66] Elsasser W M 1933 *J. Phys. Radium* **4** 549; 1934 *J. Phys. Radium* **5** 389

[67] Guggenheimer J 1934 *J. Phys. Radium* **5** 253, 475

[68] Goeppert Mayer M 1949 *Phys. Rev.* **75** 1969

[69] Haxel O, Jensen J H D and Suess H E 1949 *Phys. Rev.* **75** 1766

[70] Goeppert Mayer M 1973 *Nobel Lectures: Physics, 1963–72* (Amsterdam: Elsevier) p 20

[71] Jensen J H D 1973 *Nobel Lectures: Physics, 1963–72* (Amsterdam: Elsevier) p 40

[72] Kurath D 1990 *Nucl. Phys.* A **507** 1c

[73] Weidenmüller H A 1990 *Nucl. Phys.* A **507** 5c

[74] Goeppert Mayer M 1948 *Phys. Rev.* **74** 235

[75] Bohr N 1977 Comments on atomic and nuclear constitution *Niels Bohr: Collected Works* vol 9 (Amsterdam: North-Holland) p 523

[76] Feshbach H, Porter C and Weisskopf V F 1954 *Phys. Rev.* **96** 448

[77] Stapp H P, Ypsilantis T J and Metropolis N 1957 *Phys. Rev.* **105** 302

[78] Talmi I 1990 *Nucl. Phys.* A **507** 295c

[79] Baldwin G C and Klaiber G S 1947 *Phys. Rev.* **71** 3; 1948 *Phys. Rev.* **73** 1156

[80] Goldhaber M and Teller E 1948 *Phys. Rev.* **74** 1046
      Migdal A 1944 *J. Phys. (Moscow)* **8** 331

[81] Wilkinson D H 1959 *Ann. Rev. Nucl. Sci.* **9** 1

[82] Brown G E and Bolsterli M 1959 *Phys. Rev. Lett.* **3** 472

[83] Thouless D J 1960 *Nucl. Phys.* **21** 225; 1961 *Nucl. Phys.* **22** 78

[84] Lewis M B and Bertrand F E 1972 *Nucl. Phys.* A **196** 337
      Pitthan R and Walcher Th 1971 *Phys. Lett.* B **36** 563
      Fukuda S and Torizuka Y 1972 *Phys. Rev. Lett.* **29** 1109

[85] Bertrand F 1981 *Nucl. Phys.* A **354** 129c
      Bertsch G F 1981 *Nucl. Phys.* A **354** 157c

[86] Rainwater J 1950 *Phys. Rev.* **79** 432

[87] Rainwater J 1976 Background for the spheroidal nuclear model proposal

(Nobel Lecture) *Rev. Mod. Phys.* **48** 385

[88]   Bohr A 1951 *Phys. Rev.* **81** 134
[89]   Alder K and Winther A 1966 *Coulomb Excitation* (New York: Academic) (a collection of reprints with an introductory review)
[90]   Goldhaber M and Hill R D 1952 *Rev. Mod. Phys.* **24** 179
[91]   Bohr A and Mottelson B 1953 *Phys. Rev.* **89** 316
[92]   Bohr A and Mottelson B 1953 *Phys. Rev.* **90** 717
[93]   Bohr A and Mottelson B R 1953 *Dan. Mater. Fys. Medd.* **27** No. 16
[94]   Asaro F and Perlman I 1953 *Phys. Rev.* **91** 763; 1954 *Phys. Rev.* **93** 1423. See also, 1953 *Phys. Rev.* **92** 694, 1495
[95]   Huus T and Zupancic C 1953 *Dan. Mater. Fys. Medd.* **28** No. 1
[96]   McLelland C L and Goodman C 1953 *Phys. Rev.* **91** 51
[97]   Temmer G M and Heydenburg N P 1954 *Phys. Rev.* **94** 426
[98]   McGowan F K and Stelson P H 1955 *Phys. Rev.* **99** 127
[99]   Alder K, Bohr A, Huus T, Mottelson B and Winther A 1976 *Rev. Mod. Phys.* **28** 432
[100] Bohr A 1976 Rotational motion in nuclei (Nobel Lecture) *Rev. Mod. Phys.* **48** 365
[101] Mottelson B 1976 Elementary modes of excitation in the nucleus (Nobel Lecture) *Rev. Mod. Phys.* **48** 375
[102] Nilsson S G 1955 *Mater.-Fys. Medd. Dan. Vidensk. Selsk.* **29** No. 16
[103] Inglis D R 1954 *Phys. Rev.* **96** 1059
       Thouless D J and Valatin J G 1962 *Nucl. Phys.* **31** 211
[104] Strutinsky V M 1966 *Yad. Fiz.* **3** 614; 1967 *Nucl. Phys.* A **95** 420
[105] Bjornholm S and Lynn J E 1980 *Rev. Mod. Phys.* **52** 725
[106] Arima A and Iachello F 1975 *Phys. Rev. Lett.* **35** 1069
       Arima A, Otsuka T, Iachello F and Talmi I 1977 *Phys. Lett.* **66** B 205
[107] Weisskopf V F 1937 *Phys. Rev.* **52** 295
       Weisskopf V F and Ewing D H 1940 *Phys. Rev.* **57** 472, 935
[108] Hauser W and Feshbach H 1952 *Phys. Rev.* **87** 366
[109] Kapur P L and Peierls R E 1938 *Proc. R. Soc.* A **166** 277
[110] Wigner E P 1946 *Phys. Rev.* **70** 606
       Wigner E P and Eisenbud L 1947 *Phys. Rev.* **72** 29
[111] Lane A M and Thomas R G 1958 *Rev. Mod. Phys.* **30** 257
[112] Lawrence E O, Livingston M S and Lewis G N 1933 *Phys. Rev.* **44** 56
[113] Oppenheimer J R and Phillips M 1935 *Phys. Rev.* **47** 845; 1935 *Phys. Rev.* **48** 500
[114] Serber R 1947 *Phys. Rev.* **72** 1008
[115] Butler S T 1950 *Phys. Rev.* **80** 1095; 1951 *Proc. R. Soc.* A **208** 559
[116] Levinson C A and Banerjee M K 1957 *Ann. Phys., NY* **2** 471, 499; 1958 *Ann. Phys., NY* **3** 67
[117] McDaniels D K, Blair J S, Chen S Y and Farwell G W 1960 *Nucl. Phys.* **17** 614
[118] Blair J S 1959 *Phys. Rev.* **115** 928
[119] Feshbach H 1958 *Ann. Phys., NY* **5** 357; 1962 *Ann. Phys., NY* **19** 287; 1967 *Ann. Phys., NY* **43** 410
[120] Hofstadter R, Fechter H R and McIntyre J A 1953 *Phys. Rev.* **91** 422
[121] Hofstadter R 1956 *Rev. Mod. Phys.* **28** 214
[122] Hyde E K, Perlman I and Seaborg G T 1971 *Nuclear Properties of the Heavy Elements* (New York: Dover) ch 9 p 745
[123] Hyde E K, Perlman I and Seaborg G T 1971 *The Nuclear Properties of the Heavy Elements* vol II (New York: Dover) p 745
[124] Seaborg G T and Loveland W D 1984 *Treatise on Heavy Ion Science* vol 4, ed D A Bromley (New York: Plenum) p 254

[125] McMillan E and Abelson P H 1940 *Phys. Rev.* **57** 1186
[126] Kennedy J W, Seaborg G T, Segrè E and Wahl A C 1946 *Phys. Rev.* **69** 555
(The original report was written in May 1941 but only published after the World War II.)
[127] Fields P R *et al* 1956 Transplutonium elements in thermonuclear test debris *Phys. Rev.* **102** 180
[128] Barber R C *et al* 1992 Discovery of the transfermium elements *Prog. Part. Nucl. Phys.* **29** 453 (This is the report of a working group set up by the IUPAP and IUPAC to consider questions of priority in the discovery of elements with nuclear charge greater than 100.)
[129] Hofmann S, Nivor V, Hessberger E P, Armbruster P, Folger H, Munzenberg G, Schölt H J, Popeko A C, Yeremin A V, Andreyev A N, Saro S, Janik R and Leino M 1995 *Z. Phys.* A (January)
[130] Bethe H A 1973 *Nobel Lectures: Physics, 1963–72* (Amsterdam: Elsevier)
[131] Fowler W A 1983 Nobel Prize Lecture *Nobel Lectures in Physics 1981–90* ed G Ekspong (Singapore: World Scientific) p 172
[132] Rolfs C E and Rodney W S 1988 *Cauldrons of the Universe* (Chicago, IL: Chicago University Press)
[133] Clayton D D 1968 *Principles of Stellar Evolution and Nucleosynthesis* (New York: McGraw-Hill)
[134] Rutherford E 1929 *Nature* **123** 313
[135] Gamow G 1936 *Atomic Nuclei and Nuclear Transformations* (Oxford: Oxford University Press)
[136] Gamow G 1948 *Nature* **162** 680
Alpher R and Herman R C 1950 *Rev. Mod. Phys.* **22** 153
[137] Staub H and Stephens W E 1939 *Phys. Rev.* **55** 131
[138] Hemmendinger A 1948 *Phys. Rev.* **73** 806; 1949 *Phys. Rev.* **74** 1267
[139] Hoyle F 1946 *Mon. Not. R. Astron. Soc.* **106** 343; *Astrophys. J. Suppl.* **1** 121
[140] Salpeter E E 1952 *Astrophys. J.* **115** 326
[141] Dunbar D N F, Pixley R E, Wenzel W A and Whaling W 1953 *Phys. Rev.* **92** 649
[142] Burbidge E M, Burbidge G R, Fowler W A and Hoyle F 1957 *Rev. Mod. Phys.* **29** 547
[143] Fowler W A 1958 *Astrophys. J.* **127** 551
Cameron A G W 1958 *Ann. Rev. Nucl. Sci.* **8** 249
[144] Baade W, Burbidge G R, Hoyle F, Burbidge E M, Christy R F and Fowler W A 1956 Supernovae and californium 254 *Proc. Astron. Soc. Pacific* **68** 296
[145] Baade W 1943 *Astron. J.* **97** 119; 1945 *Astron. J.* **102** 309
[146] Hoyle F 1975 *Astronomy and Cosmology* (San Francisco, CA: Freeman) p 383

Chapter 16

UNITS, STANDARDS AND CONSTANTS

Arlie Bailey

16.1. Introduction

The central theme of this chapter is the application of new discoveries in physics to replace arbitrary material standards of measurement by physical constants. This has not been an accidental unplanned process. Metrology is a branch of science which has been consistently sponsored by government since standards of weight, length and capacity were issued by royal decree in Babylon about 2500 BC. It is generally accepted that maintenance and dissemination of standard weights and measures for use in trade ('legal metrology'), often in association with the coinage, is a responsibility which only government can carry†.

However, advanced technology is involved and in the eighteenth century scientific interest in accurate measurement began to develop. The Royal Society in Britain and the Academie des Sciences in France took part in experiments aimed at improving standards, including international comparisons. One conclusion drawn from this was that there would be advantage in having standards related to physical constants which could be reproduced wherever they were required. The length of the seconds pendulum was considered as a possibility but rejected because of its dependence on the value of gravity. When, in 1790, Talleyrand proposed to the French National Assembly the setting up of the metric system, the standards chosen were one ten-millionth of the quadrant of the Earth's meridian for the metre and the mass of one cubic decimetre of water at 4 °C for the kilogram. Much effort

† *In this chapter the word 'standard' is used exclusively in the sense of 'measurement standard' (French 'étalon'), not as a 'specification standard' (as in 'British standard'— French 'norme').*

went into determining these, and practical standards were made which became known as the 'Metre and Kilogram of the Archives'. Improved measurement techniques revealed that they were not exact realizations of the defined values so the definitions were abandoned: but the material standards remained in use.

In the nineteenth century, the growth of technology-based industry led to requirements for a wider range of standards and for national and international organizations to support them. In Britain this was largely undertaken by the British Association for the Advancement of Science, founded in 1841, which set up facilities for the testing of thermometers, clocks, lenses and other instruments, and took the initiative in defining electrical units and developing standards to meet the needs of the electricity supply industry.

In 1867 a number of scientists attending the Paris Exhibition discussed the needs of industry for internationally accepted standards. As a result, the French Government called a conference in 1875 at which twenty countries were represented and which led to the signing of the Convention of the Metre. This established the International Weights and Measures Organization including:

(i) The International Bureau of Weights and Measures (BIPM), the laboratory at Sèvres, near Paris, which establishes basic standards, carries out international comparisons, and makes determinations of fundamental physical constants, in cooperation with national laboratories [1].

(ii) The General Conference on Weights and Measures (CGPM), which oversees the administration and financing of BIPM and ratifies important decisions of CIPM†.

(iii) The International Committee of Weights and Measures (CIPM), the scientific committee which executes the decisions of CGPM, recommends changes in the international system of units and standards when needed, provides a forum for the coordination of programmes in national laboratories, and directly supervises the operation of BIPM.

(iv) A number of Consultative Committees and Working Groups which advise CIPM on specific topics and arrange international comparisons.

National laboratories with responsibility for the development and maintenance of standards were set up following the signing of the Convention of the Metre. The first of these, in 1887, was the Physikalisch-Technische Reichsanstalt (now Bundesanstalt, PTB) in Germany. This was followed by the National Physical Laboratory (NPL) in Britain [2],

---

† *Records of the meetings of the General Conference of Weights and Measures (CGPM), the International Committee of Weights and Measures (CIPM) and of its Consultative Committees and Working Groups are published by Le Bureau International des Poids et Mesures, Pavillon de Breteuil F-92313, Sèvres Cedex, France.*

the National Bureau of Standards (NBS—now the National Institute of Standards and Technology, NIST) in the USA [3], and a number of similar laboratories in other countries [4,5]. They have played a crucial part in exploring new physical phenomena which can be used in the realization of standards and the establishment of systems of units.

The metric system slowly found acceptance not only in France but in other countries in Europe and Latin America where it was in use by the mid-nineteenth century: for day-to-day purposes it is of course still by no means universally welcomed in English-speaking countries, but a major factor in its use for scientific purposes was the need for consistent measurement of electrical quantities brought about by the growth of the electric telegraph and supply industries. In 1851 W E Weber proposed a coherent system of units, based on the centimetre, gramme and second as base units, and taking two forms for electrostatic and electromagnetic quantities, based respectively on the inverse-square laws of force between electric charges or magnetic poles. Maxwell showed that the ratio of these units was determined by the speed of light; the first example of the linking of units to a fundamental physical constant. The British Association committee concerned with electrical standards recommended the adoption of the CGS units in 1863.

However, with the expansion of the supply industry it became clear that the magnitudes of the units were not convenient for practical use and in 1881 there was international agreement to use practical units, the volt, the ohm and the ampere, which were respectively 10^8, 10^9 and 10^{-1} times their CGS magnetic equivalents. They were still not easy to realize in terms of their definitions and in 1908 the International Congress in London defined the International Units in terms of material standards— the International Ohm as the resistance of a specified column of mercury, the International Ampere as the current depositing silver from solution at a specified rate and the International Volt as their product or as a given fraction of the voltage of a Weston Cell. These remained in use until 1948 when improved methods of absolute determination allowed the CGPM to replace them with the present definition of the ampere.

Meanwhile, in 1902 G Giorgi made the radical proposal which led to the adoption of the MKSA system in which there are four base units—the metre, the kilogram, the second and the ampere. This gave units of a magnitude more convenient for practical purposes, and the adoption of the ampere removed the distinction between electromagnetic and electrostatic units. The system was coherent, in the sense that other units could all be derived from the base units without multipliers other than unity. The CGPM became increasingly involved in the definition of units as other base units, thermodynamic temperature, luminous intensity and amount of substance, were identified. After much discussion the Eleventh CGPM in 1960 adopted the International System of Units, SI, which is now the recognized basis for physical measurement [6]. In this chapter we first look briefly at how the base units (other than the mole, which is not a physical quantity in the same sense as the others) and certain derived units have evolved. We then consider the determination of fundamental physical constants and how accurate knowledge of their values has reacted back on the process of measurement itself.

16.2. Units and standards

16.2.1. Mass

16.2.1.1. The kilogram. The kilogram is unique. It is the only base unit whose definition has not changed in the past hundred years and the only unit still defined in terms of a material standard. In 1878, following the signing of the Metre Convention, three prototype kilograms were made in the form of cylinders of an alloy of 90% platinum, 10% iridium [1]. They were compared with the kilogram of the archives at the Paris Observatory: the one closest in mass was chosen as the International Kilogram and has been held as such by BIPM ever since (figure 16.1). It was formally adopted by the first CGPM in 1889. A further forty cylinders were ordered in 1882: their masses were adjusted to bring them within 1 mg of the mass of the International Kilogram and 34 of them were allocated for use as national standards by the member states of the Metre Convention. Others have since been made. They are kept in specially protected conditions and used only rarely. There have been two formal international comparisons of the standards, reported to the CGPM in 1913 and 1954.

During the past century a number of special balances for the comparison of masses have been developed, mainly in national laboratories. Key points have been finding ways of avoiding changing loads on the knife-edges, keeping the system thermally stable, and eliminating ground tremors. The best system is a single-arm balance developed by the US National Bureau of Standards [1] in which reference standards are compared in turn with a counterweight. One of these is now in use by BIPM. By this means, standard kilograms can be compared with a precision of about 1 μg, or 1 part in 10^9.

Weighing is, of course, an important operation in trade as well as in physics and secondary standards are maintained nationally for this purpose. In Britain, for example, the National Weights and Measures Laboratory maintains sets of standards for metric, avoirdupois and troy weights, all traceable to the NPL kilogram.

16.2.1.2. Possible alternatives. Much effort has been put into exploring ways in which the kilogram might be expressed in terms of fundamental physical constants rather than as the mass of a material standard. This might be done through the electrical units: the ampere is defined in terms

Figure 16.1. *The kilogram. The kilogram is now the only physical unit defined in terms of a material standard—this platinum–iridium cylinder was constructed in 1878 and is held by BIPM.*

of the force between a pair of current-carrying conductors and is thus related to the kilogram. If the process were reversed and the ampere were defined in terms of fundamental constants, the kilogram could be derived.

Determination of the values of the fundamental constants is discussed in section 16.3. Possibilities which have been explored include the use of the gyromagnetic ratio of the proton [7, 8], the Faraday constant and Avogadro's number [9]. At present these all have accuracies in the region of 1 in 10^6, which is about three orders of magnitude short of what is needed to replace the material standard.

The most promising method now appears to be that devised by

See also
p 1260

Figure 16.2. *The NPL moving-coil experiment—an ongoing attempt to replace the definition of the kilogram in terms of a material standard by relating its value to electrical phenomena.*

Kibble at NPL in which a coil suspended from a balance is placed partly in the field of a permanent magnet (figure 16.2). Two measurements are made: first, a known current flowing through the stationary coil is balanced against a known mass and, second, the coil is moved through the field with known velocity and the induced voltage measured. From these the mass can be related to the electrical quantities, without direct involvement of the fundamental constants and, since the same equipment is used, a number of uncertain factors cancel out [10, 11], but there is still much work to be done before this method is fully evaluated.

16.2.2. Length

16.2.2.1. The International Metre. The metre started in the same way as the kilogram, defined in terms of a material standard, but its subsequent history has been different: after seventy years it was redefined in terms of a physical quantity, wavelength, and subsequently in terms of a fundamental constant, the speed of light.

Using the same platinum–iridium alloy as for the kilogram, a number of prototypes of the metre were constructed and one of these was selected as the international standard metre and ratified by the CGPM at its first meeting in 1889. However, the metre was not formally defined as the length of the international prototype until the Seventh CGPM in 1927. A number of copies were made and distributed to national laboratories and BIPM as reference standards. An international comparison to verify the values of the national metres started in 1921 and continued for some fifteen years, the results being published in 1940.

The metre prototypes consisted of a Pt–Ir bar of x-shaped cross-section of overall length about 120 cm. On the flat central strip three transverse lines were engraved at each end, the distance between the two middle lines defining the metre [1]. The conditions of use had to be closely specified and the 1927 definition in full stated [6]

> The unit of length is the metre, defined by the distance, at 0°, between the axes of the two central lines marked on the bar of platinum–iridium kept at the BIPM and declared Prototype of the metre by the 1st CGPM, this bar being subject to standard atmospheric pressure and supported on two cylinders of at least one centimetre diameter, symmetrically placed in the same horizontal plane at a distance of 571 mm from each other.

BIPM was involved, from its early days, in the calibration of scales for the measurement of lengths other than one metre and of surveying tapes. A 'geodetic baseline' was installed for the latter purpose, using a number of microscopes and travelling scales and capable of calibrating a 24 m wire with an uncertainty of about 10 μm. A number of techniques for accurately comparing the lengths of metre standards were developed in the early part of this century: they are described in some detail by Johnson [12]. Using the Brunner comparator, BIPM could compare two standards with a precision of 0.1 μm [1].

16.2.2.2. Wavelength standards. The idea that the wavelength of light could be used as a measure of length is perhaps not quite as old as the wave theory itself but certainly dates back to the early years of the nineteenth century. In 1859 Maxwell suggested that the sodium yellow line could be used as a standard. BIPM used interferometers to monitor length standards and in 1893 Michelson provided one which he used at

BIPM to measure the wavelength of the cadmium red line, which later (1927) became the length standard for spectroscopy and the basis for the angstrom. In the period up to 1940 some nine determinations of the metre in terms of wavelengths were made in various laboratories. In the process much was learned about the detailed characteristics of various spectral lines and what features had to be taken into account for the most accurate measurements. Doppler effect, hyperfine structure and separation of isotopes were just some of the problems which had to be tackled. After World War II the idea of a wavelength standard was given much consideration by CIPM and its Consultative Committee for the Definition of the Metre. About 1950 the choice of a suitable isotope was narrowed down to cadmium-114, mercury-198 and krypton-84 or -86. BIPM examined in detail the profiles of lines from these and ways in which they could be generated. In the end, CCDM recommended use of the krypton-86 orange line, produced by a cold-cathode discharge lamp cooled to liquid nitrogen temperature [13].

16.2.2.3. The speed of light. Section 16.3.2 describes how the measurement of the value of the speed of light developed until it became one of the most accurately known physical constants. Then in 1983 the Seventeenth CGPM redefined the metre in the following terms

See also
p 1262

> The metre is the length of the path travelled by light in vacuum during a time interval of 1/299 792 458 of a second.

Since laser interferometry had played a large part in leading up to this decision, most of the facilities needed to relate reference standards to the definition were already in existence and the change enabled improved precision to be achieved with relatively minor alterations in procedures.

16.2.2.4. Applications. There are, of course, innumerable applications where accurate measurement of length is needed. For trade purposes line standards are widely used. In engineering, gauge blocks and similar instruments have to be calibrated. For surveying there are tapes. All these and other requirements have to be met by national services in legal metrology with measurements traceable through the national standards laboratory.

In the early part of the century, optical comparators were in general use [12]. With the change to wavelength standards, interferometric methods arrived and have been adapted to use lasers. In some cases they have influenced the requirements. At BIPM a technique for standardizing 24 m baselines for tape measurement was developed using Fabry–Perot étalons [14], and distance-measuring instruments for surveying have been developed: the first of these was the mekometer [15] which uses light polarization—modulated at typically 500 MHz. By comparing the phases of the transmitted and reflected waves, accuracies of the

Samuel Wesley Stratton

(American, 1861–1931)

Dr Stratton was born in Litchfield, Illinois, the son of a farmer. He became interested in farm machinery and other mechanical devices and went to the University of Illinois where he read mechanical engineering, followed by research. In 1892 he moved to the University of Chicago at the invitation of Albert Michelson, with whom he did research on instrumentation. He also became interested in defence matters and served as a US Navy Lieutenant during the Spanish–American war of 1898.

He was appointed first Director of the National Bureau of Standards when it was founded in 1901. He was responsible for finding a site for the Bureau, for planning the laboratories and for deciding its programme of work. This required constant interaction with Congress and he acquired a reputation as 'a scientific politician'. World War I made great demands on NBS and Stratton's defence background was invaluable in deciding the projects to be undertaken, including aeronautical research and work for the Navy.

After the war, Stratton used his contacts with industry to redirect the Bureau's programme along lines of more commercial interest: this brought him into contact with the Secretary of Commerce, Herbert Hoover, who, in 1923, arranged for Stratton to become President of the Massachusetts Institute of Technology. Again he took a very active interest in steering the programme of work, while maintaining his interest in NBS as a member of the Visiting Committee. In October 1931 he died from a heart attack while dictating an article on Thomas Edison. His broad background in science and engineering, his sound judgement of national needs for research and his political sense made him the ideal first Director of NBS.

order of a millimetre per kilometre are possible. This, and its further developments such as the Georan, which uses two wavelengths, have found applications in surveying, civil engineering and geodesy [16].

For scientific purposes there has been a continuing demand for spectral lines of accurately known wavelength. In 1965 the International Astronomical Union recognized the wavelength of the Kr-86 line used to define the metre and four other wavelengths each from Kr-86, Hg-198 and Cd-114, covering the wavelength range from 435 to 645 nm with uncertainties from 2 to 7 parts in 10^8 [17]. With the development of lasers other more accurate sources have become available. Following the redefinition of the metre in 1983, the CIPM recommended five stabilized laser standards [18], and recognized standards are now available throughout the infrared region.

For specialized purposes, other units have been used. The Astronomical Unit (AU—approximately the radius of the Earth's orbit round the Sun) is the distance at which the Gaussian gravitational constant, k, has the value 0.017 202 098 95. The parsec (from a *parallax* of one *second* of arc) is the distance at which 1 AU subtends an angle of 1 second: it is 3.086×10^{16} m. The light-year (9.46×10^{15} m) is perhaps more easily understood and is used in popular publications on astronomy.

At the other end of the scale we have the angstrom used in spectroscopy and named after the Swedish physicist, A J Ångström who, in 1868, published a diagram of the 'normal solar spectrum' in which the wavelengths were given in ten-millionths of a millimetre. It was defined in terms of the cadmium red line whose wavelength in normal air was 6438.4696 Å. After the change of the definition of the metre to that using the speed of light, the angstrom was redefined as exactly 10^{-10} m. The former 'X unit' or 'Siegbahn unit', about 10^{-13} m, which was formerly used to express the wavelength of x-rays, was abandoned in 1948 in favour of the angstrom.

CIPM does not, of course, encourage the use of units outside the SI system. The X-unit is included in the list of 'units generally deprecated' and the angstrom as 'in use temporarily', though the use of the nanometre (10 Å) is preferred.

16.2.3. Time and frequency

16.2.3.1. Background. The measurement of time dates from prehistory [19]. The earliest clocks were forms of sundials and water clocks. Oscillating systems, linking time to frequency, were not introduced until the middle ages. The earliest of these was the foliot, from about 1300, in which a pivoted bar moved to and fro to allow a toothed wheel to rotate: this was not a sinusoidal oscillator, but perhaps more in the nature of a multivibrator. The pendulum clock, a true oscillator, was invented about

1660 and the spring-controlled balance wheel twenty years later. These have remained the principles of mechanical clocks.

Until the 1960s time scales were considered to be almost exclusively the province of astronomers, although some pragmatic steps were taken to meet everyday needs. Local time was in general use until the mid-nineteenth century, so the time in Bristol, for example, lagged that of London by ten minutes. The coming of the railways led to a common 'railway time' and in 1880 'Greenwich Time' became the legal standard for Britain. At an international conference in Washington in 1884 it was agreed that the world should be split into a number of time zones based on Greenwich time: these have survived with little change to the present day.

The growth of industry in the nineteenth century led to a demand for checking the accuracy of clocks and watches which astronomers were not equipped to meet. In Britain, the British Association took over the disused Kew Observatory and set up facilities for testing these and other instruments. Responsibility for this passed to NPL on its founding in 1900. It is perhaps interesting to note that NPL tested 200 watches for the London Olympic Games in 1948.

16.2.3.2. Time keepers and frequency standards. The first pendulum clocks had a variation of about ten seconds a day in their time-keeping. More accurate operation was needed, particularly for the purposes of navigation on long sea voyages. Sources of error included temperature changes with their effect on the length of the pendulum, the effect of barometric pressure, and the need to couple mechanically or electrically to the pendulum to keep it swinging and to drive the rest of the mechanism. By 1900 much work had been done on these and stabilities of about 0.01 second per day had been achieved. The ultimate development of the pendulum clock was the Shortt clock which reached its final form in 1924 (figure 16.3). It used two pendulums: one was free-swinging except for a minimum period every half-minute when a driving impulse was applied and a synchronizing signal provided for the second, 'slave' pendulum which drove the escapement. This made possible a variation of only about 0.001 second per day. One of these clocks was used at the National Physical Laboratory to provide time signals interpolating between signals from astronomical observatories: this was used for accurate time interval measurement purposes, as in the checking of chronometers.

Another form of clock based on an accurate frequency was developed at NPL using a tuning fork [20], starting in 1922 with an improved version in 1931. It used an elinvar fork between the coils of an electromagnet driven by a thermionic valve. In an airtight enclosure and at constant temperature, it had a stability of about 5 in 10^8 over one

Figure 16.3. *The Shortt clock—the ultimate pendulum clock with master and slave pendulums used by NPL in the 1920s for time interval measurements.*

hour and 3 in 10^7 over a week. It remained in use until superseded by the quartz clock.

The use of the piezoelectric properties of quartz to control accurately the frequency of an oscillatory circuit is well known. Early work was done by Cady in 1921 and the first application to time measurement was made by Marrison, of the Bell Telephone Laboratories, in 1927. Work at NPL led to the development of the 'Essen Ring' which became widely used as a very stable standard of time and frequency. Stabilities of 4 in 10^{10} over one hour and 1 in 10^8 over a month were achieved [21].

Even though they have been superseded by atomic clocks for the most accurate purposes, quartz oscillators are still in widespread use for a great variety of applications, scientific, industrial, and even for cheap wrist watches which are stable to perhaps 10 seconds in a year. The properties of quartz change slightly with age. It has been proposed that this might be overcome by using an external reference. In 1949 H Lyons at NBS developed a system of this kind using an absorption line in ammonia as the reference; a long-term frequency stability of about 1 in 10^8 was demonstrated [22]. This was, in effect, the first atomic clock (figure 16.4). Since then many other possible systems have been investigated [23–26].

In the 1920s and 1930s a number of experiments showed that state transitions could occur in beams of atoms passing through magnetic and

Figure 16.4. *The first atomic clock—constructed at NBS in 1949, this used a microwave absorption line in the ammonia molecule to control the frequency of an oscillator to about 1 part in 20 million.*

radio-frequency fields [23]. In 1938 Rabi and his colleagues demonstrated a resonance curve for the transition between two states in the lithium nucleus [27], and in 1950 Ramsey showed that a much sharper resonance could be obtained by using two separate regions of RF field with a drift space between them [28]. This principle was used in the first operational caesium clock developed at NPL by Essen and his colleagues and brought into use in 1955 [29]. In the caesium clock, the ground state is split into two components with an energy difference corresponding to a microwave

Figure 16.5. *The NPL caesium clock of the 1970s.*

frequency of about 9192 MHz (figure 16.5). In the Essen version, the beam of atoms from an oven passed through a magnet which separated the two states. The atoms in one state then passed through a microwave cavity about 50 cm long with a collimating slit in the middle and a longitudinal magnetic field, and finally through a second cavity, coupled to the first and a further magnet designed to deflect any atoms whose state had changed on to a detector. By varying the frequency of the applied microwave field the resonance pattern could be explored: the central peak was found to have a width in the region of 500 Hz. By designing a circuit to lock on to this, a stable source was obtained with a standard deviation of frequency of 1 in 10^{10}.

There have been many further developments in caesium clocks and, as well as being used in national laboratories, they are widely available commercially, some in portable form. The best standards now have accuracies of about 1 in 10^{14} [30–32]. Particularly interesting work has been carried out at NRC in Canada and PTB in Germany [33,34].

Many other atomic systems have also been studied [26]. These include rubidium standards, which are commercially available, thallium and hydrogen beam systems and a variety of masers and lasers. The hydrogen maser has very good short-term stability and is in use in many laboratories, often supplementing caesium standards.

16.2.3.3. Time scales and the second. At the beginning of this century time scales were based entirely on astronomical observations and the second was defined as 1/86 400 of the mean solar day. Because of the movement of the Earth in its orbit round the Sun, the length of the true solar day, the interval between successive crossings of the meridian by the Sun, varies during the year by some seconds, and the time scale based on it differs from a scale based on the mean solar day, averaged over the year, by a maximum of sixteen minutes in November. The difference is defined by 'the equation of time' [35].

For astronomical purposes the position of the Sun is less important and sidereal time is used. This is based on the time at which stars cross the meridian. There are corrections for minor changes in the Earth's axis of rotation: as measurement accuracy improved it was observed that there can be unpredictable changes in the Earth's rate of rotation caused by structural changes which affect its moment of inertia.

A variety of other time scales have been used for specific purposes. In 1952 the International Astronomical Union (IAU) introduced Ephemeris Time for dynamical calculations. Effectively it is based on the length of the tropical year 1900 and its disadvantage for general use is that it is accurately known only years after the event because of the need to process the observations [36–38]. Other scales, following the introduction of the atomic clock, are mentioned below.

As the accuracy of time measurement improved, first with quartz clocks and later with caesium clocks, it became clear that the definition of the second in terms of the mean solar year was no longer satisfactory. In 1956 the CIPM set up its Consultative Committee on the Definition of the Second (CCDS) to study the problem, and after much discussion the CGPM, in 1967–8, adopted the following definition

> The second is the duration of 9 192 631 770 periods of the radiation corresponding to the transition between the two hyperfine levels of the ground state of the caesium 133 atom.

This remains the definition of the second.

From 1955 the Bureau International de l'Heure (BIH) maintained a mean atomic scale known as AM by comparing clocks in different countries by means of low-frequency radio signals. In 1970 the name was changed to International Atomic Time (TAI). By then mean sidereal time had been improved by correcting certain irregularities due to Earth motion and was known as Universal Time (UT). To maintain continuity it was arranged that TAI and UT would coincide on 1 January 1958. For everyday use it is necessary to take account of the Earth's rotation and from 1 January 1972 Coordinated Universal Time (UTC) was established with the second markers coinciding with those of TAI but 'leap seconds' added or subtracted from time to time to keep the hours and minutes aligned with the solar day. This is the scale now generally used for such purposes as broadcast time signals.

BIPM had of course long been concerned with the realization of the second as a unit but after the introduction of TAI it became increasingly concerned also with time scales. CCDS was deeply involved in advising BIH and in 1971 the Fourteenth CGPM endorsed TAI and gave a formal definition of it. The Fifteenth CGPM in 1975 endorsed the use of UTC. In 1987 responsibility for time scales was transferred to BIPM and BIH ceased to exist. The other part of its programme, concerned with Earth rotation, was retained at the Paris Observatory under the new title of the International Earth Rotation Service (IERS). BIPM is continuing to maintain and disseminate TAI [39] and has now set up its own time-keeping unit using caesium clocks.

16.2.3.4. The dissemination of time and frequency. A very important role of observatories has been to make time scales available to those who need to use them. New communication techniques have always had the transmission of time signals as an early task. Just before 1800 the British Admiralty set up a chain of semaphore stations to provide a link from London to the dockyard at Portsmouth and a daily time signal was transmitted (one wonders how they managed on foggy days). When the electric telegraph was developed it was widely used to transmit time signals. Radio was also used at an early stage. In 1908 the Bureau des Longitudes in France proposed that the radiotelegraph station which had been set up on the Eiffel Tower for military purposes should be used to transmit time hourly. At the 1912 conference which led to the setting up of the BIH a programme of time and weather signals from radio stations in various parts of the world was started. This enabled ships at sea to check their chronometers. The stations chosen all operated on the long wavelength of 2500 metres, thus ensuring reception over long distances [40].

These services were primarily for military or technical purposes. From the very early days of broadcasting, time signals were transmitted for reception by the general public. In Britain, the Royal Greenwich Observatory provided the timing for the BBC's 'six pips', referred to its quartz clocks after 1942; and a number of broadcast transmitters in different parts of the world have accurately controlled carrier frequencies. These include the BBC 198 kHz transmitter at Droitwich whose frequency is controlled to within 2 in 10^{12}.

Since the 1920s a number of countries have provided special radio transmissions for accurate time and frequency signals, on a range of frequencies from VLF to HF. But as alternative methods have evolved many of these have been terminated. One development which depends on accurate timing and has provided useful signals has been the Loran-C navigation system. This uses a number of stations in different locations transmitting accurately synchronized pulses on 100 kHz. By receiving these pulses and comparing their times of arrival, a receiver can determine its position with an error of 50–500 m, to a distance of 2000 km or more. At a known position, the signals provide time comparisons to about 1 μs, equivalent to an uncertainty in frequency of about 1 in 10^{12}. This method has been used to combine national scales to form TAI.

To a large extent radio methods have been replaced by satellites as a means of time comparison. Travelling clocks have been used by the US Naval Observatory [41] and it is much involved in GPS, the US Navy's Global Positioning Satellite System, in which each satellite carries atomic clocks and from which time can be derived to about 10 ns. There has also been recent work at NBS [42].

16.2.4. Temperature

16.2.4.1. Background. The first thermometers were invented about 1600 and many different types have followed [43]. These were of practical use but the scientific significance of temperature did not become clear until the nineteenth century. In the words of the Ruhemanns 'Not until Lord Kelvin showed that the second law of thermodynamics supplied us with an absolute scale of temperature was a thermometer anything more than an arbitrary definition of a zero point and a degree' [44]. Kelvin showed that the Carnot cycle could be used to define a thermodynamic temperature scale independent of the material used, and it was later shown that a thermometer using a 'perfect' gas at zero pressure would measure thermodynamic temperature. By the end of the nineteenth century the use of the thermodynamic scale and of the gas thermometer for its realization had been accepted. The present century has seen a wide variety of developments in techniques of temperature measurement and their applications [43]. We have space only to consider a few of the fundamental aspects here.

16.2.4.2. Practical temperature scales. From its foundation in 1875, BIPM was actively involved in thermometry, mainly for the practical reason that its standards of the kilogram and the metre needed to be kept at accurately known temperatures. The hydrogen thermometer was selected as giving the best approximation to the thermodynamic scale and in 1887 the CIPM adopted this for the standard, using as fixed points the temperatures of melting ice and boiling water in specified conditions.

With the development of national standards laboratories early in this century other requirements emerged, in particular the need for a greater temperature range. In 1911 it was proposed that the thermodynamic scale should be officially adopted with a number of additional fixed points to extend the range. The outbreak of World War I delayed matters but, after long discussion, the Seventh CGPM authorized the

International Temperature Scale for temperatures above −190 °C. While accepting the thermodynamic scale as formally defining temperature, it was agreed that a practical way of realizing it should be stated. Six fixed points were specified, from the boiling point of oxygen to the freezing point of gold, together with the instruments to be used (platinum resistance thermometer and Pt/PtRh thermocouple) with their interpolation formulae.

This scale, which became known as ITS-27, was widely welcomed and came into general use. In time the introduction of new techniques showed that some revision was desirable and in 1939 the CIPM set up a Consultative Committee on Thermometry (CCT) to take advice and consider what changes should be made [46]. Again, war intervened, but in 1948 the Ninth CGPM adopted the revised International Practical Temperature Scale, IPTS-48. It also decided to change the name of the unit from the 'degree Centigrade' to the 'degree Celsius'. The scale was very similar to the earlier one, with a few refinements. The differences between them were insignificant below about 600 °C but rose considerably at higher temperatures.

It was realized that for the thermodynamic scale it was anomalous to have a unit based on the interval between two fixed points: all that is needed is an absolute zero and one fixed point. After much discussion the Tenth CGPM decided in 1954

> to define the thermodynamic temperature scale by choosing the triple point of water as the fundamental fixed point, and assigning to it the temperature 273.16 degrees Kelvin, exactly.

This made the degree Kelvin (now simply the kelvin) as nearly equal to the degree Celsius as possible.

In 1968 the CIPM recommended a further revision of the scale and IPTS-68 was formally adopted by the Fifteenth CGPM in 1975 [47,48]. New fixed points were added to extend the range and the interpolation formulae were revised to align the practical scale with the thermodynamic scale as closely as new techniques had made possible.

In 1990 yet another scale was introduced [49], known as ITS-90. The lowest temperature on the scale was reduced to 0.65 K and the temperature assigned to the sixteen fixed points were adjusted again to match thermodynamic temperatures as closely as possible. The structure of the scale illustrates how recent developments in physics have influenced the measurement process:

(i) between 0.65 and 5.0 K, the temperature, T_{90}, is defined in terms of the law relating vapour pressure and temperature for helium-3 and helium-4;

(ii) between 3.0 K and the triple point of neon (25 K), it is defined in terms of a helium gas thermometer;

(iii) between the triple point of hydrogen (14 K) and the freezing point of silver (1235 K) it uses platinum resistance thermometers;

(iv) above the freezing point of silver it uses the Planck radiation law.

It will be observed that some of the defining ranges overlap: this allows consistency of measures over limited ranges without sharp discontinuities. The differences between T_{90} and T_{68} are within 0.02 K up to 200 °C and within 0.3 K to 1000 °C.

16.2.5. *Photometry*

Even more than thermometry, photometric measurements have application mainly for industrial and domestic purposes and have not made an important contribution to the science of physics. However, the major national standards laboratories have been involved in these measurements from their early days and BIPM more recently: the candela is the SI base unit of luminous intensity. So we will give a very brief account of the way photometry units and standards have developed.

The earliest official standard of illumination was probably the British Parliamentary Candle introduced by the Metropolitan Gas Referees in 1860 for testing gas lighting. Other countries developed different standards and there was little coordination between them, perhaps because what was being sought was not a standard simply of a physical quantity but of its subjective effect. In 1909 agreement was reached between NBS, NPL and LCIE (the French Laboratoire Centrale d'Electricité) to use a common unit based on carbon filament lamps. At a meeting of the Commission Internationale de l'Eclairage (CIE) in 1921, Belgium, Italy, Spain and Switzerland joined in the agreement and the unit was named the 'International Candle'.

Further developments led to CIPM establishing a Consultative Committee for Photometry (CCP) in 1937. They recommended the adoption of a unit, the 'new candle' based on the brightness of a black-body radiator. In 1948 this was formally approved by the Ninth CGPM, with the name changed to 'candela'. The CGPM also approved the definition of the lumen and when the SI units were adopted in 1960, they included the candela as the base unit and the lumen and the lux as derived units [50–52].

As photometric techniques improved it became possible to relate measures of illumination to objective radiometric measurements and the disadvantages of the black-body source became apparent. In 1979 the Sixteenth CGPM decided that a complete change was needed and adopted an entirely new definition of the candela

The candela is the luminous intensity, in a given direction,
of a source that emits monochromatic radiation of frequency

540 × 10^{12} Hz and that has a radiant intensity in that direction of 1/683 watt per steradian.

The frequency of 540 × 10^{12} Hz, corresponding to a wavelength of 555 nm, was chosen as the peak of the sensitivity curve of the human eye [53].

16.2.6. *Electrical units and standards*

16.2.6.1. Background. The definitions of the electrical units and the availability of standards for their realization have been closely linked with the development of the science and its industrial applications. The phenomena of electromagnetism were first recognized in the sixteenth century by Gilbert, but it was not until Faraday's experiments were followed by Maxwell's equations in the nineteenth century that it really became a major branch of physics. The electric telegraph was introduced in the 1830s and the first supplies of power for electric lighting in the 1880s. In 1861 the British Association for the Advancement of Science set up an Electrical Standards Committee which played the leading part in defining a rational system of units and developing practical standards. It was received with some scepticism by the telegraph industry who felt that 'The gentlemen who constitute the Committee are but little connected with practical telegraphy ... and may be induced simply to recommend the adoption of Weber's absolute units, or some other units of a magnitude ill adapted to the peculiar and various requirements of the electric telegraph.' [54]. In fact the Committee showed great perception in distinguishing between the roles of units and standards and in building a useful practical system on a sound scientific base.

As there was at that time no coherent system for electrical quantities it gave this matter high priority. About ten years earlier Weber had proposed the centimetre–gram–second system and the Committee accepted this as the base while recognizing that multiples of the units to match practical needs would be desirable. It reported back to the Association in 1863 and the practical units were defined as the ohm, the ampere and the volt, being respectively 10^9, 10^{-1} and 10^8 times the CGS electromagnetic units.

The British Association also sponsored a large programme of work to construct reference standards and to make absolute determinations of their value. The first wire-wound resistors were constructed and their values determined in 1863 by Maxwell, Fleeming Jenkin and Balfour Stewart. The method used had been devised by Weber and used a coil rotating in the Earth's field which both caused a current to pass through the resistor and deflected a magnet. The ampere was determined in a similar manner using Weber's electrodynamometer. After some years, the main activity was transferred to the Cavendish Laboratory

Richard Tetley Glazebrook

(British, 1854–1935)

Sir Richard Glazebrook was born in Liverpool. After taking his degree in mathematics at Trinity College, Cambridge, he undertook research in optics at the Cavendish Laboratory, under Maxwell. In 1883 the British Association appointed him secretary of its Committee on Electrical Standards. He was responsible for maintaining the BA standard resistors and capacitors and for absolute determinations of the units.

In 1900, after a year as the Principal of University College, Liverpool, he became the first Director of the National Physical Laboratory. His experience, his clearsightedness and his strategic abilities enabled him to lead NPL successfully through its early years and to impress the Government with the vital part it had to play in developing new technologies, including aerodynamics, ship testing, roads and radio. About 1917 he negotiated with the Treasury new arrangements for the control of NPL as a government establishment in DSIR, the Department of Scientific and Industrial Research.

In 1919 he retired from NPL but continued to advise on its programme. He became Zaharoff Professor of Aviation at Imperial College. He was knighted in 1917 and received many other honours and awards. He was a major figure in setting the pattern of twentieth century research in applied physics in Britain and internationally.

in Cambridge and Dr Glazebrook, later Sir Richard and the first Director of NPL, took charge of it [55, 56].

In 1881, at a congress in Paris, the BA system of units was accepted internationally. It was agreed that a column of mercury should be adopted as a reference standard for the ohm, as had been proposed by Siemens, with specified dimensions. This was known as the 'legal ohm'.

Although BIPM was not officially involved in electrical standards at that time, Dr Benoit of the Bureau made a number of copies of the legal ohm for the French government. But, as the value differed slightly from the BA value, it was never recognized in Britain.

This was the first step towards setting up a practical system of standards which could be used in industry and would not be affected by the occasional changes in absolute values which continuing research was revealing. In 1893 an international conference in Chicago took this process further and it culminated in the London conference of 1908. It was here agreed that International Electrical Units would be defined in terms of material standards:

(i) the International Ohm as the resistance of a column of mercury at the temperature of melting ice, 14.4521 grams in mass, of constant cross-sectional area, and of a length of 106.300 centimetres;

(ii) the International Ampere as the current which, when passed through a solution of silver nitrate in water deposits silver at a rate of 0.001 118 00 of a gram per second;

(iii) the International Volt as the pressure to produce a current of 1 ampere through 1 ohm; subsequently an alternative definition was accepted on the basis that a Weston cell at a temperature of 20 °C has an EMF of 1.0183 International Volts.

The London Conference very effectively cleared the air. The International Units provided a satisfactory practical system for industry and other users for many years, while allowing research on absolute units to carry on unhindered.

16.2.6.2. The role of the national laboratories. Although BIPM maintained some facilities for electrical measurements, which in 1907 again provided standards for the French government, they were not formally involved in electrical standards until later. In 1921 the Sixth CGPM amended the Convention of the Metre to include electrical units. In 1927 the Seventh CGPM set up the Consultative Committee for Electricity (CCE): this met the following year and recommended that BIPM should play a full part in international electrical standards.

Up to that time, the national laboratories had been responsible for establishing and coordinating standards. Indeed, the demand for standards to meet the need of the rapidly growing electrical industry played a major part in the decisions to set up national laboratories. The first of these, the Physikalisch-Technische Reichsanstalt (PTR) in Germany, was founded in 1883 with strong support from Siemens and with Helmholtz as its first director.

In Britain the Electric Lighting Acts of 1882 and 1888 made the Board of Trade (originally set up in 1696 to look after the trade of the American colonies and later charged with the oversight of weights and measures) responsible for the regulation of the electrical supply industry. This included the provision of standards and the type approval of electricity meters. It gave legal endorsement to the International Units, provided national standards based on the BA standards and set up a laboratory for the verification of instruments. But it was recognized that it would not be in a position to develop new standards or take part in international activities, so when NPL had established its programme, the Board of Trade standards were handed over to it in 1920: for legal reasons they were maintained as separate standards until about 1950.

Similar developments have taken place in other countries. In the United States, Congress was reluctant to sanction the setting up of a national laboratory, though after Edison established his research laboratory at Menlo Park in 1876 and others followed (but electrical instruments for export had to be sent to PTR in Germany for calibration) the case became overwhelming and NBS was established in 1901. In some countries the work is done jointly by industry and government: in France, for example, LCIE the Laboratoire Centrale des Industries Electriques, is jointly supported.

16.2.6.3. Absolute determinations. While the International Units met the needs of industry they were less satisfactory from the scientific point of view and national laboratories continued to look for better ways of realizing them from their CGS definitions. In the early 1900s, PTR, NBS, NPL and LCIE undertook a coordinated programme using a variety of current balances to determine the ampere in terms of the force between current-carrying coils (figure 16.6). The ohm was measured by the Lorenz machine (in which a conducting disc rotates in the field from a coil of known dimensions) and by Campbell's inductor whose mutual inductance is calculable from its dimensions. The results of these measurements were used to calibrate mercury resistance standards and Weston cells which had travelled between the participating laboratories [55]. Continuing work in this area after World War I led to BIPM's involvement in electrical standards and eventually to the SI system (section 16.2.6.4.).

After World War II further developments took place; of these two of the most important originated in NML in Australia. In 1956 the Thomson–Lampard theorem was published [57,58]. They showed that if the wall of a conducting cylinder of arbitrary cross-section is divided parallel to the axis into four parts with negligible gaps, the cross-capacitance between opposite pairs of faces is calculable and if they are made symmetrical the capacitance per unit length is given by

$$C = \varepsilon_0 (\ln 2)/\pi \ \text{F m}^{-1}.$$
$$= 1.953\,549\,04 \ \text{pF m}^{-1}.$$

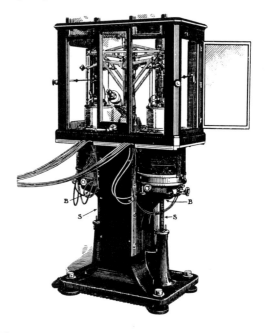

Figure 16.6. *The Ayrton–Jones current balance for the absolute determination of the ampere–initiated by the British Association and completed at NPL in 1907.*

Capacitors using this principle were constructed in NML, NBS, ETL and NPL and used by means of a quad-bridge relating capacitance and resistance to determine the absolute value of the ohm with an uncertainty in the region of 1 part in 10^7, rather better than an order of magnitude improvement on previous determinations [59].

An absolute determination of the volt can be made by measuring the force of attraction between two plane electrodes at different voltages. The force per unit area between two infinite planes with a voltage V between them and d metres apart in vacuum is given by

$$\varepsilon_0 V^2/(4\pi d) \text{ N m}^{-1}.$$

A number of measurements based on this were made using metal plates and a balance at NBS and ETL. In 1965 Clothier, of NML, proposed the use of a plate over a pool of mercury: the rise in the surface level of the mercury when a voltage was applied could be measured with great accuracy. This has achieved an uncertainty in the region of 3 in 10^7 [60, 61].

16.2.6.4. Later developments of electrical units. Following the recommendation of the CCE, BIPM in 1929 set up reference standards of the volt and the ohm and in 1932 the first international comparison took place when four national laboratories sent standards to BIPM. Further comparisons, with a growing number of participants, took place at intervals of two years until they were interrupted by World War II. They resumed later, but from 1957 only took place every three years.

The 1928 CCE concluded that the International Volt, Ohm and Ampere should not be perpetuated but that they should be replaced by an absolute system as soon as the relations between the two sets of units could be established with adequate accuracy. After further work, CIPM decided that the absolute system should come into use on 1 January 1948. The basis was the modified CGS system proposed by Giorgi in 1901 [62, 63]. In this, the MKSA system, the units are of a magnitude more suitable for practical measurements and the awkward 4π was removed from the definition of the ampere. The definitions ratified by the Ninth CGPM are of the ampere in terms of the force between two parallel conductors, the volt as the potential difference between two points of a conductor carrying one ampere and dissipating one watt, and the ohm as the resistance in which one volt produces one ampere [6]. Other units were also defined and they were all taken over into the SI system when that was formally approved by the Eleventh CGPM in 1960, with the ampere designated as the base unit for electrical quantities.

However, the definition of the ampere does not lend itself to easy practical realization, so intercomparisons with BIPM continued and it was accepted by convention that the values of the BIPM standards would be taken as the volt and the ohm. By 1970 some ten national laboratories were sending their standards to BIPM for intercomparison every three years: the spreads in value were about 1 in 10^6 for the ohm and two or three times this for the volt [64]. In the light of the results, national laboratories have sometimes adjusted the values of their standards and BIPM also made an adjustment in 1968.

With the growing workload for BIPM and the limitation of accuracy caused by having to transport reference standards, particularly Weston cells, it became clear that an alternative procedure was desirable. As the work described in section 16.3 of this chapter proceeded, it became clear that physical constants could potentially replace material standards. There have been many suggestions for ways in which electrical standards might be derived. At one time the most promising appeared to be through the gyromagnetic ratio of the proton [7]. In practice, however, this had not achieved any significant improvement and the two effects which are now used are the Josephson effect for the volt and the quantum Hall effect for the ohm.

In 1962 Josephson predicted the effect which now bears his name [65]. If two superconductors are linked by a thin insulating gap irradiated by a frequency f, an applied DC voltage will produce a current and the current–voltage characteristic has voltage steps given by

See also
p 941

$$\Delta V = (h/2e)f$$

where h is Planck's constant and e the charge on the electron. This gave rise to a great deal of interest in metrological laboratories because it clearly offered a way of determining the volt in terms of frequency, accurately measurable, and fundamental constants. The difficulty was that our knowledge of the constants was not as precise as the precision with which voltage standards can be compared. But if we write the relation as

$$\Delta V = K_J f$$

where K_J, the Josephson constant, has a value to be agreed, this still, though not now an absolute method, provides a useful way of monitoring voltage standards and comparing them without the need to transport Weston cells. A number of national laboratories made measurements of the Josephson constant in terms of their volts and used it to monitor their standards. At its meeting in 1972, CCE took note of these developments and concluded that the BIPM volt on 1 January 1969, V_{69-BI}, was equal within half a part in a million to the voltage which would be produced by the Josephson effect in a junction irradiated at the frequency of 483 594.0 GHz (figure 16.7). Not all countries were prepared to adopt this value and it became clear that it was not in fact consistent with the SI value. In 1986 CCE set up a Working Group to review the evidence and after it reported back CCE recommended that the value

$$K_{J-90} = 483\,597.9 \text{ GHz V}^{-1}$$

should be adopted from 1 January 1990 and used where possible for monitoring national standards, with an uncertainty estimated to be 4 in 10^7. CCE made it clear that this represented a reference standard and did not imply any change in the SI definition of the units or in the values of the fundamental constants. It expressed the view that no change in the recommended value would be necessary in the foreseeable future.

One practical problem was the small voltage produced by a Josephson junction—about 10 mV for a frequency of 5 GHz, for example. This made it less than easy to monitor a 1 V standard, but work at NIST has since led to the development of arrays of up to 20000 junctions in series, which gets round this problem [66].

Monitoring the ohm is basically similar. If a semiconductor carries a current I in a direction at right angles to a magnetic field, a voltage V_H is generated at right angles to both. This is known as the Hall effect and the ratio

$$R_H = V_H / I$$

is known as the Hall resistance. It normally depends on the geometry and the strength of the field, but in 1980 von Klitzing demonstrated that in certain semiconductors where the current flow is effectively limited to two dimensions the Hall resistance is quantized and has the value

$$R_H = R_K / i$$

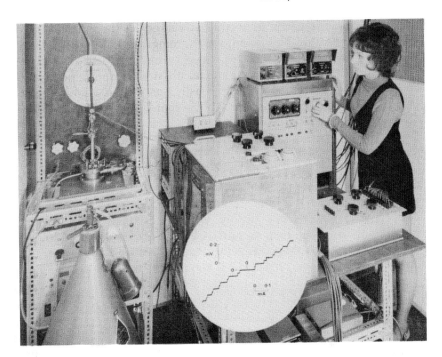

Figure 16.7. *The Josephson junction voltage standard at NPL—inset shows the current–voltage characteristic.*

where $i = 1, 2, 3, \ldots$ (depending on the field strength) and R_K is a constant known as the von Klitzing constant. Its theoretical value is h/e^2 [67–70]. Using this effect it is possible to set up a standard of resistance which can be used to monitor national ohms. As with Josephson, the equipment is not simple—it has to operate at liquid helium temperatures, for example—so that it can hardly be used as an everyday reference, but for comparing standards at the highest level without the need to transport them it has been very useful.

In 1986 the CCE set up another Working Group to consider the quantum Hall effect. After it reported in 1988, CCE recommended that the von Klitzing constant should be used from 1 January 1990 to monitor the values of national ohms and that its value should be

$$R_{K-90} = 25\,812.807 \text{ ohms}.$$

CCE again emphasized that this implied no change in the definition of the units or in the values of h and e but that it was simply a reference value to which no change was likely to be needed in the foreseeable future. The estimated uncertainty was 2 in 10^7.

The changes in connection with the volt and the ohm were announced by BIPM a year in advance of the date when they were to take place [71].

1259

At the same time the new temperature scale, ITS-90, was announced. Advice was given in some detail on the new procedures to be used and on how results should be expressed. This involved the national laboratories in a considerable amount of work not only in changing their own procedures but also in making sure that all laboratories dependent on them for calibrations were aware of the situation. Almost all national volts and ohms were effectively changed in value by some parts in a million and this had to be taken into account in maintaining calibration histories.

In fact, the process seems to have gone reasonably smoothly and we look forward to a more tranquil period for electrical measurements. It is somewhat intriguing that a major shake-up of electrical measurements seems to take place every twenty years: in 1908 we adopted the International Units, in 1928 the first CCE met, in 1948 the 'absolute' units were adopted, in 1968 a major readjustment of the values of the national volts and ohms was decided and in 1988 we were committed to Josephson and von Klitzing. One wonders what 2008 will bring.

These changes have all been accompanied by reductions in uncertainty. At the time of writing we seem to be ahead of the user demand, but this is unlikely to remain so. The periodic review of the values of the fundamental constants will continue and there are still possible new techniques to be explored. A useful survey of the present situation and future trends has been given by Kibble [10].

The other intriguing point is that although the thrust has repeatedly been towards standards based on absolute values and physical constants, we are in fact ending the century as we began it with electrical quantities referred to standards chosen by an international committee.

16.3. Physical constants

16.3.1. *Types of constant*
We come now to an application of measurement techniques which is fundamental to physics and which reacts back on the process of measurement itself, as we have just seen in relation to the electrical units.

Physical theory depends on a large number of quantities which have values which we believe to be constant over the whole of time and the whole of the universe (although this belief has occasionally been questioned). It is convenient to divide them into several categories:

(a) Mathematical constants, such as e and π whose values are fixed and are calculable from formulae to as many places of decimals as may be needed.

(b) Defined constants, such as the speed of light in vacuum, $c = 299\,792\,458$ m s^{-1} exactly, and the permeability of free space

$$\mu_0 = 4\pi \times 10^{-7} = 1.256\,637\ldots \times 10^{-6} \text{ H m}^{-1}.$$

Table 16.1. *Fundamental constants of physics (reference [59] p 2).*

Velocity of electromagnetic radiation in free space	c
Elementary charge	e
Mass of electron at rest	m_e
Mass of proton at rest	m_p
Avogadro constant	N_A
Planck constant	h
Universal gravitational constant	G
Boltzmann constant	k

From these may be derived the permittivity of free space

$$\varepsilon_0 = 1/\mu_0 c^2 = 8.854\,19\ldots \times 10^{-12}\ \text{F m}^{-1}$$

and the characteristic impedance of free space

$$Z_0 = (\mu_0/\varepsilon_0)^{1/2} = 376.730\ldots \Omega.$$

(c) Measured constants, which comprise the great majority of physical constants, including, for example, the gravitational constant, G, the Planck constant, h, the charge of the electron, e (not to be confused with $e = 2.718\ldots$ in (a) above), and many others. Their values are determined by measurement and many of them are closely interrelated. For this reason the most accurate values are derived not from individual measurements but by a least-squares treatment of all the available data. This is undertaken from time to time by CODATA, the Committee on Data for Science and Technology of the International Council of Scientific Unions (ICSU). The latest set of data on some hundred quantities was published in 1986 [72]: some changes have already been proposed [73].

These constants are often referred to, by CODATA for instance, as the fundamental physical constants. This can be somewhat confusing, as the links between them mean that the number of independent constants is very much less than the total number of those whose values are published. They are all perhaps fundamental in a sense in some branch of physics, but one may be forgiven for thinking that some are more fundamental than others. There is merit in the list of the fundamental constants of physics given by Petley (reference [59] p 2) and shown in table 16.1.

It is important to remember that constants may not be forever assigned to a particular class. Today π is for us a mathematical constant calculable from a formula to any accuracy we wish. This was not so in the third century BC when Archimedes 'measured the circle by inscribing and circumscribing polygons, increasing the number of sides till the polygons nearly met on the circle' [74]. In this way he arrived at a value between $3\frac{1}{7}$ and $3\frac{10}{71}$ (within three decimal places of the exact value). More recently the speed of light has changed from being a measured constant to a defined constant: after many determinations of its value, the CGPM in 1983 redefined the metre, thus fixing the speed of light (see section 16.2.2.3). Further measurements of its value would thus be meaningless. It is important to distinguish between this case and the apparently similar cases of the Josephson and von Klitzing constants for which CCE gave recommended values in 1988 for monitoring the volt and the ohm (section 16.2.6.4), but no change of definition of the units was involved, the constants remain measurable, and improved values may be recommended at some time in the future.

It is also worth noting that the values for the constants derived from coordinated experiments have not always been universally accepted. For example, Eddington derived a theoretical value of exactly 137 for the fine structure constant, α^{-1}, on the basis of the number of independent components of 'a complete energy tensor' [75]. Experimental values have been in the region of 137 but not exactly equal to it: the latest CODATA figure is 137.035 989 61 [72]. A more profound challenge is offered by cosmological theories which assume physical 'constants' in fact change over periods of time.

In this section we consider first the two 'classical' constants, c and G, whose values have been measured over several hundred years and are not closely linked to those of other constants. Then we discuss the ways in which the determination of linked groups of constants has become organized and finally we give examples of the coordinated measurements which have contributed to this. Many hundreds of such determinations have taken place in this century and clearly we do not have space to discuss all the experimental techniques in detail. We do, however, give some examples of the most significant developments and list references where further details can be found.

16.3.2. The speed of light

We have already mentioned how measurements of the speed of light eventually led to its becoming a defined quantity. These measurements have a long history.

In classical times it was believed that light travelled with an infinite velocity. Galileo doubted this and attempted to measure the velocity by covering and uncovering lanterns a couple of miles apart but of course this did not give a result. The first effective measurement was made by Roemer in 1676 by observing the time of the eclipses of the satellites of Jupiter from different points in the Earth's orbit. In 1849 Fizeau made the first successful measurement on Earth, using a rotating wheel as a timing shutter on a beam of light reflected from a mirror some miles distant. Foucault made similar measurements. But one of the most significant developments about this time was Maxwell's theory of electromagnetic waves, with his conclusion that the speed of light is given by

$$c = (\mu_0 \varepsilon_0)^{-1/2} = \nu\lambda$$

where ν is the frequency, λ the wavelength, and μ_0 and ε_0 are the permeability and permittivity of free space, as before. This opened the way to a variety of methods of measuring c which we will discuss later. Towards the end of the century there was much debate on the existence of the aether. It was expected that if the Earth was moving through the aether the speed of light would depend on the direction of transmission. The Michelson–Morley experiment of 1887 showed that this was not the case. This led to the idea of the Fitzgerald–Lorentz contraction and ultimately to Einstein's theory of relativity in 1905. So, even before the start of the present century, measurement had a profound effect on physical theory.

Throughout the twentieth century there has been a continuing interest in the value of the speed of light and many careful determinations of its value have been made, far too many for us to describe them all in detail here. A variety of methods have been used. In the first half of the century, Michelson made further measurements using the rotating mirror method [76], Mittelstaedt [77, 78], Hüttel [79] and Anderson [80, 81] used a basically similar method with a Kerr cell as the timing shutter, and Mercier [82] found the velocity of electromagnetic waves at about 50 MHz by observing standing waves on parallel transmission lines. But the most original method was that used by Rosa and Dorsey [83] in 1907 from the ratio of the electrostatic and electromagnetic electric units, which gave much better precision than any other method at that time. The results up to 1940 were reviewed by Birge [84]. He was somewhat critical of the error estimates in some of the earlier measurements and after detailed analysis of the results arrived at

$$c = 299\,776 \pm 4 \text{ km s}^{-1}$$

as the best available value.

The new technologies developed for defence purposes during World War II had profound influence on many aspects of metrology. Radio navigation results showed the need for better values of c and the opening up of new areas of the spectrum in the microwave and infrared regions made new measurements possible. Essen and his colleagues at NPL used a microwave cavity resonator to measure the wavelength of radiation of known frequency [85, 86]. His colleague, Froome, made similar measurements at higher frequencies (24 and 72 GHz) using a microwave interferometer [87, 88] (figure 16.8). Further Kerr cell measurements were made by a number of people, radar was used, and other geodetic

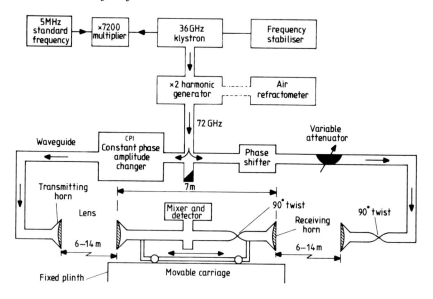

Figure 16.8. *The Froome interferometer for measurement of the speed of electromagnetic radiation at microwave frequencies.*

methods. Froome's results were the most accurate, with a quoted uncertainty of 0.1 km s^{-1} for his 1958 result of 299 792.5 km s^{-1}.

After 1970, the development of lasers with very high spectral purity made even better accuracies possible (figure 16.9). Frequencies were compared with the caesium standard using harmonic mixing and wavelengths were measured in terms of the krypton-86 length standard by means of intermediate calibrated lasers. A coordinated programme of work at NBS [89, 90], NPL [91, 92] and NRC [93] led to results with standard deviations in the region of 1 m s^{-1}. These measurements, and earlier results, are discussed in some detail by Petley [59]. They provide a clear example of how new technology can make possible a significant improvement in measurement accuracy—about two orders of magnitude in this case.

These measurements had a profound effect on the SI system of units, as we have already seen in section 16.2.2. In 1975, after a review of the results available, the CGPM recommended that the value $c = 299\,792\,458$ m s^{-1} should be used for the speed of light, with an uncertainty of $\pm 4 \times 10^{-9}$. Then in 1983 the Seventeenth CGPM used this value to redefine the metre as the length of the path travelled in vacuum in $1/299\,792\,458$ of a second. It also invited the CIPM to draw up rules for the practical realization of the metre, and this was done in terms of lasers stabilized by certain specified molecular transitions. Thus the speed of light ceased to be a measurable constant. No further determinations of its value will be made.

Figure 16.9. *The array of lasers used to measure the speed of light at NPL in the mid-1970s.*

16.3.3. *The constant of gravitation*

The other constant of which independent determinations have been made over several centuries is the gravitational constant, G. In the latter part of the seventeenth century, Newton developed his law of gravitation, starting reputedly from the relation between the fall of an apple and the motion of the planets and published in *Principia*. We can express the law as

$$F = Gm_1m_2r^{-2}$$

where F is the force between two masses m_1 and m_2 a distance r apart. Over the years much work has gone into determining the value of G and whether it really is a constant. Its value is not only of theoretical interest, but it provides a way of determining the mass, and hence the density, of the Earth.

Early determinations of G were made by measuring the deflection from the vertical of a plumbline near a large mountain. The first attempt to do this in 1740 at Mount Chimborazo in Peru by Bouguer was ruined by atrocious weather conditions, and although it was clear that the effect occurred, the results were not numerically significant. Better results were obtained by Maskelyne, then Astronomer Royal, in 1774 on Mount Schiehallion in Perthshire. Others measured G by comparing the values of the acceleration due to gravity at the surface of the Earth and the bottom of a deep mine, but it was clear that none of these methods could produce very accurate results and they were superseded by laboratory measurements of the attractive force between two masses. The first of these used a torsion balance; it was started by Michell, who died before the construction of the apparatus was complete. Cavendish took over the equipment and carried out the experiment, publishing his results in 1798. He obtained the value of 5.448 for the density of the Earth, corresponding to

$$G = 6.754 \times 10^{-11} \text{ m}^3 \text{ kg}^{-1} \text{ s}^{-2}.$$

Many other determinations followed, particularly in the latter half of last century, including the work of Boys, who used a torsion balance with an arm rotating about a vertical axis, similar to that of Cavendish, and of Poynting, who used a balance with an arm rotating in the vertical plane. Two large masses were attached to the ends of the arm, with a central knife-edge so arranged that gravity provided the restoring force. A very large mass was mounted on a turntable so that it could be positioned under either of the two masses. By moving it from one position to the other and observing the deflections, the forces, and hence G, could be calculated. Useful surveys of these measurements of G were given by Boys [94] and by Poynting and Thomson [95]. The former gives a table of the values achieved at the end of last century.

In the present century there have been many further determinations of G, all using some form of the torsion balance. Heyl [96] used an oscillation method and Zahradnicek [97] a resonance method. More recently there have been further results published by Karagioz [98] and Sagitov [99] in the USSR and Luther and Towler [100] in the USA. The results have been reviewed by Lowry *et al* [101] and by Petley (reference [59] pp 226–36). The CODATA recommended value is now

$$G = 6.67259(85) \times 10^{-11} \text{ m}^3 \text{ kg}^{-1} \text{ s}^{-2}$$

with an uncertainty of about 1 part in 10^4 [73].

Interest in the gravitational constant is not only in its absolute value. Gravitational theory is central to studies of the development of the universe. Newton's simple inverse-square law has in this century been called into question in a number of ways. It is a fascinating story but unfortunately we do not have the space to go into it here in detail. We will try to indicate some of the main directions which have been explored and the measurements which have affected them, and indicate where further details can be found. A major change was brought about by Einstein's theory of relativity which led to the law of gravitation based on tensors [102]. There have also been attempts to develop a quantum theory of gravitation [103].

There has been much interest in trying to determine whether G really is constant over vast distances and long periods of time, an

See also p 286

interest stimulated by the publication in 1937 by Dirac of his theory of multiplicative creation cosmology, in which G and other 'constants' are changing. On this theory the change in G might be in the region of 10^{-11} per year—quite out of range of direct measurements, of course. This led to many attempts to estimate limits to any variation in G, mainly using palaeontological evidence of changes in the Earth's temperature, changes in the orbits of the Earth and other planets, continental drift, clustering of galaxies, and also more recently using radar and lidar measurements of the distances of the Moon and planets, and other phenomena. A survey of the researches and a summary of the results are given by Petley (reference [59] pp 47–51). The recent results have ranged from an uncertainty of less than 10^{-12} per year to about 10^{-10}. Clearly, in relation to Dirac's theory, this does not prove that G is definitely constant, but it is fair to say that it provides no evidence to the contrary.

Another question has been whether gravitational force is a function solely of mass or whether it depends on the nature of the material involved. The classical experiments to investigate this were those of Eötvös [104]. Using a wide variety of substances, he found no evidence to suggest that the composition had any effect and subsequent work has confirmed this (reference [59], pp 241–6). There have also been checks of the inverse-square law and on the possible variation of G with other physical parameters, such as quantum state. So far no clear cut dependence has been found, but further experiments are continuing and seem likely to do so indefinitely. Some surveys of the present situation and recent results are given in reference [105].

16.3.4. *Linked sets of constants*

The two constants we have just discussed, c and G, are of significance in classical physics and their values have been determined mainly by direct measurement. The other constants that we have to consider are almost all involved in atomic or quantum theory and their values are generally related. It has therefore become the practice to determine the values not just of single constants but of linked groups from series of measurements.

The need for this began to be appreciated early in this century, but the first formal recognition was provided by the publication in the early 1920s of the International Critical Tables. These gave the values of nine 'Accepted Basic Constants' and of a number of other constants derived from them. Raymond Birge, of the University of California, Berkeley, was not entirely happy with the derivation of some of the published values and undertook an extensive critical evaluation of all the available evidence. The results were published in his classic paper of 1929 [106] 'Probable values of the general physical constants (as of January 1, 1929)'. He stressed the importance of using the least-squares method to calculate probable errors. In 1930 and 1931, W N Bond published papers [107, 108] which discussed Birge's results and built on them and other data to give results which he claimed to be '... definitely the most accurate so far obtained by any method'. These were criticized by Birge in his next paper in 1932 'Probable values of e, h, e/m and α' [109]. But Bond had made a very important contribution by using a graphical presentation, consistent with the least-squares method, to display the results of a series of related measurements and to assist in evaluating the optimum values. This was adopted by Birge in his 1932 paper and has become known as the 'Birge–Bond Diagram': an example, with a detailed explanation, is given by Petley (reference [59] p 296). In 1941, Birge published a further review of 'The general physical constants' [84], giving the latest values of some dozen principal constants and a number of derived constants. His principal constants included the speed of light, the gravitational constant, the litre (in cm^3), the volume of ideal gas, the International Ohm and Ampere, some atomic weights, the standard atmosphere, the ice point, the Joule equivalent, the Faraday constant, the Avogadro number, the electronic charge and the Planck constant. It is interesting to note that some of these, the litre and the International Ohm and Ampere, are no longer measurable constants.

In 1969 Taylor, Parker and Langenberg published their paper 'Determination of e/h, using macroscopic quantum phase coherence in superconductors: implications for quantum electrodynamics and the fundamental physical constants' [110]. In this they showed that the improved accuracy in the measurement of e/h as a result of using the Josephson effect made it necessary to recalculate the values of all the related constants. It became clear that revaluations were necessary from time to time. The International Union of Pure and Applied Physics (IUPAP) had set up a Commission on Atomic Masses and Related Constants following a conference in 1956. Further conferences followed at intervals of a few years and, in 1972, the emphasis was changed to Atomic Masses and Fundamental Constants [111]. The parent body of IUPAP, ICSU, has CODATA as a committee now publishing least-squares revisions of the fundamental constants from time to time, as we have seen [72].

16.3.5. Some examples of linked determinations

16.3.5.1. F, R, N_A and k. During the greater part of the nineteenth century, ideas were developing of the atomic and molecular structure of matter and the kinetic theory of gases, together with Maxwell's electromagnetic theory. In particular, it was realized that some properties could usefully be expressed per mole of substance, i.e. in terms of the gram-molecular weight. The quantities in this group all have their roots in this period and precede the discovery of the electron. They are:

F, the Faraday, the electric charge carried by one gram-equivalent of any element;

R, the gas constant, given by $PV = RT$ for one mole of an ideal gas;

N_A, the Avogadro constant, the number of molecules per mole; and

k, the Boltzmann constant, given by $k = R/N_A$.

Direct measurements of the Faraday have been carried out by electrolysis, measuring the quantity of material deposited from an electrolyte when a known current is passed for a known time. The early measurements, including one by Rayleigh and Sidgwick [112], were made using silver. This has the disadvantage that silver does not normally consist of a single isotope. Iodine does and for this reason the measurements made by Washburn and Bates in 1912 using iodine were of particular importance [113]. Subsequent measurements have also used organic compounds, but the reported uncertainties have not greatly changed (reference [59], section 4.7). The currently accepted value is [73]

$$F = 96\,485.309 \text{ C mol}^{-1}.$$

The difficulty in measuring the gas constant, R, is of course that no gas is an ideal gas and Boyle's law is not strictly obeyed. The traditional method has been to measure the density of a real gas, usually oxygen, at various pressures and to extrapolate to zero pressure. More recently, Quinn and his colleagues have used an acoustic interferometer maintained at an accurately known temperature to determine R [114] (figure 16.10). The results have been reviewed by Colclough [115] and the latest value from CODATA is [73]

$$R = 8.314\,510 \text{ J mol}^{-1} \text{ K}^{-1}.$$

In the mid-nineteenth century, Loschmidt attempted to measure Avogadro's number, but the result was wildly in error. Early in the present century, Rutherford measured the charge of an alpha particle by counting the number of particles given off by a sample of radium and the total charge. A little later Millikan measured the charge on the electron directly (see following section). From the known values of ionic charges in electrolytes, this made it possible to calculate N_A. In 1900 Planck had stated his law of black-body radiation which involved k and h. From experimental results, he calculated the values of k and h and derived $N_A = R/k$. More recently, N_A has been determined by accurate measurement of the density of silicon together with its lattice spacing. When allowance is made for the isotopic composition of the samples, this method has provided probably the best values of N_A so far [116]. The latest CODATA value is [73]

$$N_A = 6.022\,1367 \times 10^{23} \text{ mol}^{-1}.$$

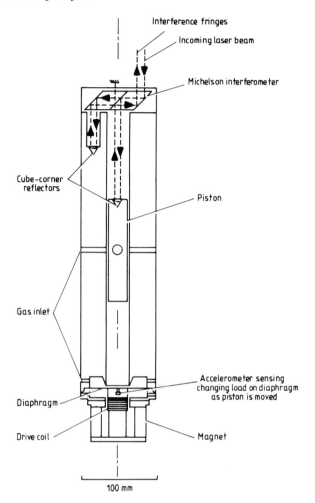

Interference fringes

Incoming laser beam

Michelson interferometer

Cube-corner reflectors

Piston

Gas inlet

Accelerometer sensing changing load on diaphragm as piston is moved

Diaphragm

Drive coil

Magnet

100 mm

Figure 16.10. *The acoustic interferometer for measurement of the universal gas constant used by Quinn et al [114] in the early 1970s.*

Although Planck derived a value of the Boltzmann constant, k, from radiation measurements, it is now calculated from the values of R and N_A, which are more accurately known by other means, as we have just seen. The CODATA value is [73]

$$k = 1.380\,658 \times 10^{-23} \text{ J K}^{-1}.$$

16.3.5.2. e, h, e/m and α. This group of constants forms the core of Birge's least-squares adjustment of 1932 [109]. They are related to the properties of the electron and indeed measurement of some of them

1270

Table 16.2. *Latest CODATA values for physical constants.*

Elementary charge	e	$1.602\,177\,33 \times 10^{-19}$ C
Planck constant	h	$6.626\,0755 \times 10^{-34}$ J s
Electron charge/mass ratio	e/m_e	$1.758\,819\,62 \times 10^{11}$ C kg^{-1}
Fine structure constant	α	$7.297\,353\,08 \times 10^{-3}$
Rydberg constant	R_∞	$10\,973\,731.534$ m^{-1}
Electron mass	m_e	$9.109\,3897 \times 10^{-31}$ kg

led to the identification of the electron in the first place. By the mid-nineteenth century, ionic conduction in electrolytes was well understood and when, after their discovery by Röngten in 1895, x-rays were found to induce conduction in gases, it was easy to identify this as caused by ionic conduction in the gaseous state, but cathode rays, which had first been produced around 1870, presented more of a problem. They were first thought to be waves similar to light, but then Crookes showed that they were deflected by a magnetic field and in 1890 Schuster used this effect to show that they must be particles with a charge-to-mass ratio, e/m, many times that of the hydrogen ion. Many further measurements followed: among the most notable were Thomson's measurement of the deflection by electric as well as magnetic fields [117] and Millikan's determinations of the charge by the oil drop method [118] which continued into the 1920s. These measurements established the identity of the electron as a particle with a mass about one two-thousandth of that of the hydrogen ion and an equal but opposite charge [119].

Later determinations of the properties of the electron have made use of spectroscopic and indirect methods, usually involving the values of some other constants including Planck's constant, h, the fine structure constant, α $(= \mu_0 c e^2/2h)$, the Rydberg constant R_∞ $(= m_e c \alpha^2/2h)$, the Faraday constant and the Avogadro constant. The relations between these were discussed in some detail by Birge [84, 109]. More recently, the discovery of the Josephson and von Klitzing effects has provided alternative routes. The Josephson constant is given by

$$K_J = 2e/h$$

and the von Klitzing constant by

$$R_K = h/e^2 = \mu_0 c/2$$

where μ_0 is the permeability of free space.

The need to provide reference standards for the volt and the ohm (see section 16.2.6.4) led to a concentrated and coordinated programme of measurement of these two constants to the best possible accuracy. The results were reviewed by two working parties of the CCE [120]. Recommended values were agreed by the CIPM. The latest CODATA values for the constants mentioned above are shown in table 16.2 [73].

16.3.5.3. Other constants. The CODATA list gives values of over 100 constants. We have been able to discuss only a minority of these. Most of the others are of a specialized nature: about 60, for example, give values for detailed properties of fundamental particles—the electron, the muon, the proton, the neutron and the deuteron. A few are of more general interest—the Bohr magneton, the Stefan–Boltzmann constant, the first and second radiation constants, and the Wien displacement law constant, for example. One constant which has been the subject of some interest recently has been the gyromagnetic ratio of the proton; it was considered that this might provide an alternative and accurate means for the absolute determination of the ampere and a number of measurements were undertaken with this end in view [8, 9].

Determination of the values of the physical constants is an ongoing process, likely to continue indefinitely as measurement techniques improve. A new least-squares adjustment is scheduled for completion in 1995 [73]. A new set of recommended values will then presumably be issued by CODATA.

16.4. Applications

In this chapter we have looked in some detail at the derivations of the base units of the SI system and of the standards for maintaining them, and also at the determinations of the fundamental physical constants. These are the areas of metrology most relevant to the science of physics and where international cooperation is most deeply involved.

In addition, national standards laboratories have the responsibility of maintaining secondary standards of the base units and of derived units for the support of trade, industry and domestic purposes and of disseminating these through calibration services. In Britain, for example, the National Measurement Accreditation Service, NAMAS, which is a service of the National Physical Laboratory, accredits over a thousand calibration and testing laboratories in industry and elsewhere so that measurements made are traceable to the national and international standards. For the most part there is little basic physics involved in this and details of the services provided in most countries are readily available [4, 5].

There are two areas of derived quantities where the science of metrology has developed mainly in this century, where the establishment of standards has been vital and where BIPM has played a major part in establishing them. These are radioactivity and microwaves. Radioactivity was discovered by Becquerel in 1896 when he observed that uranium salts could affect a photographic plate. Thorium and radium were soon also discovered and in 1911 Marie Curie made the first radium standard which was deposited with BIPM and became the first international standard. The need for further standards eventually led to BIPM setting up new laboratories, opened in 1964 and responsible for the preparation of standard sources, establishment of x-ray and γ-ray dosimetry standards and neutron measurements [1]. They have also led the way to the definition of the SI units of radioactivity including the gray and the sievert [121].

See also p 1863

Microwave applications to communications, radar and navigation grew enormously during World War II. After the war it soon became clear that there was a need for international harmonization for standards of measurement. In 1952 URSI, the International Union of Radio Science, sponsored international comparisons of microwave power, followed by other quantities including noise, attenuation and permittivity. A more formal arrangement than URSI could provide was clearly desirable and in 1965 CCE appointed a Working Group on RF Measurements, GT-RF, which organizes international comparisons. More than a dozen laboratories take part. Frequency bands up to 18 GHz are covered for most quantities, with a few measurements up to about 100 GHz and laser power at 10.6 and 1.06 μm [122, 123]. Lists of quantities offered by national laboratories are given in the URSI Register [5].

Metrology is an interesting example of a subject which serves both pure science and everyday needs. National systems of weights and measures date from at least 3000 BC and have been internationally coordinated at least since Roman times. In the early eighteenth century the Royal Society in Britain and the Academy of Science in France arranged intercomparisons and began to look at the possibility of relating practical standards to physical constants. Since then the development of metrology as a science and its applications in trade and everyday life have gone hand in hand. There can be few branches of science where international cooperation has been so active and harmonious over so many years and where the results have influenced so wide a range of human activity. In this chapter we have emphasized its importance to new developments in physics and we end with references to papers by Cook which give a comprehensive and profound view of this topic [124–126].

References

[1] Page C H and Vigoureux P 1975 *The International Bureau of Weights and Measures 1875–1975* (Washington, DC: US Department of Commerce)

[2] Pyatt E C 1983 *The National Physical Laboratory—A History* (Bristol: Hilger)

[3] Cochrane R C 1966 *Measures for Progress—A History of the National Bureau of Standards* (Washington, DC: US Department of Commerce)

[4] Dobbie B, Darrell J, Poulter K and Hobbs R 1987 *Review of DTI Work on Measurement Standards* (London: Department of Trade and Industry) ch 20

[5] Bailey A E (ed) 1990 *URSI Register of National Standards Laboratories for Electromagnetic Metrology* (Bristol: Hilger)

[6]   Bell R J (ed) 1993 *SI—The International System of Units* 6th edn (London: HMSO)

[7]   Vigoureux P 1971 The gyromagnetic ratio of the proton—a survey *Precision Measurement and Fundamental Constants* (Gaithersburg, MD: NBS)

[8]   Kibble B P and Hunt G J 1979 A measurement of the gyromagnetic ratio of the proton in a strong magnetic field *Metrologia* **15** 5–30

[9]   Taylor B N 1973 Determining the Avogadro constant to high accuracy via improved measurements of the absolute ampere *Metrologia* **9** 21–3

[10]   Kibble B P 1991 Present state of the electrical units *Proc. IEE* A **138** 187–97

[11]   Kibble B P, Robinson I A and Belliss J H 1988 A realisation of the SI watt by the NPL moving-coil balance *NPL Report* DES 88

[12]   Johnson W H 1923 Comparators *and* Line standards of length *Dictionary of Applied Physics, III* ed R T Glazebrook (London: Macmillan) pp 232–57, 465–77

[13]   1960 Comptes Rendues *11th CGPM Resolution* 6

[14]   Carré P and Hamon J 1966 Mesure interférentielle de la base géodésique du BIPM *Metrologia* **2** 143–50

[15]   Froome K D and Bradsell R H Distance measurements by means of a modulated light beam yet independent of the speed of light *Symposium on Electronic Distance Measurement (Oxford, September, 1965)* (London: Hilger-Watts)

[16]   Laurila S H 1976 *Electronic Surveying and Navigation* (New York: Wiley)

[17]   1965 *Transactions of the International Astronomical Union* vol XII A

[18]   1983 Procès-Verbaux *CIPM Recommendation* 1

[19]   Ward F A B 1970 *Time Measurement—Historical Review* (London: Science Museum)

[20]   Dye D W and Essen L 1934 *Proc. R. Soc.* A **143** 285

[21]   Essen L 1938 *Proc. Phys. Soc.* **50** 413

[22]   Lyons H 1949 Microwave spectroscopic frequency and time standards *Electron. Eng.* **68** 251

[23]   Beehler R E 1967 A historical review of atomic frequency standards *Proc. IEEE* **55** 792–805

[24]   Audoin C and Vanier J 1976 Atomic frequency standards and clocks *J. Phys. E: Sci. Instrum.* **9** 697–720

[25]   Terrien J 1976 Standards of length and time *Rep. Prog. Phys.* **39** 1067–108

[26]   Vanier J and Audoin C 1989 *The Quantum Physics of Atomic Frequency Standards* (Bristol: Hilger)

[27]   Rabi I I, Zacharias J R, Millman S and Kusch P 1938 A new method of measuring nuclear magnetic moment *Phys. Rev.* **53** 318

[28]   Ramsey N F 1950 A molecular beam resonance method with separated oscillatory fields *Phys. Rev.* **78** 695–9

[29]   Essen L and Parry J V L 1957 The caesium resonator as a standard of frequency and time *Phil. Trans.* A **250** 45–69

[30]   Bowhill S A (ed) 1984 Brief reviews of developments in frequency standards and in time scale formation and comparisons *Review of Radio Science 1981–83* (Brussels: URSI (International Union of Radio Science))

[31]   Mungall A G 1986 Frequency and time—national standards *Proc. IEEE* **74** 132–6

[32]   See several other papers in the 1986 special issue on radio measurement methods and standards *Proc. IEEE* **74** 137–82.

[33]   Mungall A G, Daams H and Boulanger J-S 1980 Design and performance of the new 1-m NRC primary caesium clocks *IEEE Trans. Instrum. Meas.* **IM-29** 291–7

[34]   Becker G 1977 Performance of the primary Cs-standard of the

Physikalisch-Technische Bundesanstalt *Metrologia* **13** 99–104

[35] Ward F A B 1970 *Time Measurement—Historical Review* (London: Science Museum) ch I

[36] Guinot G and Seidelmann P K 1988 Time scales: their history, definition and interpretation *Astron. Astrophys.* **194** 304–8

[37] Winkler G M R and Van Flandern T C 1977 Ephemeris time, relativity, etc *Astron. J.* **82** 84–92

[38] See also Kaye and Laby 1992 (reprint) Astronomical and atomic time systems *Tables of Physical and Chemical Constants* 15th edn (London: Longmans) section 1.9.1

[39] Quinn T J 1991 The BIPM and the accurate measurement of time *Proc. IEEE* **79** 894–905

[40] Fleming J A 1913 *The Wonders of Wireless Telegraphy* (London: SPCK)

[41] Winkler G M R 1986 Changes at USNO in global timekeeping *Proc. IEEE* **74** 151–5

[42] Beehler R E and Allan D W 1986 Recent trends in NBS time and frequency distribution services *Proc. IEEE* **74** 155–7

[43] Quinn T J and Compton J P 1975 The foundations of thermometry *Rep. Prog. Phys.* **38** 151–239

[44] Ruhemann M and Ruhemann B 1937 *Low Temperature Physics* (Cambridge: Cambridge University Press) p 52

[45] Hall J A 1967 The early history of the International Practical Scale of Temperature *Metrologia* **3** 25

[46] Hall J A and Barber C R 1967 The evolution of the International Practical Temperature Scale *Metrologia* **3** 78

[47] Barber C R 1969 International Practical Temperature Scale of 1968 *Nature* **222** 929

[48] *The International Practical Temperature Scale of 1968* (London: National Physical Laboratory/HMSO) (amended edn 1975)

[49] 1990 The International Temperature Scale of 1990 (ITS-90) *NPL Leaflet*. See also Kaye and Laby 1992 (reprint) The International Temperature Scale of 1990 (ITS-90) *Tables of Physical and Chemical Constants* 15th edn (London: Longmans) section 1.5.1

[50] Preston J S 1961 Photometric standards and the unit of light *NPL Notes on Applied Science No 24* (London: HMSO)

[51] Crawford B H 1962 Physical photometry *NPL Notes on Applied Science No 29* (London: HMSO)

[52] Jones O C 1975 Optical radiation scales *Contemp. Phys.* **16** 287–310

[53] Jones O C 1978 Proposed changes to the SI system of photometric units *Lighting Res. Technol.* **10** 37–40

[54] Clark Latimer 1862 Letter to the editor *The Electrician* 17 January p 1862

[55] Glazebrook R T 1913 The Ohm, the Ampere and the Volt—a memory of fifty years, 1862–1912: The Fourth Kelvin Lecture, 27 February 1913 *Proc. IEE* **50** 560–92

[56] Paul R W 1936 Electrical measurements before 1886 *J. Sci. Instrum.* **13** 1–8

[57] Thomson A M and Lampard D G 1956 A new theorem in electrostatics with applications to calculable standards of capacitance *Nature* **177** 888

[58] Lampard D G 1957 A new theorem in electrostatics with applications to calculable standards of capacitance *Proc. IEE* **104** C 271–80

[59] Petley B W 1985 *The Fundamental Physical Constants and the Frontier of Measurement* (Bristol: Hilger) pp 142–5

[60] Clothier W K 1965 A proposal for an absolute liquid electrometer *Metrologia* **1** 181–4

[61] Clothier W K, Sloggett G J, Bairnsfather H, Currey M F and Benjamin D J

1989 A determination of the volt with improved accuracy *Metrologia* **26** 9–46

[62] Giorgi G 1901 Unita razionali di elettromagnetismo *Atti Assoc. Elettrotecn.* **5** 402–18

[63] Vigoureux P 1989 Eighty-eight years of Giorgi's MKS units *J. Phys. E: Sci. Instrum.* **22** 671–3

[64] Dix C H and Bailey A E 1975 Electrical standards of measurement, Part 1 DC and low-frequency standards *Proc. IEE* **122** 1018–36

[65] Josephson B D 1962 Supercurrents through barriers *Phys. Lett.* **1** 251. See also 1964 *Rev. Mod. Phys.* **26** 216; 1965 *Adv. Phys.* **14** 419

[66] Burroughs C J and Hamilton C A 1990 Voltage calibration systems using Josephson junction arrays *IEEE Instrumentation and Measurement Conference (13–15 February, 1990)* (New York: IEEE) pp 291–4

[67] Von Klitzing K, Dorda G and Pepper M 1980 New method for high accuracy α determination based on quantised Hall resistance *Phys. Rev. Lett.* **45** 494–7

[68] Von Klitzing K and Ebert G 1985 Application of the quantum Hall effect in metrology *Metrologia* **21** 12–18

[69] Taylor B N 1987 History of the present value of $2e/h$ commonly used for defining national units of voltage and possible changes in national units of voltage and resistance *IEEE Trans. Instrum. Meas.* **IM-36** 659–64

[70] Taylor B N and Witt T J 1989 New international electrical reference standards based on the Josephson and quantum Hall effects *Metrologia* **26** 47–62

[71] Quinn T J 1989 News from the BIPM *Metrologia* **26** 69–74

[72] The 1986 adjustment of the fundamental physical constants: report of the CODATA Task Group on Fundamental Constants *CODATA Bulletin 63* (Oxford: Pergamon). See also 1973 Recommended consistent values of the fundamental physical constants *CODATA Bulletin 11* (Paris: CODATA Secretariat)

[73] Cohen E R and Taylor B N 1992 The fundamental physical constants *Phys. Today* August BG 9–13

[74] Dampier-Whetham W C D 1930 *A History of Science* (Cambridge: Cambridge University Press) p 47

[75] Eddington A S 1946 *Fundamental Theory* (Cambridge: Cambridge University Press)

[76] Michelson A A, Pease F G and Pearson F 1935 *Astrophys. J.* **82** 26

[77] Mittelstaedt O 1929 *Ann. Phys., Lpz* **2** 285

[78] Mittelstaedt O 1929 *Phys. Z.* **30** 165

[79] Hüttel A 1940 *Ann. Phys., Lpz* **37** 365

[80] Anderson W C 1937 *Rev. Sci. Instrum.* **8** 239

[81] Anderson W C 1941 *J. Opt. Soc. Am.* **31** 187

[82] Mercier J 1924 *J. Phys. Radium* **5** 168

[83] Rosa E B and Dorsey N E 1907 *Bur. Stand. Bull.* **3** 433

[84] Birge R T 1942 *Rep. Prog. Phys.* **8** 90

[85] Essen L and Gordon-Smith A C 1948 *Proc. R. Soc.* A **194** 348

[86] Essen L and Froome K D 1951 *Proc. Phys. Soc.* B **64** 862

[87] Froome K D 1958 *Proc. R. Soc.* A **247** 109

[88] Froome K D and Essen L 1969 *The Velocity of Light and Radio Waves* (New York: Academic)

[89] Bay Z and White J A 1972 *Phys. Rev.* D **5** 796

[90] Evenson K M, Wells J S, Peterson F R, Danielson B L, Day G W, Barger R L and Hall J L 1972 *Phys. Rev. Lett.* **29** 1349

[91] Blaney T G, Bradley C C, Edwards G J, Jolliffe B W, Knight D J E,

Rowley W R C, Shotton K C and Woods P T 1977 *Proc. R. Soc.* A **355** 61

[92]   Woods P T and Jolliffe B W 1976 *J. Phys. E: Sci. Instrum.* **9** 395

[93]   Baird K M, Smith D S and Witford B G 1980 *Opt. Commun.* **12** 367

[94]   Boys C V 1923 Earth, the mean density of the, *Dictionary of Applied Physics* ed R T Glazebrook (London: Macmillan)

[95]   Poynting J H and Thomson J J 1934 *Textbook of Physics; Properties of Matter* (London: Griffin)

[96]   Heyl P R 1930 *Bur. Stand. J. Res.* **5** 1243

         Heyl P R and Chrzanowski P 1942 *J. Res. NBS* **29** 1

[97]   Zahradnicek J 1933 *Phys. Z.* **34** 126

[98]   Karagioz O V, Ismaylov V P, Agafonov N L, Kocheryan E G and Tarakanov Yu A 1976 *Izv. Akad. Sci. USSR* **12** 351

[99]   Sagitov M V, Milyukov V R, Monakhov E A, Nazhidinov V S and Tadzhidinov Kh G 1977 *Dokl. Akad. Nauk.* **245** 567

[100]   Luther G G and Towler W R 1982 *Phys. Rev. Lett.* **48** 121

[101]   Lowry R A, Towler W R, Parker H M, Kuhlthau A R and Beams J W 1972 The gravitational constant *G Atomic Masses and Fundamental Constants* 4 ed J H Sanders and A H Wapstra (New York: Plenum) p 521

[102]   See, for example, Nunn T P 1923 *Relativity and Gravitation* (London: University of London Press); Hawking S W and Israel W 1979 *General Relativity—an Einstein Centenary Survey* (Cambridge: Cambridge University Press)

[103]   Ashtekar A and Geroch R 1974 Quantum theory of gravitation *Rep. Prog. Phys.* **37** 1211–56

[104]   Eötvös R V, Pekar D and Feteke E 1922 *Ann. Phys., Lpz* **68** 11

[105]   See, for example, Petley B W 1985 *The Fundamental Physical Constants and the Frontier of Measurement* (Bristol: Hilger) ch 7

         Gillies G T 1982 *BIPM Rapport* BPM-82/9

         Will C M 1984 The confrontation between general relativity and experiment, an update *Phys. Rep.* **113** 345–422

[106]   Birge R T 1929 *Rev. Mod. Phys.* **1** 1

[107]   Bond W N 1930 *Phil. Mag.* **10** 994

[108]   Bond W N 1931 *Phil. Mag.* **12** 632

[109]   Birge R T 1932 *Phys. Rev.* **40** 228

[110]   Taylor B N, Parker W H and Langenberg D N 1969 *Rev. Mod. Phys.* **41** 375

[111]   Sanders J H and Wapstra A H (ed) 1972 *Atomic Masses and Fundamental Constants 4* (New York: Plenum)

[112]   Lord Rayleigh and Mrs Sidgwick 1884 *Phil. Trans.* **175** 411

[113]   Washburn E W and Bates S J 1912 *J. Am. Chem. Soc.* **34** 1341, 1515

[114]   Quinn T J, Colclough A R and Chandler T R D 1976 *Phil. Trans.* **283** 367

[115]   Colclough A R 1981 *Precision Measurement and Fundamental Constants 2, NBS Special Publication 635* (Gaithersburg, MD: NBS) p 263

[116]   Deslattes R D 1980 *Ann. Rev. Phys. Chem.* **31** 435

[117]   Thomson J J 1897 *Phil. Mag.* **44** 293

[118]   Millikan R A 1912 *Trans. Am. Electrochem. Soc.* **21** 185

[119]   Millikan R A 1917 *The Electron—Its Isolation and Measurement and the Determination of some of its Properties* (Chicago: University of Chicago Press) (2nd edn 1924)

[120]   Comité Consultatif d'Electricité 1988 *Report of the 18th meeting*, Appendix E2, *Report from the Working Group on the Josephson Effect*; and Appendix E3, *Report from the Working Group on the Quantum Hall Effect BIPM*. See also Hartland A 1988 Quantum standards for electrical units *Contemp. Phys.* **29** 477

[121]   International Commission on Radiological Protection 1980 *Radiation Quantities and Units ICRP Report 33* (Oxford: ICRP)

[122]   Bailey A E 1980 International harmonisation of microwave standards *Proc. IEE H* **127** 70–3

[123]   Bailey A E, Hellwig H W, Nemoto T and Okamura S 1986 International organisation in electromagnetic metrology and international comparison of RF and microwave standards *Proc. IEEE* **74** 9–14

[124]   Cook A H 1972 Quantum metrology–standards of measurement based on atomic and quantum phenomena *Rep. Prog. Phys.* **35** 463

[125]   Cook A H 1975 The importance of precise measurement in physics *Contemp. Phys.* **16** 395

[126]   Cook A H 1977 Standards of measurement and the structure of physical knowledge *Contemp. Phys.* **18** 393

ILLUSTRATION ACKNOWLEDGMENTS

Chapter 9

R P Feynman—AIP Meggers Gallery of Nobel Laureates

R R Wilson—Courtesy of Fermilab

9.1 Reprinted with permission from *Phys. Rev.* **72**, 241–3. Copyright 1995 The American Physical Society

9.2 AIP Emilio Segrè Visual Archives

9.3 Instituts Internationaux de Physique et de Chimie, Bruxelles

9.5 Reprinted with permission from *Nature* **160** p855. Copyright 1995 Macmillan Magazines Ltd

9.7 Reprinted with permission from *Phys. Rev. Lett.* **12**, 204–6. Copyright 1964 The American Physical Society

9.8 Photograph by Alan W Richards. Courtesy AIP Emilio Segrè Visual Archives

9.9 M Goldhaber, L Grodzins and A W Sunyar *Phys. Rev.* **109**, 1015–7. Copyright 1958 The American Physical Society

9.10 Courtesy of Brookhaven National Laboratory

9.11 J W Cronin, V L Fitch and R Turlay *Phys. Rev. Lett.* **13**, 138–40. Copyright 1964 The American Physical Society

9.12 Courtesy of Fermilab

9.16 Courtesy of Stanford Linear Accelerator Center and US Department of Energy

9.18 From M Riordan, *The Hunting of the Quark*, Simon & Schuster, 1987

9.19 S Glashow and A Salam: photographs courtesy of AIP. Photograph of S Weinberg courtesy of Fermilab

9.21 (*a*) Brookhaven National Laboratory. Courtesy AIP Emilio Segrè Visual Archives (*b*) Reprinted with permission from *Phys. Rev. Lett.* **33** 1404–6. Copyright 1974 The American Physical Society

9.22 Left: (*a*) Reprinted with permission from *Phys. Rev. Lett.* **33** 1406–8. Copyright 1974 The American Physical Society. (*b*) Reprinted with permission from *Phys. Rev. Lett.* **33** 1406–8. Copyright 1974 The American Physical Society. (*c*) Reprinted with permission from *Phys. Rev. Lett.* **33** 1406–8. Copyright 1974 The American Physical Society

9.22 Right: Stanford Linear Accelerator Center and US Department of Energy

9.24 Reprinted with permission from *Phys. Lett.* **122B** 398–410. Copyright 1983 Elsevier Science BV, Amsterdam

9.25 Courtesy of Fermilab

9.26 Top: reproduced with

permission from *Annual Review of Nuclear and Particle Science* Vol 41 © 1991, by Annual Reviews, Inc.

9.28 Reprinted with permission from *Phys. Rev.* **D50** 1173-825. Copyright 1994 The American Physical Society

9.30 Reprinted with permission from *Phys. Rev.* **D45** S1–S584. Copyright 1992 The American Physical Society

9.31 (*a*) Reprinted with permission from *Phys. Rev. Lett.* **39** 1240–2. Copyright The American Physical Society. (*b*) Reprinted with permission from *Rev. Mod. Phys.* **61** 547–60. Copyright 1989 The American Physical Society

9.33 Top: DESY, Hamburg. Bottom: CESR, Cornell University

9.34 AIP Niels Bohr Library

9.35 Courtesy of Fermilab

9.37 Bottom: courtesy of Joe Stancampiano and Karl Luttrell

9.39 Reproduced with permission from *Proceedings DPF 94 Meeting (Albuquerque, NM, August 1994)*, World Scientific, Singapore

Chapter 10

L Prandtl—AIP Emilio Segrè Visual Archives, Lande Collection

G I Taylor—Cambridge University Library. Courtesy AIP Emilio Segrè Visual Archives

Chapter 11

H Kamerlingh Onnes—Burndy Library. Courtesy AIP Emilio Segrè Visual Archives

J Bardeen—AIP Emilio Segrè Visual Archives

11.1 From *Low Temperature Physics* by L C Jackson, John Wiley (US) and Methuen (UK)

11.4 Courtesy of Prof J F Allen FRS

11.9 From *Helium-3* by W P Halperin and L P Pitaevskii, North-Holland, Amsterdam

11.11 From *Superconductivity* by P F Dahl, AIP Press, New York

11.12 Reprinted with permission from *Appl. Phys. Lett.* **51** 57. Copyright 1987 The American Physical Society

11.13 Reprinted with permission from *Phys. Rev.* B **38**, 2477. Copyright 1988 The American Physical Society

Chapter 12

P Debye—AIP Emilio Segrè Visual Archives. Photograph by Francis Simon

M Born—AIP Emilio Segrè Visual Archives, Lande Collection

12.3 (*a*) Reprinted with permission from *Physica* **14** 139. Copyright 1948 Elsevier Science. (*b*) Reprinted with permission from *Physica* **14** 510. Copyright 1948 Elsevier Science

12.5 R Berman, 1953, *Advances in Physics* **2** 103, Taylor and Francis

12.7 (*a*) Reprinted with permission from M Blackman, *Proc. R. Soc.* A**148** 365. Copyright 1935 The Royal Society

12.8 Figure 1: reprinted with permission from E W Kellerman, *Phil. Trans. R. Soc.* **238** 513. Copyright 1940 The Royal Society. Figure 2: reprinted with permission from E W Kellerman, *Proc. R. Soc.* A**178** 17. Copyright 1941 The Royal Society

12.11 Reprinted with permission from *Phys. Rev.* **100** 756. Copyright 1955 The American Physical Society

12.12 Reprinted with permission from *Phys. Rev.* **112** 90. Copyright 1958 The American Physical Society

12.14 Reprinted with permission from *Phys. Rev.* **128** 1099.

Copyright 1962 The American Physical Society

12.15 Reprinted with permission from *Phys. Rev.* B **13** 4258. Copyright 1976 The American Physical Society

12.19 Reprinted with permission from *Phys. Rev.* **65** 117. Copyright 1944 The American Physical Society

Chapter 13

H Massey—Photograph by D H Rooks

W F Meggers—AIP Meggers Collection of Nobel Laureates

G Herzberg—AIP Meggers Collection of Nobel Laureates

13.1 Courtesy of J J McLelland, NIST

13.2 From *Physics of Atoms and Molecules* by U Fano, University of Chicago Press

13.3 T E Sharp, *Atomic Data* **2** 119–69 (1972), Academic Press, San Diego

13.4 A E Douglas and W E Jones, *Can. J. Phys.* **44** 2251 (1966), National Research Council of Canada, Ottawa

13.5 From *Physical Chemistry* by R S Berry, S A Rice and J Ross. Copyright 1980 John Wiley and Sons. Reprinted by permission of John Wiley and Sons, Inc.

13.6 A P Lukirskii *et al*, *Optics and Spectroscopy* **9** 262. Copyright 1960 The Optical Society of America

13.7 Courtesy of R D Deslattes, NIST

13.8 Courtesy of R P Madden, NIST

13.9 Reprinted with permission from *Rev. Mod. Phys.* **40** 456, figure 68. Copyright 1968 The American Physical Society

13.10 Reprinted with permission from A B Bleaney *et al*, *Proc. Roy. Soc.* **A189**. Copyright 1947 The Royal Society, London

13.11 From U Fano and A R P Rau, *Atomic Collisions and Spectra*, Academic Press, Orlando

13.12 IOP Publishing Ltd, Bristol

13.13 Reprinted with permission from *Phys. Rev. Lett.* **10** 104, figure 2. Copyright 1963 The American Physical Society

13.14 IOP Publishing Ltd, Bristol

13.15 (*a*) G Herzberg and C Jungen, 1972, *J. Mol. Spect.* **41** 425, Academic Press, Orlando

13.16 Reprinted with permission from K T Lu, *Phys. Rev.* A4, 579. Copyright 1971 The American Physical Society

13.17 (*a*) Reprinted with permission from *Phys. Rev.* **128** 662, figure 7. Copyright 1962 The American Physical Society. (*b*) Reprinted with permission from P F Ziemba and E Everhart, *Phys. Rev. Lett.* **2** 299, figure 1. Copyright 1959 The American Physical Society

13.18 Reprinted with permission from S U Ovchinnokov and E A Solov'ev, *Sov. Phys. JETP* **63** 538, figure 2. Copyright 1986 American Institute of Physics

13.19 Reprinted with permission from R Villa and R D Deslattes, *J. Chem. Phys.* **44** 4399. Copyright 1966 American Institute of Physics

13.20 B W Petley, *The Fundamental Physical Constants*, Adam Hilger

13.21 Reprinted with permission from K von Klitzing *et al*, *Phys. Rev. Lett.* **45** 494, figure 1. Copyright 1980 The American Physical Society

13.22 IOP Publishing Ltd, Bristol

Chapter 14

P E Weiss—AIP Emilio Segrè Visual Archives, Goudsmit Collection

J H Van Vleck—AIP Meggers Gallery of Nobel Laureates

14.2 From *Magnetic Bubbles*, p84, by A H Bobeck and H E D Scovil. Copyright © June 1971 Scientific American, Inc. All Rights reserved

14.7   From R M Bozorth
       Ferromagnetism D Van Nostrand
       Co., Inc. Copyright Bell
       Telephones Laboratories, Inc.
14.10 Reprinted with permission
       from A R Mackintosh, *Physics
       Today*, June 1977, figure 6,
       p28. Copyright 1977 American
       Institute of Physics
14.11 Reprinted with permission
       from Griffiths, *Nature* **158** 670.
       Copyright 1946 Macmillan
       Magazines Ltd
14.12 From J K Standley, *Oxide
       Magnetic Materials*, figure 9.7,
       p183, Clarendon Press, Oxford
14.14 Reprinted with permission
       from A R Mackintosh, *Physics
       Today*, June 1977, figure 4,
       p26. Copyright 1977 American
       Institute of Physics
14.15 From A B Pippard, *Physics
       of Vibration*. Reprinted with
       the permission of Cambridge
       University Press, Cambridge
14.16 Courtesy of Professor B S
       Worthington

Chapter 15

E O Lawrence—Lawrence Radiation
       Laboratory. Courtesy AIP
       Emilio Segrè Visual Archives
E Fermi—AIP Emilio Segrè Visual
       Archives
15.1   Niels Bohr Institute. Courtesy
       AIP Emilio Segrè Visual
       Archives
15.2   Courtesy of Max-Plank-Institut
       für Kernphysik
15.3   Argonne National Laboratory.
       Courtesy AIP Emilio Segrè
       Visual Archives

15.4   Lawrence Berkeley Laboratory.
       Courtesy of AIP Emilio Segrè
       Visual Archives
15.5   Department of Terrestrial
       Magnetism, Carnegie Institute
       of Washington. Courtesy AIP
       Emilio Segrè Visual Archives
15.7   Oakland Tribune. Courtesy AIP
       Emilio Segrè Visual Archives

Chapter 16

R T Glazebrook—Courtesy of Mrs
       Marguerite Pyatt, Hungerford,
       Berks
16.2   *Measurement at the Frontiers of
       Science*, 1991, National Physical
       Laboratory
16.3   E Pyatt, *National Physical
       Laboratory—A History*, 1983,
       Adam Hilger
16.4   Cochrane, *Measures for Progress*,
       NIST
16.5   E Pyatt, *National Physical
       Laboratory—A History*, 1983,
       Adam Hilger
16.6   E Pyatt, *National Physical
       Laboratory—A History*, 1983,
       Adam Hilger
16.7   E Pyatt, *National Physical
       Laboratory—A History*, 1983,
       Adam Hilger
16.8   B W Petley, *The Fundamental
       Constants and the Frontier of
       Measurement*, 1988, Adam
       Hilger
16.9   E Pyatt, *National Physical
       Laboratory—A History*, 1983,
       Adam Hilger
16.10 B W Petley, *The Fundamental
       Constants and the Frontier of
       Measurement*, 1988, Adam
       Hilger

SUBJECT INDEX

Note—volume and page numbers in **bold type** refer to items in boxes

Georan, II.1242
Geospace, III.1981, III.1997–2001
  future study propsects, III.2011
  pictorial models, III.2010
Germanium, I.75, I.470, II.982,
    II.1208, III.1281, III.1301,
    III.1319, III.1322, III.1324,
    III.1332–6, III.1340, III.1544,
    III.1749
*Gesellschaft deutscher Naturforscher
    und Ärzte*, I.5
Giant branch, III.1698
Giant dipole resonance (GDR),
    II.1213–14
Giant molecular clouds, III.1765
Giant stars, III.1696, III.1698,
    III.1708–9
Gibbs ensembles, I.531–3
Gibbs free energy, III.1543, III.1548
Gibbs function, I.549
Gibbs Paradox, I.528
Ginzburg–Landau theory, I.568,
    II.934, II.935, II.940
Glashow, Iliopoulos and Maiani
    (GIM) mechanism, II.700
Glashow model, II.694
Glashow–Weinberg–Salam theory,
    II.695, II.697, II.698, II.707
Glass industry, III.1517
Glasses, III.1533, III.1600–2
  phonons in, II.1004–7
Glitches, III.1754
Global Positioning Satellite
    System, II.1249
Glomar Challenger, III.1976, III.1980
Gluons, II.636, II.637, II.684, II.714,
    II.716
Glutamine, structure of, I.449
God Particle, The, III.2020
God principle, III.2019–20
Gold, III.1515, III.1543
Gold-198, III.1905
Goodyear, III.1593, III.1597
Göttingen quantum mechanics,
    I.183–96

Gradient-index lenses, III.1391
Grain boundaries, III.1518–20,
    III.1528
Grain size distribution, III.1549
Grand canonical ensemble, I.532
Grand partition function (GPF),
    I.533
Grand unified theory, I.323,
    II.763–4
Grating spectroscopy, III.1391
Gravitation, II.757–62, III.2037
  and quantum theory, I.318
  in special theory of relativity,
    I.283–4
  Newtonian constant of, I.28
  theory of, I.283–4, II.762
Gravitational collapse, I.306–12
Gravitational constant (G),
    II.1265–7
Gravitational energy, I.300–1
Gravitational field equations, I.253
Gravitational radiation, I.312–15
Gravitational waves, I.312, I.315
Gravity fields
  Moon, II.888
  Sun, II.888
Gray (unit), III.1867
Grazing collisions by ions and
    atoms, II.1041–2
Great Debate, III.1716–21
Green statistics, II.687
Greenhouse effect, II.899
Green's function, I.617, I.623–4
Ground-state energy, II.1045
Group-theoretical model, I.211–15
Group velocity, II.856
Guanine, I.504
Guided wave optics, III.1458
Guy's Hospital, III.1864
Gyromagnetic effect, II.1153
Gyromagnetic ratio, II.1237, II.1257
Gyroscope
  femtosecond-pulsed dye-laser
    ring, III.1433

in diagnosis, III.1859–63
in medical physics, III.1857
see also Radiotherapy

Yang–Mills theory, II.692–4, II.713,
    II.714, II.763
$YBa_2Cu_3O_7$, II.959
$YBa_2Cu_3O_{7-\delta}$, I.490–1
Yerkes system, III.1698
Yield-stress, III.1520–1, III.1523
Yin-Yang coil, III.1664
Young's double-slit experiment,
    III.1402
Yttrium–iron–garnet, II.1159
Yukawa meson theory, I.378–83,
    II.651–2

Z-bosons, II.637, II.706–9
Zeeman effect, I.7, I.36, I.52, I.54,
    I.99–103, I.160–2, I.182, I.183,
    I.184, I.211, II.1114, II.1115,
    II.1149
Zeeman splittings, II.1120

Zeeman sub-state, II.953
Zeitschrift für Kristallographie, I.511
Zeitschrift für Metallkunde, III.1556
Zenith angle, III.1992
Zernike's circle polynomials,
    III.1409
Zero-point motion, II.973–5,
    II.1004
Zero-resistance state, II.961
Zero sound, II.950, III.1364
ZETA (Zero-Energy
    Thermonuclear Apparatus)
    toroidal pinch experiment,
    III.1641–2, III.1645, III.1657
Zhukovski's theorem, II.822
Zinc, III.1360
Zinc oxide, III.1537
Zinc sulphide scintillators, II.1200
Zincblende, III.1514
Zone-levelling, III.1543, III.1544
Zone-refining, III.1543, III.1544,
    III.1545
Zweig's rule, II.685

NAME INDEX

Note—volume and page numbers in **bold type** refer to items in boxes

Karagioz, O V, gravitation, II.1266

Karle, I L, structure of arginine dihydrate, I.449–50

Karle, J
determinantal inequalities, I.448
structure of arginine dihydrate, I.449–50
tangent formula, I.449
The Solution of the Phase Problem. I. The Centrosymmetric Crystal, I.449

Karpushko, F V, optical bistability, III.1450

Kasai, C, flow imaging, III.1914

Kasper, J S
crystal structure of dekaborane, I.447
inequality relations, I.447

Kasteleyn, P W, bond percolation model, I.573

Kastler, A, optical pumping, II.1097, III.1415, III.1447

Kasuya, T, RKKY interaction, II.1164

Kay, W, scintillation counting, I.113

Kaya, S, ferromagnetism, II.1134

Kaye, radiation protection, III.1890

Kayser, H
handbook of spectroscopy, I.76
history of spectroscopy, I.24

Keck, D B, chemical vapour deposition, III.1447

Keeler, J, spiral nebulae, III.1950

Keenan, P C
MKK system, III.1698
radio astronomy, III.1736

Keesom, A P, liquid helium, II.914

Keesom, W H
atomic vibrations, III.1287
liquid helium, II.914, II.922
Seebeck coefficient, III.1310

Keffer, F, spin-wave modes, II.1151

Kekulé, A, Karlsruhe conference, I.45

Keldysh, L V, photo-assisted tunnelling, III.1423

Kelker, H, liquid crystals, III.1540

Kellar, J N, line broadening, I.472

Keller, A, polyethylene, III.1552

Kellermann, E W, NaCl, II.987–9

Kellers, C F, λ-point of liquid He^4, I.559

Kelley, P L, photodetection, III.1441

Kellman, E, MKK system, III.1698

Kelly, A, Walter Rosenhain, III.1517–19

Kelly, M, cuprous oxide rectifiers, III.1323

Kelvin, Lord
conducting layer, III.1984
cooling Earth, III.1945
Earth structure, III.1946
Earth's rigidity, III.1953
gaseous molecule, I.48
law of equal partition of energy, I.143–4
Maxwell's Demon, I.576
papers and patents, I.2
radium atoms, I.62
second law of thermodynamics, I.521
second wrangler, I.11
Seebeck effect, III.1284
ship–wave pattern, II.883
Sun's time-scale, III.1694
temperature scales, II.1249

Kemmer, N
meson theory, I.394
new particles, I.393
quantum field theory interaction, I.378
symmetric theory, I.403
U-particles, I.393

Kendrew, J C
globular proteins, I.501

Mackintosh, A R, exchange
  interactions, II.1164
McLennan, liquid helium, II.922
McMahon, R G, redshift quasars,
  III.1798
McMillan, E M
  electron synchrotron, II.731
  frequency modulation, II.1202
  ion acceleration, II.730
  neptunium, II.1224
  60″ cyclotron, II.1202
  transuranium element, II.1223
Madansky, L, spark counters,
  II.753
Madelung, E
  electron vibration, II.983–4
  quantum theory, I.227
Madey, J M J, free-electron laser,
  III.1456
Maeda, K, ionosphere, III.1984
Maiani, L
  charmed quark, II.700, II.705
  neutral currents, II.700
Maiman, T H, **III.1425**
  ruby laser, III.1424–6, III.1894
Majorana, E
  deuteron, I.376
  nuclear stability, II.1190
Maker, P D, pump-probe
  configuration, III.1445
Maki, A G, HCN laser line,
  III.1443
Maki, Z, quark–lepton analogy,
  II.700
Mallard, J R, gamma camera,
  III.1907
Malmberg, J H,
  magneto-electrostatic
  confinement, III.1681
Mandel, L
  coherence theory, III.1433
  optical coherence, III.1392
  photon antibunching, III.1460
  photon beams, III.1424

quantum theory of coherence,
  III.1439
sub-Poissonian statistics, III.1460
Mandelbrot, B
  fractals, I.574
  self-similarity, I.574
Mandelstam, L I
  first-order Raman effect, II.996
  Mandelstam–Brillouin
  scattering, III.1397
  Raman effect (combination
  scattering), III.1404–6
  Raman scattering, II.996
  secondary scattered radiation,
  I.172
Mandelstam, S, Twelfth Solvay
  Conference, Brussels, II.646
Manley, J M, non-linear optics,
  III.1424
Mansfield, P
  magnetic resonance imaging,
  III.1924
  nuclear magnetic resonance,
  II.1175
Mansur, L K, nuclear materials
  research, III.1534
Marconi, G
  electromagnetic waves, I.29
  radio waves, III.1995
  telegraph signals, I.54
  wireless transmission,
  III.1619–20, III.1986
Marechal, A, coherent imaging,
  III.1418
Margon, B, binary sources, III.1756
Mark, H
  biomolecular structures, I.497
  single crystals, III.1527
Marsden, E
  α-particles, I.113
  α-scattering, II.1200
  medical physics, III.1864
Marshak, R E
  *Conceptual Foundations of Modern
  Particle Physics*, II.770

Nishikawa, S, electron diffraction, I.463

Nishina, Y
Dirac equation, I.218
letter, I.212
new particles, I.389

Nishizawa, J, semiconductor maser, III.1423

Niu, K, charmed particles, II.700

Noether, F, rigid body, I.279

Nordheim, L
disordered materials, III.1365
Schrödinger equations, I.205

Nordsieck, A
First Shelter Island Conference, II.645
Master-type equation, I.605
nuclear exchange potential, I.376

Nordström, G, special-relativistic theories, I.284

Northrop, G A, thermal conductivity, II.982

Northrop, T G
adiabatic confinement, III.1679–80
trapped particles in magnetic-mirror fields, III.1683

Nouchi, photon, III.1451–2

Novikov, I D
black holes, III.1779, III.1780
structure in the Universe, III.1801

Nowick, A, anelastic relations, III.1523

Nuckolls, J
direct-drive approach to inertial fusion, III.1676
fusion power, III.1673
radiation implosion, III.1674

Obreimow, crystal growth, III.1280

Occhialini, G P S
cloud chambers, I.371, II.651
electron–positron pairs, I.231

O'Dell, C R, helium abundance, III.1790

Oehme, R
parity violation in β-decay, II.665
singularities in complex angular momentum, II.682

Oersted, H C
electric current, III.1958
electromagnetic interactions, III.1953

Ohkawa, T, multipole geometry, III.1645

Ohnuki, Y, quark–lepton analogy, II.700

Okayama, T, slow mesotrons, I.406

Oke, B, supernovae, III.1759

Oke, I, energy in the Sun, III.1748

O'Keefe, J A, optical microscopy, III.1406

Oken, L, *Gesellschaft deutscher Naturforscher und Ärzt*, I.5

Okubo, S, mass relations among baryons and mesons, II.661

Oldham, R D
earthquake waves, III.1956
seismic wave paths, III.1954

Olszewski, K, liquefied gases, I.27

O'Neill, E L, communications theory, III.1417

O'Neill, G K, *Storage-ring synchrotron: device for high-energy physics research*, II.739

Onsager, L, **I.551**
entropy relations, I.600–2
Fermi surface, III.1351
ferroelectric transition in KH_2PO_4, I.571
hypothetical vortices, II.931
Ising model, I.554–61, II.1018
Onsager relations, I.616–17, I.626
quantum theory, III.1350

Oort, J H
cosmological density, III.1799

DATE DUE